T0311931

Processes and Ore Deposits of Ultramafic-Mafic Magmas through Space and Time

Processes and Ore Deposits of Ultramafic-Mafic Magmas through Space and Time

Edited by

Sisir K. Mondal

Jadavpur University, Kolkata, West Bengal, India

William L. Griffin

Macquarie University, Sydney, NSW, Australia

Elsevier
Radarweg 29, PO Box 211, 1000 AE Amsterdam, Netherlands
The Boulevard, Langford Lane, Kidlington, Oxford OX5 1GB, United Kingdom
50 Hampshire Street, 5th Floor, Cambridge, MA 02139, United States

Notices

Knowledge and best practice in this field are constantly changing. As new research and experience broaden our understanding,
changes in research methods, professional practices, or medical treatment may become necessary.

Practitioners and researchers must always rely on their own experience and knowledge in evaluating and using any information,
methods, compounds, or experiments described herein. In using such information or methods they should be mindful of their own
safety and the safety of others, including parties for whom they have a professional responsibility.

To the fullest extent of the law, neither the Publisher nor the authors, contributors, or editors, assume any liability for any injury
and/or damage to persons or property as a matter of products liability, negligence or otherwise, or from any use or operation of any
methods, products, instructions, or ideas contained in the material herein.

British Library Cataloguing-in-Publication Data
A catalogue record for this book is available from the British Library

Library of Congress Cataloging-in-Publication Data
A catalog record for this book is available from the Library of Congress

ISBN: 978-0-12-811159-8

For Information on all Elsevier publications
visit our website at https://www.elsevier.com/books-and-journals

Working together
to grow libraries in
developing countries

www.elsevier.com • www.bookaid.org

Publisher: Candice Janco
Acquisition Editor: Amy Shapiro
Editorial Project Manager: Tasha Frank
Production Project Manager: Anitha Sivaraj
Cover Designer: Victoria Pearson

Typeset by MPS Limited, Chennai, India

Contents

List of Contributors

Ahmed H. Ahmed
Helwan University, Cairo, Egypt; King Abdulaziz University, Jedda, Saudi Arabia

Shoji Arai
Kanazawa University, Kanazawa, Japan

Stephen J. Barnes
CSIRO/Mineral Resources, Kensington, Perth, WA, Australia

Graham C. Begg
Macquarie University, Sydney, NSW, Australia; Minerals Targeting International PL, West Perth, WA, Australia

Elena A. Belousova
Macquarie University, Sydney, NSW, Australia

Dmitriy A. Chareev
Institute of Experimental Mineralogy RAS, Chernogolovka, Russia

Marco L. Fiorentini
The University of Western Australia, Perth, WA, Australia

William L. Griffin
Macquarie University, Sydney, NSW, Australia

Jon M.A. Hronsky
Western Mining Services PL, West Perth, WA, Australia; The University of Western Australia, Crawley, WA, Australia

Rais M. Latypov
University of the Witwatersrand, Wits, South Africa

Margaux Le Vaillant
The University of Western Australia, Perth, WA, Australia; CSIRO/Mineral Resources, Kensington, Perth, WA, Australia

Chusi Li
Indiana University, Bloomington, IN, United States

Kreshimir N. Malitch
Institute of Geology and Geochemistry, Ural Branch of Russian Academy of Sciences (UB RAS), Ekaterinburg, Russia; Macquarie University, Sydney, NSW, Australia

Sisir K. Mondal
Jadavpur University, Kolkata, West Bengal, India

Ria Mukherjee
University of the Witwatersrand, Johannesburg, South Africa

Suzanne Y. O'Reilly
Macquarie University, Sydney, NSW, Australia

Kwan-Nang Pang
Academia Sinica, Taipei, Taiwan

Edward M. Ripley
Indiana University, Bloomington, IN, United States

J. Gregory Shellnutt
National Taiwan Normal University, Taipei, Taiwan

Sergey F. Sluzhenikin
Institute of Geology of Ore Deposits, Petrography, Mineralogy, and Geochemistry, Russian Academy of Sciences (IGEM RAS), Moscow, Russia

Joyashish Thakurta
Western Michigan University, Kalamazoo, MI, United States

Jill A. VanTongeren
Rutgers University, New Brunswick, NJ, United States

Anna Vymazalová
Czech Geological Survey, Prague, Czech Republic

Inna Yu Badanina
Institute of Geology and Geochemistry, Ural Branch of Russian Academy of Sciences (UB RAS), Ekaterinburg, Russia

Acknowledgements

We would like to thank all the authors for their efforts and patience regarding this edited book. We wish to acknowledge Tasha Frank and Amy Shapiro of Elsevier for their help to bring out the book. We are grateful to the following persons for their reviews of various chapters for this special issue:

Bernard Charlier (University of Liege, Belgium)
Chusi Li (Indiana University, United States)
Dejan Milidragovic (British Columbia Geological Survey, Canada)
Edward M. Ripley (Indiana University, United States)
Emil Makovicky (Copenhagen University, Denmark)
Fernando Gervilla (Universidad de Granada, Spain)
Ilya Veksler (GFZ German Research Centre for Geosciences, Germany)
Louis J. Cabri (Cabri Consulting Inc., Canada)
N. Alex Zirakparvar (American Museum of Natural History, United States)
Shoji Arai (Kanazawa University, Japan)
Svetlana Tessalina (Curtin University, Australia)
Tim Kusky (China University of Geosciences, China)
Wolf-Gang Maier (Cardiff University, United Kingdom)
Yan Wang (Guangzhou Institute of Geochemistry, China)
Cover photography credits: Sisir K. Mondal and William L. Griffin

Introduction

Economic deposits of Ni-Cu sulfides, platinum-group elements (PGEs), chromite, and Ti-V-bearing magnetite are formed from ultramafic-mafic magmas at high temperature after emplacement in a crustal magma chamber. Chromite deposits are also formed in the conduits/fissures of the upper mantle section of the oceanic lithosphere at mid-ocean ridges or in back-arc rift settings. The petrogenetic evolution of the magmatic ore deposits are directly linked to the evolution of the parental ultramafic-mafic magmas through space and time. Like any other ores, magmatic ore deposits reflect the convergence of a set of favorable circumstances. For magmatic deposits, ore-forming elements are present in ultramafic-mafic magmas; therefore, the characteristics of the different mantle source reservoirs and the partial melting processes are important for the formation of these deposits. After emplacement of the parental ultramafic-mafic magmas in the crustal chamber, silicate melt-crystal equilibria play a vital role in the formation of the magmatic ore deposits. In addition, the magma chamber dynamics and the addition of crustal materials, including sulfur, to the parental ultramafic-mafic magmas play significant roles in the formation of Ni-Cu sulfide and PGE deposits. The host ultramafic-mafic rocks of magmatic ore deposits may occur as synvolcanic bodies in Archean and Proterozoic greenstone belts (e.g., komatiitic suite, tholeiitic suite, and anorthosites), intra-cratonic intrusions (e.g., mafic layered intrusions or subvolcanic intrusions in the conduits of the flood basalts), or bodies related to orogenesis (e.g., ophiolites, Alaskan-type zoned complexes). Understanding of the known magmatic ore deposits in terms of spatial and temporal distribution, and their characteristic features and genesis, gives clues for the exploration and exploitation of similar deposits.

This volume is a collection of state-of-the-art articles covering fundamental processes and the current state of knowledge on magmatic ore deposits. We expect that the primary audience for this book will be geologists, geochemists, exploration geologists, and volcanologists, as well as practitioners from the mining and metallurgical industry, ore metal processing, and mineral beneficiation groups. This volume has been edited with the idea in mind that it can be used as a reference book both for research work and regular course work, extending the audience both to research scholars and dissertation students. We also expect that the edited volume will provide knowledge and new concepts helpful to exploration geologists targeting new metal prospects. The volume also contains updated knowledge on the fundamental processes responsible for magmatic ore deposits, the relationships between tectonic settings and magmatic ore mineralization, links between magmatic ore mineralization and processes in the mantle (both shallow and deep) versus the crust, as well as the changing patterns in magmatic ore mineralization through time, and their bearing on the chemical evolution of the Earth's mantle and crust.

The book covers a wide range of topics, which are organized thematically. Begg et al. (Chapter 1: Global- to Deposit-Scale Controls on Orthomagmatic Ni-Cu(-PGE) and PGE Reef Ore Formation, pp. 1−46) present an overview on global- to deposit-scale controls on magmatic Ni-Cu (-PGE) and reef type PGE ore formation and translate the associated characteristics and processes into factors that can be useful for predictive exploration targeting. Vaillant et al. (Chapter 2: Review of Predictive and Detective Exploration Tools for Magmatic Ni-Cu-(PGE) Deposits, With a

Focus on Komatiite-Related Systems in Western Australia, pp. 47−78) provide a general overview of the genetic processes responsible for Ni-Cu-PGE magmatic deposits, and describe the geochemical and geophysical techniques that are currently used, addressing how these exploration tools relate to each other and could be integrated at different scales. Ripley and Li (Chapter 3: Metallic Ore Deposits Associated With Mafic to Ultramafic Igneous Rocks, pp. 79−111) deliver a general review on the metallic ore deposits associated with mafic to ultramafic igneous rocks. They briefly describe the deposit types and key geological-lithological features, existing genetic theories for the origin of the deposits, and finally they discuss concentration mechanisms of metals where similar processes may be involved for very different types of deposits. VanTongeren (Chapter 4: Mixing and Unmixing in the Bushveld Complex Magma Chamber, pp. 113−138) reviews the evidence for new magma inputs and mixing within the magma chamber stratigraphy, as well as the present state of understanding regarding the source magmas of the Bushveld Complex (South Africa). In addition, VanTongeren presents a review of the physical and chemical signatures of immiscibility and magma unmixing in the Bushveld Complex. Arai and Ahmed (Chapter 5: Secular Change of Chromite Concentration Processes From the Archean to the Phanerozoic, pp. 139−157) re-examine the secular change of chromite concentration processes from the Archean to the Phanerozoic to evaluate changes in magmatic processes in the mantle, which in turn have possibly controlled the distribution of Cr in the upper mantle and the crust. The chapter by Mukherjee and Mondal (Chapter 6: Petrogenetic Evolution of Chromite Deposits in the Archean Greenstone Belts of India, pp. 159−195) provides updates on the petrology, geochemistry (bulk-rock and isotope), and geochronology of the ultramafic-mafic rocks and their chromite deposits from the Indian Archean greenstone belts, and compares these occurrences with similar deposits in globally distributed greenstone belts, so as to understand their bearing on Earth's evolution in the Archean. Their chapter also explores whether the ultramafic-mafic magmatism responsible for ore mineralization was linked to amalgamation or break-up stages of a supercontinent cycle. Based on the evidence from radiogenic- and stable-isotope data, Malitch et al. (Chapter 7: New Insights on the Origin of Ultramafic-Mafic Intrusions and Associated Ni-Cu-PGE Sulfide Deposits of the Noril'sk and Taimyr Provinces, Russia: Evidence From Radiogenic- and Stable-Isotope Data, pp. 197−238) present new insights on the origin of ultramafic-mafic intrusions and associated Ni-Cu-PGE sulfide deposits of the Noril'sk and Taimyr provinces, Russia. Pang and Shellnut (Chapter 8: Magmatic Sulfide and Fe-Ti Oxide Deposits Associated With Mafic-Ultramafic Intrusions in China, pp. 239−267) present a brief review of ultramafic-mafic rocks in China, and the related magmatic ore deposits. Their review starts with background information on the geology of China, provides a short summary of the major phases of ultramafic-mafic magmatism with which the magmatic ore deposits are associated, and finally gives brief descriptions of individual ore deposits and the research work done on them. Thakurta (Chapter 9: Alaskan-Type Complexes and Their Associations With Economic Mineral Deposits, pp. 269−302) reviews the Alaskan-type complexes and their associations with economic mineral deposits. This article describes the salient characteristics of Alaskan-type complexes in terms of their petrological and geochemical features and their tectonic occurrence, and investigates the potential of these complexes as target locations for economic Cu-Ni-PGE mineralization. In the final chapter (Chapter 10: Experimental Aspects of Platinum-Group Minerals, pp. 303−354), Vymazalová and Chareev present the most updated knowledge on the experimental aspects of platinum-group minerals (PGMs). This article reviews the experimental methods that can be applied for the synthesis of platinum-group compounds and minerals, provides an overview of 26 PGMs described since 2002, and discusses their experimental aspects. In addition, they summarize the ternary PGM and corresponding systems, and compile the available data for relevant phase diagrams.

Sisir K. Mondal[1] and William L. Griffin[2]

[1]Jadavpur University, Kolkata, West Bengal, India [2]Macquarie University, Sydney, NSW, Australia

GLOBAL- TO DEPOSIT-SCALE CONTROLS ON ORTHOMAGMATIC NI-CU(-PGE) AND PGE REEF ORE FORMATION

Graham C. Begg[1,2], Jon M.A. Hronsky[3,4], William L. Griffin[1] and Suzanne Y. O'Reilly[1]

[1]*Macquarie University, Sydney, NSW, Australia* [2]*Minerals Targeting International PL, West Perth, WA, Australia*
[3]*Western Mining Services PL, West Perth, WA, Australia*
[4]*The University of Western Australia, Crawley, WA, Australia*

CHAPTER OUTLINE

Processes and Ore Deposits of Ultramafic-Mafic Magmas through Space and Time. DOI: http://dx.doi.org/10.1016/B978-0-12-811159-8.00002-0

1.1 INTRODUCTION

The declining rate of ore deposit discovery over the last few decades has reflected the gradual exhaustion of the near-surface search space. Increasingly, mineral exploration is heading deeper, and into more remote, covered terranes in the quest for new productive areas. This focus on deposits with limited surface expression has emphasized the importance of exploration targeting in the discovery process. Targeting for ore deposits relies on well-understood genetic models and on the ability to translate these models into key parameters that can be identified in the available datasets (Hronsky and Groves, 2008). In recognition of the dynamic nature of the ore-forming process, and the broad range of influencing factors across space and time, ore deposit genesis is now commonly described in terms of a mineral system, a set of Earth processes that mobilize and concentrate elements. More specifically, a mineral system is a temporally and spatially hierarchical and interconnected set of processes that results in deposit formation and modification. Ore deposits formed from mafic-ultramafic magmas are the products of mineral systems involving various degrees of partial melting within the Upper Mantle. The magnitude of this melting, the concentration and focusing of the magmas, and their subsequent physical and chemical interaction with the lithosphere, ultimately determine the quality (size, grade) and style (geometry, sulfide-rich versus sulfide-poor, massive or disseminated, etc.) of mineralization.

Ni-Cu(-PGE)-sulfide deposits and platinum-group-element (PGE) reefs are the most economically important types of orthomagmatic Ni-Cu or PGE mineralization. The former are a globally important source of Ni and PGE production, while the latter are the dominant source of PGEs. The aim of this chapter is to provide an overview of the global- to district-scale controls on the genesis of such deposits, and to translate the attendant features and processes into factors that can be useful for exploration targeting.

1.2 DEPOSITS AT A GLANCE

1.2.1 OVERVIEW

Economically significant Ni-Cu(-PGE)-sulfide and PGE reef deposits are found within the continental lithosphere; they are mostly hosted by cratons, and particularly their peripheries (e.g., Begg et al., 2010, Maier and Groves, 2011; Table 1.1). Few are known in modern-day equatorial regions (Fig. 1.1), perhaps largely due to the difficulty of exploring for deeply weathered sulfide orebodies in a well developed regolith. Deposits found in major orogenic belts are rarely significant in size, and no major economic deposit is cospatial with a contemporaneous magmatic arc. It will be argued below that both genetic and preservational factors are likely to be important. Deposit ages range from mid-Mesoarchean to Tertiary.

The convective mantle is the primary source of melts relevant for the formation of deposits associated with mafic and ultramafic rocks. In the case of orthomagmatic Ni-Cu(-PGE) and PGE reef deposits, the most significant deposits show a strong association with Large Igneous Provinces (LIPs; Naldrett, 1989, 2004; Pirajno, 2000; Kerrich et al., 2000, 2005; Maier, 2005; Ernst, 2007; Pirajno et al., 2009; Begg et al., 2010; Dobretsov et al., 2010; Maier and Groves, 2011) and/or komatiites. The attendant mafic to ultramafic magmas require hot ($\geq 1500°C$) upwelling mantle,

Table 1.1 Lithospheric Setting and Magmatic Association of Significant Ni-Cu(-PGE) and PGE Camps and Deposits (> 100 kt Ni equiv., and/or >1 Moz PGE)

Approx. Deposit Age (Ga)	Age (Ga)	Camps or Deposits [giant; bold]	Coeval Permissive LIP?	Associated Craton / ON (<25 km) Margin or NEAR (<120 km) Margin	Lithospheric Setting of Camps/Deposits	MgO Group; PGE Status	Selected References
0.06—0.03	0.03	Brady's Glacier	No	No (microcontinent / ON)	Convergent margin	LM	Himmelberg and Loney (1981)
		Other Cordilleran[a]	No	No (microcontinent / ON)	Convergent margin	LM	
	0.056	**Skaergaard**	Yes	Rae / ON	Intracontinental/Small marginal basin	LM; PGE Reef	Wotzlaw et al. (2012), Nielson et al. (2015)
0.34—0.22	0.217	Hongqilin	Yes	North China / NEAR	Intracontinental	LM	Lightfoot and Evans-Lamswood (2015)
	0.232	Wellgreen	Yes	No (microcontinent / ON)	Intracontinental rift?	LM; PGE-rich	Mortensen and Hulbert (1991), Barkov et al. (2002)
	0.251	**Norilsk**	Yes	Siberian / NEAR	**Intracontinental**	LM; PGE-rich	Kamo et al. (2003), Lightfoot and Evans-Lamswood (2015)
	0.258	Ban Phuc	Yes	South China / ON	Intracontinental	LM	Lightfoot and Evans-Lamswood (2015)
	0.259	Yangliuping	Yes	South China / ON	Intracontinental	LM	Song et al. (2003)
	0.269	Huangshan	Yes	Tarim / ON	Intracontinental	LM	Zhou et al. (2004); Lightfoot and Evans-Lamswood (2015)
	~0.282	**Huangshandong**	Yes	Tarim / ON	**Intracontinental**	LM	Mao et al. (2002), Lightfoot and Evans-Lamswood (2015)
	0.285	Xiangshan	Yes	Tarim / ON	Intracontinental	LM	Qin et al. (2003), Lightfoot and Evans-Lamswood (2015)
	0.287	Kalatongke	Yes	No (microcontinent / ON)	Intracontinental	LM	Zhang et al. (2009), Lightfoot and Evans-Lamswood (2015)
	0.344	Aguablanca	Yes	No (microcontinent / ON)	Intracontinental	LM; PGE-rich?	Ordóñez-Casado et al. (2008)
0.73	~0.726	Kingash	Yes?	? (microcontinent / ON)	?	LM	Ernst et al. (2012), Ariskin et al. (2016)
	0.728	Yoko-Dovyren	Yes?	Siberian / ON?	Small marginal basin?	LM; PGE-rich	Ariskin et al. (2013)

(Continued)

Table 1.1 Lithospheric Setting and Magmatic Association of Significant Ni-Cu(-PGE) and/or >1 Moz PGE Camps and Deposits (>100 kt Ni equiv., and/or >1 Moz PGE) *Continued*

Approx. Deposit Age (Ga)	Age (Ga)	Camps or Deposits [giant; bold]	Coeval Permissive LIP?	Associated Craton / ON (<25 km) Margin or NEAR (<120 km) Margin	Lithospheric Setting of Camps/Deposits	MgO Group; PGE Status	Selected References
0.86–0.80	~0.80?	Limoeiro	?	No (microcontinent / ON)	Intracontinental	LM	Mota-e-Silva et al. (2013)
	0.832	**Jinchuan**	?	**North China / ON**	?	LM	Zhang et al. (2010), Lightfoot and Evans-Lamswood (2015)
	~0.855	Munali	?	Congo / NEAR	Intracontinental	LM	Johnson et al. (2006), Evans (2011)
1.11–1.07	?	Ntaka Hill	?	No	Intracontinental	LM?	
	1.068	**West Musgrave**	Yes	**SAC, WAC, NAC / ON**	Intracontinental	LM	Seat et al. (2007, 2011)
	1.099	Duluth	Yes	**Superior / NEAR**	Intracontinental rift	LM; PGE-rich	Paces and Miller (1993)
	1.106	Tamarack	Yes	Superior / NEAR	Intracontinental	LM	Goldner (2011), Lightfoot and Evans-Lamswood (2015)
	1.107	Eagle	Yes	Superior / NEAR	Intracontinental	LM; PGE-rich	Lightfoot and Evans-Lamswood (2015)
1.40 – 1.30	1.11?	Jacomynspan	Yes?	Kalahari / ON	Intracontinental	LM	Maier et al. (2016a)
	~1.30	Nova-Bollinger	Yes?	Yilgarn / ON	Intracontinental rift	LM	Lightfoot and Evans-Lamswood (2015)
	1.334	**Voisey's Bay**	Yes	**Nain / ON**	**Intracontinental**	LM	Maier et al. (2007, 2010)
1.88–1.84	1.403	**Kabanga**	Yes?	**Tanzania / ON**	**Intracontinental rift**	LM; PGE-rich	Hoatson and Blake (2000)
	1.844	Savannah	No	Kimberley/ON	Small marginal basin	LM	Lightfoot et al. (2001)
	1.85	**Sudbury**	Yes	**Superior / NEAR**	**Intracontinental**	LM; PGE-rich	Page and Hoatson (2000)
	1.856	Panton Sill	No	Kimberley/ON	Small marginal basin	LM; PGE Reef	Hulbert et al. (2005), Machado et al. (2010)
	1.88	**Thompson**	Yes	Superior / ON	**Peri-cratonic basin**	HM	Lesher (2007)
	1.88	**Raglan**	Yes	**Superior / ON**	**Small marginal basin**	HM; PGE-rich	Gaal (1980)
	1.883	Kotalahti	Yes	Karelia / ON	Intracontinental rift	LM	
	?	Lynn Lake	?	Sask ? / NEAR	?	LM	

1.98 - 1.92	1.92	Allarechka	Yes	Kola / ON	Intracontinental	LM	Skuf'in and Bayanova (2006)
	1.98	**Pechenga Camp**	**Yes**	**Kola / NEAR**	**Intracontinental rift**	**LM; PGE-rich**	Skuf'in and Bayanova (2006), Hanski et al. (2011)
2.09—2.05	2.05	Nkomati	Yes	Paleo-Kaapvaal / ON	Intracontinental rift	LM	Maier (2005)
	2.056	**Bushveld PGE Reefs**	**Yes**	**Paleo-Kaapvaal / NEAR**	**Intracontinental**	**LM; PGE Reef**	Zeh et al. (2015)
	2.06	**Platreef (Flatreef)**	**Yes**	**Paleo-Kaapvaal / NEAR**	**Intracontinental**	**LM; PGE-rich**	Maier et al. (2008a,b)
	2.06?	**Sakatti**	**Yes?**	**Karelia / ON**	**Intracontinental rift**	**LM**	Mutanen and Huhma (2001); Le Vaillant et al. (2016)
	2.06	Keivitsa	Yes?	Karelia / ON	Intracontinental rift	LM	Mutanen and Huhma (2001); Le Vaillant et al. (2016)
	~2.06	Santa Rita	Yes?	Sao Francisco / ON	Intracontinental rift	LM	Barnes et al. (2001)
	2.07	Voronezh	?	Paleo-Sarmatia / NEAR	Intracontinental	LM	Chernyshov et al. (2012)
	2.09	Samapleu	?	Kenema-Man / NEAR	Intracontinental	LM	Gouedji et al. (2014)
2.58—2.44	2.44	**Portimo Complex**	**Yes**	**No**	**Intracontinental**	**LM; PGE-Reef**	Maier (2005)
	2.44	Penikat	Yes	No	Intracontinental rift	LM; PGE-Reef	Maier (2005)
	2.49	Monchegorsk	Yes	Kola / NEAR	Intracontinental	LM; PGE-rich	Maier (2005), Sharkov and Chistyakov (2012)
2.74—2.69	2.575	**Great Dyke**	**Yes**	**Paleo?-Zimbabwe / NEAR**	**Intracontinental**	**LM; PGE Reef**	Oberthür et al. (2002)
	2.689	Lac des Iles	Yes	Paleo-Superior/ NEAR	Intracontinental	LM; PGE-rich	Barnes and Gomwe (2011)
	~2.7	Tati	Yes	Paleo-Zimbabwe / NEAR	Intracontinental rift	LM; PGE-rich	Maier et al. (2008a,b)
	~2.70	Shangani, etc	Yes	Paleo?-Zimbabwe / NEAR	Intracontinental rift?	HM	Hofmann et al. (2014), Prendergast and Wilson (2015)
	2.7?	**Selebi-Phikwe**	**Yes**	**Zimbabwe / ON**	**?**	**LM; PGE-rich**	Maier et al. (2008a,b)
	2.702	Montcalm	Yes	No?	Intracontinental rift	HM	Barrie et al. (1990)
	2.704	**Stillwater**	**Yes**	**Paleo-Wyoming / NEAR**	**Intracontinental**	**LM; PGE Reef**	Premo et al. (1990)
	2.705	**Kambalda**	**Yes**	**Paleo-Yilgarn / NEAR**	**Intracontinental**	**HM**	Barnes (2006)
	2.705	**Agnew-Wiluna Belt**	**Yes**	**Paleo-Yilgarn / NEAR**	**Intracontinental rift**	**HM**	Barnes (2006)
	2.705	Silver Swan	Yes	Paleo-Yilgarn / NEAR	Intracontinental rift	HM	Barnes (2006)
	2.72	Shebandowan	Yes	Paleo-Superior?/ NEAR	Intracontinental rift	HM	Houlé and Lesher (2011)
	~2.73?	Dumont	Yes	No?	Intracontinental rift	HM	Houlé and Lesher (2011)

(Continued)

Table 1.1 Lithospheric Setting and Magmatic Association of Significant Ni-Cu(-PGE) and PGE Camps and Deposits (>100 kt Ni equiv., and/or >1 Moz PGE) *Continued*

Approx. Deposit Age (Ga)	Age (Ga)	Camps or Deposits [giant; bold]	Coeval Permissive LIP?	Associated Craton / ON (<25 km) Margin or NEAR (<120 km) Margin	Lithospheric Setting of Camps/ Deposits	MgO Group; PGE Status	Selected References
	2.73?	Grasset	Yes	Paleo-Superior/ NEAR	Intracontinental	LM; PGE-rich	Barnes et al. (2007), Houlé and Lesher (2011)
	2.734	Ferguson Lake	Yes	Hearne/ NEAR	Intracontinental rift	LM; PGE-rich	Campos-Alvarez et al. (2012)
	2.735	Eagle's Nest	Yes	Paleo-Superior/ ON	Intracontinental rift	LM; PGE-rich	Mungall et al. (2010)
2.92 ~ 2.8		Windarra	Yes	East (Paleo-) Yilgarn / NEAR	Intracontinental rift	HM	Kositcin et al. (2008)
	<2.808	Lake Johnston	Yes	Paleo-Yilgarn /ON	Intracontinental rift	HM	Romano et al. (2014), Barnes (2006), Mole et al. (2013)
	2.92?	Forrestania	Yes	Paleo-Yilgarn /ON	Intracontinental rift	HM	Wang et al. (1996), Barnes (2006), Mole et al. (2013)
	2.925	Munni Munni	?	Paleo-Pilbara / ON	Intracontinental rift	LM; PGE Reef	Hoatson and Keays (1989)
3.03	3.033	Stella	?	Kimberley Block / NEAR	Intracontinental rift	LM; PGE Reef	Maier et al. (2003)

Giant camps/deposits (larger bold type) are those deposits whose ultimate size will likely satisfy the >1.3 Mt of Ni metal criterion as outlined by Schodde and Hronsky (2006), or >6 Moz PGE. LIP, large Igneous Province, following the definition of Ernst (2014); the listed LIPs are those that at the time of eruption/intrusion satisfied, or were likely to have satisfied, this definition (i.e., not necessarily well preserved or exposed). HM, High-MgO; LM, Low-MgO. SAC, WAC, NAC = south, west, and north Australian cratons.
[a]Tertiary Ni-Cu deposits of the Canadian and Alaskan cordillera, including Bohemia Basin, Cogburn, and Victor.

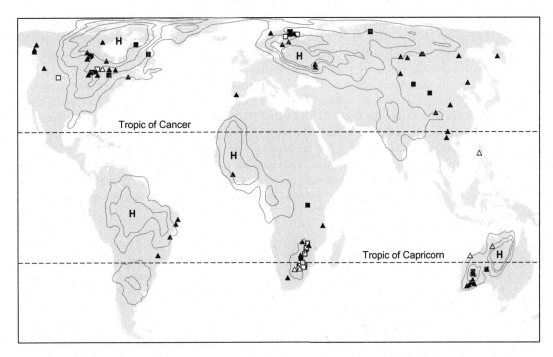

FIGURE 1.1

Global distribution of major (triangles, >100 kt Ni equiv., >1 Moz PGE) and giant (squares, >1 Mt Ni equiv., >6 Moz PGE) Ni-Cu(-PGE) (black) and PGE (white) deposits associated with mafic to ultramafic complexes. Contours depict velocity highs at the $100-175$ km depth range from a 2011 version of the global bodywave shear wave seismic tomography model of Steve Grand (created as per Grand, 2002; processed as described in Begg et al., 2009). Dashed lines bracket the tropics.

and therefore have been linked to high-degree melting in rapidly ascending mantle plumes arriving at depths of 100 km or less (Herzberg and O'Hara, 2002; e.g., Figure 3 of Begg et al., 2010). This situation will be limited to areas of relatively thin lithosphere, with melts rising to the upper crust via translithospheric faults.

As an alternative to a plume-related origin, some workers have invoked delamination of a lithospheric root following (micro)continental collision or magma injection (e.g., Elkins-Tanton, 2005). While delamination may also result in hot mantle upwellings, the scale of late melting in postcollisional settings appears to be relatively limited compared to the extremely rapid injection of massive volumes of melt ($>1 \times 10^6$ km^3 in approximately $1-2$ Myr) involved with some mineralized LIPs (Siberian Traps, Reichow et al., 2009; Bushveld Complex, Zeh et al., 2015). In addition, some deposit-related LIPs occur in settings where either the timing or the lithospheric situation is not suited to a delamination scenario. For example, the c. 1.3 Ga Nain Plutonic Suite (NPS) in Canada, which hosts the Voisey's Bay Ni-Cu-PGE deposit, intruded almost 0.5 Gyr after the collision that formed the Trans-Hudson Orogen. The sutures and orogenic belts proximal to the 2.05 Ga Bushveld Complex, which hosts most of the world's PGE resource, are much older (> 0.5 Gyr) than the LIP. Mantle xenoliths and xenocrysts in much younger kimberlites that intrude the Bushveld Complex clearly indicate the continued presence of thick Archean Subcontinental Lithospheric Mantle (SCLM) in the region (Griffin et al., 2003). It is unlikely that lithospheric delamination can be the cause of magmatism in deposits that are precollisional (e.g., Agnew Wiluna Belt deposits, Kambalda, Raglan, Thompson, Kabanga, Pechenga), or significantly postdate (>100 Myr) collision (e.g., Voisey's Bay, Duluth). Finally, paleomagnetically-reconstructed sites of major LIP magmatism over the last 300 Myr exhibit a strong spatial connection with distinctive velocity features adjacent to the core-mantle boundary (CMB; see discussion on temporal distribution of deposits and supercontinent cycles in Section 1.5). If this is a feature typical of earlier supercontinent cycles, whilst a connection of some smaller deposits (and their associated relatively small melt volumes) to lithospheric delamination cannot be ruled out, for the reasons described above we favor a connection of the larger deposits and their attendant LIPs, to deep-sourced mantle plumes.

1.2.2 KEY ELEMENTS OF THE MAGMATIC SYSTEM

For a detailed treatment of the magmatic processes involved in the generation of Ni-Cu(-PGE)-sulfide and PGE reef deposits, the reader is referred to Lesher and Keays (2002), Naldrett (1997, 2004, 2010) and references therein, Barnes and Lightfoot (2005), Cawthorn et al. (2005), Maier and Groves (2011), and other papers in this volume.

1.2.2.1 Ni-Cu(-PGE) Deposits

The generic magmatic process that produces sulfide-rich Ni-Cu(-PGE) deposits involves high-degree partial melting in the mantle, ascent of the melts into the crust, sulfide-melt saturation and incorporation of chalcophile elements, concentration of the sulfide-melt and upgrading of metal contents, and emplacement of the sulfide melts in the deposit environment. Sulfide-melt saturation may occur at a range of crustal depths, including the upper-crustal site of deposit formation and preservation. Immiscibility of sulfide melts in their host ultramafic-mafic silicate magma leads to segregation of the sulfide melts, after which they are free to migrate according to the dynamics of the tectonomagmatic system. Sulfide melts increase their metal content through interaction with large volumes of metal-undepleted silicate magma (e.g., Naldrett, 2004, 2010). Cooling leads to sulfide exsolution. Ores tend to be sulfide-rich (10%−90% sulfide), with assemblages dominated by pyrrhotite, (and generally lesser) pentlandite, chalcopyrite and pyrite. Ore tenors depend on initial composition of the magma and how much metal partitions into the sulfide liquid; PGE contents (Os, Ir, Ru, Rh, Pt, Pd) can be highly variable.

One of the key mechanisms required for ore formation is sulfide-melt saturation. Sulfide solubility in magmas increases with decreasing pressure (Mavrogenes and O'Neill, 1999), so that melts formed at mantle depths are strongly undersaturated in sulfide at upper-crustal levels. For a sulfide melt to form and segregate from the magma, either: (1) sulfur must be added from an external source, (2) the sulfide solubility must be reduced by compositional or physical changes in the magma, or (3) the volume of silicate liquid must be reduced through crystallization of silicate minerals (Naldrett, 1989, 2004, 2010; Lesher and Keays, 2002; Ripley and Li, 2013). Contamination of injected melts by S-rich (meta)sediments is commonly invoked as the cause of sulfide-melt saturation (Lesher and Groves, 1986; Lesher and Keays, 2002; Ripley et al., 2002;

Lightfoot and Keays, 2005; Keays and Lightfoot, 2004; Bekker et al., 2009). While this may be true in some, or perhaps many cases, there are others where this process apparently has not occurred, despite the presence of such sediments in the host rocks (e.g., Pechenga deposit, Hanski et al., 2011).

Classification schemes based on deposit type become increasingly complex with each new variant that is described or discovered. For the purpose of greenfields exploration targeting there is very little practical utility in such models. Naldrett (2004, 2010) instead produced a classification scheme based on a combination of the composition of the associated magmas (also a reflection of the active tectonic environment), orebody morphology and mineralogy, and genetic origin of the deposit (e.g., impact-related). Whilst useful for understanding details of deposit style, this scheme does not easily translate into a predictive model. Our aim is to produce a predictive holistic model that can encompass both Ni-Cu(-PGE) deposits and PGE deposits. For simplicity, we have used the composition of the inferred parental magmas of cogenetic rocks (after Griffin et al., 2013) to divide the Ni-Cu(-PGE) deposits and PGE deposits into two end-member types: High-MgO (ultramafic; komatiitic to picritic magmas with >18 wt% MgO) and Low-MgO (mafic; mostly basaltic magmas with <12 wt% MgO). Although there is a full spectrum of deposits between these end-members (titular Medium-MgO deposits), we elect to classify these deposits into the end-member group that they most resemble, utilizing both the MgO content and the geometrical form of the deposits. PGE reef deposits and most PGE-rich deposits, fall into the Low MgO group. The majority of High MgO deposits are (mostly Neo-)Archean, with the youngest being Paleoproterozoic, whereas Low MgO deposits span all eras from the Neoarchean onwards.

High-MgO (or ultramafic) Ni-Cu(-PGE) deposits are typically associated with either extrusive or shallow subvolcanic environments which have been developed in thinned crust (generally rift basins) and in relatively extensional to neutral tectonic conditions. They are spatially associated with well-developed fault corridors; Archean examples are hosted by narrow, regionally continuous broadly linear greenstone belts. The best examples are associated with (paleo)craton boundaries (see Section 1.3.2). The relatively unfractionated mantle melts appear to have followed translithospheric structures that facilitated easy access to the upper crust, without much interaction with either SCLM or the deeper crust en route.

Such direct, relatively unimpeded access to the upper crust ensures that melts arrive in the (sub)volcanic environment undersaturated with respect to sulfide melt (Mavrogenes and O'Neill, 1999). To attain sulfide-melt saturation these melts typically have assimilated sulfides from sediments within the host upper-crustal volcanic-sedimentary pile. Classic examples are the Kambalda deposits of Western Australia, and the Raglan deposits in Canada (Lesher and Keays, 2002). Massive sulfide ores are common, typically a meter to several meters thick, and formed by gravitational settling and accumulation of sulfides at the base of lava channels. Well-preserved mineralized zones have a massive sulfide base that transitions upwards into a zone of matrix sulfides and eventually into a zone of disseminated sulfides within a cumulate-rich channel. Economically less-important disseminated ores generally occur in layers 1–10 m thick directly above, or peripheral to, the massive ores. Deposits are invariably deformed by later shortening events. The reconstructed predeformational form of individual orebodies is dominantly highly linear, near horizontal, and typically several hundred meters to more than a kilometer long (e.g., Perring, 2015).

Low-MgO (mafic) Ni-Cu(-PGE) deposits may form in areas of thinned crust (such as a marginal/rift basin), but more generally are found in areas of average to above- average crustal thickness. They form at a range of upper-crustal levels, mostly deeper than for High-MgO deposits, and are genetically and spatially associated with intrusions rather than volcanics. They are ubiquitously associated with (paleo)craton boundaries and/or active orogenic belts, and form under a wide range of tectonic conditions, with the exception of regional strong extension (see Section 1.3.2). They have a strong spatial association with translithospheric faults, fault intersections, and fault-defined corridors. Deposits are either hosted within, or have a close spatial and genetic relationship with, interconnected networks of variably differentiated mafic and ultramafic intrusions, generally in the roots of associated LIPs. Selected parts of these intrusion networks act as conduits for the injection and accumulation of sulfide melts. Collectively, these conduits are referred to as chonoliths. Lightfoot and Evans-Lamswood (2015) identify three different geometrical forms of chonoliths associated with deposits:

1. Classic chonoliths are pipe-shaped intrusions typically hundreds of meters across and $1-10$ km in length that may be either cylindrical, oblate, or ribbon-like in cross section, and often occur in clusters. Examples include Baimazhai, Tongdongzi, Talnakh, Kharaelakh, Noril'sk I, Noril'sk II, Chernagorsk, Maslovskoe, Tamarack, Current Lake, Babel−Nebo, Nkomati, Limoeiro, Chibasong, Wellgreen, and Voronezh;
2. Intrusions with rhomboid-shaped outcrop patterns and funnel-shaped cross sections. Examples include Jinchuan, Hongqiling Number 1, Jingbulake, Kalatongke, Huangshan, Huangshandong, Limahe, Qingquanshan, Lengshuiqing, Zhubu, Ban Phuc, the Ovoid, Discovery Hill, Eastern Deeps, Eagle, Double Eagle, Aguablanca, Maksut, Santa Rita, and Suwar;
3. Widened pipe-like zones within dykes, and sills, and structural discontinuities in the dykes and sills. Examples include Reid Brook and Eastern Deeps dykes (Voisey's Bay), the Worthington and Copper Cliff Offsets of the Sudbury Igneous Complex, Hongqiling Number 7 Intrusion (China), and Pechenga (Russia).

More rarely, Ni-Cu(-PGE) deposits occur as broad zones (of mostly, or entirely, disseminated mineralization) within large (tabular) layered intrusive complexes characterized by multiple pulses of mafic-ultramafic intrusion (e.g., Duluth). The Platreef is a 100 km long, <250 m wide tabular, pyroxenitic intrusion with a roughly 40 degrees westerly dip, transgressing basement rocks at the base of the Bushveld Complex. It hosts several horizons of disseminated Ni-Cu-rich PGE mineralization, and is the possibly the world's largest PGE resource. Unique in the spectrum of Low-MgO deposits are the ores associated with the Sudbury Igneous Province in Canada, which collectively comprise the world's second largest Ni-Cu(-PGE) camp.

The mafic-ultramafic intrusions genetically associated with Low-MgO Ni-Cu(-PGE) deposits are fractionated products of ultramafic magmas, which in all but one case (Sudbury) are accepted to be mantle-derived. The Sudbury Igneous Complex apparently formed due to impact-induced melting of preexisting, mostly mafic (LIP) crust (Keays and Lightfoot, 1999; Lightfoot et al., 2001; Mungall et al., 2004), caused by a c. 1.86 Ga bolide/comet impact (Naldrett, 1997). However, geochemical techniques cannot rule out a 20%−30% contribution from an impact-triggered injection of mantle melt (P. Lightfoot, pers. comm.). Given the Sudbury impact occurred towards the margin of the Superior Craton, and at a time close to the formation of other giant Ni-Cu(-PGE) camps around the periphery of the craton (Thompson, Raglan), such a mantle melt contribution may be likely. There is no other known incidence of significant Ni-Cu(-PGE) mineralization associated with an impact event.

Most of the largest Low-MgO Ni-Cu(-PGE) deposits are associated with intrusions that have experienced significant crustal contamination, which may have occurred at any crustal level. While some ores show isotopic evidence for the assimilation of *S*-bearing country rocks (Ripley and Li, 2003; Maier et al., 2010), others show no addition of crustal sulfur (e.g., Babel−Nebo, Seat et al., 2011). Other possibilities are that sulfide saturation is induced by assimilation of silica from country rocks, or by melt fractionation (Ripley and Li, 2013). Country rocks vary widely from (meta) sedimentary to (meta)igneous, and have generally been subject to earlier deformation and metamorphism. Most deposits also have experienced significant deformation, either immediately after their formation (e.g., Raglan), or much later. Ores may be massive (mostly 1−10 m thick, but may exceed 100 m) or disseminated (10−100 m thick); the former generally are found towards the base of intrusions, and the latter as broad zones within the intrusive host. Some massive ores extend beyond their associated intrusion as veins penetrating into adjacent country rocks (e.g., ores associated with the Talnakh intrusion at the Noril'sk camp). Individual orebodies may be pod-like, ribbon-like, or cylindrical in shape; they may be continuous on the 100 m-km scale, and are generally strongly influenced by the shape of the associated/hosting intrusion. Within the context of the larger related magmatic system/LIP, deposits are spatially associated with potential "feeder" intrusions such as chonoliths, and/or "feeder" positions with respect to the main entry point of magmas into a larger (differentiated) intrusion/chamber. This may be apparent at the scale of the upper-crustal expression of the magmatic system, or at the scale of an individual camp. Both relationships are well illustrated at the camp scale by the linked orebodies of the Voisey's Bay Ni-Cu(-PGE) camp (Lightfoot et al., 2012).

The most dramatic difference between end-member High-MgO ultramafic and Low-MgO mafic-ultramafic magmas is their viscosity (e.g., Bhattacharji, 1967; Persikov, 1991). Low-viscosity High-MgO lavas are able to flow quickly, comparable to an oily fluid. As a consequence, ores related to High-MgO volcanism typically have linear, continuous (100 m-km) forms consistent with gravitational sulfide settling from a magma flowing in a conduit/channel. By contrast, Low-MgO magmas are orders of magnitude more viscous. Their sulfide-rich ores never display stratified, gravitational settling textures, but rather often display textures suggestive of high-pressure injection (breccias, fracture and vein arrays) late in the intrusive history (e.g., Noril'sk, Urvantsev, 1971; Voisey's Bay, Evans-Lamswood et al., 2000; Lightfoot and Evans-Lamswood, 2015). One proposed explanation is that breccia-textured massive sulfide ores form as sulfide infiltration breccias (e.g., Barnes et al., 2016), however this must be a second-order effect as these can logically only form once there has been a significant local accumulation of sulfide. In many cases, a close association between massive sulfide and abundant, variably assimilated, wallrock xenoliths is observed (e.g., Voisey's Bay; Evans-Lamswood et al., 2000) which suggests that there may be an important role for xenolith-rich hybrid magmas (likely to be anomalously buoyant) in transporting sulfides upwards through these Low-MgO systems. Magmas with intermediate MgO compositions (e.g., 12−18 wt% MgO) will have viscosities between these two end-members. As a result, associated orebodies are likely to have pod-like form (e.g. Raglan deposit), rather than linear and continuous geometries.

The seemingly dynamic injection of massive ores in the Low-MgO systems raises the possibility that sulfide saturation occurred at depth (Ripley and Li, 2013), with highly pressurized (volatile-rich?) sulfide melts expelled from deeper magma chambers at a late stage in the crystallization of the host melts, akin to the fault-valve behavior common in hydrothermal systems (e.g., Sibson and Scott, 1998; Lightfoot and Evans-Lamswood, 2015). Such fluid (or in this case, sulfide melt-bearing magma) release episodes, where overpressure combined with the prevailing stress conditions overcome some form of physical threshold barrier, are a hallmark of Self-Organized Critical Systems (SOCS; Hronsky, 2011; McCuaig and Hronsky, 2014). It is highly informative that whilst associated LIP durations of c. 1–20 Myr can generally be established with precise radiometric age dating, discriminating between different mineralizing pulses of magma within individual camps is not resolvable with these techniques. This is despite field evidence for an established sequence of mineralized intrusive pulses. This points to dynamic injection of mineralized magmas in short-lived transient events. On the scale of the associated LIP or major magmatic event, the volume of the host intrusions is tiny by comparison, yet these small intrusions host large quantities of sulfides, far outstripping (by 1–3 orders of magnitude) the sulfide dissolving capacity of such small volumes of silicate magma (e.g., Naldrett et al., 1992; Lightfoot and Evans-Lamswood, 2015). Combined, these observations are strong evidence for deep (perhaps as much as kilometers below the deposit level, but still mid- to upper- crustal) accumulation of magmatic sulfide melts, prior to dynamic injection into the upper-crustal deposit environment, along with relatively small volumes of mafic-ultramafic melt. Subsequent gravity-inspired downward migration of sulfide melt to the base of an intrusive magma batch will be favored by any decrease in magma flow velocity, highlighting the importance of changes in magma conduit geometries in the accumulation of sulfide melt and ultimate generation of massive sulfide ores.

1.2.2.2 PGE Reef Deposits

The generic magmatic process for the generation of PGE reef deposits differs in that the ores tend to be sulfide-poor, with 0.5%–5% sulfide. PGEs may be hosted within sulfides or occur as Platinum Group Minerals (PGMs). The deposits generally occur as thin (centimeters- to meters-scale) spatially continuous (one to tens of kilometer) tabular bodies/layers within major differentiated layered intrusion complexes or dykes, rather than associated with feeder intrusions like their Ni-Cu counterparts. These PGE reefs are invariably associated with major stratigraphic discontinuities in the composition of their host intrusions that are likely to represent major magma unconformity events during the formation of these magma chambers. These reefs are commonly, but not always, associated with chromite- (and in some cases, magnetite-) rich layers, attesting to the anomalous processes associated with their formation (e.g., Naldrett et al., 2012). The overall form of the host intrusion is generally either lopolithic or dyke-like; the largest deposits are associated with intrusive complexes with longest dimensions measured in tens to hundreds of kilometer. The intrusive complexes generally straddle, or are adjacent to, large translithospheric faults. However, within the intrusions individual ore deposits may have only a broad spatial correlation with these faults. In comparison to Ni-Cu(-PGE) sulfide deposits, PGE reef deposits and their associated intrusions commonly formed further from syn-mineral craton boundaries, but typically close to paleo-craton boundaries internal to the craton. Classic deposit examples are the narrow reef ores of the Bushveld Complex in South Africa, the Stillwater Complex in the USA, the Great Dyke in Zimbabwe, and the Skaergaard intrusion in Greenland. An exception to the m-scale reef thickness typical of most of these deposits is the Flatreef deposit below the northern Bushveld Complex, which comprises thick zones (tens of meters) of Ni-Cu-rich PGE mineralization within a separate tabular ultramafic intrusion close to the base of the complex. The Flatreef is the deeper extension of the steeper 100 km-long Platreef intrusion that hosts extensive, generally narrower, zones of

Ni-Cu-rich PGE mineralization. The geometry of this intrusion at the base of the complex, and adjacent to a paleocraton boundary represented by the Thabazimbi-Murchison Lineament (TML), is essentially equivalent to that of a feeder intrusion typical of a Low-MgO Ni-Cu(-PGE) deposit environment. Importantly, Platreef magmas are considered as a likely separate pulse of compositionally equivalent magmas to those of the Critical Zone of the Bushveld, which hosts the PGE reefs (McDonald et al., 2005). Thus it serves as a possible connection between the deposit styles, and demonstrates that to a first order the lopolithic layered intrusions are the potential magmatic nonsurficial discharge equivalent of thick volcanic piles found in some LIPs associated with Ni-Cu (-PGE) deposits (e.g., Norilsk). Magmas that form PGE reef deposits appear to enter the layered intrusion as S-poor, sulfide-undersaturated melts, and reach sulfide saturation via processes such as fractionation and/or magma mixing within the intrusion (Cawthorn et al., 2005; Naldrett, 2010; Maier and Groves, 2011). As a result, they form laterally extensive thin horizons, generally at the transition from ultramafic to overlying mafic parts of the intrusion.

1.3 LITHOSPHERIC, TECTONIC, AND GEODYNAMIC SETTING: CONTINENT TO CRATON SCALE

1.3.1 LITHOSPHERIC RESILIENCE: PRESERVATION VERSUS GENETIC FACTORS

Mapping of the architecture and tectonothermal history of the SCLM and overlying crust reveals that continents are largely comprised (perhaps >70% by area in the SCLM) of Archean and reworked Archean lithosphere (e.g., Begg et al., 2009; Griffin et al., 2009). This is consistent with Re-Os dating of sulfide phases in SCLM xenoliths (Griffin et al., 2004; Griffin and O'Reilly, 2007) and the Hf isotope record of crustal zircons globally (Belousova et al., 2010; Dhuime et al., 2012). The presence of a dominantly Archean SCLM implies that most juvenile SCLM generated in ocean basins and island arcs is ultimately returned to the convecting mantle, though some will be stored on continents as obducted lithosphere or as accreted arc terranes, until demolished by continent-continent collision. The key mechanisms for the return of juvenile lithosphere to the mantle include subduction, syn/postcollisional lithospheric delamination, and ablation/thinning at the base of the SCLM.

Domains recognized as cratons existing today are simply the best-preserved fragments of the Archean lithosphere, stitched together during Archean and Proterozoic accretionary and collisional events. Thus, many cratons have had a prominent Proterozoic tectonomagmatic history. If there is no significant post-Archean tectonothermal overprint evident in the crust, the SCLM beneath these regions is termed *archon*. Paleo- to Meso-Proterozoic SCLM is termed *proton*, and Neoproterozoic to Phanerozoic SCLM is termed *tecton* (terminology modified from Janse, 1994). However, investigation of the tectonothermal history of the crust and SCLM indicates that many protons and tectons are in fact former archons that have suffered tectonothermal overprinting (e.g., Begg et al., 2009) during one or more supercontinent cycles. These "reworked" archons are classified as proton/archon, tecton/archon and so on.

As a first-order global observation, increased lithospheric reworking is inversely correlated with lithospheric thickness. Areas of the thickest lithosphere are dominated by archons and proton/archons (e.g., Artemieva et al., 2009; Begg et al., 2009; Griffin et al., 2009). At SCLM depths (e.g., 50−175 km) the seismic velocity of these regions is anomalously high. Variation in the velocity is dominantly linked (inversely) to variations in temperature, but is also correlated (positively) with the degree of melt depletion (essentially the removal of Fe) in the mantle rocks (Afonso and Schutt, 2012). The most depleted mantle roots (archons and proton/archons) typically also have low geotherms and hence will have the highest seismic velocities (Griffin et al., 2009). Successive waves of tectonothermal reworking have the effect of thinning (and probably weakening) the SCLM (Begg et al., 2009), and increasing its Fe content through the interaction with melts introduced from the convecting mantle. This process is termed refertilization (Griffin et al., 2009). Volumetrically, arc magmatism is the dominant cause of refertilization, and the water-rich nature of these magmas contributes to weakening and ultimately thinning of the lithospheric mantle.

The interpretation of geophysical data in terms of lithospheric structure and composition, combined with multiscale and multidisciplinary geoscientific data including crustal geology, magnetics, gravity, various seismic-related data, geochronology, mantle petrology and geochemistry, and isotopic data (e.g., Re-Os, Hf, Nd, Pb), is the basis of Lithospheric Architecture Mapping. This involves the delineation of Upper Lithospheric Domains (ULDs), comprising both the upper SCLM (mapped at nominal 80−100 km depth) and the autochthonous crust.

Fig. 1.2 shows the location of major Ni-Cu(-PGE) and PGE deposits in Africa with respect to the tectonothermal history of its ULDs. There is a clear relationship between the deposits and regions with the best-preserved (generally thickest) old lithosphere, namely archons and proton/ archons. Given that such areas should also exhibit the highest seismic velocity at SCLM depths, then at a broad (continental) scale we should observe a correlation of deposits with (the edges of) regions of higher velocity in seismic data. This correlation is clearly visible in Fig. 1.1, suggesting that this relationship has global application. Note, however, that some deposits lie in lower-velocity regions (e.g., western United States, eastern China, eastern Africa, eastern Brazil). This distribution generally reflects postmineralization mantle dynamics, resulting in thinning/reworking/refertilization of the ancient SCLM and consequently higher present day geotherms (Griffin et al., 2013). Our mapping of the tectonothermal history of the continental lithosphere beyond Africa is at an advanced stage, and to date supports the African findings (Begg et al., 2009). Most of the larger orthomagmatic Ni-Cu(-PGE) sulfide and PGE reef deposits are hosted by archons and proton/archons that have experienced only one, or at most two, significant episodes of post-Archean reworking. These regions generally feature thick, higher-velocity SCLM roots.

We use the term *lithospheric resilience* to describe the degree to which thick SCLM is preserved. The SCLM of the most resilient lithosphere is most likely to be archon, or proton/archon with only one post-Archean tectonothermal overprint. High resilience seems to be favorable for deposits on two fronts. Firstly, the transition from thick (resilient) to thin lithosphere provides a good focus for mantle melting, and such transitions are most likely to be found around the peripheries of the most resilient (thickest) lithosphere. Secondly, highly resilient lithosphere has in general experienced less tectonism than more reworked lithosphere, and has probably sheltered its crust and hence any contained deposits. From a tectonic perspective, multiply reworked ancient lithosphere is unlikely to preserve intact deposits, and will often only have the potential to preserve deposits younger than the youngest major tectonothermal event. Thus a low-resilience tecton/proton/archon will be less likely to host deposits, because: (1) unless it is juxtaposed to highly resilient lithosphere, there is unlikely to be a prominent step (from thick to thin lithosphere) to focus mantle melting, and; (2) multiple tectonothermal overprints are likely to have destroyed (e.g., by uplift and erosion) or compromised (e.g., by structural dismemberment) any deposits that predate the youngest tectonothermal event.

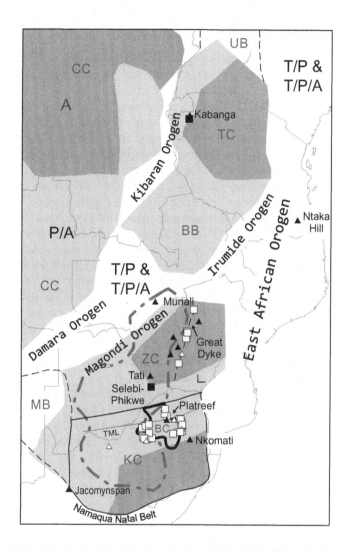

FIGURE 1.2

Deposits and lithospheric resilience. Relationship of giant (squares) and major (triangles) Ni-Cu(-PGE; black) and PGE (white) deposits of Africa with respect to archons (A), proton/archons (P/A), and intervening areas of tecton/ proton (T/P) and tecton/proton/archon (T/P/A). Deposits are preferentially located within, or at the margins of the most resilient lithosphere (A, P/A). Thin black lines outline the Kaapvaal Craton and the internal paleocraton boundary represented by the TML. The high velocity SCLM root at 100−175 km depth (thick gray dash-dot line) is taken from the updated 2011 seismic tomography model of Grand (2002). Note the position of the Great Dyke (dashed line) along the edge of this anomaly, suggestive of a paleocraton boundary. *BB*, Bangweleu Block; *BC*, Bushveld Complex (thick black outline); *CC*, Congo Craton; *KC*, Kaapvaal Craton; *L*, Limpopo Belt; *MB*, Maltahohe Block; *UB*, Ugandan Block; *ZC*, Zimbabwe Craton.

1.3.2 CRATONIC ADDRESS AND THE IMPORTANCE OF CRATON MARGINS

Most Ni-Cu(-PGE) deposits fit the craton-margin model employed by some mineral exploration groups over the last two to three decades. This model was partially recognized by Groves et al. (1987), briefly described by Kerrich et al. (2005), fully articulated by Begg et al. (2010), and reemphasized by Maier and Groves (2011). The core component of the model is that deposits dominantly form at major lithospheric domain boundaries at the margins of cratons as they exist at the time of the mineralizing event. On the other hand, while some PGE reef deposits are located on craton margins (Skaergaard, Panton Sill), the majority occur more internal to cratons (e.g., Maier and Groves, 2011). We will investigate this more fully below.

The craton-margin model recognizes that most Ni-Cu(-PGE) deposits are found near the margins of large lithospheric blocks (dimensions mostly measured in hundreds of kilometers), and in particular near the margins of cratons (sometimes more than 1000 km across). There are two important concepts regarding craton margins. Firstly, cratons are amalgams of multiple lithospheric blocks, and blocks can be added or detached by plate tectonics over time, resulting in a shift of the craton margin. In other words, the spatial extent of a craton is time-sensitive. Multiple cratons may join to form a larger craton. The adjoining boundary/suture then becomes a paleocraton boundary. Secondly, the term *craton margin* should be used loosely to capture the family of major translithospheric faults that may be susceptible to tectonic reactivation around the edges of a craton. This zone may extend inboard from the titular "craton boundary/edge" by more than 100 km, and outboard by generally a lesser distance. It is often sensitive to the presence of a paleocraton boundary (e.g., former suture zone) that may be (sub)parallel with the current craton boundary and situated along the edge of an older, more central part of the craton.

Table 1.1 summarizes the lithospheric position and setting of most significant Ni-Cu(-PGE) and PGE reef deposits in space and time. Some key examples are illustrated on Fig. 1.3. We classify the spatial relationship as *on* a craton margin/edge if the deposit occurs within 25 km of a mapped outer edge, and *near* the margin/edge if it is 25−120 km distant. In some instances it is difficult to assess the craton-margin relationship, particularly for deposits with poor age constraints, or where there has been major postmineral displacement of previously adjacent lithospheric blocks (e.g., Jinchuan deposit, China). The boundaries shown in Fig. 1.3 rely to various degrees on published outlines, but in general are simplified versions of more detailed unpublished lithospheric maps, constructed as described above in Section 1.3.1. The accuracy of the lithospheric mapping obviously depends on the quality and spatial resolution of the available data. In general we attempt to depict the position of the lithospheric boundaries as they would be either at, or directly below, the base of the crust. That is, it is the edge of SCLM blocks that we regard as the most spatially important and durable boundary. Above such edges are a range of structures connecting to the upper crust. Mapping of these SCLM block/domain edges attempts to overcome the misleading impact of shallow-dipping (imbricate) crustal structures, and to more accurately portray the positions that are more likely to focus any injection of mantle-derived melts into the overlying crust. Subsequent migration of melts to the upper crust will often be spatially associated with major relatively steep-dipping (translithospheric) faults.

The deposits depicted in Fig. 1.3 (and to a lesser extent Fig. 1.2) dominantly lie either on or directly adjacent to craton margins. These include: Huangshan camp and Sayan Area deposits in northwest China and Siberia, Fig. 1.3A; Thompson, Raglan, and Voisey's Bay deposits in Canada,

Fig. 1.3B; almost all significant deposits in China, including Jinchuan, Fig. 1.3C; all post-Archean deposits in Australia, including West Musgraves camp, Nova, Savannah and Panton Sill, Fig. 1.3D, and; Selebi-Phikwe, Kabanga, and Jacomynspan in central and southern Africa, Fig. 1.2. Other deposits lie on structures slightly inboard of the margin (e.g., Norilsk, Fig. 1.3A; various deposits of the Mid-Continent Rift area, Fig. 1.3B). Yet others are demonstrated to lie on or close to paleo-craton boundaries, including: The Neoarchean High-MgO deposits fringing the eastern margin of the Western Yilgarn paleocraton (Fig. 1.3D; for details see Begg et al., 2010; Mole et al., 2013); the Bushveld Complex PGE deposits and associated Nkomati Ni-Cu(PGE) deposit (discussed further in this section; Fig. 1.2), and; the Neoproterozoic Dovyren and Chaya deposits of the southern Siberian Craton (Fig. 1.3A). The c. 1.1Ga mafic-ultramafic intrusive complexes and deposits associated with the Mid-Continent Rift (Fig. 1.3B) are associated with large translithospheric structures around the periphery of the rift. The c. 1106 Ma Tamarack and c. 1107 Ma Eagle deposits, developed during the incipient rifting stage of the MCR (Lightfoot and Evans-Lamswood, 2015), lie along a major E-W fault corridor parallel to, but inboard from the margin of the Superior Craton. It is not understood why the MCR developed its peculiar curved plan-view shape, taking a "bite" of the edge of the craton, but the possibility remains that it exploited one or more preexisting structures.

There is geological uncertainty as to the position of some deposits with respect to craton margins at the time of mineralization. For instance, whilst the giant c. 827 Ma Jinchuan Ni-Cu(-PGE) deposit (Li et al., 2005) lies adjacent to the North China Craton margin, the syn-mineralization lithospheric context of this margin is not well understood. Similarly, the c. 725 Ma Kingash Ni-Cu (-PGE; Ernst et al., 2012) deposit lies outside the Siberian craton, and is genetically linked to the margin of the adjacent Tomsk microcontinent (Fig. 1.3A). Given that a younger magmatic arc environment appears to have existed between the position of Kingash and the Siberian craton, there is some doubt as to the relative positions of the craton and microcontinent at 725 Ma, despite the presence of similar-aged mineralization (e.g., c. 728 Ma; Dovyren deposit; Ariskin et al., 2013) further to the east on the southern flank of the craton.

The 67 deposits/camps in Table 1.1 have been assessed for their relationship to craton margins. Of these, 29 have been classed as on a craton margin, whilst another 26 are classed as near, collectively making up >80% of those evaluated. Of the 12 remaining deposits/camps, 7 (Kingash, Limoeiro, Aguablanca, Kalatongke, Wellgreen, Brady's Glacier and "Other Cordilleran") occur on the margin of microcontinental blocks in Proterozoic or Phanerozoic orogenic belts. Uncertainty surrounds the position of Kingash with respect to the Siberian Craton at the time of mineralization (see above). The c. 287 Ma Kalatongke deposit in western China (Zhang et al., 2009) could be part of the Huangshan camp (including the giant c. 282 Ma Huangshandong deposit; Mao et al., 2002), but displaced approximately 700 km by sinistral movement on the adjacent Irtysh shear zone. The region was subject to extensive Permian strikeslip tectonics following closure of oceans within the CAOB in the Late Carboniferous. Therefore, Kalatongke may have been either on or near the margin of the Tarim craton at the time of mineralization. This seems logical, as the deposit is the same age as the c. 290−274 Ma Tarim LIP mapped on the craton (Liu et al., 2016 and references therein). Similarly, the large low-grade (currently subeconomic) Wellgreen deposit, on the edge of the exotic Alexander terrane (microcontinent) in the Alaskan cordillera, is associated with the Triassic Nicolai LIP. This LIP is interpreted to have erupted synchronously with rifting and separation of the Alexander microcontinent from its Arctic (possibly northern Siberian; cf. Nelson and Colpron, 2007) parent continent. The exact nature of this syn-mineralization lithospheric setting is

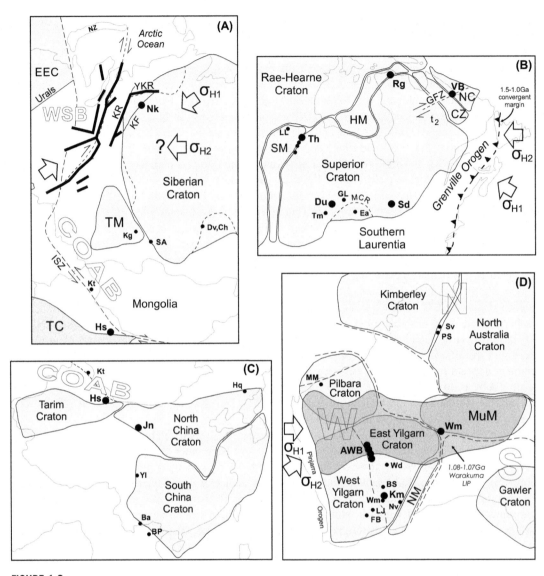

FIGURE 1.3

Examples of the lithospheric, tectonic, and geodynamic setting of significant Ni-Cu(-PGE) and PGE deposits (giant, large circles), showing cratonic lithosphere (gray), reworked lithosphere (white) and selected major structures/fault zones (dashed lines). Large broad arrows denote principal horizontal stress direction, σ_H. See text for details. (A) Northern Central Asian Orogenic Belt (COAB) and Siberia. Thick black lines depict Late Permian-Early Triassic rifts (transtensional fault zones) beneath the Jurassic-Tertiary West Siberia Basin (WSB). A NNE- to NE-directed regional maximum horizontal stress direction (σ_{H1}) during ongoing indentation of the Siberian Craton favored transtensional opening of these zones during Siberian Traps flood-basalt volcanism. Note the favorable orientation of the Noril'sk-Kharaelakh Fault (KF). Ni-Cu-PGE mineralization at the Norilsk (Nk)

(Continued)

unclear. On the other hand, the relatively large, but low grade Brady's Glacier and other smaller "Other Cordilleran" deposits in the Cordillera of Canada and Alaska all occur on the margins of microcontinents, and appear to be coeval with the adjacent convergent-margin magmatic arc. None are currently economic or operating. Similar arc-related, modest-sized Ni-Cu(-PGE) deposits are known elsewhere, but are not economically significant, and are not treated further here. It appears that magmatic arc environments, with their much lower magma flux rates relative to LIPs, are not conducive to the formation of large, high quality Ni-Cu(-PGE) deposits.

There are 19 deposits in Table 1.1 that formed on or near a paleocraton margin within the larger host craton. Most of these are Archean deposits, and all predate 2.0 Ga. Many of the larger PGE-rich (>1 g/t PGE) deposits are included within this subset, including most of the PGE reef deposits/camps (the Bushveld PGE reefs, the Great Dyke, Stillwater, Munni Munni, Lac des Iles). Two deposits/camps that do not appear to lie on or near a craton margin are Portimo and Penikat within the Fennoscandian Shield. Both are associated with PGE reefs within intrusions at the unconformity between Archean basement and overlying sedimentary rocks of the adjacent Peräpohja Basin, a Paleoproterozoic intracratonic rift. At first observation the NNE-trending Great

---

◄ camp occurs when there is a switch from this stress orientation (e.g., to σ_{H2}). See Section 1.3.4.1 for details. Major Permian sinistral slip on the Irtysh Shear Zone (ISZ) may have moved the early Permian Kalatongke deposit (Kt) from a position much closer to the coeval Huangshan camp (Hs). The Neoproterozoic Dovyren (Dv) and Chaya (Ch) deposits lie along a prominent Neoproterozoic rifted embayment on the margin of the Siberian Craton. Other deposits/camps: *Kg*, Kingash; *SA*, Sayan Area. Other features: *EEC*, East European Craton; *KR*, Khudosei Rift; *NZ*, Novaya Zemlya; *TC*, Tarim Craton; *TM*, Tomsk Microcontinent; *YKR*, Yenisei-Khatanga Rift. (B) Proterozoic Ni-Cu(-PGE) deposits of the Eastern Canadian Shield. Note that the Voisey's Bay (VB) mineralization occurs following a stress switch from NW-directed (σ_{H1}) during intrusion of the NPS along the favorably-oriented boundary between the Nain Craton (NC) and the Core Zone (CZ) microcontinent, to more westerly-directed (σ_{H2}), resulting in dextral movement along the preexisting Garder Fracture Zone (GFZ) synchronous with injection of mafic-ultramafic intrusions and magmatic sulfides. The stress switch is assigned to events at the Grenville convergent margin. See Section 1.3.4.2 for details. Deposits/Camps: *Du*, Duluth; *Ea*, Eagle; *GL*, Great Lakes; *LL*, Lyn Lake; *Rg*, Raglan; *Sd*, Sudbury; *Th*, Thompson; *Tm*, Tamarak. *HM*, Hudson Microcontinent; *MCR*, axis of 1.1 Ga Mid-Continent Rift; *SM*, Sask Microcontinent. (C) China. *Ba*, Baimazhai; *BP*, Bhan Phuc; *Hs*, Huangshan camp; *Hq*, Hongqilin; *Jn*, Jinchuan; *Kt*, Kalatongke; *Yl*, Yangliuping. (D) Central and western Australia. Archean deposits within the Yilgarn and Pilbara craton lie on intracratonic rifts, with Munni Munni (MM) and the largest Neoarchean Yilgarn deposits (Agnew Wiluna Belt, AWB; Kambalda, Km; Widgiemooltha, Wm; Black Swan, BS) either on or near major paleocraton boundaries (see dashed lines). Proterozoic deposits Savannah (Sv), Panton Sill (PS), and most likely Nova (Nv), formed during closure of their host peri-cratonic rifts during collisional events. East-directed shortening during c. 1.09−1.07 Ga collision along the western margin of Australia (Pinjarra Orogen) favored intrusion of the c. 1.08 Ga Warakurna LIP (dark gray) along east-west cratonic boundaries as major intrusive complexes and sills. The main focus of magmatic injection is at the cratonic triple junction in the West Musgraves, where subsequent weakening of the crust resulted in uplift and bimodal volcanism. A c. 1.07 Ga stress switch to a more NE-directed shortening triggered injection of mineralizing chonoliths within the West Musgraves camp (Wm). The stress switch is related to events at the collisional margin. See Section 1.3.4.3 for details. Other deposits: *LJ*, Lake Johnson; *FB*, Forrestania Belt; *Wd*, Windarra. *MuM*, Mulga Microcontinent; *NM*, Nornalup Microcontinent; The greater Northern- (N), Southern- (S), and Western- (W) Australian cratons outlined by the dashed lines.

Dyke cuts across the center of the Zimbabwe craton, and appears to reach its margins at a high angle—visible in magnetics at the southern margin, but obscured by younger rocks of the Zambezi Belt at its northern margin. It is therefore intracratonic to the Zimbabwe craton and seemingly not lying on or near a craton margin. However, over most of its length the Great Dyke runs subparallel to, and slightly offset (<50 km) from, several Meso- to Neoarchean greenstone belts, including Sinoia-Sipolilo, Salisbury-Shamwa, Midland, Manisi, Selukwe and Belingwe. In addition, the seismic tomography model summarized in Figs. 1.1 and 1.2 reveals that the Great Dyke follows the eastern edge of a regional high velocity SCLM feature beneath the western half of the host Zimbabwe craton. This observation is supported by the more detailed seismic tomography model produced by the Kaapvaal experiment (e.g., Fig. 2b in Griffin et al., 2013). The suggestion is therefore that the dyke follows a major physical break in the SCLM, possibly representing a paleocraton boundary.

The Nkomati Ni-Cu deposit, and the collective Bushveld PGE reef deposits, actually occur within the coeval Kaapvaal craton. However, both appear to be spatially associated with tectonically reactivated paleocraton margins internal to the Kaapvaal craton. Nkomati lies adjacent to the same translithospheric structure that localized the Paleo- to Meso-Archean Barberton greenstone belt. It is not clear whether this structure has experienced Paleoproterozoic reactivation. The Bushveld PGE reefs, hosted within the Bushveld Igneous Complex, are adjacent to the TML (Fig. 1.2), an internal Kaapvaal paleocratonic boundary separating the Murchison domain to the north from the rest of the craton. Reactivation of the TML under a NW far-field horizontal principal stress regime was synchronous with the intrusion of the Bushveld Complex (Good and De Wit, 1997; Clarke et al., 2009), and the TML has effectively acted as a syn-mineralization craton boundary. The PGE deposits richest in Ni-Cu occur closest to the TML, further emphasizing the importance of this structure as an injection point for mantle magmas. This relatively consistent spatial relationship between PGE reef deposits, tectonically reactivated translithospheric faults, and positions generally more inboard of craton margins than typical Ni-Cu(-PGE) sulfide deposits, suggests that a more inboard setting favors the development of a (ponded) layered intrusion rather than the magmatic (high flux?) conduits hosting most craton-margin deposits. The formation of large (lopolithic) layered intrusions appears to require (sub)horizontal layering within the crust, with injection of high-pressure magmas at favorable mechanical interfaces. Suitable geological environments include aerially extensive shallow-dipping unconformities between crystalline basement and overlying relatively undeformed sedimentary packages (e.g., 2.45 Ga Portimo and Penikat intrusions, Finland; 56 Ma Skaergaard intrusion, Greenland), and thick sequences of broadly flat-lying platform sediments (Bushveld Complex).

An additional and important question then becomes: Is the more inboard setting intrinsically favorable for PGE accumulation due to the dynamics of layered intrusion magma chambers (particularly given that many PGE reefs are interpreted as having scavenged PGE from very large magma volumes; Naldrett, 2010), or is there something else influencing the PGE metal budget? This will be discussed further in Section 1.3.5.

1.3.3 CONTINENTS AND THEIR MARGINS

As documented in Table 1.1, most significant Ni-Cu(-PGE) and PGE reef deposits formed on or near (paleo)craton margins in intracontinental settings, many significantly distant (e.g., several hundred to more than a thousand kilometers) from the nearest passive- or convergent- margin. Such a setting is in line with the timing of many of the deposits, which are often associated with the existence of supercontinents (see Section 1.5). The second most common setting, particularly for Ni-Cu(-PGE) deposits, is within pericratonic intracontinental rifts, mostly in the form of Archean greenstone belts. None of these rifts widened to the point of production of oceanic lithosphere, though some may have come close, such as the now-inverted rifts hosting the Nova-Bollinger deposit in the Albany Fraser Orogen along the southeastern margin of the Yilgarn craton, Australia, and the Kabanga camp in the Kibaran Orogen on the western margin of the Tanzania craton. Some peri-cratonic rifts form narrow basins close to the continental margin, but isolated from the major ocean by an intervening microcontinent. The c. 1.3 Ga Nova-Bollinger deposit (Maier et al., 2016a) formed during closure of such a basin between the Yilgarn Craton and the Nornalup microcontinent (Fig. 1.3D). The c. 1.85−1.84 Ga Panton Sill PGE reef and Savannah Ni-Cu(-PGE) deposits in Western Australia, may too have formed in this setting, adjacent to the North Australian Craton (NAC), but on the cusp of (or during?) the c. 1.85−1.84Ga collisional event that sutured the NAC to the adjacent Kimberley Craton (Fig. 1.3D). The giant Thompson Camp appears to have formed in a peri-cratonic rift between the Superior Craton and the outboard Sask microcontinent (Fig. 1.3B). With further rifting, some peri-cratonic rifts transition into small (peri-cratonic) marginal basins. Marginal basins are defined here as basins that have widened enough to generate some oceanic lithosphere, but not a major ocean. They are generally the result of rifting of a microcontinental ribbon/fragment or craton from an existing continent. The circum-Superior Craton Raglan Camp formed in such a setting (Fig. 1.3B). In most cases such ribbons are reaccreted to their original host continent during major ocean-closing events. Both the Thompson Belt and Raglan deposits formed prior to rift closure, and on the continental side of the rift, directly adjacent to the continental lithosphere. Limited basin width also limited the amount of crustal thickening during collisional closure of these basins. Therefore, while the Thompson Belt (in particular) and Raglan deposits are deformed, they are preserved in the eroded remnants of the resultant orogenic belts.

There are few deposits known to exist adjacent to genuine passive margins that faced large oceans. The notable exception is the c. 56 Ma Skaergaard Intrusion PGE-Reef deposit in eastern Greenland adjacent to the modern post 55 Ma North Atlantic passive margin. The intrusion occurs at the unconformity between underlying Precambrian gneiss and overlying Paleocene tholeiitic volcanics of the North Atlantic Igneous Province, and formed at the onset of seafloor spreading and opening of the North Atlantic (Mosar et al., 2002), with subsequent development of the passive margin. The intrusion is roughly 280 km inboard of the edge of the continental shelf, and actually lies on a major translithospheric structure that trends perpendicular to the coastline. This structure is at the northernmost limit of the Paleoproterozoic Nagssugtoqidian orogen (possible eastern continuation of the Disko Bugt suture and/or equivalent to the Baffin suture on Baffin Island; St-Onge et al., 2009), and we interpret it as a Paleoproterozoic craton margin separating the Rae craton to the north and the Aasiaat microcontinent to the south. Although it is at the inboard limit of the Atlantic passive margin, it is not clear if the deposit would survive the effects of crustal thickening, uplift, and erosion, were the Atlantic Ocean to close. The total lack of such deposits in the ancient rock record therefore suggests that surviving the closure of a significant width ocean is problematic.

1.3.4 **KINEMATIC/DYNAMIC FACTORS**

A critical component necessary for deposit formation is that the related translithospheric structures must be either tectonically active or at least favorably-oriented under the prevailing stress field for syn-intrusion reactivation. Lightfoot and Evans-Lamswood (2015) document the strong structural control on the form of mafic-ultramafic intrusions and the nature of the conduits that facilitate injection of the massive sulfide melts. They conclude that most deposits occur within structural corridors that experience syn-mineral strike-slip motion, either in the form of transtension or transpression.

Here we discuss a few examples of the larger, plate-scale controls on the genesis of Ni-Cu (-PGE) and PGE deposits related to mafic and ultramafic magmas.

1.3.4.1 Noril'sk Camp

The largely Paleozoic magmatic arcs of the COAB were terminated in a series of Late Paleozoic accretionary and collisional events and aided by relative convergence between the Siberia Craton and the East European Craton (Baltica; e.g., Windley et al., 2007). Synchronous with the eruption of the Siberian Traps LIP at c. 251 Ma, the region now overlain by the younger West Siberian Basin experienced right lateral slip on N-S faults, with a modest transtensional component creating narrow N-S rifts (Allen et al., 2006). A series of temporally-related en echelon NE-SW- to NNE-SSW- oriented narrow rifts connected to the N-S rifts indicates that the maximum horizontal stress (σ_{H1}) was oriented NE-SW to NNE-SSW during their development (Fig. 1.3A). Whilst the regional basaltic LIP volcanism attests to an extensional component of this transtension, the amount of extension appears to have been relatively modest, since volcanism across the region now occupied by the West Siberian Basin erupted subaerially through this time (Holt et al., 2012). The majority of Ni-Cu-PGE mineralization in the Noril'sk camp occurs adjacent to the Noril'sk-Kharaelakh fault within the Noril'sk-Kharaelakh Trough, a <10 km deep depression in the basement that hosts Neoproterozoic to Permian sediments, overlain by the thickest accumulation of mafic volcanics (up to 3.5 km-thick) belonging to the Siberian Traps on the craton. This NNE-trending trough traverses the NW corner of the Siberian craton, and was optimally oriented for reactivation with respect to a Late Permian NE- or NNE-directed principal horizontal stress field. We interpret the Noril'sk-Kharaelakh fault structure as the reactivated expression of a probably Early Proterozoic (or Archean?) suture between the main body of the craton and a microcontinental block to the NW. On-craton along the trend of the Noril'sk-Kharaelakh fault, the earliest Siberian Traps lavas erupted into a shallow lacustrine or lagoonal environment, in contrast to the subaerial eruption of most of the lavas elsewhere on the craton (Czamanske et al., 1998). Most of the mineralization is in sill-like intrusions (chonoliths) hosted by Paleozoic sediments in the footwall of the volcanic pile. The Traps sequence on the NW part of the craton is divided into three main assemblages from oldest to youngest. The upper portion of the second assemblage is comprised of the strongly metal-depleted (~99% of precious metals, and ~80% of Ni, Cu, Co removed) basaltic rocks of the Nadezhdinsky Formation (Lightfoot and Keays, 2005). Thickness isopachs of the first two assemblages are parallel with the trend of the trough, whereas thickness isopachs for the third (youngest) assemblage lie at an angle of 90 degrees to the trough (Fedorenko, 1994), indicative of a switch in regional stress conditions (e.g., to σ_{H2}, Fig. 1.3A; precise orientation unknown) away from the prevailing NNE- to NE- directed maximum horizontal stress direction. Crucially, the observed timing of the Ni-Cu-PGE mineralizing intrusions is during the early stages of the youngest assemblage (Fedorenko, 1994), consistent with the possibility that a sudden regional tectonic stress switch served as the trigger for release of magmas rich in magmatic sulfide melt from deeper within the magma conduit network.

1.3.4.2 Voisey's Bay

The tectonic setting of the Voisey's Bay Ni-Cu deposit is depicted in Fig. 1.3B. The deposit is associated with the large intrusive 1.35−1.29 Ga NPS, which intruded at the intersection of the margin of the Nain Craton and the cross-cutting E-W Gardar Fracture Zone (GFZ; Myers, 2008). The dominant trend of the NPS is parallel to the NNW-trending craton margin. Oriented almost orthogonal to the c. 1.5−1.0 Ga Grenville convergent margin, tectonic reactivation of the Nain craton margin is to be expected during periods of broadly N-S maximum horizontal stress (σ_{H1}), and is likely to have facilitated intrusion of mantle-derived magmas of the NPS. At a late stage of NPS magmatism (c. 1334 Ma; Lightfoot and Evans-Lamswood, 2015), synchronous with the onset of dextral transtension along the GFZ, a set of narrow NPS-related mafic dykes and associated magmatic sulfide melts intruded the GFZ adjacent to the craton margin. The majority of the economic mineralization injected as "sulfide and fragment-laden batches of magma" along conduits within the dyke network as the last event within the mafic intrusion (Lightfoot and Evans-Lamswood, 2015). The cause of the onset of dextral transtension (at time t_2) was likely a stress switch associated with events at the Grenville convergent margin (σ_{H2}), which served as the trigger for the release of major accumulations of massive sulfide melt from deeper-seated regions of the intrusive network.

1.3.4.3 West Musgraves (Nebo-Babel)

The c. 1.08 Ga Warakurna LIP in Australia extends as a series of sills, volcanics and intrusive complexes following major cratonic boundaries for more than 1500 km in an E-W direction (Fig. 1.3D). The largest concentration of associated mafic-ultramafic magmatism, interpreted to likely represent the location of a mantle plume impact, is in the West Musgraves region at the triple junction of the Western- and Southern-Australian cratons with the Mulga Microcontinent, a lithospheric block attached to the southern margin of the greater North Australian Craton. Intrusion of thick, c. 1.08 Ga layered mafic complexes (Giles Complex) in the West Musgrave, are followed by uplift and subsequent bimodal volcanism of the Bentley Supergroup (Smithies et al., 2015). West Musgraves Ni-Cu mineralization is associated with broadly NE-trending c. 1068 Ma chonolith intrusions containing multiple magmatic pulses, and seemingly later than, or slightly overlapping with, the bimodal volcanism (Seat et al., 2007, 2011). A series of c. 1.09−1.02 Ga high-grade metamorphic and polyphase deformation and magmatic events within the Pinjarra Orogen on the Western edge of the Yilgarn Craton, but not significantly affecting the craton, are consistent with an oblique (dextral) collisional event on this margin (Ksienzyk et al., 2012 and references therein). We interpret that synchronous with Warakurna plume impact close to the cratonic triple junction, broadly E-W shortening (σ_{H1}) at an early (c. 1.09−1.07 Ga) stage of this orogenic event is responsible for the LIP selecting the E-W cratonic margins as the preferred site of crustal injection of magma. At the West Musgraves, voluminous mafic intrusions weakened an already warm, thick orogenic crust (Smithies et al., 2015), resulting in extension and uplift, followed by the bimodal volcanism. We interpret that the Pinjarra orogen reached an advanced stage of oblique collision at c. 1.07 Ga, manifested by a switch to dextral strike slip along the western margin of the Yilgarn craton (Darling Fault). This resulted in transfer of a NE-directed maximum stress (σ_{H2}) through the continental lithosphere. We invoke this stress switch as the mechanism for releasing magmatic sulfide melts from deeper-seated regions of the magma conduit network, and their dynamic injection into the evolving multiphase bodies of the broadly NE-trending upper-crustal chonoliths of the West Musgraves camp.

1.3.4.4 High-MgO Versus Low-MgO (Ultramafic vs Mafic-Ultramafic)

High-MgO, ultramafic-dominant, end-member Ni-Cu(-PGE) deposits are exclusively Archean to Paleoproterozoic in age (Table 1.1), consistent with the higher melting potential of mantle plumes in the early Earth (modeling indicates a decline of about 150°C over the last 3 Ga; Labrosse and Jaupart, 2007). Archean examples are hosted by intracratonic greenstone-belt terranes, mostly along paleocraton margins (e.g., Agnew-Wiluna Belt camp, Yilgarn Craton; Fig. 1.3D). Paleoproterozoic examples (Thompson, Raglan) occur in peri-cratonic rifts close to the continental margin (discussed above; Fig. 1.3B). This setting coincides with a transition from thick to thinner lithosphere, which can focus a dynamic mantle upwelling such as a mantle plume, resulting in high-degree partial melting of the upwelling mantle (Begg et al., 2010; Barnes et al., 2012; Griffin et al., 2013). A combination of thinned lithosphere and neutral to extensional tectonic conditions will facilitate ascent of the magma into the rift. The high MgO compositions of these komatiitic melts associated with deposits indicate that those melts experienced minimal fractionation en route to the upper crust. The ultramafic rocks hosting the deposits, however, show evidence of extensive contamination by upper crustal sulfur-rich rocks, an essential step in deposit formation, as described in Section 1.2.2.1. Archean rift settings, with their attendant bimodal magmatism, provided a sedimentary environment rich in exhalative sulfides. The ingestion of these sulfides by the ultramafic intrusions and flows induced sulfide saturation in the melts, allowing gravitational accumulation of the metal-rich sulfide melt at the base of flows where it ultimately crystallized to form accumulations of massive sulfide (e.g., Lesher and Keays, 2002).

The rift architecture is likely to have provided a first order constraint on localization of the ultramafic intrusions and flows as well as their geometry. Intersections of steep transfer faults with the rift axis would provide a particularly favorable translithospheric conduit (Perring, 2015). Such sites may well have localized the preceding volcanism, as well as any associated felsic volcanic centers and thick accumulations of marine exhalative sulfides. The fundamental plumbing site therefore effectively localizes the key ingredients—high volumes of ultramafic magma, plus large quantities of sulfidic sediment—in the same location. After rising to volcanic or subvolcanic depths ultramafic lavas and sills are likely to have flowed/injected along the submarine rift axis, parallel to bounding rift-related graben edges. As a result, despite multiple overprinting shortening events (Beresford et al., 2002, 2005; Kositcin et al., 2008; Perring, 2015), the orebodies are mostly oriented subparallel to the paleocraton boundary and former rift axis.

In contrast, the Low-MgO (mafic-ultramafic rock associated) Ni-Cu(-PGE) deposits are dominantly post-Archean in age. As described in Sections 1.2.2.1 and 1.3.2, their typical setting is along craton margins, occasionally following a period of extension and rift development, but more commonly where injecting magmas have been introduced into relatively thick orogenic crust. Magmas intruding thick crust are more likely to have a complex ascent path involving multiple interconnected dykes and sills. As a result, they are likely to experience significant fractionation and there is strong potential for contamination by the enclosing country rocks. The affinity of such deposits with craton margins is favored not just by a transition from thick to thinner SCLM, and the presence of translithospheric faults as the plumbing system, but also the predilection of craton margins to undergo tectonic reactivation during regional tectonic events. The scarcity of postcratonization deposits close to the centers of cratons is a testament to the internal structural integrity of these ancient lithospheric blocks.

1.3.5 IMPACT ON METALLOGENY: PGE-RICH VERSUS PGE-POOR

Multiple generations of mantle-derived magmas from across the Kaapvaal craton are PGE-rich, indicating that the PGE content of mineralized intrusions is not just a function of crustal processes (Maier and Groves, 2011). Hughes et al. (2016) show that lamprophyric dykes cutting the western Bushveld Complex have Pt/Pd ratios that fall within the range of the hosting western Bushveld Complex. Given an SCLM origin for the lamprophyre magmas, the finding is consistent with a possible SCLM contribution of PGE metals to the Bushveld melts. Similarly, the mafic intrusions on some cratons or parts of cratons appear to be particularly well endowed with PGE, while others are poorly endowed (e.g., Yilgarn craton, Australia). These observations are consistent with the possibility of a control in the source region. Xenoliths derived from cratonic roots commonly contain sulfide, and these sulfides carry PGEs (Griffin et al., 2002). Ascending melts coming into contact with the lithospheric root would be able to dissolve sulfides (Mavrogenes and O'Neill, 1999), suggesting that the SCLM could contribute PGE to the magmas. Melting of a mantle plume enriched in a component of incompletely equilibrated (meteorite-derived) late veneer, or in a component from the Earth's outer core, is proposed as a possible alternative source of the PGEs by Maier et al. (2016b).

Conventional models of asthenospheric melting tend to treat most magmas as equally endowed with PGEs (and Ni). However, this makes it difficult to explain the spatial distribution of both Ni-Cu(-PGE) deposits and PGE reef deposits. LIPs and komatiites in areas without significant SCLM roots apparently do not contain significant deposits of these types (Zhang et al., 2008), consistent with the finding above that most major deposits and endowment are associated with the most resilient (best-preserved, thick) SCLM. Flood basalts from provinces that lack Ni-sulfide deposits have lower Os contents than basalts from provinces containing such deposits (Fig. 1.4A). Thus the high-Os signature of "fertile" flood basalts appears, empirically, to reflect their interaction with old SCLM, which also has been metasomatically enriched in sulfur over time (discussed further in this section).

Izokh et al. (2016) examined the PGE contents of mafic rocks related to the Permo-Triassic Siberian LIP across the Siberian craton. They observed that early magmas in the central part of the LIP, and those peripheral to the craton and main magmatic outpouring of the LIP center and likely plume head, have lower PGE contents than later magmas closer to the plume head. Drawing on analogous findings for other LIPs, they contend that elevated K contents and high PGE contents of magmas in the mantle plume head are sourced from the core-mantle boundary. In support of this argument, they note that the komatiites of the Emeishan LIP have $^{187}Os/^{188}Os$ ratios similar to chondrites (Hanski et al., 2004), and conclude that this indicates that the high PGE contents were brought from the core-mantle boundary by "thermochemical plumes."

However, such arguments ignore abundant evidence on the time-integrated behavior of sulfides and PGEs in the SCLM (Griffin et al., 2002, 2004). In-situ analysis of PGEs and Re-Os systematics in mantle-derived xenoliths, mainly from the Siberia and Kaapvaal cratons, shows that the SCLM is a major reservoir for PGEs, and that their distribution is controlled by sulfides. During the extensive melt depletion involved in the formation of cratonic roots, Os and Ir behaved as compatible

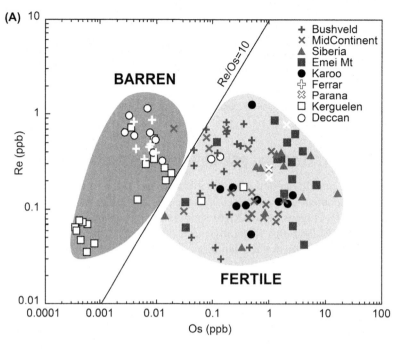

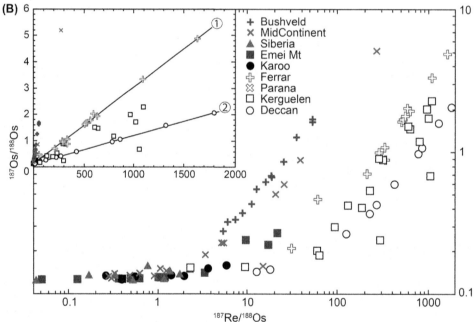

FIGURE 1.4

Re-Os data for LIPs and other magmas. (A) Re and Os contents of mafic rocks from provinces with known Ni-Cu (-PGE) sulfide deposits (*dark symbols*) and provinces lacking known deposits (*white symbols*). (B) $^{187}Re/^{188}Os$ versus $^{187}Os/^{188}Os$ in flood-basalt suites. Main figure, log−log scale; inset, linear scale. The plots in the inset correspond to: (1) Ferrar dolerites, 177 ± 2 Myr ago, intercept 0.125 ± 0.033 (γOs = −0.6 ± 0.26); (2) Deccan traps, 65.6 ± 0.3 Myr ago, intercept 0.12843 (γOs = 1.5 ± 0.3).

After Griffin, W.L., Begg, G.C., O'Reilly, S.Y., 2013. Continental-root control on the genesis of magmatic ore deposits.

Nat. Geosci. 6, 905–910

elements, relative to elements such as Pt and Pd, and concentrated in the residues, probably as micronuggets (see Lorand et al., 2013 for a review of PGE systematics). The oldest Os model ages are found in high-Os + Ir sulfides that may be residual, or may have formed by reaction between introduced sulfide and such micronuggets. The significance of this is that mantle-derived sulfides from beneath ancient cratons have significantly higher mean Os contents than sulfides in xenoliths and massif peridotites from areas with younger mantle lithosphere, where the PGE content is derived largely from asthenospheric melts (Griffin et al., 2002, 2004; Fig. 1.4B).

Most peridotite xenoliths from cratonic roots carry several generations of sulfides; some of these have model ages that correspond to major crustal orogenic events, suggesting that the SCLM is periodically traversed by asthenosphere-derived melts that introduce more sulfide. Many xenoliths contain >1 generation of sulfide (e.g., Griffin et al., 2002, 2004, 2011), and younger generations commonly have highly radiogenic Os (high ^{187}Os/^{188}Os, giving "future" model ages), reflecting scavenging of PGEs from silicate minerals with high Re/Os. These magmatic sulfides also tend to be enriched in Pt + Pd, relative to Os + Ir. Thus the ^{187}Os/^{188}Os ratios of the bulk sulfide component of the SCLM will, at any time, give a model age that is significantly younger than its age of formation (Griffin et al., 2004). The chondritic ^{187}Os/^{188}Os of the Emeishan komatiites (Hanski et al., 2004) therefore can represent simply the signature of a strongly refertilized SCLM, rather than the core-mantle boundary.

Sequential refertilization of the SCLM by asthenosphere-derived melts and fluids, especially around craton margins and the base of the deep cratonic roots, thus leads to enrichment in both sulfide and incompatible elements, including K (O'Reilly and Griffin, 2012). The pattern observed by Izokh et al. (2016) in the Siberian Traps, with coupled PGE and K contents, therefore is what would be predicted for interaction of a high-T melt with the deep SCLM: enrichment of the melts in K, and in sulfide with high Pt/Os, like those observed in Norilsk and other PGE-rich LIP-related deposits. These patterns imply a significant role for the SCLM in the localization and PGE fertility of mantle-derived magmas involved in the formation of Ni-Cu-PGE and PGE reef deposits.

1.4 GENERAL MODEL

A lithosphere-scale general model (The Sudbury camp is not considered here due to the low probability of other similar bolide impact-related events resulting in a metallogenically significant outcome.) for the location of Ni-Cu(-PGE) sulfide and PGE reef deposits is shown in Fig. 1.5. The salient features generic to the variants presented are:

1. Impact of a dynamic upwelling of convecting mantle (generally a mantle plume) onto the base of thick (>150 km) cratonic lithosphere, followed by flow of the upwelling/plume to regions of thinner lithosphere at the craton, or paleocraton, margin.
2. At roughly 100 km depth or less, decompression-related high-degree partial melting of the upwelling/plume generates large volumes of upper mantle/asthenospheric melt.
3. Variable interaction of the upper mantle/asthenospheric melts with SCLM. Melting of previously refertilized/metasomatized regions of SCLM may contribute significant quantities of PGE to the melt. This appears to be particularly favored in the region of paleocraton boundaries internal to the larger craton.
4. Transfer of melts to the upper crust via tectonically active translithospheric faults and an interconnected network of intrusive magmatic conduits. Melts may interact with wallrocks at all

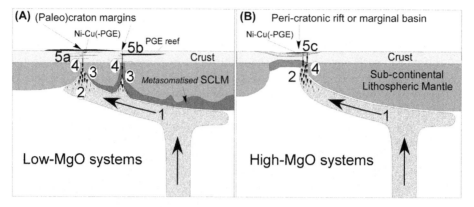

FIGURE 1.5

Generalized model for the formation of Ni-Cu(-PGE) and PGE deposits. (A) Low-MgO systems within a collage of juxtaposed lithospheric blocks. High-MgO deposits do not form in this setting. (B) High-MgO systems and rifted lithosphere. Occurrence of Low-MgO deposits in this setting is rare. (1) Impact of a mantle upwelling (plume) and flowage to regions of thinner lithosphere; (2) Adiabatic melting of the upwelling mantle; (3) Melts may incorporate PGE by melting refertilized/metasomatized SCLM. This process seems to be particularly favored at paleocraton boundaries inboard of the craton margin. (4) Transfer of melts into the crust via active translithospheric faults and an interconnected intrusion (sills and dykes) network. (5a) Low-MgO Ni-Cu(-PGE) systems; (i) Sulfide melt saturation via either crustal contamination or system dynamics (including fractionation); (ii) accumulation of magmatic sulfides in deeper parts of the (upper crustal?) conduit system; (iii) dynamic injection of sulfide-rich melts into upper-crustal conduits (e.g., chonoliths) and intrusions following breach of a mechanical threshold barrier, sometimes induced by switches in far-field stress conditions. (5b) PGE reef systems: (i) Injection of generally S-poor, sulfide-undersaturated ultramafic-mafic magma into a large layered intrusion; (ii) sulfide melt saturation and gravitational settling to form narrow laterally-extensive PGE-rich reefs. (5c) High-MgO Ni-Cu(-PGE) systems: (i) Eruption or injection of sulfide-undersaturated ultramafic magma into volcanic or subvolcanic channels; (ii) Extensive contamination of magma with sulfidic sediments results in magmatic sulfide melt saturation; (iii) accumulation of sulfide melt at the base of the channel by gravitational settling.

    points of the journey (SCLM, lower crust, upper crust) unless isolated by earlier pulses of intrusions. The intrusive/extrusive network varies depending on the composition, form, rheology and structural framework of upper-crustal units, coupled with the prevailing stress conditions.

5. Variable interaction of melts with crust, and/or the dynamics of the magmatic system, combine to induce sulfide (Ni, Cu, PGE) saturation within the magma, and/or PGM precipitation. Dependent on tectonic setting, far-field stress conditions and the nature of the country rocks, this may occur either *en route* within the plumbing system (Low-MgO Ni-Cu(-PGE)), in situ within a ponded intrusive mass (e.g., layered intrusion; PGE reef), or in extrusive flow or subvolcanic channels (High-MgO Ni-Cu(-PGE)). Within large Low-MgO magmatic systems, accumulation of concentrated magmatic sulfide melts may occur in the magmatic conduit system (e.g. at Noril'sk during Nadezhdinsky Formation magmatism) below the level of final emplacement. The injection of these sulfide melts into shallow level upper-crustal intrusive networks frequently follows a stress switch, and/or the breaching of a physical threshold barrier, akin to the behavior of a SOCS.

1.5 DEPOSIT TEMPORAL DISTRIBUTION, THE SUPERCONTINENT CYCLE, AND DYNAMIC EARTH MODEL

The temporal distribution of significant Ni-Cu(-PGE) and PGE reef deposits, and/or regions with deposits of approximately equivalent age, is broadly similar (Table 1.1; Fig. 1.6); neither are uniformly distributed through time. This is consistent with the strong genetic connection between these deposit types. The majority are distinctively clustered in time into 4 prominent groupings, Late Archean, Middle to Late Paleoproterozoic, Middle Mesoproterozoic to Early Neoproterozoic, and Late Paleozoic.

Also shown on Fig. 1.6 are the lifespans of the four supercontinents of Kenorland, Nuna, Rodinia and Pangea, exhibiting a cyclicity with a period of approximately 0.8 Ga. Each exists within a continuum of formation and breakup under the agency of plate tectonics. We have broken each supercontinent cycle into three stages: *amalgamation*, *peak*, and *breakup* (Fig. 1.6). Each stage is defined by geological observation from our 15 year program of analysis under the umbrella of the Global Lithospheric Architecture Mapping Project (GLAM, see also Section 1.3.1; e.g., Begg et al., 2009). The amalgamation period is when the first significant collisions between lithospheric domains begin to build a new continent, and often result in rapid clustering of major continental masses such as the Late Neoproterozoic Gondwana, regarded here as merely a staging post on the way to the final supercontinent. The beginning of the peak period begins with the final collisions and resultant orogenic belts that stitch the majority (>85%) of the global continental lithosphere into a single entity, the supercontinent. The breakup period encompasses the full spectrum from rifting and development of new oceans as the supercontinent begins to fragment, through to the final separation of the major land masses into a multitude of continents and microcontinents such as experienced in the present day. The last three (and possibly all) supercontinents appear to display a smooth transition between final breakup of one supercontinent and the beginning of amalgamation of the next. We employ the following breakdown of each supercontinent cycle: Kenorland, amalgamation from 2.93 Ga, peak from 2.7 Ga, breakup from 2.45 to 2.25 Ga; Nuna, amalgamation from 2.15 Ga, peak from 1.9 Ga, breakup from 1.65 to 1.40 Ga; Rodinia, amalgamation from 1.38 Ga, peak from 1.1 Ga, breakup from 0.82 to 0.59 Ga; Pangea, amalgamation from 0.59 Ga, peak from 0.3 Ga, breakup from 0.18 to 0.06 Ga.

The four broad peaks of Ni-Cu(-PGE) and PGE deposit formation described above display a strong temporal connection with the supercontinent cycle, in keeping with observations made by earlier workers (e.g., Kerrich et al., 2005). The dominant period of ore formation for each supercontinent is during the later stage of the amalgamation and early stages of the peak. Interestingly, this is coincident with times of maximum structural disturbance and shortening of the lithosphere, as large (micro)continental blocks are sutured together. This may seem at odds with the formation of deposits that require major production of mafic-ultramafic mantle-derived magmas and their injection into the crust, both of which intuitively may more easily be imagined as occurring during periods of regional extension. But the truth is stranger than the imagination. The late amalgamation and peak periods of supercontinent formation are most favored for two crucial reasons: (1) coincidence with cospatial mantle plume activity (see below), and; (2) the presence of translithospheric fault networks that are either active or at the point of activation, and therefore can be exploited by the mantle melts to develop conduits to the upper crust. In addition, delamination of juvenile lithosphere caught in the vice formed by colliding blocks of more durable continental lithosphere

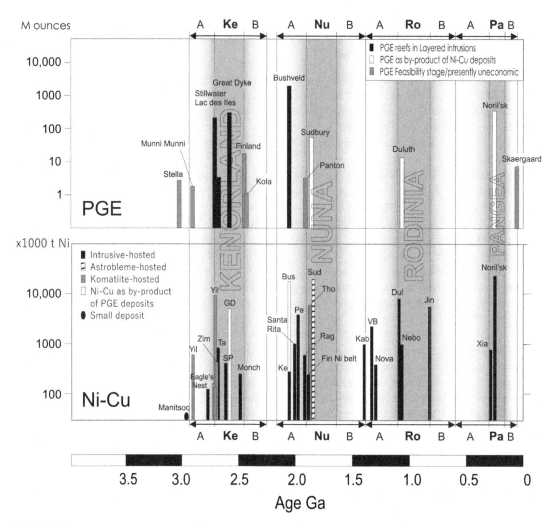

FIGURE 1.6

Secular distribution of PGE and Ni-Cu deposits and key elements of the supercontinent cycle (modified after Maier, W.D., Groves, D.I., 2011, Temporal and spatial controls on the formation of magmatic PGE and Ni–Cu deposits. Mineral. Deposita 46, 841–857). Supercontinent peaks Kenorland, Ke, Nuna, Nu, Rodinia, Ro, and Pangea, Pa, are preceded by periods of amalgamation, A, and dispersed during periods of breakup, B (see text for details). Komatiitic-hosted (High-MgO) deposits are restricted to the earlier two supercontinent cycles. Note the clustering of deposits about the amalgamation and peak stages. *Bus*, Bushveld; *Dul*, Duluth; *Fin Ni belt*, Finnish Ni belt; *GD*, Great Dyke; *Jin*, Jinchuan; *Kab*, Kabanga; *Ke*, Keivitsa; *Monch*, Monchegorsk; *Nebo*, Nebo-Babel; *Pe*, Pechenga; *Rag*, Raglan; *SP*, Selebi Phikwe; *Sud*, Sudbury; *Ta*, Tati; *Tho*, Thompson; *VB*, Voisey's Bay; *Xia*, Xinjiang Province deposits (including Huangshandong, Huangshan, Kalatongke, Xiangshan); *Yil*, Yilgarn; *Zim*, Zimbabwe.

(e.g., archons and reworked archons) is likely to be more common during this time period. Such delamination may result in mantle upwellings capable of forming some of the smaller deposits (though this has yet to be confirmed for any significant deposit), but as discussed in Section 1.2.1 this mechanism is not a plausible explanation for the formation of the larger deposits. This time period also favors deposit preservation, as larger blocks of durable continental lithosphere are available to host the resultant deposits, in contrast to isolated microcontinental blocks that may be prone to major tectonic reworking (and deposit loss) during their collision.

The coincidence of mantle plumes and the late almalgamation and peak periods of a supercontinent cycle can be reconciled with a model of whole-Earth processes whereby mantle plumes are generated within or directly above the D″ region at the base of the lower mantle. The D″ is a seismically-anomalous zone above the Core-Mantle Boundary (CMB) that contains the remains of subducted slabs of ocean lithosphere. Seismic tomography data image some slabs traversing the whole mantle (e.g., Simmons et al., 2015), with the coldest slabs most likely to penetrate all the way to D″. The slabs are the probable origin of the higher velocity regions of D″ known as the slab graveyards (Richards and Engebretson, 1992). Seismic images of D″ reveal two large areas of relatively low velocity, surrounded by the higher-velocity slab graveyards (Fig. 1.7). These two large regions of low velocity, rising to more than 1500 km above the CMB, and tapering towards their tops, are known as the African and Pacific Large Low Shear Velocity Provinces (LLSVPs; Lay, 2005). They are interpreted as regions of relatively hot mantle, perhaps also with a distinctive Fe-rich composition (e.g., Tackley, 2012).

Using paleomagnetic data, and building on the work of Burke and Torsvik (2004) and Thorne et al. (2004), Torsvik et al. (2006, 2008) reconstructed the paleoeruption position of LIPs formed in the last 300 Myr. They demonstrated that the LIPs, in addition to modern-day hotspots, formed directly over the edges of the LLSVP regions, providing strong evidence that mantle plumes rise from the base of the mantle in response to the dynamic geometrical organization of mantle features (Fig. 1.7). Numerical modeling supports this finding (Tackley, 2011; Steinberger and Torsvik, 2012); Tackley (2011; following Olson and Kincaid, 1991) demonstrates that when slabs in the graveyard reach a critical temperature, the harzburgitic parts of the slabs become buoyant and rise from the D″ region as mantle plumes. The LLSVPs also coincide with areas of elevation in the geoid; the +10 m contour of the geoid coincides with the edge regions of the LLSVPs when projected vertically upward (Burke et al., 2008). The geoid anomaly gives support to the idea of a hot, buoyant region of lower mantle with a component of upflow, relative to broad regions of downflow elsewhere associated with subducted slabs.

Fig. 1.7 illustrates the position of the Pangea supercontinent (after Scotese, 2001) relative to the African LLSVP and the reconstructed positions of <300 Ma LIPs and hotspots (note that the 297 Ma Skagerrak-Centered LIP is the oldest considered; Torsvik et al., 2008). The supercontinent is positioned directly above the African LLSVP, and in the path of the plumes responsible for most of the LIPs. This superposition of features points to a direct connection between supercontinent formation and the genesis of an underlying LLSVP. We propose that with closure of the major oceans, and gradual switch of subduction focus to the peripheries of the growing supercontinent, portions of the slab graveyard beneath the nascent supercontinent become sufficiently buoyant to initiate the rise of large plumes. In this way, the area of mantle that had been the focus of subduction to build the supercontinent, becomes the focus of subsequent plume production. The ensuing succession of plumes (a plume "bloom") effectively bombard the base of the supercontinent where the topography of the Lithosphere-Asthenosphere Boundary (LAB) focuses plume melting into spatially

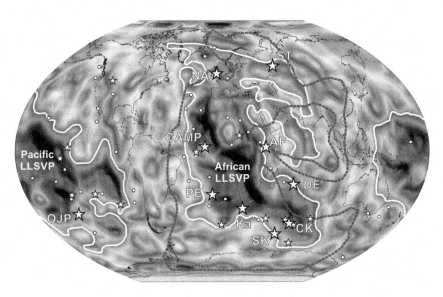

FIGURE 1.7

Relationship between LIPs, hotspots, supercontinents, and the lower mantle. Paleomagnetically reconstructed positions of <300 Ma LIPs (stars) and hotspots (dots) at the time of their initial eruption (after Torsvik et al., 2006, 2008) relative to the zones of anomalously low shear wave velocity (LLSVPs; *white outline*) within the D″ layer directly above the CMB (unpublished 2011 update of Grand, 2002, seismic tomography model). The largest LIPs (large stars) are: 251 Ma Siberian Traps, ST; 201 Ma Central Atlantic Magmatic Province, CAMP; 184 Ma Karoo, Ka; 132 Ma Parana-Etendeki, PE; 120 Ma Ontong Java Plateau, OJP; 115 Ma Central Kerguelen and Southern Kerguelen, CK, SK; 65 Ma Deccan Traps, DE; 62 Ma North Atlantic Igneous Province, NA; 30 Ma Afar, Af. The approximate position of the Pangea supercontinent at 250 Ma (after Scotese, 2001) is outlined by the *dashed line*. Most of the LIPs and hotspots are spatially correlated with the LLSVPs, and particularly around their margins. In addition there is a clear correlation of the African LLSVP, its associated LIPs and hotspots, and the position of the Pangea supercontinent. ST, Ka, and NA host Ni-Cu(-PGE) and/or PGE reef deposits.

discrete regions of thinner lithosphere. The temporal coincidence with times of maximum activation of translithospheric structures provides a link with the crustal environment and the opportunity for magma migration and deposit genesis.

Ultimately, the change of overall subduction configuration, mantle convection, and dynamic topography (reflected in the geoid) above the African LLSVP, conspired to fragment and disperse the Pangea supercontinent. As breakup proceeds and oceans open above the LLSVP, the chances of large mantle plumes impacting continental lithosphere diminish significantly, a key factor in the relative lack of Ni-Cu(-PGE) and PGE deposits during these time periods.

1.6 EXPLORATION SUMMARY

The purpose of this section is to summarize the most useful pragmatic exploration targeting criteria at each level in the scale hierarchy of targeting. The emphasis is on generic parameters because these are the most useful in exploration which is often focused on relatively poorly known regions, where new major deposits are most likely to be found.

This discussion focuses mostly on sulfide-rich Ni-Cu(-PGE) deposits, not the much rarer class of sulfide-poor PGE reefs. Exploration for the latter is essentially about identifying very large mafic intrusive complexes (generally slightly inboard of craton margins; often associated with paleocraton boundaries) and then focusing on the stratigraphic positions of major magmatic unconformities. Because these reefs tend to be very continuous (i.e., many tens of km) with broadly similar levels of mineralization, they are not particularly challenging exploration targets if they are actually present.

1.6.1 GLOBAL

At the global scale there are time periods within the supercontinent cycle that are most favorable for the generation of major deposits. These periods begin during the late parts of the amalgamation stage of a supercontinent, and end during the middle stages of the peak of the supercontinent stage, as defined in Fig. 1.6. Deposits do form at other times, but much less frequently. They rarely form during the breakup stage.

1.6.2 CONTINENTAL

At a continental-scale the key ingredients in a targeting model relate to the tectonic environment and the degree of lithospheric resilience. The presence of either a LIP or magmatism permissive of an LIP is also favorable, but diagnostic information may not be available, so a permissive approach should be made. The favorable time periods identified at the global scale may be used as a proxy here. The lithospheric resilience refers to the extent to which thick SCLM is preserved; the most resilient lithosphere generally coincides with archons and with proton/archons that have experienced relatively modest post-Archean tectonothermal reworking. Craton margins surrounding such resilient blocks will have the highest priority due to the potential for focused melting of upwellings of convective mantle. In addition, deposit preservation appears to be enhanced by the close proximity to relatively resilient lithospheric domains.

From the perspective of tectonic settings the spatial distribution of major deposits is inversely correlated with the presence of a syn-mineral convergent margin. Therefore, proximity (<300 km?) to these margins should be avoided. Similarly, passive margins that fringed major oceans at the time of a potential mineralizing event should be avoided, as there are no significant deposits known to have formed in and/or survived tectonic closure of, such an ocean. As a result of the avoidance of these two tectonic settings, intracontinental regions at the time of the mineralizing event in question are deemed most favorable. There will also be certain time periods within each continent, and each part of a continent, that are more favorable due to the timing of activation of major structures, including craton margins, under the prevailing stress field. Periods dominated by strong extension should be avoided, as there is a poor correlation of such periods with deposit formation, possibly due to a lack of magmatic focus.

1.6.3 REGIONAL

Although conceptual models (see above) for Ni-Cu(-PGE) deposits relate these to LIPs, in most cases the discovery of major new deposits occurred prior to the recognition of their host LIP and its extent. One reason for this is that the exposure of most pre-Phanerozoic LIPs is fragmentary because of later deformation, erosion, and coverage by younger sediments. If evidence for a LIP exists, as is commonly the case with some younger examples, then this is a positive, but in the absence of diagnostic information, a permissive approach applies. Geochemical evidence of contamination of plume melts by metasomatized SCLM should be seen as a positive for prospectivity of a LIP for PGE-rich mineralization, however the requisite analytical (major and trace element; isotopic) data will seldom be available. Therefore, in many cases, practical regional targeting is generally not driven by the concept of a host LIP as defining the region of interest.

Instead, at the regional scale, the most fundamental targeting element is the presence of a large volume of mafic/ultramafic rocks, consistent with a site of focused magma emplacement, spatially coincident with a major translithospheric structure that is likely to be associated with the margin of a lithospheric domain (i.e., craton or paleocraton margin). These mafic/ultramafic rocks typically comprise only a small proportion of their host LIP. A prospective region is typically on the scale of 100 to a few hundreds of kilometers in strike-length (subparallel to a controlling translithospheric structure). In some cases, the concept of a prospective region may be essentially equivalent to that of a camp, as discussed below.

A good example of this for High-MgO systems is the Norseman-Wiluna belt (within the Yilgarn Craton of Western Australia), which formed on the rifted margin of the Youanmi paleocraton (Western Yilgarn in Fig. 1.3D) about 2.8−2.7 Ga. In this case, a map of c. 2.7 Ga komatiites shows a strong concentration within this belt and constitutes a first-order targeting criterion. Good examples of this prospective regional association include the Fraser belt on the Yilgarn Craton margin, the Giles Complex in the West Musgrave region (which is associated with the confluence of several major translithospheric craton boundary structures), and the NPS, associated with a structural intersection on the margin of the Nain Craton in Labrador.

Prospective regions for High-MgO deposits are likely to be dominated by volcanic or subvolcanic units, whereas for prospective Low-MgO deposits the presence of a large volume of intrusions is important. Empirically, concentrations of mafic rocks dominated by dyke swarms do not appear to be prospective for significant Ni-Cu(-PGE) deposits. This may be because mafic magmatism is less focused in such situations and dyke swarms are interpreted to represent magma flow away from a center of upwelling (e.g., Ernst and Buchan, 2002).

In many cases, the presence and nature of a prospective mafic/ultramafic concentration will need to be inferred from geophysical data, primarily magnetic surveys. Many ultramafic rocks tend to be magnetic because serpentinization of olivine (a very common alteration process) generates magnetite. In such rocks, the intensity of magnetization may be a crude proxy for the original olivine content of the rock and therefore a proxy for focused magma flux. Systems dominated by mafic intrusions also tend to have a magnetic expression but in contrast to ultramafic systems, the strongest magnetic response is usually associated with the more fractionated parts of these intrusions which crystallize titanomagnetite as a primary igneous phase. The patterns of magnetic response are important in assessing regional prospectivity for Low-MgO systems. Favorable mafic intrusive complexes are likely to have complex, ameboid geometries rather than the regular patterns of dyke swarms.

It is important to note that the spatial concentration of mafic/ultramafic rocks that define a regional area of interest need not all necessarily be coeval with and directly related to the ore-hosting magmatic event. Rather, favorable regions are commonly the loci of multiple phases of mantle-derived magmatism. The ore-hosting event may be only one (or more) emplacement event within a broader magmatic history.

Ophiolites are also commonly emplaced along major lithospheric domain boundaries and can therefore superficially resemble prospective regions for Ni-Cu(-PGE) deposits, particularly as they host minor occurrences of this style of mineralization. Probably the best practical way to eliminate these settings is on the basis of evidence for lateral continuity of ultramafic units rather than the significant lateral heterogeneity as we expect in prospective intrusive complexes.

In some cases (e.g., Jinchuan region, China) prospective mafic/ultramafic intrusions occur as a only relatively minor component along a major translithospheric structure. However, these intrusions typically show strong differentiation which is a proxy for focused magma flux. Therefore if it is possible to identify a series of volumetrically minor but strongly differentiated mafic/ultramafic intrusions along a major lithosphere domain-bounding structure, this setting should also be considered a prospective region for Ni-Cu(-PGE) deposits.

Because sulfide saturation via crustal contamination is considered to be important in the formation of many Low-MgO Ni-Cu(-PGE) deposits, and essential for the High-MgO deposits, it may be thought that the regional presence of sulfur-rich rocks (either sulfidic sediments or evaporites) in close association with a mafic/ultramafic intrusive complex may be an important factor in regional targeting. However for several reasons, this is not the case in a practical sense. The presence of sulfur-bearing supracrustals may be impossible to evaluate at the early stages of exploration (as was the case with the West Musgrave region) or sulfur-rich rocks may be so widespread their presence does not help targeting (e.g., typical komatiite-hosting greenstone belts) or the site of crustal sulfide assimilation may be below the current level of exposure (i.e., as inferred for the Platreef by Penniston-Dorland et al, 2008). In addition, other mechanisms can result in sulfide saturation (discussed in Section 1.2.2.1), and contamination by S-rich crust is not a universal feature of Low-MgO deposits. We therefore regard the presence of sulfidic sediments in the host succession as a favorable, but not essential ingredient of the exploration model at this scale.

1.6.4 CAMP

A prospective camp- (or district-) scale target area for Ni-Cu(-PGE) deposits is considered to represent a localized zone of more focused upward magma flux than laterally equivalent parts of a prospective region. It is localized by an intersection of major translithospheric structures that may be observable in regional geophysical datasets, most notably gravity.

Importantly, the camp must relate to a focus of the specific event that is directly linked to mineralization, which may be a relatively small component of a broader prospective mafic-ultramafic complex. The typical scale of a camp ranges from about 10−100 km in diameter/strike-length.

There are two different ways to target a camp within a prospective region. The first is to predict its location based on interpreted intersections of fundamental structures, based on geophysical and geological data. The second is to identify the permissive signature of a camp in geological, geochemical, and geophysical datasets.

This signature will usually relate to evidence for either a spatial concentration of intrusions/volcanics with more differentiated and/or higher-MgO compositions, or for an underlying deep-seated mafic intrusive complex in gravity data. Ideally, both of these will be observable in the data. In some cases, there may be direct geochemical evidence for mineralization, or at least anomalously primitive intrusions, in wide-spaced regional geochemical data (e.g., Nova discovery—Bennett et al., 2014).

The Norilsk region, where abundant mafic intrusions extend over at least 300 km of strike, is a good example where several camp-scale areas of magmatic focus are defined by concentrations of differentiated mafic intrusions (Yakubchuk and Nikishin, 2004). The largest of these, centered on the position of the major ore deposits of Norilsk and Talnakh, has an area of about 100×20 km. Three smaller (20–60 km in strike-length) concentrations of differentiated intrusions are associated with three additional areas of subeconomic mineralization. In High-MgO systems, more magnesian parts of the target intrusive/extrusive complex, which commonly represent areas of more focused magma emplacement, will typically have a stronger magnetic signature because of the effects of serpentinization (see above). Additionally, areas of magma focus may coincide with areas of earlier intense felsic volcanism. While potentially rich in sulfidic sediments, felsic volcanic centers may also serve as a proxy for the intersection of the rift axis with high-angle transfer faults, and are particularly useful when targeting for High-MgO deposits. In Low-MgO systems, magnetic data may help in the definition of a camp through the imaging of areas of greater complexity within a region of mafic intrusions.

The East Kimberley Ni-Cu-PGE province in Western Australia, which hosts the Savannah and Panton Sills deposits, is a good example of a camp-scale signature related to a deep-seated gravity signature. The area of mineralized and related differentiated intrusions is closely associated with a long-wavelength positive gravity anomaly with dimensions of approximately 100×50 km, which likely represents an underlying, deep mafic intrusive complex.

1.6.5 DEPOSIT

At the deposit scale, the key geological targeting concept relates to the identification of individual magma-emplacement conduits, within host intrusions or volcanics. Intrusions that have acted as significant magma conduits may be flanked by extensive zones of contact metamorphism. Changes in intrusion geometry can be important for localizing sulfides. However, typically exploration at this scale is driven by empirical direct-detection technologies focused on locating sulfide mineralization. Electromagnetic surveying is the most commonly used technique because most economic bodies of Ni-Cu(-PGE) mineralization include a significant component that is strongly electrically conductive. If the area is exposed at the surface or covered by a thin regolith amenable to reconnaissance drilling, geochemical techniques will typically also be employed. Once a significant occurrence of mineralization has been discovered the key is to define the geological controls on its host conduit zone and to follow this.

1.7 SUMMARY: FROM GLOBAL TO LOCAL

We have described above many of the features intrinsic to the broad-scale localization of magmatic Ni-Cu(-PGE) and PGE reef deposits in space (continental to camp) and time (supercontinent cycles). We have identified geodynamic, lithospheric, and tectonic factors of importance, and described how these are translated into features that may be recognized in geological, geophysical, and geochemical datasets at a range of scales. Many other studies have addressed petrological and geochemical features of these deposits and their related mafic and ultramafic intrusions.

The role of SCLM is primarily mechanical, focusing the high-degree melting of impacting mantle plumes in areas of thinner lithosphere, and partitioning tectonic strain at cratonic margins. The spatial coincidence of these two is the most important first-order control on the localization of deposits. Interaction of the dominantly asthenosphere-derived melts with the metasomatized SCLM may also increase the PGE budget. Coupled with the mechanical properties of the upper crust within stable craton interiors, this is consistent with the tendency for PGE-reef deposits to form further inboard from craton margins than their Ni-Cu(-PGE) counterparts, and heightens the exploration relevance of paleocraton boundaries that may have experienced tectonic reactivation during a major mafic-ultramafic magmatic event.

We have learned that mafic-ultramafic magmas not only dance to the tune of whole-mantle convection, mantle plumes, and supercontinent cycles. They are also given instructions via the interaction of plate tectonics with the continental lithospheric architecture. This determines the when and the where of Ni-Cu(-PGE) and PGE deposit formation. The most significant High-MgO Ni-Cu(-PGE) deposits dominantly form in narrow peri-cratonic rifts during the pre-1.8 Ga period of higher mantle temperatures. They are dependent on the architecture of the rifts to provide both the required magmatic focus, and the sulfidic crustal rocks that act as the essential contaminant to drive magmatic sulfide formation. On the other hand, Low-MgO Ni-Cu(-PGE) deposits apparently are much more dependent on dynamic factors. They appear to accumulate sulfide melts deep in the crustal conduit, below the level of deposit emplacement. The upwards escape of these sulfide melts and their dynamic injection into the deposit sites is particularly sensitive to the breaching of mechanical threshold barriers, similar to the fault-valve behavior of hydrothermal fluids (e.g., Sibson and Scott, 1998) such as associated with orogenic gold systems (e.g., Robert and Brown, 1986), or with many magmatic-hydrothermal systems (e.g., large porphyry Cu-Mo systems; Rohrlach and Loucks, 2005; Loucks, 2014). Regional tectonic stress switches have played a key role in facilitating this process in several important Ni-Cu(-PGE) deposit camps (Noril'sk, Voisey's Bay, West Musgraves). We anticipate that this interplay of tectonics with the dynamic behavior of magmatic systems will become increasingly recognized, and has important implications for deposit exploration.

We put forward a simple holistic model (Fig. 1.5) for Ni-Cu(-PGE) and PGE reef formation. This model facilitates prediction at the greenfields stage of exploration for new deposits and camps, particularly at the continental- to regional-scales. The camp-scale represents the crossover point from a combined conceptual/empirical approach to the use of direct-detection techniques.

ACKNOWLEDGEMENTS

The manuscript benefited from a review by Wolf Maier. Wolf is also thanked for sharing an updated graph of the secular distribution of PGE and Ni-Cu resources adapted for use in Fig. 1.6. This work has been supported by funding from the Australian Research Council, Macquarie University and Minerals Targeting International Pty Ltd. The analytical data were obtained using instrumentation funded by DEST Systemic Infrastructure Grants, ARC LIEF, NCRIS/AuScope, industry partners and Macquarie University. This is contribution 978 from the ARC Centre of Excellence for Core to Crust Fluid Systems (http://www.ccfs.mq.edu.au) and 1158 in the GEMOC Key Centre (http://www.gemoc.mq.edu.au).

REFERENCES

Afonso, J.C., Schutt, D.L., 2012. The effects of polybaric partial melting on density and seismic velocities of mantle restites. Lithos 134−135, 289−303.

Allen, M.B., Anderson, L., Searle, R.C., Buslov, M., 2006. Oblique rift geometry of the West Siberian Basin: tectonic setting for the Siberian flood basalts. J. Geol. Soc. London 163 (6), 901−904.

Ariskin, A.A., Kislov, E.V., Danyushevsky, L.V., Nikolaev, G.S., Fiorentini, M.L., Gilbert, S., et al., 2016. Cu−Ni−PGE fertility of the Yoko-Dovyren layered massif (northern Transbaikalia, Russia): thermodynamic modeling of sulfide compositions in low mineralized dunite based on quantitative sulfide mineralogy. Mineral. Deposita 51, 993−1011.

Ariskin, A.A., Kostitsyn, Y.A., Konnikov, E.G., Danyushevsky, L.V., Meffre, S., Nikolaev, G.S., et al., 2013. Geochronology of the Dovyren Intrusive Complex, northwestern Baikal area, Russia, in the Neoproterozoic. Geochem. Inter. 51, 859−875.

Artemieva, I.M., 2009. The continental lithosphere: reconciling thermal, seismic, and petrologic data. Lithos 109, 23−46.

Barkov, A.Y., Laflamme, J.G., Cabri, L.J., Martin, R.F., 2002. Platinum-group minerals from the Wellgreen Ni−Cu−PGE deposit, Yukon, Canada. Can. Mineral. 40 (2), 651−669.

Barnes, S.J., 2006. Komatiite-hosted nickel sulfide deposits: geology, geochemistry, and genesis. Soc. Econ. Geol. Special Publication 13, 51−118.

Barnes, S., Le Vaillant, M., Lightfoot, P., Staude, S., Pina, R., 2016. Sulfide-matrix magmatic ore breccias: their origin at infiltration-melting fronts. In: Godel, B., Barnes, S., Gonzales−Alvarez, I., Fiorentini, M., Le Vaillant, M. (Eds.), Geological Society of Western Australia Record 2016/13 − 13th International Ni-Cu-PGE symposium. Fremantle, Australia, p. 8. , Abstracts.

Barnes, S.J., Lesher, C.M., Sproule, R.A., 2007. Geochemistry of komatiites in the Eastern Goldfields Superterrane, Western Australia and the Abitibi Greenstone Belt, Canada, and implications for the distribution of associated Ni−Cu−PGE deposits. Appl. Earth Sci. 116 (4), 167−187.

Barnes, S.-J., Lightfoot, P., 2005. Formation of magmatic nickel sulfide ore deposits and processes affecting the Cu and PGE contents. Econ. Geol. 100th Anniversary volume 179−214.

Barnes, S.J., Melezhik, V.A., Sokolov, S.V., 2001. The composition and mode of formation of the Pechenga nickel deposits, Kola Peninsula, northwestern Russia. Can. Mineral. 39 (2), 447−471.

Barnes, S.J., Van Kranendonk, M.J., Sonntag, I., 2012. Geochemistry and tectonic setting of basalts from the Eastern Goldfields Superterrane. Aust. J. Earth Sci. 59 (5), 707−735.

Barnes, S.-J., Gomwe, T.S., 2011. The Pd deposits of the Lac des Iles complex, northwestern Ontario, Rev. Econ. Geol. 17, 351−370.

Barrie, C.T., Naldrett, A.J., Davis, D., 1990. Geochemical constraints on the genesis of the Montcalm Gabbroic Complex and Ni-Cu deposit, Western Abitibi Sub-province, Ontario. Can. Mineral. 28, 451−474.

Begg, G.C., Griffin, W.L., Natapov, L.M., O'Reilly, S.Y., Grand, S.P., O'Neill, C.J., et al., 2009. The lithospheric architecture of Africa: seismic tomography, mantle petrology and tectonic evolution. Geosphere 5, 23−50.

Begg, G.C., Hronsky, J.A.M., Arndt, N.T., Griffin, W.L., O'Reilly, S.Y., Hayward, N., 2010. Lithospheric, cratonic and geodynamic setting of Ni-Cu-PGE sulfide deposits. Econ. Geol. 105, 1057−1070.

Bekker, A., Barley, M.E., Fiorentini, M.L., Rouxel, O.J., Rumble, D., Beresford, S.W., 2009. Atmospheric sulfur in Archean komatiite-hosted nickel deposits. Science 326, 1086−1089.

Belousova, E.A., Kostitsyn, Y.A., Griffin, W.L., Begg, G.C., O'Reilly, S.Y., Pearson, N.J., 2010. The growth of the continental crust: constraints from zircon Hf-isotope data. Lithos 119, 457−466.

Bennett, M., Gollan, M., Staubman, M., Bartlett, J., 2014. Motive, means and opportunity: key factors in the discovery of the Nova-Bollinger magmatic nickel-copper sulfide deposits in Western Australia. In: Kelley, K.D., Golden, H.C. (Eds.), SEG Special Publication 18. Building Exploration Capability for the 21st Century. pp. 301−320.

Beresford, S.W., Cas, R.A.F., Lahaye, Y., Jane, M., 2002. Facies architecture of an Archean komatiite-hosted Ni-sulphide ore deposit, Victor, Kambalda, Western Australia: implications for komatiite lava emplacement, J. Volcanol. Geotherm. Res., 118. pp. 57−75.

Beresford, S.W., Stone, W.E., Cas, R.A.F., Lahaye, Y., Jane, M., 2005. Volcanological controls on the localisation of a komatiite-hosted Ni-Cu-(PGE) deposit, Coronet, Kambalda, Western Australia. Econ. Geol. 100, 1457−1467.

Bhattacharji, S., 1967. Mechanics of flow differentiation in ultramafic and mafic sills. J. Geol. 101−112.

Burke, K., Steinberger, B., Torsvik, T.H., Smethurst, M.A., 2008. Plume generation zones at the margins of Large Low Shear Velocity Provinces on the core-mantle boundary. Earth Planet. Sci. Lett. 265, 49−60.

Burke, K., Torsvik, T.H., 2004. Derivation of large igneous provinces of the past 200 million years from long-term heterogeneities in the deep mantle. Earth Planet. Sci. Lett. 227, 531−538.

Campos-Alvarez, N.O., Samson, I.M., Fryer, B.J., 2012. The roles of magmatic and hydrothermal processes in PGE mineralization, Ferguson Lake deposit, Nunavut, Canada. Mineral. Deposita 47, 441−465.

Cawthorn, R.G., Barnes, S.-J., Ballhaus, C., Malitch, K.N., 2005. Platinum group element, chromium and vanadium deposits in mafic and ultramafic rocks. Econ. Geol. 100th Anniversary volume 215−250.

Chernyshov, N.M., Ryborak, M.V., Al'bekov, A.Y., Bayanova, T.B., 2012. U-Pb age of granitoids from the Ol'khovskii ring pluton of the Voronezh crystalline massif (Northern Part of the conjunction zone of Sarmatia and Volgo-Uralia). Doklady Earth Sci. 444 (1), 618−620.

Clarke, B., Uken, R., Reinhardt, J., 2009. Structural and compositional constraints on the emplacement of the Bushveld Complex, South Africa. Lithos 111, 21−36.

Czamanske, G.K., Gurevitch, A.B., Fedorenko, V., Simonov, O., 1998. Demise of the Siberian plume: paleo-geographic and paleotectonic reconstruction from the prevolcanic and volcanic record, north-central Siberia. Inter. Geol. Rev. 40, 95−115.

Dobretsov, N.L., Borisenko, A.S., Izokh, A.E., Zhmodik, S.M., 2010. A thermochemical model of Eurasian Permo−Triassic mantle plumes as a basis for prediction and exploration for Cu−Ni−PGE and rare-metal ore deposits. Rus. Geol. Geophys. (Geologiya i Geofizika) 51 (9), 903−924, 1159−1187).

Dhuime, B., Hawkesworth, C.J., Cawood, P.A., Storey, C.D., 2012. A Change in the Geodynamics of Continental Growth 3 Billion Years Ago. Science 335, 1334−1336.

Elkins-Tanton, L.T., 2005. Continental magmatism caused by lithospheric delamination. Geol. Soc. Am. Special Papers 388, 449−461.

Ernst, R.E., 2007. Large igneous provinces in Canada through time and their metallogenic potential. Geol. Assoc. Can. Special Publication 5, 929−937.

Ernst, R.E., 2014. Large Igneous Provinces. Cambridge University Press, Cambridge, 653 pp.

Ernst, R.E., Buchan, K.L., 2002. Maximum size and distribution in time and space of mantle plumes: evidence from large igneous provinces. J. Geodynam. 34 (2), 309−342.

Ernst, R.E., Hamilton, M.A., Soderlund U., 2012. A proposed 725 Ma Dovyren- Kingash LIP of southern Siberia, and possible reconstruction link with the 725−715 Ma Franklin LIP of northern Laurentia. GACMAC Joint Annual Meeting B Geoscience at the Edge, May 27−29, St. John's, Newfoundland and Labrador, Canada. Abs vol. 35.

Evans, D.M., 2011. Geodynamic setting of Neoproterozoic nickel sulphide deposits in eastern Africa. Appl. Earth Sci. 120 (4), 175−186.

Evans-Lamswood, D.M., Butt, D.P., Jackson, R.S., Lee, D.V., Muggridge, M.G., Wheeler, R.I., et al., 2000. Physical controls associated with the distribution of sulfides in the Voisey's Bay Ni-Cu-Co deposit, Labrador. Econ. Geol. 95, 749−770.

Fedorenko, V.A., 1994. Evolution of magmatism as reflected in the volcanic sequence of the Norilsk region. In: Lightfoot, P.C., Naldrett, A.J. (Eds.), Proceedings Sudbury−Norilsk Symposium, Ontario Geological Survey Spec Paper, 5. pp. 171−184.

Gaál, G., 1980. Geological setting and intrusion tectonics of the Kotalahti nickel-copper deposit, Finland. Bull. Geol. Soc. Finland 52 (1), 101−128.

Goldner, B.D., 2011. Igneous petrology of the Ni-Cu-PGE Mineralized Tamarack intrusion, Aitkin and Carlton counties, Minnesota. Masters dissertation. University of Minnesota, USA.

Good, N., de Wit, M.J., 1997. The Thabazimbi-Murchison lineament of the Kaapvaal craton, South Africa: 2700 Ma of episodic deformation. J. Geol. Soc. London 154, 93–97.

Gouedji, F., Picard, C., Coulibaly, Y., Audet, M.-A., Auge, T., Goncalves, P., et al., 2014. The Samapleu mafic-ultramafic intrusion and its Ni-Cu-PGE mineralization: an Eburnean (2.09 Ga) feeder dyke to the Yacouba layered complex (Man Archean craton, western Ivory Coast). Bull. Soc. Geol. France 185 (6), 393–411.

Grand, S., 2002. Mantle shear-wave tomography and the fate of subducted slabs. Philos. Trans. Royal Soc. London A360, 2475–2491.

Griffin, W.L., Begg, G.C., O'Reilly, S.Y., 2013. Continental-root control on the genesis of magmatic ore deposits. Nat. Geosci. 6, 905–910.

Griffin, W.L., Begg, G.C., Dunn, D., O'Reilly, S.Y., Natapov, L.M., Karlstrom, K., 2011. Archean lithospheric mantle beneath Arkansas: continental growth by microcontinent accretion. Bull. Geol. Soc. Am. 123, 1763–1775.

Griffin, W.L., Graham, S., O'Reilly, S.Y., Pearson, N.J., 2004. Lithosphere evolution beneath the Kaapvaal Craton. Re-Os systematics of sulfides in mantle-derived peridotites. Chem. Geol. 208, 89–118.

Griffin, W.L., O'Reilly, S.Y., Afonso, J.C., Begg, G.C., 2009. The composition and evolution of lithospheric mantle: a re-evaluation and its tectonic implications. J. Petrol. 50, 1185–1204.

Griffin, W.L., O'Reilly, S.Y., Natapov, L.M., Ryan, C.G., 2003. The evolution of lithospheric mantle beneath the Kalahari Craton and its margins. Lithos 71, 215–241.

Griffin, W.L., O'Reilly, S.Y., 2007. The earliest subcontinental mantle. In: Van Kranendonk, M., Smithies, H., Bennett, V. (Eds.), Earth's Oldest Rocks. Elsevier, Amsterdam, pp. 1013–1035.

Griffin, W.L., Spetsius, Z.V., Pearson, N.J., O'Reilly, S.Y., 2002. In-situ Re-Os analysis of sulfide inclusions in kimberlitic olivine: new constraints on depletion events in the Siberian lithospheric mantle. Geochem. Geophys. Geosys. 3, 1069.

Groves, D.I., Ho, S.E., Rock, N.M.S., Barley, M.E., Muggeridge, M.Y., 1987. Archean cratons, diamond and platinum; evidence for coupled long-lived crust–mantle systems. Geology. 15, 801–805.

Hanski, E., Walker, R.J., Huhma, H., Polyakov, G.V., Balykin, P.A., Hoa, T.T., et al., 2004. Origin of the Permian-Triassic komatiites, northwestern Vietnam. Contrib. Mineral. Petrol. 147 (4), 453–469.

Hanski, E.J., Luo, Z.-Y., Oduro, H., Walker, R.J., 2011. The pechenga Ni-Cu sulfide deposits, northwestern russia: a review with new constraints from the feeder dikes. In: Li, C., Ripley, E.M. (Eds.), Magmatic Ni-Cu and PGE Deposits: Geology, Geochemistry, and Genesis. Rev. Econ. Geol., 17. pp. 145–162.

Herzberg, C., O'Hara, M.J., 2002. Plume-associated ultramafic magmas of Phanerozoic age. J. Petrol. 43, 1857–1883.

Himmelberg, G.R., Loney, R.A., 1981. Petrology of the ultramafic and gabbroic rocks of the Brady Glacier nickel-copper deposit. Fairweather Range, southeastern Alaska: U.S. Geological Survey Professional Paper 1195, 26.

Hoatson, D.M., Blake, D.H., 2000. Geology and economic potential of the Palaeoproterozoic layered mafic-ultramafic intrusions in the East Kimberley, Western Australia. Aust. Geol. Surv. Org. Bull. 246, 469.

Hoatson, D.M., Keays, R.R., 1989. Formation of platiniferous sulfide horizons by crystal fractionation and magma mixing in the Munni Munni layered intrusion, West Pilbara Block, Western Australia. Econ. Geol. 84, 1775–1804.

Hofmann, A., Bekker, A., Dirks, P., Gueguen, B., Rumble, D., Rouxel, O.J., 2014. Comparing orthomagmatic and hydrothermal mineralization models for komatiite-hosted nickel deposits in Zimbabwe using multiple-sulfur, iron, and nickel isotope data. Mineral. Deposita 49 (1), 75–100.

Holt, P.J., van Hunen, J., Allen, M.B., 2012. Subsidence of the West Siberian Basin: effects of a mantle plume impact. Geology 40 (8), 703–706.

Houlé, M.G., Lesher, C.M., 2011. Komatiite-associated Ni-Cu-(PGE) deposits, Abitibi greenstone belt, Superior Province, Canada. Reviews in. Econ. Geol. 17, 89–121.

Hronsky, J.M.A., 2011. Self-organized critical systems and ore formation: the key to spatial targeting? Soc. Econ. Geol. Newslett. 84, 14–16, 2011.

Hronsky, J.M.A., Groves, D.I., 2008. Science of targeting: definition, strategies, targeting and performance measurement. Aust. J. Earth Sci. 55 (1), 3–12.

Hughes, H.S.R., McDonald, I., Kinnaird, J.A., Nex, P.A.M., Bybee, G.M., 2016. The metallogenesis of the lithospheric mantle and its bearing on the Bushveld magmatic event: evidence from lamprophyric dykes and critical metals in mantle xenolith sulphides. Paper number 3061. Abstract 35th International Geological Congress, Cape Town, South Africa. (http://www.americangeosciences.org/information/igc).

Hulbert, L.J., Hamilton, M.A., Horan, M.F., Scoates, R.F.J., 2005. U-Pb zircon and Re-Os isotope geochronology of mineralized ultramafic intrusions and associated nickel ores from the Thompson Nickel Belt, Manitoba, Canada. Econ. Geol. 100 (1), 29–41.

Izokh, A.E., Medvedev, A., Ya., Fedoseev, G.S., Polyakov, G.V., Nikolaeva, I.V., et al., 2016. Distribution of PGE in permo-triassic basalts of the siberian large igneous province. Rus. Geol. Geophys. 57, 809–821.

Janse, A.J.A., 1994. Is Clifford's Rule still valid? Affirmative examples from around the world. In: Meyer, H.O.A., Leonardos, O.H. (Eds.), Diamonds: Characteriza- tion, genesis and exploration: Brazilia, Departement Nacional da Production Mineralia. pp. 215–235.

Johnson, S.P., Waele, D.B., Evans, D., Tembo, F., Banda, W., 2006. A record of Neoproterozoic divergent processes along the southern margin of the Congo Craton, paper presented at 21st Colloquium of African Geology, Maputo, Mozambique. Geolical Survey of South Africa 3-5.

Kamo, S.L., Czamanske, G.K., Amelin, Y., Fedorenko, V.A., Davis, D.W., Trofimov, V.R., 2003. Rapid eruption of Siberian flood-volcanic rocks and evidence for coincidence with the Permian–Triassic boundary and mass extinction at 251 Ma. Earth. Planet. Sci. Lett. 214 (1), 75–91.

Keays, R.R., Lightfoot, P.C., 1999. The role of meteorite impact, source rocks, protores and mafic magmas in the genesis of the Sudbury Ni–Cu–PGE sulfide ore deposits. In: Keays, R.R., Lesher, C.M., Lightfoot, P.C., Farrow, C.E.G. (Eds.), Dynamic processes in magmatic ore deposits and their application to mineral exploration. Geological Association of Canada Short Course Notes, 13. pp. 329–366.

Keays, R.R., Lightfoot, P.C., 2004. Formation of Ni–Cu–platinum group element sulphide mineralisation in the Sudbury impact melt sheet. Mineral. Petrol. 82, 217–258.

Kerrich, R., Goldfarb, R.J., Groves, D.I., Garwin, S., 2000. The geodynamics of world-class gold deposits: characteristics, space–time distribution, and origins. Rev. Econ. Geol. 13, 501–552.

Kerrich, R., Goldfarb, R.J., Richards, J.P., 2005. Metallogenic provinces in an evolving geodynamic framework. Econ. Geol. 100th Anniversary Volume 1097–1136.

Kositcin, N., Brown, S.J.A., Barley, M.E., Krapez, B., Cassidy, K.F., Champion, D.C., 2008. SHRIMP U-Pb age constraints on the Late Archaean tectonostratigraphic architecture of the Eastern Goldfields superterrane, Yilgarn craton, Western Australia. Precambrian. Res. 161, 5–33.

Ksienzyk, A.K., Jacobs, J., Boger, S.D., Košler, J., Sircombe, K.N., Whitehouse, M.J., 2012. U–Pb ages of metamorphic monazite and detrital zircon from the Northampton Complex: evidence of two orogenic cycles in Western Australia. Precambrian. Res. 198, 37–50.

Labrosse, S., Jaupart, C., 2007. Thermal evolution of the Earth: secular changes and fluctuations in plate characteristics. Earth. Planet. Sci. Lett. 260, 465–481.

Lay, T., 2005. The deep mantle thermo-chemical boundary layer: the putative mantle plume source. In: Foulger, G.R., Natland, J.N., Presnall, D.C., Anderson, D.L. (Eds.), Plates, Plumes and Paradigms, Geological Society of America Special, vol. 388. pp. 193–205.

Le Vaillant, M., Barnes, S.J., Fiorentini, M.L., Santaguida, F., Törmänen, T., 2016. Effects of hydrous alteration on the distribution of base metals and platinum group elements within the Kevitsa magmatic nickel sulphide deposit. Ore Geol. Rev. 72 (7), 128–148.

Lesher, C.M., 2007. Ni-Cu-(PGE) deposits in the Raglan area, Cape Smith Belt, New Quebec. Mineral Deposits of Canada: a synthesis of major deposit-types, district metallogeny, the evolution of geological provinces, and exploration Methods. Geol. Assoc. Can. 5, 351–386.

Lesher, C.M., Groves, D.I., 1986. Controls on the formation of komatiite-associated nickel-copper sulfide deposits. In: Genkin, A.D., Naldrett, A.J., Ridge, J.D., Sillitoe, R.H., Vokes, F.M. (Eds.), Geology and Metallogeny of Copper Deposits. Springer-Verlag, Heidelberg, pp. 43–62.

Lesher, C.M., Keays, R.R., 2002. Komatiite-associated Ni–Cu-(PGE) deposits: geology, mineralogy, geochemistry and genesis. In: Cabri, L. (Ed.), The geology, geochemistry, mineralogy and beneficiation of the platinum-group elements. Canadian Institute of Mining, Metallurgy and Petroleum, 54. pp. 579–618.

Li, X.H., Su, L., Chung, S.L., Li, Z.X., Liu, Y., Song, B., et al., 2005. Formation of the Jinchuan ultramafic intrusion and the world's third largest Ni-Cu sulfide deposit: associated with the ∼825 Ma south China mantle plume?. Geochem. Geophys. Geosys. 6, 11. Available from: http://dx.doi.org/10.1029/2005GC001006.

Lightfoot, P.C., Evans-Lamswood, D., 2015. Structural controls on the primary distribution of mafic–ultramafic intrusions containing Ni–Cu–Co–(PGE) sulfide mineralization in the roots of large igneous provinces. Ore Geol. Rev. 64, 354–386.

Lightfoot, P.C., Keays, R.R., 2005. Siderophile and chalcophile metal variations in flood basalts from the Siberian Trap, Noril'sk Region: implications for the origin of the Ni–Cu-PGE sulfide ores. Econ. Geol. 100, 439–462.

Lightfoot, P.C., Keays, R.R., Doherty, W., 2001. Chemical evolution and origin of nickel sulfide mineralization in the sudbury igneous complex, Ontario, Canada. Econ. Geol. 96, 1855–1875.

Lightfoot, P.C., Keays, R.R., Evans-Lamswood, D., Wheeler, R., 2012. S saturation history of the Nain Plutonic Suite mafic intrusions: origin of the Voisey's Bay Ni-Cu-Co sulfide deposit, Labrador, Canada. Mineral. Deposita 47, 23–50.

Liu, Y., Lü, X., Wu, C., Hu, X., Duan, Z., Deng, G., et al., 2016. The migration of Tarim plume magma toward the northeast in Early Permian and its significance for the exploration of PGE-Cu–Ni magmatic sulfide deposits in Xinjiang, NW China: as suggested by Sr–Nd–Hf isotopes, sedimentology and geophysical data. Ore Geol. Rev. 72 (1), 538–545.

Lorand, J.P., Luguet, A., Alard, O., 2013. Platinum-group element systematics and petrogenetic processing of the continental upper mantle: a review. Lithos 164, 2–21.

Loucks, R.R., 2014. Distinctive composition of copper-ore-forming arc magmas. Aust. J. Earth Sci. 61, 5–16.

Machado, N., Gapais, D., Potrel, A., Gauthier, G., Hallot, E., 2010. Chronology of transpression, magmatism, and sedimentation in the Thompson Nickel Belt (Manitoba, Canada) and timing of Trans-Hudson Orogen-Superior Province collision This article is one of a series of papers published in this Special Issue on the theme of Geochronology in honour of Tom Krogh. Can. J. Earth. Sci. 48 (2), 295–324.

Maier, W.D., 2005. Platinum-group element (PGE) deposits and occurrences: mineralization styles, genetic concepts, and exploration criteria. J. Afr. Earth Sci. 41, 165–191.

Maier, W.D., Barnes, S.J., Gartz, V., Andrews, G., 2003. Pt-Pd reefs in magnetitites of the Stella layered intrusion, South Africa: a world of new exploration opportunities for platinum group elements. Geology. 31 (10), 885–888.

Maier, W.D., Peltonen, P., Livesey, T., 2007. The ages of the Kabanga North and Kapalagulu intrusions, western Tanzania: a reconnaissance study. Econ. Geol. 102 (1), 147–154.

Maier, W.D., Barnes, S.J., Chinyepi, G., Barton Jr., J.M., Eglington, B., Setshedi, I., 2008a. The composition of magmatic Ni–Cu–(PGE) sulfide deposits in the Tati and Selebi-Phikwe belts of eastern Botswana. Mineral. Deposita 43 (1), 37–60.

Maier, W.D., de Klerk, L., Blaine, J., Manyeruke, T., Barnes, S.-J., Stevens, M.V.A., et al., 2008b. Petrogenesis of contact-style PGE mineralization in the northern lobe of the Bushveld Complex: comparison of data from the farms Rooipoort, Townlands, Drenthe and Nonnenwet. Mineral. Deposit 43, 255–280.

Maier, W.D., Barnes, S.-J., Sarkar, A., Ripley, E., Li, C., Livesey, T., 2010. The Kabanga Ni sulfide deposit, Tanzania: I. Geology, petrography, silicate rock geochemistry, and sulfur and oxygen isotopes. Mineral. Deposita. 45, 419–441.

Maier, W.D., Groves, D.I., 2011. Temporal and spatial controls on the formation of magmatic PGE and Ni–Cu deposits. Mineral. Deposita 46, 841–857.

Maier, W.D., Smithies, R.H., Spaggiari, C.V., Barnes, S.J., Kirkland, C.L., Yang, S., et al., 2016a. Petrogenesis and Ni-Cu sulphide potential of mafic-ultramafic rocks in the Mesoproterozoic Fraser Zone within the Albany-Fraser Orogen, Western Australia. Precambrian. Res. 281, 27–46.

Maier, W.D., Barnes, S.J., Karykowski, B.T., 2016b. A chilled margin of komatiite and Mg-rich basaltic andesite in the western Bushveld Complex, South Africa. Contrib. Mineral. Petrol. 171 (6), 1–22.

Mao, J.W., Yang, J.M., Qu, W.J., Du, A.D., Wang, Z.L., Han, C.M., 2002. Re−Os age of Cu−Ni ores from the Huangshandong Cu−Ni sulfide deposit in the East Tianshan Mountains and its implication for geodynamic processes. Mineral. Deposita 21, 323–330, in Chinese.

Mavrogenes, J.A., O'Neill, H.St.C., 1999. The relative effects of pressure, temperature and oxygen fugacity on the solubility of sulfide in mafic magmas. Geochim. Cosmochim. Acta 63, 1173–1180.

McCuaig, T.C., Hronsky, J.M.A., 2014. The mineral system concept: the key to exploration targeting. Soc. Econ. Geol. Special Publication 18, 153–176.

McDonald, I., Holwell, D.A., Armitage, P.E.B., 2005. Geochemistry and mineralogy of the Platreef and "Critical Zone" of the northern lobe of the Bushveld Complex, South Africa: implications for Bushveld stratigraphy and the development of PGE mineralisation, Mineral. Deposita, 40. p. 526.

Mole, D.R., Fiorentini, M.L., Cassidy, K.F., Kirkland, C.L., Thebaud, N., McCuaig, T.C., et al., 2013. Crustal evolution, intra-cratonic architecture and the metallogeny of an Archaean craton, Geol. Soc. London, Special Publications, 393. pp. SP393–SP398.

Mortensen, J.K., Hulbert, L.J., 1991. A U-Pb zircon age for a Maple Creek gabbro sill, Tatamagouche Creek area, southwestern Yukon Territory. Rad. Age Isotopic Stud. Rep. 5, 91–92.

Mosar, J., Eide, E.A., Osmundsen, P.T., Sommaruga, A., Torsvik, T.H., 2002. Greenland-Norway separation: a geodynamic model for the North Atlantic. Norw. J. Geol. 82, 281–298.

Mota-e-Silva, J., Ferreira Filho, C.F., Della Giustina, M.E.S., 2013. The Limoeiro deposit: Ni-Cu-PGE sulfide mineralization hosted within an Ultramafic tubular magma conduit in the Borborema Province, Northeastern Brazil. Econ. Geol. 108 (7), 1753–1771.

Mungall, J.E., Ames, D.E., Hanley, J.J., 2004. Geochemical evidence from the Sudbury structure for crustal redistribution by large bolide impacts. Nature. 429, 546–548.

Mungall, J.E., Harvey, J.D., Balch, S.J., Azar, B., Atkinson, J., Hamilton, M.A., 2010. Eagle's nest: a magmatic Ni-sulfide deposit in the James Bay Lowlands, Ontario, Canada. Soc. Econ. Geol. Special Publication 15, 539–557.

Mutanen, T., Huhma, H., 2001. U-Pb geochronology of the Koitelainen, Akanvaara and Keivitsa layered intrusions and related rocks. Special paper, Geol. Surv. Finland 229–246.

Myers, J.S., Voordouw, R.J., Tettelaar, T.A., 2008. Proterozoic anorthositegranite Nain batholith: structure and intrusion processes in an active lithosphere- scale fault zone, northern Labrador. Can. J. Earth. Sci. 45, 909–934.

Naldrett, A.J., 1989. Ores associated with flood basalts. In: Whitney, J.A., Naldrett, A.J. (Eds.), Ore Deposition Associated With Magmas: Society of Economic Geologists: Dordrecht. pp. 103–118.

Naldrett, A.J., 1997. Key factors in the genesis of Noril'sk, Sudbury, Jinchuan, Voisey's Bay and other world-class Ni−Cu−PGE deposits: implications for exploration. Aust. J. Earth Sci. 44, 283–315.

Naldrett, A.J., 2004. Magmatic Sulfide Deposits: Geology, Geochemistry, and Exploration. Springer Verlag, Heidelberg, Berlin, p. 727.

Naldrett, A.J., 2010. Secular variation of magmatic sulfide deposits and their source magmas. Econ. Geol. 105, 669–688.

Naldrett, A.J., Lightfoot, P.C., Fedorenko, V.A., Doherty, W., Gorbachev, N.S., 1992. Geology and geochemistry of intrusions and flood basalt of the Noril_sk region, USSR, with implication for the origin of the Ni−Cu ores. Econ. Geol. 87, 975–1004.

Naldrett, A.J., Wilson, A., Kinnaird, J., Yudovskaya, M., Chunnett, G., 2012. The origin of chromitites and related PGE mineralization in the Bushveld Complex: new mineralogical and petrological constraints. Mineral. Deposita 47, 209–232.

Nelson, J., Colpron, M., 2007. Tectonics and metallogeny of the British Columbia, Yukon and Alaskan Cordillera, 1.8 Ga to the present. In: Goodfellow, W.D. (Ed.), Mineral Deposits of Canada: A Synthesis of Major Deposit-Types, District Metallogeny, the Evolution of Geological Provinces, and Exploration Methods: Geological Association of Canada, Mineral Deposits Division, Special Publication No, 5. pp. 755−791.

Nielsen, T.F., Andersen, J.Ø., Holness, M.B., Keiding, J.K., Rudashevsky, N.S., Rudashevsky, V.N., et al., 2015. The Skaergaard PGE and gold deposit: the result of in situ fractionation, sulphide saturation, and magma chamber-scale precious metal redistribution by immiscible Fe-rich melt. J. Petrol. 56 (8), 1643−1676.

Oberthür, T., Davis, D.W., Blenkinsop, T.G., Höhndorf, A., 2002. Precise U-Pb mineral ages, Rb-Sr and Sm-Nd systematics for the Great Dyke, Zimbabwe—constraints on late Archean events in the Zimbabwe craton and Limpopo belt. Precambrian. Res. 113, 293−305.

Olson, P., Kincaid, C., 1991. Experiments on the interaction of thermal convection and compositional layering at the base of the mantle. J. Geophys. Res. 96, 4347−4354.

Ordóñez-Casado, B., Martin-Izard, A., García-Nieto, J., 2008. SHRIMP-zircon U−Pb dating of the Ni−Cu−PGE mineralized Aguablanca gabbro and Santa Olalla granodiorite: confirmation of an Early Carboniferous metallogenic epoch in the Variscan Massif of the Iberian Peninsula. Ore Geol. Rev. 34 (3), 343−353.

O'Reilly, S.Y., Griffin, W.L., 2012. Mantle metasomatism. In: Harlov, D.E., Austrheim, H. (Eds.), Metasomatism and the Chemical Transformation of Rock, Lecture Notes in Earth System Sciences. Springer-Verlag, Berlin Heidelberg, pp. 467−528. , DOI 10.1007/978-3-642-28394-9_12.

Paces, J.B., Miller Jr., J.D., 1993. Precise U-Pb ages of Duluth Complex and related mafic intrusions, northeastern Minnesota: new insights for physical, petrogenetic, paleomagnetic and tectono-magmatic processes associated with 1.1 Ga Midcontinent rifting. J. Geophys. Res. 98B, 13997−14013.

Page, R.W., Hoatson, D.M., 2000. Geochronology of the mafic-ultramafic intrusions. In: Hoatson, D.M., Blake, D.H. (Eds.), Geology and Economic Potential of the Palaeoproterozoic Layered Mafic-Ultramafic Intrusions in the East Kimberley, Western Australia, 246. Australian Geological Survey Organisation, Bulletin, pp. 163−172.

Penniston-Dorland, S.C., Wing, B.A., Nex, P.A., Kinnaird, J.A., Farquhar, J., Brown, M., et al., 2008. Multiple sulfur isotopes reveal a magmatic origin for the Platreef platinum group element deposit, Bushveld Complex, South Africa. Geology 36 (12), 979−982.

Perring, C.S., 2015. A 3-D geological and structural synthesis of the leinster area of the agnew-wiluna belt, Yilgarn Craton, Western Australia, with special reference to the volcanological setting of komatiite-associated nickel sulfide deposits. Econ. Geol. 110, 469−503.

Persikov, E.S., 1991. The viscosity of magmatic liquids: experiment, generalized patterns. A model for calculation and prediction. Applications.. In: Perchuk, L.L., Kushiro, I. (Eds.), Physical Chemistry of Magmas. Springer, New York, pp. 1−40.

Pirajno, F., 2000. Ore Deposits and Mantle Plumes. Kluwer Academic Publishers, Dordrecht, 556 p.

Pirajno, F., Ernst, R.E., Borisenko, A.S., Fedoseev, G., Naumov, E.A., 2009. Intraplate magmatism in Central Asia and China and associated metallogeny. Ore Geol. Rev. 35 (2), 114−136.

Premo, W.R., Helz, R.T., Zientek, M.L., Langston, R.B., 1990. U-Pb and Sm-Nd ages for the Stillwater Complex and its associated sills and dikes, Beartooth Mountains, Montana: identification of a parent magma?. Geology 18 (11), 1065−1068.

Prendergast, M.D., Wilson, A.H., 2015. The nickeliferous archean madziwa igneous complex, Northern Zimbabwe: petrological evolution, magmatic architecture, and ore genesis. Econ. Geol. 110 (5), 1295−1312.

Qin, K.Z., Zhang, L.C., Xiao, W.J., Xu, X.W., Yan, Z., Mao, J.W., 2003. Overview of major Au, Cu, Ni and Fe deposits and metallogenic evolution of the eastern Tianshan Mountains, Northwestern China. Tectonic evolution and metallogeny of the Chinese Altay and Tianshan (London) 227−249.

Reichow, M.K., Pringle, M.S., Al'Mukhamedov, A.I., Allen, M.B., Andreichev, V.L., Buslov, M.M., et al., 2009. The timing and extent of the eruption of the Siberian Traps large igneous province: implications for the end-Permian environmental crisis. Earth. Planet. Sci. Lett. 277, 9−20.

Richards, M.A., Engebretson, D.C., 1992. Large scale mantle convection and the history of subduction. Nature 355, 437−440.

Ripley, E.M., Li, C., Shin, D., 2002. Paragneiss assimilation in the genesis of magmatic Ni−Cu-Co sulfide mineralization at Voisey's Bay, Labrador: d34S, d13C and S/Se evidence. Econ. Geol. 97, 1307−1318.

Ripley, E.M., Li, C., 2003. Sulfur isotopic exchange and metal enrichment in the formation of magmatic Cu-Ni-(PGE) deposits. Econ. Geol. 98, 635−641.

Ripley, E.M., Li, C., 2013. Sulfide saturation in mafic magmas: is external sulfur required for magmatic Ni-Cu-(PGE) ore genesis? Econ. Geol. 108, 45−58.

Robert, F., Brown, A.C., 1986. Archean gold-quartz veins at the Sigma mine, Abitibi greenstone belt, Quebec. Part I: geologic relations and formation of the vein system. Econ. Geol. 81, 578−592.

Rohrlach, B.D., Loucks, R.R., 2005. Multi-million-year cyclic ramp-up of volatiles in a lower-crustal magma reservoir trapped below the Tampakan copper gold deposit by Mio−Pliocene crustal compression in the southern Philippines. In: Porter, T.M. (Ed.), Super Porphyry Copper & Gold Deposits—A Global Perspective, 2. pp. 369−407. PCG Publishing, Adelaide.

Romano, S.S.N., Thébaud, N.J.M., Mole, D.R., Wingate, M.T.D., Kirkland, C.L., Doublier, M.P., 2014. Geochronological constraints on nickel metallogeny in the Lake Johnston belt, Southern Cross Domain. Aust. J. Earth Sci. 61 (1), 143−157.

Schodde, R., Hronsky, J.M.A., 2006. The role of world-class mines in wealth creation. Soc. Econ. Geol. Special Publication 12, 71−90.

Scotese, C.R., 2001. Digital paleogeographic map archive on CD-ROM. PALEOMAP Project, Arlington, Texas.

Seat, Z., Beresford, S.W., Grguric, B.A., Waugh, R.S., Hronsky, J.M.A., Gee, M.M.A., et al., 2007. Architecture and emplacement of the Nebo-Babel gabbronorite-hosted magmatic Ni-Cu-PGE sulfide deposit, West Musgrave, Western Australia. Mineral. Deposita 42, 551−581.

Seat, Z., Mary Gee, M.A., Grguric, B.A., Beresford, S.W., Grassineau, N.V., 2011. The Nebo-Babel Ni-Cu-PGE Sulfide Deposit (West Musgrave, Australia): Pt. 1. U/Pb zircon ages, whole-rock and mineral chemistry, and O-Sr-Nd isotope compositions of the intrusion, with constraints on petrogenesis. Econ. Geol. 106, 527−556.

Sharkov, E.V., Chistyakov, A.V., 2012. The Early Paleoproterozoic Monchegorsk layered mafite-ultramafite massif in the Kola Peninsula: geology, petrology, and ore potential. Petrology 20 (7), 607−639.

Sibson, R.H., Scott, J., 1998. Stress/fault controls on the containment and release of overpressured fluids: examples from gold-quartz vein systems in Juneau, Alaska, Victoria, Australia, and Otago, New Zealand. Ore Geol. Rev. 13, 293−306.

Simmons, N.A., Myers, S.C., Johannesson, G., Matzel, E., Grand, S.P., 2015. Evidence for long-lived subduction of an ancient tectonic plate beneath the southern Indian Ocean. Geophys. Res. Lett. 42 (21), 9270−9278.

Skuf'in, P.K., Bayanova, T.B., 2006. Early Proterozoic central-type volcano in the Pechenga structure and its relation to the ore-bearing gabbro-wehrlite complex of the Kola Peninsula. Petrology 14 (6), 609−627.

Smithies, R.H., Kirkland, C.L., Korhonen, F.J., Aitken, A.R.A., Howard, H.M., Maier, W.D., et al., 2015. The Mesoproterozoic thermal evolution of the Musgrave Province in central Australia—Plume vs. the geological record. Gondwana Res. 27 (4), 1419−1429.

Song, X.Y., Zhou, M.F., Cao, Z.M., Sun, M., Wang, Y.L., 2003. Ni−Cu−(PGE) magmatic sulfide deposits in the Yangliuping area, Permian Emeishan igneous province, SW China. Mineral. Deposita 38 (7), 831−843.

Steinberger, B., Torsvik, T.H., 2012. A geodynamic model of plumes from the margins of Large Low Shear Velocity Provinces. Geochem. Geophys. Geosys. 13 (1).

St-Onge, M.R., Van Gool, J.A.M., Garde, A.A., Scott, D.J., 2009. Correlation of Archaean and Palaeoproterozoic units between northeastern Canada and western Greenland: constraining the pre-collisional upper plate accretionary history of the Trans-Hudson orogen. In: Cawood, P.A., Kroner, A. (Eds.), Earth Accretionary Systems in Space and Time. The Geological Society, London, Special Publications, 318. pp. 193−235.

Tackley, P.J., 2011. Living dead slabs in 3-D: the dynamics of compositionally-stratified slabs entering a "slab graveyard" above the core−mantle boundary. Phys. Earth Planet. Interiors 188, 150−162.

Tackley, P.J., 2012. Dynamics and evolution of the deep mantle resulting from thermal, chemical, phase and melting effects. Earth Sci. Rev. 110, 1−25.

Torsvik, T.H., Smethurst, M.A., Burke, K., Steinberger, B., 2006. Large igneous provinces generated from the margins of the large low-velocity provinces in the deep mantle. Geophys. J. Inter. 167, 1447−1460.

Torsvik, T.H., Smethurst, M.A., Burke, K., Steinberger, B., 2008. Long term stability in deep mantle structure: evidence from the ∼ 300 Ma Skagerrak-Centered Large Igneous Province (the SCLIP). Earth. Planet. Sci. Lett. 267, 444−452.

Thorne, M.S., Grand, S.P., Garnero, E.J., 2004. Geographic correlation between hot spots and deep mantle lateral shear-wave velocity gradients. Phys. Earth Planet. Interiors 146, 47−63.

Urvantsev, N.N., 1971. Genetic characteristics of copper-nickel deposits of the Noril'sk region as criteria for predicting the massive ores of this type within the Yeniseisky nickel-bearing province. Geology and deposits of the Noril'sk region: Noril'sk 234−238, in Russian.

Wang, Q., Schiøtte, L., Campbell, I.H., 1996. Geochronological constraints on the age of komatiites and nickel mineralisation in the Lake Johnston greenstone belt, Yilgarn Craton, Western Australia. Aust. J. Earth Sci. 43 (4), 381−385.

Windley, B.F., Alexeiev, D.V., Xiao, W., Kröner, A., Badarch, G., 2007. Tectonic models for accretion of the Central Asian Orogenic belt. J. Geol. Soc. London 164, 31−47.

Wotzlaw, J.-F., Bindeman, I.N., Schaltegger, U., Brooks, C.K., Naslund, H.R., 2012. High-resolution insights into episodes of crystallization, hydrothermal alteration and remelting in the Skaergaard intrusive complex. Earth. Planet. Sci. Lett. 355, 199−212.

Yakubchuk, A., Nikishin, A., 2004. Noril'sk−Talnakh Cu−Ni−PGE deposits: a revised tectonic model. Mineral. Deposita 39, 125−142.

Zeh, A., Ovtcharova, M., Wilson, A.H., Schaltegger, U., 2015. The Bushveld Complex was emplaced and cooled in less than one million years - results of zirconology, and geotectonic implications. Earth. Planet. Sci. Lett. 418, 103−114.

Zhang, M., Kamo, S.L., Li, C., Hu, P., Ripley, E.M., 2010. Erratum to: precise U−Pb zircon−baddeleyite age of the Jinchuan sulfide ore-bearing ultramafic intrusion, western China. Mineral. Deposita 45, 215.

Zhang, M., O'Reilly, S.Y., Wang, K.L., Hronsky, J., Griffin, W.L., 2008. Flood basalts and metallogeny: the lithospheric mantle connection. Earth-Sci. Rev. 86 (1), 145−174.

Zhang, Z., Mao, J., Chai, F., Yan, S., Chen, B., Pirajno, F., 2009. Geochemistry of the Permian Kalatongke mafic intrusions, northern Xinjiang, northwest China: implications for the genesis of magmatic Ni-Cu sulfide deposits. Econ. Geol. 104 (2), 185−203.

Zhou, M.F., Lesher, C.M., Yang, Z.X., Li, J.W., Sun, M., 2004. Geochemistry and petrogenesis of 270 Ma Ni−Cu−(PGE) sulfide-bearing mafic intrusions in the Huangshan district, Eastern Xinjiang, Northwest China: implications for the tectonic evolution of the Central Asian orogenic belt. Chem. Geol. 209, 233−257.

FURTHER READING

Hoatson, D.M., Jaireth, S., Jaques, A.L., 2006. Nickel sulfide deposits in Australia: characteristics, resources, and potential. Ore Geol. Rev. 29 (3), 177−241.

REVIEW OF PREDICTIVE AND DETECTIVE EXPLORATION TOOLS FOR MAGMATIC NI-CU-(PGE) DEPOSITS, WITH A FOCUS ON KOMATIITE-RELATED SYSTEMS IN WESTERN AUSTRALIA

2

Margaux Le Vaillant[1,2]**, Marco L. Fiorentini**[1] **and Stephen J. Barnes**[2]

[1]*The University of Western Australia, Perth, WA, Australia* [2]*CSIRO/Mineral Resources, Kensington, Perth, WA, Australia*

CHAPTER OUTLINE

2.1 INTRODUCTION

About 60% of the world's nickel is currently produced from the exploitation and recovery of Ni-Cu-(PGE)-bearing sulfides associated with magmatic systems, mainly mafic and ultramafic magmatic intrusions and lava flows (Naldrett, 2004; Mudd, 2010). The remaining nickel production is almost entirely from limonitic and saprolitic laterite deposits (Golightly, 1981; Barnes and Lightfoot, 2005). The existing exploration problem for magmatic systems lies in the fact that they are extremely difficult targets. Indeed, the highly dynamic conduits where mineralization is

Processes and Ore Deposits of Ultramafic-Mafic Magmas through Space and Time. DOI: http://dx.doi.org/10.1016/B978-0-12-811159-8.00003-2

commonly found, which are part of lithospheric plumbing systems and/or extensive volcanic fields where magma is transported, do not generally form large detectable footprints that can be easily followed up during exploration.

The purpose of this study is to review the application of exploration methods to search for magmatic systems, with a focus on komatiite-hosted systems where the most recent work has been carried out with a spatially constrained approach. The footprints, which are not necessarily geochemical in nature, are generally cryptic to most datasets, but can be revealed if the integration of various techniques is carried out at the appropriate scale. In this chapter, we will provide a general overview on the genetic process for orthomagmatic Ni-Cu-(PGE) systems with a mineral system approach (McCuaig et al., 2010). We will then describe the geochemical and geophysical techniques that are currently utilized, addressing how these exploration tools relate to each other and could be integrated at various scales.

2.2 NI-CU-(PGE) ORE-FORMING PROCESSES AND ASSOCIATED EXPLORATION TOOLS

2.2.1 ORE FORMING PROCESSES

A widely accepted working model for the genesis of magmatic Ni-Cu-(PGE) mineral systems involves the presence of a fertile magma (1), which reaches sulfide saturation (2) at various stages during emplacement and crystallization at different crustal levels; the sulfide liquid is then chemically enriched in metals and physically transported and accumulated in specific environments depending on a range of physical conditions (3). Barnes et al. (2016) comprehensively discussed the mineral system framework for magmatic mineral systems (Fig. 2.1).

The magma is the primary source of metals. Its fertility largely depends on the nature and degree of partial melting of the mantle source that generated it. Significant Ni-Cu-(PGE) deposits are known to be associated with a wide range of magma types, ranging from komatiites, ferropicrites, picrites, all the way to tholeiitic and alkali basalts. The varying nature of the magmas is not a controlling factor on the prospectivity of the given magmatic province but rather affects the metal content of the resulting sulfide mineralization.

In order to form nickel-rich sulfides, it is vital for the magmatic system to reach sulfide saturation prior to extensive olivine crystallization. Therefore, thermo-mechanical assimilation of sulfur-bearing lithologies during magmatic emplacement appears to be the most efficient way to form metal-rich sulfides. While this model is well established for komatiites, for which sulfide-rich exhalative and/or sedimentary rocks located proximal to the volcanic vent are the sulfur source (Lesher, 1983; Huppert et al., 1984; Lesher et al., 1984; Groves et al., 1986; Lesher and Groves, 1986b; Lesher and Arndt, 1995; Bekker et al., 2009; Fiorentini et al., 2012a), the question is still debated for the associated intrusive systems, for which isotope evidence is inconclusive as to the relative proportion in sulfur contribution between crustal and mantle reservoirs (e.g., Fiorentini et al., 2012b).

Once sulfide supersaturation is attained and a sulfide liquid is formed, chalcophile elements in the silicate melt partition into newly formed sulfide droplets. In order to be concentrated at an economic level, the sulfide droplets need to react with a sufficient amount of magma (Campbell and

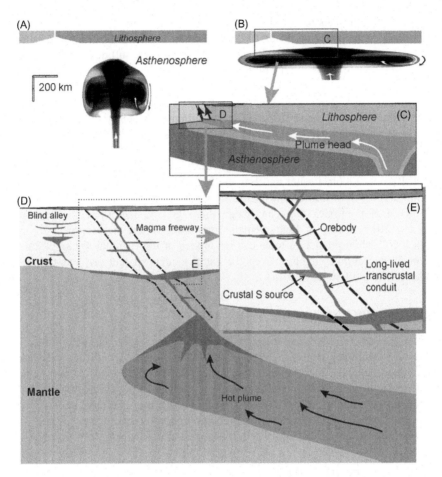

FIGURE 2.1

Schematic illustration of the lithospheric scale ore forming processes for the genesis of magmatic Ni-Cu-(PGE) deposits. (A) Starting plume ascending beneath an Archaean craton, within a few hundred kilometers of an original craton boundary. (B) Impingement and flattening of plume head beneath lithosphere. (C) Channeling of plume head and tail to thinnest lithosphere at craton margin, generation of continental rifting centered on original suture, and onset of high-Mg, low-T melts production. (D) Development of favorable and unfavorable environments for mineralization above the melting zone, showing the combination of long-lived mantle-tapping structure and high magma production giving rise to high flux "magma freeways" with potential for assimilation of crustal S, transport and deposition of magmatic sulfide ores.

Naldrett, 1979; Naldrett, 1999) and to be efficiently stirred to enhance equilibration (Lesher and Campbell, 1993; Robertson et al., 2015). Thus, a common feature for Ni-Cu-(PGE) deposits is the presence or proximity of a magma conduit or flow pathway (Lesher et al., 1984; Lesher, 1989; Barnes et al., 2016).

Finally, these sulfide droplets need to be concentrated in important quantities in a restricted locality to form ore bodies (Campbell and Naldrett, 1979; Naldrett, 1999). In komatiites, this concentration is generally thought to be achieved by gravitational settling of dense sulfide droplets to the bottom of lava flows (Naldrett, 1966; Ewers and Hudson, 1972), or freezing of dense sulfide liquid layers entrained at the base of the magmatic flow (Lesher and Campbell, 1993). In some intrusive systems, physical concentration may occur in response to changing geometry of the flow pathway and/or through dynamic interaction of the sulfide melt with surrounding country rocks, forming sulfide-rich breccias of variable nature (e.g., Robertson et al., 2015; Barnes et al., 2016; Locmelis et al., 2016).

Important primary magmatic processes leading to the genesis of Ni-Cu-(PGE) mineralization have been extensively studied and reviewed (e.g., Lesher and Keays, 1984, 2002; Lesher, 1989; Naldrett, 2004; Barnes, 2006b; Lesher and Barnes, 2009; Barnes et al., 2015). In order to translate the magmatic mineral system framework into an effective series of exploration targeting criteria, it is important to develop proxies that can image these genetic factors at the appropriate scale (McCuaig et al., 2010). Scale is fundamentally important in the development of an effective exploration protocol: while predictive tools can be applied at the regional to camp scale, detection techniques can be successfully applied from the camp to deposit and ore shoot scale (McCuaig et al., 2010; McCuaig and Hronsky, 2014).

2.2.2 CURRENT EXPLORATION METHODS

Predictive techniques are the main targeting methodology when exploring at the craton- to terrane-scale, based on favorable tectonic setting and volcanic environment (McCuaig et al., 2010; McCuaig and Hronsky, 2014). Detection techniques, both geophysical and geochemical in nature, are increasingly more effective as exploration evolves from regional scale all the way to camp, prospect, deposit, and ore shoot scales. However, both the predictive and detective techniques that have been developed to explore for magmatic mineral systems over the last four decades have only had limited success. A summary of all these techniques is presented in Table 2.1.

2.2.3 PREDICTIVE TECHNIQUES AT REGIONAL SCALE

When exploring at the regional scale, explorers look for (1) favorable tectonic environments along with (2) the most prospective geological setting. For magmatic Ni-Cu-(PGE) systems, favorable tectonic environments are represented by large crustal scale structures favoring magma flow from the mantle to the crust, such as craton boundaries at the time of plume impingement (Beresford et al., 2007; Begg et al., 2010; McCuaig et al., 2010; Mole et al., 2013, 2014) and continental rift zones subsequently inverted by polyphase deformation (Beresford et al., 2007; McCuaig et al., 2010; Fiorentini et al., 2012b), as shown in Fig. 2.2. In order to image trans-lithospheric faults, numerous geophysical techniques are in use, such as magnetotelluric, aeromagnetic, seismic, and gravity profiles.

However, geophysical techniques only provide a snapshot of the current lithospheric architecture of any given lithospheric block, whereas the location of the highly prospective paleo-margins of the terranes that assembled to form the craton would be blind to such techniques. In order to image the paleo-margins and reveal the presence of intra-cratonic prospective areas, it is possible to make use of spatially constrained isotopic terrane mapping. In fact, maps of whole-rock Sm-Nd (εNd) model ages of granitoids indicate the presence of ancient craton boundaries (Fig. 2.2). The addition of

Table 2.1 Summary of the Various Exploration Tools Described in this Paper

Regional Scale		Camp to Prospect Scale		Deposit Scale	
Favorable Tectonic Settings:		*Channelized Volcanic Environments:*		*Evidences of Sulfide Accumulation and/or Extraction:*	
Craton boundaries	Beresford et al., 2007; Begg et al., 2010; McCuaig et al., 2010; Mole et al., 2013, 2014	Prediction of prospective volcanic environments	*Barnes and Brand, 1999; Barnes et al., 2004a; Lesher et al., 2001; Fiorentini et al., 2012b; Le Vaillant et al., 2014*	Positive and negative anomalies in chalccophile elements (Ni, Cu, Co and PGE) Whole rock and Ni in olivine	*Barnes et al., 1988a; Barnes and Picard, 1993; Lesher et al., 2001*
Inverted continental rift zones	*Beresford et al., 2007; McCuaig et al., 2010; Fiorentini et al., 2012b*	Exploration tool for komatilte-hosted systems: Potentially using pXRF data	Ni/Ti vs Ni/Cr diagrams, *(Le Vaillant et al. 2014)*	Ruthenium depletion in chromite grains (komatilte hosted systems) — possible applications in lateritic terranes	*Fiorentini et al. 2008; Locmelis et al., 2011, 2013;*
		Evidence of Crustal Contamination:			
		Sulfur isotope analyses to test crustal assimilation models	*Fiorentini et al., 2012a*		
Exploration tools:	Magnetotellurics, Aeromagnetics, Seismics, Gravity, Sm-Nd model Age maps, Lu-Hf isotopes on zircons *(Deen et al., 2006; Champion and Cassidy, 2007, 2008; Begg et al., 2009; McCuaig et al., 2010; Mole et al., 2012, 2014; Perring et al., 2015a, 2015b)*	Anomalous enrichment in incompatible elements such as Zr, Th, LREE	*Lesher et al., 2001; Barnes and Hill, 2004b; Fiorentini et al., 2012b*	Exploration tool:	Whole rock analyses, Laser ablation ICP-MS Analyses of olivine and chromite to evaluate their Ni or Ru content
		Possible pervasive contamination signals in certain cases (e.g., Black Swan, Perseverance)	*Barnes et al., 1988c, 2004a, 2007; Barnes and Fiorentini, 2012;*	*Example: Subtle PGE enrichments and depletion signals in host komatiite units at the Long Victor and Maggie Hays deposits, Western Australia, up to 400 m away from massive sulfides (Barnes et al., 2013; Heggie et al., 2012).*	
Areas of Enhanced Magmatic Flux:		Exploration tools:	Laboratory geochemical analyses, portable XRF (pXRF) analyses potentially directly on the field *(Le Vaillant et al., 2014)*	*Empirical Detection Tools:*	
High proportions of high MgO komatiite magmas, and abundance of strongly adcumulate olivine-rich cumulates	*Barnes and Fiorentini, 2012*			Use of Ni/ Cr to delineate ore-related channels in komatiite hosted systems	*Barnes and Brand, 1999; Barnes et al., 2013*

(Continued)

Table 2.1 Summary of the Various Exploration Tools Described in this Paper *Continued*

Regional Scale		Camp to Prospect Scale		Deposit Scale
Favorable Tectonic Settings:		*Channelized Volcanic Environments:*		*Evidences of Sulfide Accumulation and/or Extraction:*
		Example: Presence of a pervasice contamination signal in komatiite units of the Black Swan komatiite complex, as well as the Perseverance ultramafic complex, Western Australia (Barnes et al., 1988b, 2004a)		*Example: delineation of ore shoots around the Kambalda Dome, Western Australia (Barnes and Brand, 1999)* Airborne and down hole electromagnetic surveys
Exploration tools:	Regional airborne aeromagnetic Surveys *(Barnes et al., 2004a; Grguric and Riley, 2006; Fiorentini et al., 2007)*	*Evidences of Sulfide Accumulation and/or Extraction:*		*Example: The Nova discovery (Western Australia) was mode when testing a large and strong EM anomaly (ASX Announcement – 26.07.2012)*
		Positive and negative anomalies in chalccophile elements (Ni, Cu, Co and PGE) Whole rock and Ni in olivine *Exploration tool:*	Barnes et al., 1988a; Barnes and Picard, 1993; Lesher et al., 2001 Whole rock analyses, Laser ablation ICP-MS analyses of olivines to evaluate their nickel content	*Hydrothermal Ni-As-PGE Haloes:* Presence of nickel arsenides (gersdoffite mainly) in veins or within a plan of foliation creating a geochemical signal (enrichment in Ni, As, Pd and Pt) extending up to 1,780 m away from massive nickel-sulfides. Shown in Komatiite hosted system, applicable to any type of magmatic Ni-Cu-(PGE) deposit.
Example: The Nebo Babel deposit in the West Mustgrave, Western Australia, was discovered using lithospheric Architectural targeting combining areas presenting evidences of major focused mantle-derived magmatism with anomalous lithospheric architecture (boundary of the Yilgam Craton, and convergence of multiple faults including the Mundrobilla fault) (Pers. Comm. John Hronsky)				
		Examples: LowNi contents of oivines in the Kambalda Dome komatiite flows and in the Perseverance camp, Western Australia (Lesher et al. 1981; Barnes et al., 1988a); Anoumalous positive and negative PGE concentrations in the komatiite basalt of the Raglan camp, Northern Quebec (Barnes and Picard,1993; Lesher et al., 2001)		*Le Vaillant, 2014; Le Vaillant et al. 2015a,b)* *Example: Hydrothermal geochemical halo observed surrounding the Miitel and the Sarah's Find deposits, Western Australia (Le Vaillant et al. 2015a,b)*
Modified from Le Vaillant, M., Fiorentini, M.L., Barnes, S.J., 2016b. Review of lithogeochemical exploration tools for komatiite-hosted Ni-Cu-(PGE) deposits. J. Geochem. Explor. 168, 1–19.				

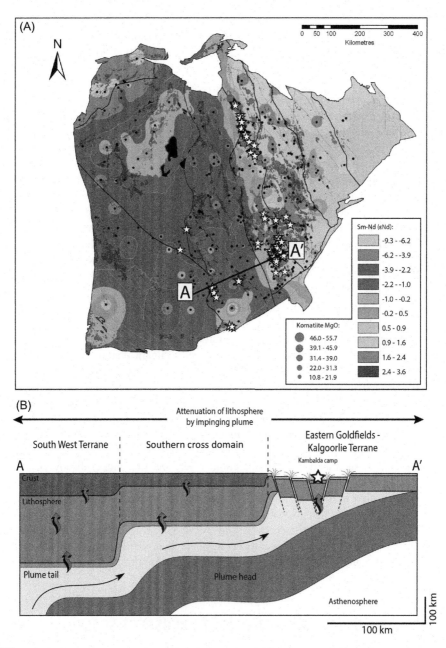

FIGURE 2.2

(A) Sm-Nd (εNd) isotopic map (geometric interval) highlighting the internal architecture of the Yilgarn craton, Western Australia, and showing the location of komatiite-hosted nickel deposits along its paleo-margins (white stars). (B) Interpreted lithospheric cross-section based on the Sm-Nd isotopic mapping.

Adapted from Mole, D.R., Fiorentini, M.L., Cassidy, K.F., Kirkland, C.L., Thebaud, N., McCuaig, T.C., et al., 2013. Crustal Evolution, Intra-Cratonic Architecture and the Metallogeny of an Archaean Craton. Geological Society, London, Special Publications 393.

Lu-Hf isotopic data from well-dated zircons can add an additional dimension of time combined with information on the magmatic source, allowing craton margin positions to be tracked through time (Deen et al., 2006; Champion and Cassidy, 2007, 2008; Begg et al., 2009; Mole et al., 2012, 2014).

Currently, aeromagnetic and geological maps are still the main tools in practice to identify favorable tectonic settings and host lithologies, at a range of scales. For example, to assist in exploration targeting for komatiite systems, Perring et al. (2015a, 2015b) have proposed the existence of accretionary growth faults, recognizable in regional magnetic and gravity patterns, as primary controls on the location of mineralized dunite channels in the Agnew-Wiluna greenstone belt of Western Australia. However, in poorly exposed terranes typical of Archaean greenstone belts, there is a high rate of false positive anomalies created when interpreting mapped lithologies (i.e., structures interpreted as mantle-tapping, which are in fact limited to the upper crust; McCuaig et al., 2010).

When exploring for magmatic Ni-Cu-(PGE) deposits, explorers look for areas of enhanced magmatic flux, which are reflected for example in the presence of large sulfide globules, in excess of 1 cm, indicating that the transporting magma was capable of generating a massive sulfide accumulation (Barnes et al., 2017). In the case of komatiites, proxies for enhanced flux are also high MgO concentrations and abundance of strongly olivine mesocumulate-adcumulate textured rocks (Lesher et al., 1984; Lesher, 1989; Hill et al., 1995; Lesher and Keays, 2002; Arndt et al., 2008; Barnes and Fiorentini, 2012). Regional airborne aeromagnetic data allow the identification of these favorable lithologies. This was one of the main exploration tools during the nickel boom in 1966−71 in Western Australia (Woodall and Travis, 1969; Ross and Travis, 1981; Marston, 1984; Hronsky and Schodde, 2006). Where high MgO rocks are serpentinized, they generate strong bulls-eye magnetic anomalies, such as at the Mount Keith deposit, Agnew-Wiluna greenstone belt, Yilgarn craton, Western Australia (Burt and Sheppy, 1975; Grguric and Riley, 2006; Fiorentini et al., 2007). However, this detection technique fails where the ultramafic rocks are converted to non-magnetic talc-carbonate assemblages, as they very commonly are in Archaean settings (such as deposits around the Widgiemooltha and Kambalda Domes, Norseman-Wiluna greenstone belt, Western Australia (Barnes et al., 2004b).

Combination of a favorable tectonic environment such as a trans-lithospheric fault with areas of high magmatic flux within a large igneous province is a highly favorable indicator of nickel endowed mining camps, such as the Kambalda camp where the first nickel discoveries were made in the mid 1960s (Hronsky and Schodde, 2006). Such settings are commonly associated with craton margins, as first identified in the Cape Smith (Raglan) and Thompson komatiite belts that surround the Superior Craton (Baragar and Scoates, 1981). The recent world-class discoveries at Nova Bollinger and Nebo Babel in the Fraser belt and West Mustgrave areas of Western Australia, respectively, are good examples of mineralized systems that were discovered using predictive targeting at the craton scale, looking for areas with evidence of major focused mantle-derived magmatism with anomalous lithospheric architecture.

2.2.4 LITHOGEOCHEMICAL TOOLS AT CAMP TO PROSPECT SCALE

Numerous lithogeochemical exploration techniques, both predictive and detective, have been developed and successfully used over the years at the camp to prospect scale, such as:

1. The search for evidence of crustal contamination of the magma, for instance anomalous enrichments in highly incompatible lithophile elements (Zr, Th, and LREE) (Lesher et al.,

2001; Barnes et al., 2004b; Fiorentini et al., 2012b) represents a predictive exploration technique used to identify prospective magmatic units that may have reached sulfide saturation due to crustal assimilation. As an example, Zr and Ti are relatively immobile and incompatible in crystallizing olivine, and Zr is more highly concentrated in crustal rocks relative to mantle melts. Hence the ratio Zr/Ti can be used as a contamination indicator, with the advantages that it is not greatly affected by alteration, metamorphism, nor fractional crystallization (Sun and Nesbitt, 1977; Huppert and Sparks, 1985) and can be measured directly in the field using portable X-Ray fluoresence tools (pXRF; Le Vaillant et al., 2014).

Contamination signals will be detected when important amounts of fractionated (felsic) material has been assimilated by the magma (Lesher and Arndt, 1995). This methodology has been successfully applied in komatiite exploration, where pervasive crustal contamination signals are observed over several km of strike, such at the Black Swan komatiite complex (Barnes et al., 2004a) as well as at the Perseverance and Mount Keith ultramafic complexes (Barnes et al., 1988b), both in the Yilgarn craton of Western Australia, where komatiites were erupted onto and within sulfide-bearing felsic country rocks (e.g., Fiorentini et al., 2007).

Barnes and Fiorentini (2012) showed that, at the scale of greenstone belts and terranes, prospective belts tend to contain significantly higher proportions of contaminated komatiites. The mineralized portions of the Abitibi greenstone belt show distinct regional patchy anomalies in the presence of contamination signals relative to the remainder of the belt, such as elevated ratios of strongly incompatible (Th, La, Ce, Zr) to less incompatible (HREE, Ti, Nb) lithophile trace elements (Sproule et al., 2002; Barnes et al., 2007). However, in cases where komatiites were erupted onto basaltic substrates with volumetrically minor sulfidic sediments, as at Kambalda, signals of contamination can be weak and spatially limited even in well-mineralized flows (Lesher and Arndt, 1995; LaFlamme et al., 2016), where the contamination seems to be limited to the flanking sheet flow facies, and absent within the most active, central part of the the channel where the ore-forming magma has been flushed out by ongoing flow (Lesher et al., 2001; Barnes et al., 2013b).

However, the application of this geochemical approach may lead to the identification of a high rate of false positive targets. First of all, the nature of the geochemical nature is very dependent upon the type of contaminant, some of which may not leave behind an easily detectable footprint. Furthermore, it is clear that not all magmatic conduits that were emplaced dynamically along any plumbing system are mineralized (e.g., Paringa Balsalts, Kambalda; Redman and Keays, 1985). However, combining evidence of crustal contamination with other observations can help identify these false positives.

2. When exploring for komatiite-hosted deposits, channelized volcanic environments are the most prospective settings (Barnes and Brand, 1999; Lesher et al., 2001; Barnes et al., 2004b; Fiorentini et al., 2012b). These can be detected with a series of geochemical criteria (Fig. 2.3), which are particularly useful where original rock textures are unrecognizable. For example, a Ni/Ti vs Ni/Cr diagram can be used to delineate favorable volcanic environments; Ni/Ti will correlate with original olivine content constraining the silicate nickel background and highlighting subtle sulfide-related nickel enrichment. This can be combined with the Ni/Cr ratio, which gives information on the volcanic environment.

Empirical observations show that in komatiite systems prospective olivine-rich channel facies rocks tend to have higher Ni/Cr ratios than unprospective non-channel facies rocks of

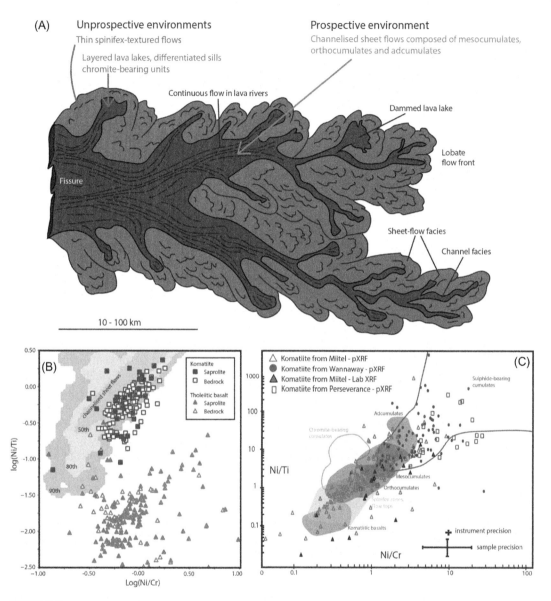

FIGURE 2.3

(A) Schematic illustration of a regional komatiite flow field, modified from Hill et al. (1995), (B) comparison of Ni/Ti and Ni/Cr ratios between fresh bedrock and "top of fresh rock" saprolite in the Agnew area in Western Australia, data compiled by Barnes et al. (2014), and (C) Plot showing the potential use of pXRF in evaluating the prospectivity of a komatiite unit using Ni/TI and Ni/Cr ratios.

Modified from Le Vaillant, M., Barnes, S.J., Fisher, L., Fiorentini, M.L., Caruso, S., 2014. Use and calibration of portable X-Ray fluorescence analysers: application to lithogeochemical exploration for komatiite-hosted nickel sulphide deposits. Geochem. Explor. Environ. Anal. 14, 199–209.

otherwise similar composition (Barnes, 1998), such as for example the mineralized Kambalda-style ore environments. These are characterized by high Ni and low Cr contents of sulfide-poor rocks (Barnes and Brand, 1999): around the Kambalda Dome variation of the Ni/Cr ratio has been used with some success in delineating fertile channels (Woolrich et al., 1981). These variations in Ni, Cr and Ti contents form part of the rationale for the use of combined Ni/Cr and Ni/Ti ratios to map favorable sulfide deposition sites within komatiite flow fields (Barnes et al., 2004b), as discussed further below. However, it is now known that this approach applies only to true komatiites and not to deposits hosted within komatiitic basalt sequences, such as those in the Raglan camp, Cape Smith belt, Northern Quebec (Barnes and Picard, 1993; Lesher et al., 2001), where the magmas were pervasively chromite saturated (Lesher and Stone, 1996; Lesher, 2007). These ratios can also potentially be calculated using concentrations collected directly on the field using pXRF (Le Vaillant et al., 2014), as shown in Fig. 2.3C.

3. Chalcophile element anomalies associated with magmatic sulfide extraction or accumulation, such as Ni, Cu, Co and platinum group elements (PGE), can be used as a detective exploration tool (Duke and Naldrett, 1978) usually restricted to near-deposit settings, but there are examples of more pervasive deposit scale signals. For example, in the komatiite-hosted Perseverance deposit, Agnew Wiluna greenstone belt, Western Australia, the observed extensive depletion of nickel in olivine is indicative of sulfide segregation, along with crustal contamination signals, over several km of strike (Barnes et al., 1988a). In addition, pervasive anomalies in PGEs, both positive and negative, have been identified at camp scale in the komatiitic basalts of the Raglan camp. However, some mineralized komatiite belts such as Forrestania, Southern Cross Province, Western Australia, and some regional-scale flow fields, such as the Silver Lake Member of the Kambalda Komatiite Formation of the Kalgoorlie Terrane, Western Australia, show no significant chalcophile element anomalies beyond the immediate deposit scale (Lesher et al., 2001; Barnes and Fiorentini, 2012; Barnes et al., 2013b). The dynamics and size of the komatiite flows must therefore be kept in mind when using these geochemical tools.

   The komatiite-hosted Perseverance deposit represents an extremely dynamic magmatic system with a lasting magma flow or multiple flows without solidification in between, resulting in pervasive felsic footwall assimilation. Extensive olivine-sulfide equilibration in such a dynamic system gave rise to Ni depletion in olivine in a very large volume of preserved cumulates. On the other hand, although the Kambalda magmatic system as a whole contains comparable volumes of sulfide ore as Perseverance, mineralization at Kambalda occurred within relatively low volume, strongly channelized magma flows with a highly localized external S source. In the Kambalda deposit, Ni depletion in olivine is restricted to the immediate surrounding rocks. Furthermore, the extent of Ni depletion depends strongly on the R factor, i.e. the mass ratio of silicate magma to sulfide liquid that equilibrate. Where R factors are relatively low, as at Perseverance, ore tenors are lower, and Ni depletion in silicate magma and olivine correspondingly is greater, relative to high R factor deposits with high Ni tenors and only minor Ni depletion in olivine (Barnes et al., 2013a).

4. At the camp to prospect scale, multiple sulfur isotopes represent a very informative tool both for prediction and detection of magmatic mineral systems. From a predictive perspective, multiple sulfur isotopes support the working hypothesis that crustal assimilation is key to sulfide saturation at least in komatiite systems (Lesher and Groves, 1986a), where the source can vary from exhalative dominated reservoirs—mostly associated with felsic volcanism in

VMS type settings (Fiorentini et al., 2012a)—all the way to sulfidic sediments in abyssal oceanic planes (La Flamme et al., 2016). In mafic hosted systems, sulfur isotopes have been used to either advocate for a mantle source of sulfur or/and to show the extreme complexity of the ore forming process, with assimilation of different crustal sulfur reservoirs at variable depths over the emplacement and crystallization history of any given magmatic system (Seat et al., 2009; Ripley, 2013).

2.2.5 DETECTIVE TECHNIQUES AT DEPOSIT SCALE

As the scale of exploration closes in to the camp and deposit scales, the availability of detective techniques increases (McCuaig et al., 2010). Over the years, several detection techniques have been developed for magmatic Ni-Cu-(PGE) systems. However, their efficiency is still limited, mainly due to the small size and ribbon-like geometry of most Ni-Cu-(PGE) sulfide ore bodies, especially the ones associated with komatiite magmas (Fig. 2.3). The main available exploration tools for magmatic sulfide systems that have been developed for the use at the deposit scale are as follows:

1. The detection of anomalous whole-rock enrichment and/or depletion of chalcophile elements (Ni, Cu, Co, PGEs) as an indication of sulfide segregation and accumulation (e.g., Lesher et al., 2001; Barnes et al., 2004b; Arndt, 2005). Fiorentini et al. (2010, 2011) suggested that enrichment and depletion should not be considered in absolute terms, as they may vary according to different magmatic provinces and magma types. They investigated the PGE background concentrations of basalts, ferropicrites and komatiites, in order to define the various controls for PGE variation, and subsequently isolate PGE variability related to sulfide saturation. Platinum-group variability has been utilized to refine ore genetic concepts (e.g., Norilsk-Talnakh deposit; Arndt, 2005) or predict the mineralization potential of different magmatic provinces (e.g., Deccan Traps; Krishnamurthy, 2015). However, the most detailed spatially constrained studies were carried out for komatiite systems. At the Long Victor (Kambalda) and Maggie Hays komatiite-hosted nickel-sulfide deposits, Western Australia, Barnes et al. (2013b) and Heggie et al. (2012) observed localized subtle PGE enrichments and/ or depletions extending up to 400 m away from mineralization, thus considerably enlarging the footprint of those notably small targets. Similarly to PGEs, the study of nickel contents in olivine also represents a useful tool (Duke and Naldrett, 1978; Duke, 1979) that has nonetheless limitations, owing to (1) the relatively rare preservation of fresh olivine in magmatic systems, especially Archean komatiites, which are almost universally pervasively altered, and to (2) the lower sensitivity of Ni relative to PGEs to sulfide-related depletion due to its much lower partition coefficient (Duke and Naldrett, 1978; Duke, 1979). On the other hand, while minero-chemical analyses for Ni in olivine are very affordable, PGE-based techniques generally require expensive high-precision analyses.
2. A similar limitation applies to the use of Ni/Cr ratios, which can delineate ore-related channels within broad komatiitic volcanic fields (Barnes and Brand, 1999), through picking up Ni enrichment directly and also by discriminating cumulates formed in high-temperature lava channels. Olivine cumulates in ore-bearing komatiite lava channels tend to have higher Ni/Cr even in samples that are not mineralized. There are two reasons for this: the magmas tend to be hotter, and therefore less likely to be chromite-saturated; and even when the magmas were

chromite-saturated (e.g., Mt Keith; Barnes et al., 2011), chromite crystallization seems to be suppressed in high-flux channel settings, especially where sulfide is present (Woolrich et al., 1981; Barnes, 1998). However, the Ni/Cr ratio is not significantly more effective than nickel anomalies alone (Barnes et al., 2013b), as discussed below in the section comparing various exploration tools.

Barnes et al (2013b) and Heggie et al (2012) compared the effectiveness of variations in Ni/Cr, nickel alone and Pt/Ti as vectors towards ore bodies within olivine cumulate-rich komatiites (Fig. 2.4). Both in extrusive (e.g., Long-Victor deposit at Kambalda) and intrusive systems (e.g., Maggie Hays, Lake Johnson greenstone belt), variations in Pt/Ti and Pd/Ti can be detected in essentially sulfide-free rocks some 400 m away from orebody. In the Kambalda case, this takes the form both of PGE depletion in flanking flow tops and PGE enrichment closer to orebody within the channel. Both the depleted and enriched signals extend up to 450 m away from the 0.4 wt% Ni grade shell (the effective limit of detectable disseminated sulfides) and are much clearer than the very subtle variation in Ni/Cr between channel and flank facies (Fig. 2.4A,B). Using the Ni/Cr ratios does not appear to have advantage over using the elemental nickel abundance by itself, although Ni/Cr can be used in mildly weathered or extensively altered rocks where absolute whole-rock Ni and Cr may have been diluted or enhanced. At Maggie Hays, a subtle increase in Ni/Cr and a much more pronounced increase in Pt/Ti and Pd/Ti appear at about 300 m distance from the disseminated sulfide shell, but the Pt/Ti and Pd/Ti define particularly clear vectors of systematic increase towards orebody. Values above 1 (mantle normalized value) appear up to 650 m away with a systematically increasing trend developed within the inner 250 m (Fig. 2.4).

3. The detection of ruthenium depletion in chromite grains, obtained by laser-ablation ICP-MS analysis of chromite grains either in situ or from mineral separates, shown as characteristic of mineralized komatiites by Locmelis et al. (2013), can potentially be used to discriminate between mineralized and barren komatiite flows (Fiorentini et al., 2008; Locmelis et al., 2011; Locmelis et al., 2013). Care must be taken to compare rocks that cooled at similar rates, as cooling rate seems to have an effect on Iridium-group PGE (IPGE) concentrations in chromites (Pagé and Barnes, 2016).

4. Finally, the evaluation of the thickness of sedimentary rock units at basal komatiite contacts could potentially be used for exploration, as a tight mutually exclusive relationship between the distribution of sulfidic sedimentary rocks and Ni-Cu-(PGE) sulfide orebody has been demonstrated at Kambalda (Bavinton, 1981; Paterson, 1984; Lesher and Groves, 1986a), and was used extensively as an exploration guide during the early delineation of the orebodies (Gresham and Loftus Hill, 1981).

Most of the exploration tools described above do not extend over 50–100 m away from massive sulfides, mainly because of the dynamic nature of the ore-forming channels and removal of early ore-forming magma by the later magma pulses. Moreover, these techniques rely on the present geometry being similar to the one upon emplacement, whereas komatiite-hosted Ni-Cu-(PGE) systems have commonly undergone polyphase deformation (e.g., Duuring et al., 2007; Layton-Matthews et al., 2007; Duuring et al., 2010; Duuring et al., 2012), rendering these detection tools largely ineffective.

Finally, geophysical methods are of great importance to identify prospective areas and define drilling targets, particularly when integrated with other datasets. Electromagnetic (EM) techniques

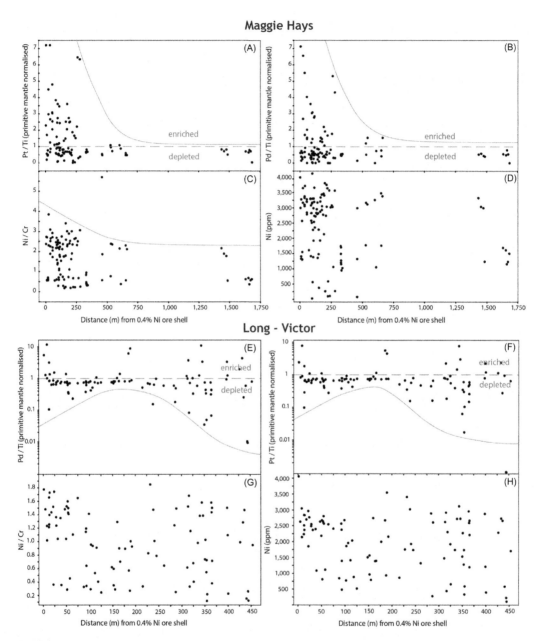

FIGURE 2.4

Comparison between the spatial extent of anomalous Ni/Cr and PGE haloes at Maggie Hays and Long Victor.

Modified from Barnes, S.J., Heggie, G.J., Fiorentini, M.L., 2013b. Spatial variation in platinum group element concentrations in ore-
bearing komatiite at the Long-Victor deposit, Kambalda Dome, Western Australia: enlarging the footprint of nickel sulphide
orebodies. Econ. Geol. 108, 913–933.

are generally very efficient to detect massive nickel-sulfide ore bodies, particularly in low conductive backgrounds. For example, the recently discovered Nova deposit in the Fraser Range Domain, Western Australia, was first intersected by a drill hole meant to test a large and strong EM anomaly (Bennett et al., 2014). The distance of detection for down-hole EM (DHEM), which is one of the standard techniques, mainly depends on the shape and size of the target, and varies between 0.5−1.5 times the smaller dimensions of the ore body, therefore generally extending 50−150 m from massive sulfides (Peters, 2006), making it a highly effective tool to follow brownfield ore body extensions. However, in Australia particularly, large amounts of saline groundwater and salt are present in the cover, challenging EM interpretations (González-Álvarez et al., 2016). And for komatiite-hosted Ni-Cu-(PGE) deposits, the following characteristics give rise to several limitations for geophysical methods: ore bodies are commonly small and deep seated, barren sedimentary and exhalative sulfide units are commonly in close spatial association with the ore bodies. Finally, the main drawback of DHEM is the high false positive rate due to EM anomalies associated with barren sulfidic or graphitic metasediments, which makes it difficult to discriminate nickel-rich from barren sulfide bodies (Peters, 2006).

All exploration tools described above are summarized in Table 2.1. In brief, magmatic Ni-Cu-PGE deposits remain very difficult exploration targets. Their general location can be reasonably well predicted owing to an extensive knowledge of their genetic processes, but to this date there are no effective predictive or detection techniques (geochemical or geophysical) that allow explorers to define and prioritize targets from sparse drilling. For example in Western Australia the rate of discovery of magmatic mineral systems has slowed down dramatically since the initial surge of exploration success for komatiite-hosted deposits between 1966 and 1973 (Hronsky and Schodde, 2006). Undiscovered deposits are highly likely to exist at depth, even in mature well-explored terranes, where they are likely to be deformed, altered and offset from their original position. A new approach is needed to aid exploration in brownfields terranes by enlarging the detectable footprints of undiscovered deposits. Secondary hydrothermal haloes surrounding primary magmatic deposits have potential to be useful signals.

2.3 HYDROTHERMAL REMOBILIZATION AND GEOCHEMICAL HALOES

Most magmatic Ni-Cu-(PGE) deposits have been altered and modified to some degree as a result of interaction with post-magmatic (or syn-magmatic) hydrothermal and metamorphic fluids. The interaction between these fluids and massive nickel-sulfide bodies has the potential to create a relatively large dispersive footprint with specific mineralogical and lithogeochemical characteristics. Few previous studies looking at hydrothermal haloes around magmatic Ni-Cu-(PGE) deposits show promising results. At the Barnet property, part of the Sudbury Cu-Ni-(PGE) camp (Canada), wide scale mobility of nickel in hydrothermal solution in the footwall of the deposit has been highlighted by the presence of elevated concentrations of nickel in secondary amphiboles (Hanley and Bray, 2009). This significant mobility (up to 700 m) was probably facilitated in that system by extreme impact-related fracture permeability. Another study by Layton-Matthews et al. (2007) on Ni-Cu-(PGE) deposits of the Thompson Nickel Belt highlights enrichment in Ni, Au, Pd, and Cu within sedimentary sulfide units adjacent to the nickel-sulfide deposits. At Thompson, this enrichment is interpreted as being either created during the mobilization of fluids generated by the metamorphism of both the ore zones and their host rocks (Bleeker, 1990) or via syn-magmatic diffusion of metals (Burnham et al., 2003). Finally, according to a study by Barrie et al. (2007), hydrothermal haloes are present around the River Valley, Fergusson Lake, and Kabanga deposits. Their results show subtle anomalies in combined metal and/or transition elements in country rocks, extending up to several hundred meters away from massive sulfides.

In addition, there are an increasing number of studies on hydrothermal nickel and/or PGE accumulations. In the Sudbury camp, the low-sulfide, Pt- and Pd-rich haloes (150 m away from massive sulfides) around vein-type Ni-Cu-(PGE) ores in the footwall of the Sudbury Igneous Complex (SIC), have been interpreted as the result of remobilization of PGE (mainly Pt and Pd) from differentiated sulfide liquids by late magmatic-deuteric and/or hydrothermal fluids by some researchers (Farrow and Watkinson, 1996; Molnár et al., 1999; Molnar and Watkinson, 2001; Molnár et al., 2001; Hanley and Mungall, 2003; Mossman et al., 2003; Hanley et al., 2005; Péntek et al., 2008; Hanley et al., 2011; Péntek et al., 2011; Molnár, 2013; Péntek et al., 2013; Tuba et al., 2014).

Finally, recent studies indicate the possibility of remobilization of large amounts of nickel from sulfide sources, on a scale of hundreds of meters up to kilometers in some specific cases (González-Álvarez et al., 2010, 2013a, 2013b; Keays and Jowitt, 2013; Pirajno and González-Álvarez, 2013); and some hydrothermal PGE-(Au) deposits have also been reported (Dillon-Leitch et al., 1986; Nyman et al., 1990; Olivo and Theyer, 2004; De Almeida et al., 2007; Bursztyn and Olivo, 2010). All these results, combined with recent studies on the behavior of Ni, Pd, and Pt in hydrothermal fluids (Barnes and Liu, 2012; Yuan et al., 2015; Le Vaillant et al., 2016a) show that, despite the widespread belief that Ni and PGE are extremely immobile elements under most circumstances, specific fluids and geological contexts have the capacity to remobilize nickel and PGE from a massive sulfide source and potentially create hydrothermal haloes useful for exploration targeting.

Le Vaillant et al. (2016b) investigated the nature of hydrothermal geochemical haloes around komatiite-hosted Ni-Cu-PGE deposits located in the Archean Norseman-Wiluna greenstone belt, which is the supracrustal component of the Kalgoorlie Terrane, within the Eastern Goldfields Superterrane (Cassidy et al., 2006), Yilgarn Craton, Western Australia (Fig. 2.5). The same methodology was applied at five komatiite hosted localities (Fig. 2.5). Results obtained for the various "hydrothermal halo" case studies are compared in Table 2.2. While the Miitel and the Sarah's find deposits display large Ni-As-PGE (Pd, Pt) hydrothermal haloes (Figs. 2.6 and 2.7), the Otter—Juan, Durkin and Perseverance ore bodies are devoid of any detectable halo, even though all these deposits have undergone similar phases of deformation and alteration. It is argued that these systems did not attain the necessary conditions allowing Ni and PGEs to be either incorporated in hydrothermal fluids and mobilized or even just re-deposited (Le Vaillant et al., 2016b).

The Miitel, Otter-Juan and Durkin ore shoots are directly comparable case studies as they are Kambalda-style type 1 deposits (Lesher, 1989; Hill and Gole, 1990) hosted at the basal contact between channelized komatiites and footwall basalts. In addition, all these komatiite localities have undergone similar metamorphic and alteration processes. However, while the Miitel deposit is extensively overprinted with arsenic metasomatism (values ranging from 35 to 2,405 ppm As), the Durkin and Otter-Juan deposits are devoid of it (maximum of 60 ppm As measured, with most values below detection limits). Le Vaillant et al. (2016b) put forward the hypothesis that the presence of arsenic plays a first order control in the development of a geochemical halo. According to

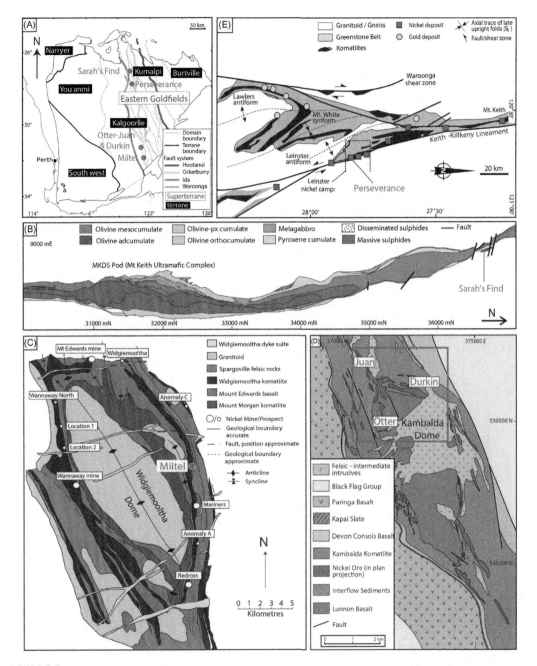

FIGURE 2.5

Geological maps locating the various case studies used within the project on hydrothermal haloes. (A) Simplified geological map of the Yilgarn Craton showing the location of the case studies, modified from Cassidy et al. (2006). (B) Location of the Sarah's Find prospect on a detailed geology of the Mount Keith ultramafic complex (from Fiorentini et al., 2007; Rosengren et al., 2007; Barnes et al., 2011). (C) Geological map of the Widgiemooltha Dome area showing existing Ni-Cu-(PGE) mines and prospects (adapted and modified after Seat et al. (2004) and McQueen (1981), originally modified after Willet et al., 1978). (D) Location map of Kambalda dome modified from Barnes (2006a). (E) Regional map of the Perseverance area, modified from Hill et al. (1995), Trofimovs et al. (2003), and Duuring et al. (2010).

Table 2.2 General information and summary of the results obtained during the study of hydrothermal haloes around various Australian komatiite-hosted nickel sulfide deposits.

Deposit	Site Stage	Greenstone Belt	Location	Contained Commodity (kt) and Grade (Ni)	Footwall type	Arsenic metasomatism	Hydrothermal halo	References
Miitel	Operating	Norseman Wiluna	42.0 km South of Kambalda X: 371457 Y:6504801	1.56 kt at 2.9% Ni	Mafic –Thick tholeiitic basalt, the Mount Edwards basalt	Yes, secondary enrichment in arsenic present and widespread	Ni-As-PGE hydrothermal halo extending up to 250 m away from massive sulfide mineralization	Cairns et al. (2003); Le Vaillant et al. (2015c), Le Vaillant (2014)
Perseverance	Care and maintenance	Agnew Wiluna	11.4 km North of Leinster X: 273997 Y: 6920833	276 kt at 2.3% Ni	Felsic-fine grained, schistose, biotite-rich volcanic-sedimentary rocks overlain by fragmental-textured rhyodacite and plagioclase-quartz phenocrysts rich rhyodacite	Elevated arsenic concentrations in some areas of the massive sulfides and areas where they have been mechanically remobilised – not widespread	No hydrothermal halo observed surrounding the massive sulfides (only mechanical remobilisation of the Sulfides)	Martin and Allchurch (1975); Binns and Groves (1976); Platt et al. (1978); Billington (1984); Gole et al. (1987); Barnes et al. (1988a, 1988b), Barnes (2006a, 2006b), Duuring et al. (2007), Duuring et al. (2010), Barnes et al. (2011), Le Vaillant (2014)

Sarah's find	Prospect	Agnew Wiluna	Northern part of the Mont Keith domain X:255200 Y:6990650	Non economic	Felsic - dacite	Yes, secondary enrichment in arsenic present and widespread	Ni-As-PGE hydrothermal halo observed extending up to 1780 m away from massive sulfide mineralization (Le Vaillant et al. 2015b)	Burt and Sheppy (1975), Dowling and Hill (1990), Hill (1995), Rosengren et al. (2005, 2007), Fiorentini et al. (2007), Le Vaillant et al. (2015b)
Durkin	Shut	Norseman Wiluna	3.9 km North from Kambalda X:372562 Y:6551055	19.17 kt at 5.1 % Ni	Mafic –thick tholeiitic basalt, the Lunnon basalt	No elevated arsenic concentrations observed – all results observed at or below background levels	No hydrothermal halo observed surrounding the massive sulfides	Marston (1984)
Otter-Juan	Care and maintenance	Norseman Wiluna	4.3 km NNW from Kambalda X:371512 Y:6551152	0.14 kt at 6.9% Ni	Mafic-thick tholeitic basalt, the Lunnon basalt	No elevated arsenic concentrations observed-all results observed at or below background levels	No hydrothermal halo observed surrounding the massive sulfides	Marston (1984)

Modified from Le Vaillant, M., Fiorentini, M.L., Barnes, S.J., 2016b. Review of lithogeochemical exploration tools for komatiite-hosted Ni-Cu-(PGE) deposits. J. Geochem. Explor. 168, 1–19.

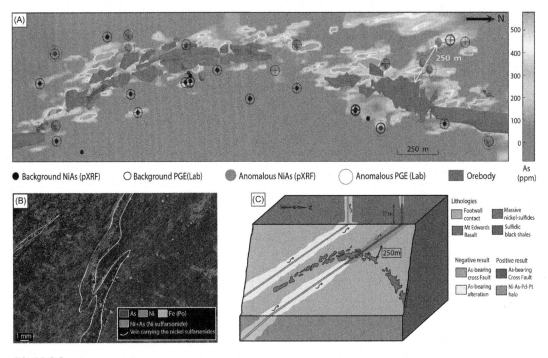

FIGURE 2.6

Summary of results from the study of the Miitel deposit. (A) Perspective view from gOcad® of a long section through the 3D model of the Miitel deposit. This image combines: (1) distribution of the arsenic in ppm at the contact between the basalt and the komatiites (model derived using Leapfrog), (2) location of pXRF analyses showing anomalously high Ni and As concentrations, and (3) location of laboratory PGE analyses highlighting samples enriched in PGE. (B) False color element concentration map (As *blue*, Ni *red*, Fe *green*), of samples DRD918-358.6 which contains nickel arsenides within small hydrothermal quartz and/or carbonate veins cross cutting the Mount Edwards footwall basalt. This map was produced using the data collected with the Maia detector array on the X-ray fluorescence microscopy beamline, at the Australian Synchrotron in Melbourne. (C) 3D block model of the Miitel system showing the possible application of the Ni-As-Pd geochemical halo to exploration targeting for nickel sulfides.

Modified from Le Vaillant, M., Barnes, S.J., Fiorentini, M.L., Miller, J., McCuaig, T.C., Muccilli, P., 2015a. A hydrothermal Ni-As-PGE geochemical halo around the miitel komatiite-hosted nickel sulphide deposit, Yilgarn Craton, Western Australia. Econ. Geol. 110, 505–530.

the model, arsenic-rich fluids, such as those commonly related to orogenic gold events (Eilu and Groves, 2001), remobilized Ni and PGEs from the massive sulfides (Wood, 2002), subsequently re-depositing them as nickel-arsenides within a geochemical halo surrounding the magmatic nickel-sulfide ore (Le Vaillant et al., 2015a).

Le Vaillant et al (2016b) investigated in detail the nature of metal remobilization and the formation of detectable hydrothermal haloes surrounding nickel-sulfide deposits, in an attempt to better document the size and geometry of these targets that effectively enlarge the footprint of primary magmatic mineralization. The study of hydrothermal haloes around selected deposits provided key insights into the environment and the conditions necessary to create Ni-As-Co-Pd geochemical

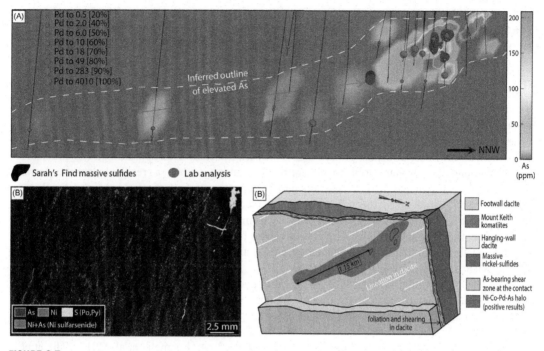

FIGURE 2.7

Summary of results from the study of the Sarah's Find deposit(A) 3D visualization of concentrations in Pd of all analyzed samples, combined with a color representation of the arsenic concentrations along the footwall contact between the Mount Keith komatiites and the Mount Keith dacite. (B) micro-XRF map of one of the sample containing nickel arsenides within the foliation in the dacite footwall. (C) Interpretative block model of the geochemical halo observed around the Sarah's Find ore body.

Modified from Le Vaillant, M., Saleem, A., Barnes, S., Fiorentini, M., Miller, J., Beresford, S., et al., 2015b. Hydrothermal remobilisation around a deformed and remobilised komatiite-hosted Ni-Cu-(PGE) deposit, Sarah's Find, Agnew Wiluna greenstone belt, Yilgarn Craton, Western Australia. Mineral. Deposita, 1–20.

haloes around magmatic Ni-Cu-(PGE) deposits. The authors concluded that arsenic evidently plays a crucial role. An important conclusion is that the absence of As-related Ni-PGE hydrothermal haloes should not be regarded as a negative indicator in exploration for Ni-Cu-(PGE) sulfides. They suggested that Ni-As-PGE "stains" are unlikely to generate false positive anomalies, but false negatives are likely (Le Vaillant et al., 2016b).

2.4 WEATHERING-RESISTANT GEOCHEMICAL SIGNALS AND INDICATOR MINERALS

Potentially mineralized systems are only rarely found in fresh rocks at the surface, particularly in regolith-dominated terrains of continental land masses that occur at tropical latitudes. It is desirable to be able to recognize the geochemical signals of fertile mineral systems in weathered rocks in any terrain that has been subjected to deep weathering (e.g., Australia, Brazil, West Africa, India). Typically, exploration programs in such terrains involve air-core or percussion drilling through shallow transported cover to sample material from the "top of fresh rock", which in many cases comprises saprolite (in-situ weathered rock; Anand and Paine, 2002). Barnes et al. (2014) compared fresh bedrock and "top of fresh rock" saprolite in the Agnew area in Western Australia and concluded that inter-element ratios of the rare-earth-elements, Zr, Ti, Cr, and Ni in saprolite were preserved from the original unweathered fresh rock. Therefore bedrock lithologies were able to be mapped using saprolite geochemistry.

Le Vaillant et al (2016b) plotted saprolite bottom-hole samples and fresh bedrock samples (from diamond drill core at depths greater than 50 m) from a study area approximately 5 km square centered around the Vivien gold mine north of Agnew, within a residual lateritic terrain (Fig. 2.3B). The Ni/Cr and Ni/Ti ratio-ratio plot is superimposed on the field defined by Barnes et al (2004b) for channelized sheet flow facies (Kambalda-style) komatiites. Komatiites from the Agnew Komatiite formation and tholeiitic basalts from the Redeemer Formation (Hayman et al., 2015) define two distinct fields that overlap closely for weathered and fresh samples. A major advantage of this particular suite of elements (Ni, Cr, and Ti) is their suitability for analysis using portable XRF devices at typical abundance levels (Le Vaillant et al., 2014). The same combination—reliable preservation through weathering, and amenability of pXRF analysis—applies to the ratio Zr/Ti that can be used as a proxy for contamination by enriched felsic material.

Mineral chemistry in lateritic terrains also has potential for exploration targeting of Ni-Cu-(PGE) sulfides. It was previously stated that Ru concentrations in chromite formed in komatiites could be used as a prospectivity indicator (Fiorentini et al., 2008; Locmelis et al., 2013). The application to weathered rocks of element ratios involving Cr in komatiites depend on the lack of independent mobility of Cr in solution during weathering (Locmelis et al., 2013). The interpretation of indicator trace element characteristics such as Ru depletion in detrital chromite grains depends very strongly on this assumption. The PGE behavior and the nature of platinum group mineral inclusions in weathered chromites has not been systematically investigated, but there is evidence that cores of large chromite crystals, particularly Cr-rich ones (Garnier et al., 2008), preserve their primary chemistry through moderate degrees of weathering (Friedrich et al., 1981; Friedrich, 1984) even in lateritic environments (Summons et al., 1981; Michailidis, 1990) and possibly within diamictite glacial sediments (Salama et al., 2016). As chromite is a widespread component of resistant heavy mineral suites, studying their composition using LA-ICP-MS has real potential applications for exploration targeting of komatiite-hosted Ni-Cu-(PGE) deposits.

2.5 CONCLUSIONS

The effectiveness of different lithogeochemical techniques depends on scale, geological history and density of available data. All the various exploration techniques described above are summarized in Table 2.1, linked to the scale at which they are most useful (Fig. 2.8). For example, indicators of favorable volcanic setting, crustal contamination processes, and magmatic sulfide formation are more likely to be used at prospect scale, but in some cases can generate regional targets. In particular, useful information can be generated using ratios of the elements Ni, Ti, Cr, and Zr. These

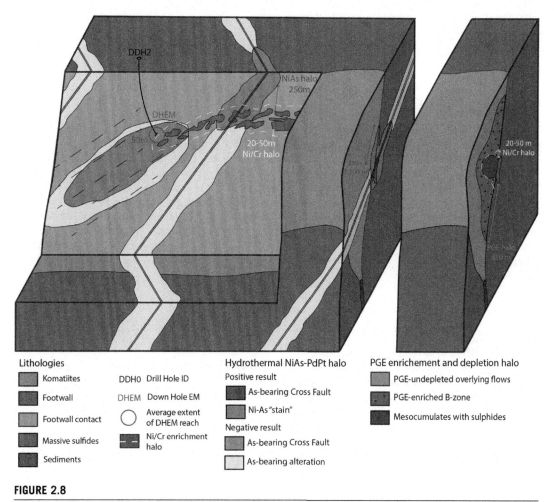

FIGURE 2.8

3D block model compiling and comparing various spatial exploration tools at the deposit scale.

elements can be measured in the field using pXRF instruments, as their ratios are reasonably robust under alteration and mild weathering. At deposit scale, the largest haloes, up to km scale, are generated by hydrothermal dispersion, but these are only formed in the presence of overprinting arsenic-bearing fluids and hence can easily give rise to false negatives. Chalcophile element anomalies, particularly PGE enrichment using proxies such as Pt/Ti, and Ni depletion in olivine or in olivine-rich cumulates, generate reliable vectors on a scale up to a few hundred meters, but are of limited value where orebodies are tectonically dismembered from their original host rocks. All of the techniques outlined here are more effective when used in combination with one another. Effective exploration at all scales relies on integration of appropriate lithogeochemical signals with geophysical data, regional and local geological understanding, along with a large element of good luck. There are no geochemical magic bullets, but lithogeochemical data can be of great value in prioritizing targets at all scales.

ACKNOWLEDGEMENTS

Financial support for this research was provided by MERIWA (project #M413), BHP Billiton Nickel West, Mincor Resources NL and First Quantum Minerals Ltd. The SIRF scholarship of the University of Western Australia and the MERIWA research grant are greatly appreciated. BHP Billiton Nickel West, Mincor Resources NL, and First Quantum Minerals Ltd are acknowledged for providing on site access and samples. This project is an output from the ARC Centre of Excellence for Core to Crust Fluid System (CCFS) and the CSIRO Minerals Resources Flagship. Marco Fiorentini also acknowledges support from the Australian Research Council through Linkage Project LP120100668 and the Future Fellowship Scheme (FT110100241). This is contribution 976 from the ARC Centre of Excellence for Core to Crust Fluid Systems (http://www.ccfs. mq.edu.au). John Miller, Ignacio Gonzalez-Alvarez, and Mike Lesher are thanked for their comments and reviews. We acknowledge financial support for this facility from the Science and Industry Endowment Fund (SIEF).

REFERENCES

Anand, R.R., Paine, M., 2002. Regolith geology of the Yilgarn Craton, Western Australia: implications for exploration. Aust. J. Earth Sci. 49, 3–162.

Arndt, N., Lesher, C.M., Czamanske, G.K., 2005. Mantle-derived magmas and magmatic Ni-Cu- (PGE) deposits. Econ. Geol. 100th Aniv. Vol. 5–24.

Arndt, N. T., Lesher, C. M., and Barnes, S. J., 2008, Komatiite: Cambridge, Cambridge University Press, 467 p.

Baragar, W., Scoates, R., 1981. The circum-Superior belt: a Proterozoic plate margin? Dev. Precambrian Geol. 4, 297–330.

Barnes, S.J., Mungall, J.E., Le Vaillant, M., Godel, B., Lesher, C.M., Holwell, D.A., et al., 2017. Sulphide-silicate textures in magmatic Ni-Cu-PGE sulphide ore deposits. 1. Disseminated and net-textured ores. Am. Mineral., online . Available from: http://dx.doi.org/10.2138/am-2016-5754.

Barnes, S.J., Fisher, L.A., Anand, R., Uemoto, T., 2014. Mapping bedrock lithologies throughin situregolith using retained element ratios: a case study from the Agnew-Lawlers area, Western Australia. Aust. J. Earth Sci. 61, 269–285.

Barnes, S.J., 1998. Chromite in komatiites, 1. Magmatic controls on crystallisation and composition. J. Petrol. 39, 1689–1720.

Barnes, S.J., 2006a. Komatiite-hosted nickel-sulphide deposits: geology, geochemistry, and genesis. Soc. Econ. Geol. Spec. Publ. 13, 51–118.

Barnes, S.J., 2006b. Komatiites: petrology, volcanology, metamorphism and geochemistry. Soc. Econ. Geol. Spec. Publ. 13, 13–49.

Barnes, S.J., Brand, N.W., 1999. The distribution of Cr, Ni, and chromite in komatiites, and application to exploration for komatiite-hosted nickel sulphide deposits. Econ. Geol. 94, 129–132.

Barnes, S.J., Fiorentini, M.L., 2012. Komatiite magmas and sulphide nickel deposits: a comparison of variably endowed archean terranes. Econ. Geol. 107, 755–780.

Barnes, S.-J., Lightfoot, P.C., 2005. Formation of magmatic nickel sulphide deposits and processes affecting their copper and platinum grounp element contents. Econ. Geol. 100th Anniv. Vol. 179–213.

Barnes, S.J., Liu, W., 2012. Pt and Pd mobility in hydrothermal fluids: evidence from komatiites and from thermodynamic modelling. Ore Geol. Rev. 44, 49–58.

Barnes, S.-J., Picard, C.P., 1993. The behaviour of platinum-group elements during partial melting, crystal fractionation, and sulphide segregation: an example from the Cape Smith Fold Belt, northern Quebec. Geochim. Cosmochim. Acta 57, 79–87.

Barnes, S.J., Gole, M.J., Hill, R.E.T., 1988a. The agnew nickel deposit, Western Australia: Part I. Structure and stratigraphy. Econ. Geol. 83, 524–536.

Barnes, S.J., Gole, M.J., Hill, R.E.T., 1988b. The Agnew nickel deposit, Western Australia: Part II. Sulphide geochemistry, with emphasis on the platinum-group elements. Econ. Geol. 83, 537–550.

Barnes, S.J., Hill, R.E.T., Evans, N.J., 2004a. Komatiites and nickel sulphide ores of the Black Swan area, Yilgarn Craton, Western Australia. 3: Komatiite geochemistry, and implications for ore forming processes. Mineral. Deposita 39, 729–751.

Barnes, S.J., Hill, R.E.T., Perring, C.S., Dowling, S.E., 2004b. Lithogeochemical exploration for komatiite-associated Ni-sulphide deposits: strategies and limitations. Mineral. Petrol. 82, 259–293.

Barnes, S.J., Lesher, C.M., Sproule, R.A., 2007. Geochemistry of komatiites in the Eastern Goldfields Superterrane, Western Australia and the Abitibi Greenstone Belt, Canada, and implications for the distribution of associated Ni–Cu–PGE deposits. Appl. Earth Sci.: IMM Trans. Section B 116, 167–187.

Barnes, S.J., Fiorentini, M.L., Duuring, P., Grguric, B.A., Perring, C.S., 2011. The Perseverance and Mount Keith Ni deposits of the Agnew-Wiluna Belt, Yilgarn Craton, Western Australia. Rev. Econ. Geol. 17, 51–88.

Barnes, S.J., Van Kranendonk, M.J., Sonntag, I., 2012. Geochemistry and tectonic setting of basalts from the Eastern Goldfields Superterrane. Aust. J. Earth Sci. 59, 707–735.

Barnes, S.J., Godel, B., Gurer, D., Brenan, J.M., Robertson, J., Paterson, D., 2013a. Sulphide-olivine Fe-Ni exchange and the origin of anomalously Ni-rich magmatic sulphides. Econ. Geol. 108, 1971–1982.

Barnes, S.J., Heggie, G.J., Fiorentini, M.L., 2013b. Spatial variation in platinum group element concentrations in ore-bearing komatiite at the Long-Victor deposit, Kambalda Dome, Western Australia: enlarging the footprint of nickel sulphide orebodies, Econ. Geol., 108. pp. 913–933.

Barnes, S.J., Cruden, A.R., Arndt, N., Saumur, B.M., 2015. The mineral system approach applied to magmatic Ni–Cu–PGE sulphide deposits: Ore Geology Reviews.

Barnes, S.J., Cruden, A.R., Arndt, N., Saumur, B.M., 2016. The mineral system approach applied to magmatic Ni–Cu–PGE sulphide deposits. Ore Geol. Rev. 76 (2016), 296–316.

Barrie, C.T., Samson, I., Price, M., Alvarez, B.C., Fryer, B., 2007. Chemical and mineralogical halos around magmatic sulphide deposits, CAMIRO Project 04E01. Canadian Mining Industry Research Organization/ Organisation Canadienne de recherche sur l'industrie miniere.

Bavinton, O.A., 1981. The nature of sulphidic metasediments at Kambalda and their broad relationships with associated ultramafic rocks and nickel ores. Econ. Geol. 76, 1606–1628.

Begg, G.C., Griffin, W.L., Natapov, L.M., O'Reilly, S.Y., Grand, S.P., O'Neill, C.J., et al., 2009. The lithospheric architecture of Africa: seismic tomography, mantle petrology, and tectonic evolution. Geosphere 5, 23–50.

Begg, G.C., Hronsky, J.A.M., Arndt, N.T., Griffin, W.L., O'Reilly, S.Y., Hayward, N., 2010. Lithospheric, cratonic, and geodynamic setting of Ni-Cu-PGE sulphide deposits. Econ. Geol. Bull. Soc. Econ. Geol. 105, 1057–1070.

Bekker, A., Barley, M.E., Fiorentini, M.L., Rouxel, O.J., Rumble, D., Beresford, S.W., 2009. Atmospheric sulphur in archean komatiite-hosted nickel deposits. Science 326, 1086–1089.

Bennett, M., Gollan, M., Staubmann, M., Bartlett, J., 2014. Motive, means, and opportunity: key factors in the discovery of the Nova-Bollinger magmatic nickel-copper sulphide deposits in Western Australia. Soc. Econ. Geol. Spec. Publ. 18, 301–320.

Beresford, S.W., Fiorentini, M., Hronsky, J.M.A., Rosengren, N., Bertuch, D.I., Saleem, A., et al., 2007. Komatiite-hosted Ni-Cu-(PGE) deposits: understanding deposit and camp footprints. In: Bierlein, F.P., Knox-Robinson, C.M. (Eds.), Proceedings of Geoconferences (WA) Inc. Kalgoorlie '07 Conference, 25−27 September 2007, 14. Geoscience Australia Record 2007, Kalgoorlie, Western Australia, pp. 159−163.

Bleeker, W., 1990. Evolution of the Thompson Nickel Belt and Its Nickel Deposits, Manitoba, Canada. University of New Brunswick, Fredericton, p. 444.

Boyd, F.R., Gurney, J.J., Richardson, S.H., 1985. Evidence for a 150-200-km thick Archaean lithosphere from diamond inclusion thermobarometry. Nature 315, 387−389.

Burnham, O.M., Halden, N., Layton-Matthews, D., Lesher, C.M., Liwanag, J., Heaman, L., et al., 2003. Geology, Stratigraphy, Petrogenesis, and Metallogenesis of the Thompson Nickel Belt, Manitoba: Final Report for CAMIRO Project 97E-02. Mineral Exploration Research centre, Sudbury, p. 410.

Bursztyn, N.E., Olivo, G.R., 2010. PGE-rich Ni-Cu sulphide mineralization in the Flin Flon Greenstone Belt, Manitoba, Canada: implications for hydrothermal remobilization of platinum group elements in basic-ultrabasic sequences. Econ. Geol. 105, 1469−1490.

Burt, D.R.L., Sheppy, N.R., 1975. Mount Keith nickel sulphide deposit. In: Knight, C.L. (Ed.), Economic Geology of Australia and Papua New Guinea, 1. Australasian Institute of Mining and Metallurgy, pp. 159−168.

Campbell, I.H., Naldrett, A.J., 1979. The influence of silicate:sulphide ratios on the geochemistry of magmatic sulphides. Econ. Geol. Bull. Soc. Econ. Geol. 74, 1503−1506.

Campbell, I.H., Griffiths, R.W., Hill, R.I., 1989. Melting in an Archaean mantle plume: heads it's basalt, tails it's komatiites. Nature 339, 697−699.

Cassidy, K.F., Champion, D.C., Krapez, B., Barley, M.E., Brown, S.J.A., Blewett, R.S., et al., 2006. A revised geological framework for the Yilgarn Craton, Western AustraliaIn: Resources, Ia (Ed.), Western Australia Geological Survey, p. 8.

Champion, D.C., Cassidy, K.F., 2007. An overview of the Yilgarn Craton and its crustal evolution, Proceedings of Geoconferences (WA) Inc. Kalgoorlie'07 Conference, pp. 25−27.

Champion, D.C., Cassidy, K.F., 2008. Geodynamics: using geochemistry and isotopic signatures of granites to aid mineral systems studies: an example from the Yilgarn craton. Geosci. Aust. Rec. 2008/09, 7−16.

De Almeida, C.M., Olivo, G.R., de-Carvalho, S.G., 2007. The Ni-Cu-PGE sulphide ores of the komatiite-hosted Fortaleza de Minas deposit, Brazil: evidence of hydrothermal remobilization. Can. Mineral. 45, 751−773.

Deen, T., Griffin, W.L., Begg, G., O'Reilly, S., Natapov, L.M., Hronsky, J., 2006. Thermal and compositional structure of the subcontinental lithospheric mantle: derivation from shear wave seismic tomography. Geochem. Geophys. Geosys. 7.

Dillon-Leitch, H.C.H., Watkinson, D.H., Coats, C.J.A., 1986. Distribution of platinum-group elements in the Donaldson West Deposit, Cape Smith Belt, Quebec. Econ. Geol. 81, 1147−1158.

Duke, J., Naldrett, A., 1978. A numerical model of the fractionation of olivine and molten sulphide from komatiite magma. Earth Planet. Sci. Lett. 39, 255−266.

Duke, J.M., 1979. Computer simulation of the fractionation of olivine and sulphide from mafic and ultramafic magmas. Can. Mineral. 17, 507−514.

Duuring, P., Bleeker, W., Beresford, S.W., 2007. Structural modification of the komatiite-associated harmony nickel sulphide deposit, Leinster, Western Australia. Econ. Geol. 102, 277−297.

Duuring, P., Bleeker, W., Beresford, S.W., Hayward, N., 2010. Towards a volcanic−structural balance: relative importance of volcanism, folding, and remobilisation of nickel sulphides at the Perseverance Ni−Cu−(PGE) deposit, Western Australia. Mineral. Deposita 45, 281−311.

Duuring, P., Bleeker, W., Beresford, S.W., Fiorentini, M.L., Rosengren, N.M., 2012. Structural evolution of the Agnew−Wiluna greenstone belt, Eastern Yilgarn Craton and implications for komatiite-hosted Ni sulphide exploration. Aust. J. Earth Sci. 59, 765−791.

Eilu, P., Groves, D.I., 2001. Primary alteration and geochemical dispersion haloes of Archaean orogenic gold deposits in the Yilgarn Craton: the pre-weathering scenario. Geochem,.Explor. Environ. Anal. 1, 183−200.

Ewers, W., Hudson, D., 1972. An interpretive study of a nickel-iron sulphide ore intersection, Lunnon Shoot, Kambalda, Western Australia. Econ. Geol. 67, 1075–1092.

Farrow, C., Watkinson, D., 1996. Geochemical evolution of the Epidote Zone, Fraser Mine, Sudbury, Ontario; Ni-Cu-OGE remobilization by saline fluids. Explor. Mining Geol 5, 17–31.

Fiorentini, M., Beresford, S., Barley, M., Duuring, P., Bekker, A., Rosengren, N., et al., 2012b. District to camp controls on the genesis of komatiite-hosted nickel sulphide deposits, Agnew-Wiluna greenstone belt, Western Australia: insights from the multiple sulphur isotopes. Econ. Geol. 107, 781–796.

Fiorentini, M.L., Bekker, A., Rouxel, O., Wing, B.A., 2012a. Multiple sulphur and iron isotope composition of magmatic Ni-Cu-(PGE) sulphide mineralization from Eastern Botswana.

Fiorentini, M.L., Barnes, S.J., Lesher, C.M., Heggie, G.J., Keays, R.R., Burnham, O.M., 2010. Platinum group element geochemistry of mineralized and nonmineralized komatiites and basalts. Econ. Geol. 105, 795–823.

Fiorentini, M.L., Barnes, S.J., Maier, W.D., Burnham, O.M., Heggie, G., 2011. Global variability in the platinum-group element contents of komatiites. J. Petrol. 52, 83–112.

Fiorentini, M.L., Beresford, S.W., Barley, M.E., 2008. Ruthernium-chromium variation: a new lithogeochemical tool in the exploration for komatiite-hosted Ni-Cu-(PGE) deposits. Econ. Geol. 103, 431–437.

Fiorentini, M.L., Rosengren, N., Beresford, S.W., Grguric, B., Barley, M.E., 2007. Controls on the emplacement and genesis of the MKD5 and Sarah's Find Ni–Cu–PGE deposits, Mount Keith, Agnew–Wiluna Greenstone Belt, Western Australia. Mineral. Deposita 42, 847–877.

Friedrich, G., 1984. Chromite-bearing limonitic laterites as weathering products of ultramafic rocks Recent researchers in geology: Products and Processes of Rock Weathering 11 16–22.

Friedrich, G., Brunemann, H.G., Wilcke, J., Stumpfl, E.F., 1981. Chrome spinels in lateritic soils and ultramafic source rocks, Acoje Mine, Zambales, Philippines. Metallogeny Mafic Ultramafic Complexes. pp. 257–278.

Garnier, J., Quantin, C., Guimaraes, E., Becquer, T., 2008. Can chromite weathering be a source of Cr in soils? Mineral. Mag. 72 (1), 49–53.

Golightly, J., 1981. Nickeliferous laterite deposits. Econ. Geol. 75, 710–735.

González-Álvarez, I., Porwal, A., Beresford, S.W., McCuaig, T.C., Maier, W.D., 2010. Hydrothermal Ni prospectivity analysis of Tasmania, Australia. Ore Geol. Rev. 38, 168–183.

González-Álvarez, I., Pirajno, F., Kerrich, R., 2013a. Hydrothermal nickel deposits: secular variation and diversity. Ore Geol. Rev. 52, 1–3.

González-Álvarez, I., Sweetapple, M., Lindley, I.D., Kirakar, J., 2013b. Hydrothermal Ni: Doriri Creek, Papua New Guinea. Ore Geol. Rev. 52, 37–57.

González-Álvarez, I., Ley-Cooper, A.Y., Salama, W., 2016. A geological assessment of airborne electromagnetics for mineral exploration through deeply weathered profiles in the southeast Yilgarn Cratonic margin, Western Australia. Ore Geol. Rev. 73, Part 3, 522–539.

Gresham, J.J., Loftus Hill, G.D., 1981. The geology of the Kambalda Nickel field, Western Australia. Econ. Geol. 76, 1373–1416.

Grguric, B., Riley, T., 2006. An integrated geometallurgical approach to optimize business outcomes at the MKD5 nickel deposit, Mount Keith, Western Australia. Econ. Geol. 12, 311.

Groves, D.I., Korkiakoski, E.A., McNaughton, N.J., Lesher, C.M., Cowden, A., 1986. Thermal erosion by komatiites at Kambalda, Western Australia and the genesis of nickel ores. Nature 319, 136–139.

Hanley, J., Ames, D., Barnes, J., Sharp, Z., Guillong, M., 2011. Interaction of magmatic fluids and silicate melt residues with saline groundwater in the footwall of the Sudbury Igneous Complex, Ontario, Canada: new evidence from bulk rock geochemistry, fluid inclusions and stable isotopes. Chem. Geol. 281, 1–25.

Hanley, J.J., Bray, C.J., 2009. The trace metal content of amphibole as proximity indicator for Cu-Ni-PGE mineralization in the footwall of the Sudbury igneous complex, Ontario, Canada. Econ. Geol. 104, 113–125.

Hanley, J.J., Mungall, J.E., 2003. Chlorine enrichment and hydrous alteration of the Sudbury Breccia hosting footwall Cu-Ni-PGE mineralization at the Fraser mine, Sudbury, Ontario, Canada. Can. Mineral. 41, 857−881.

Hanley, J.J., Mungall, J.E., Pettke, T., Spooner, E.T.C., Bray, C.J., 2005. Ore metal redistribution by hydrocarbon-brine and hydrocarbon-halide melt phases, North Range footwall of the Sudbury Igneous Complex, Ontario, Canada. Mineral. Deposita 40, 237−256.

Hayman, P.C., Thébaud, N., Pawley, M.J., Barnes, S.J., Cas, R.A.F., Amelin, Y., et al., 2015. Evolution of a ~2.7 Ga large igneous province: a volcanological, geochemical and geochronological study of the Agnew Greenstone Belt, and new regional correlations for the Kalgoorlie Terrane (Yilgarn Craton, Western Australia). Precambrian Res. 270, 334−368.

Heggie, G.J., Fiorentini, M.L., Barnes, S.J., Barley, M.E., 2012. Maggie Hays Ni Deposit: Part 2. Nickel mineralization and the spatial distribution of PGE ore-forming signatures in the Maggie Hays Ni system, Lake Johnston greenstone belt, Western Australia. Econ. Geol. 107, 817−833.

Hill, R.E.T., Gole, M.J., 1990. Nickel sulphide deposits of the Yilgarn Block. In: Hughes, F.E. (Ed.), Australasian Institute of Mining and Metallurgy. pp. 557−559.

Hill, R.E.T., Barnes, S.J., Gole, M.J., Dowling, S.E., 1995. The volcanology of komatiites as deduced from field relationships in the Norseman-Wiluna greenstone belt, Western Australia. Lithos 34, 159−188.

Hronsky, J., Schodde, R.C., 2006. Nickel exploration history of the Yilgarn Craton: from the nickel boom to today. Soc. Econ. Geol. Spec. Publ. 13, 1−11.

Huppert, H.E., Sparks, S.J., 1985. Komatiites I: eruption and flow. J. Petrol. 26, 694−725.

Huppert, H.E., Sparks, R.S.J., Tuner, J.S., Arndt, N.T., 1984. Emplacement and cooling of komatiite lavas. Nature 309, 19−22.

Keays, R.R., Jowitt, S.M., 2013. The Avebury Ni deposit, Tasmania: a case study of an unconventional nickel deposit. Ore Geol. Rev. 52, 4−17.

Krishnamurthy, P.J., 2015. Chalcophile element depletion in lower deccan trap formations and implications for Cu-Ni-PGE sulphide mineralization in the Deccan Traps, India akin to those of Norilsk-Talnakh, Siberian traps, Russia. Geol. Soc. India 85, 411. Available from: http://dx.doi.org/10.1007/s12594-015-0231-6.

LaFlamme, C., Martin, L., Jeon, H., Reddy, S.M., Selvaraja, V., et al., 2016. In situ multiple sulphur isotope analysis by SIMS of pyrite, chalcopyrite, pyrrhotite, and pentlandite to refine magmatic ore genetic models. Chem. Geol. 444 (2016), 1−15.

Layton-Matthews, D., Lesher, C.M., Burnham, O.M., Liwanag, J., Halden, N.M., Hulbert, L., et al., 2007. Magmatic Ni-Cu-Platinum-Group element deposits of the Thomson Nickel belt. In: Goodfellow, W.D. (Ed.), Mineral Deposits of Canada: A synthethis of Major Deposit-Types, District Metallogeny, the Evolution of Geological provinces, and Exploration Methods. pp. 409−432.

Le Vaillant, M., 2014. Hydrothermal remobilisation of base metals and platinum group elements around komatiite-hosted nickel sulphide deposits: applications to exploration methods, School of Earth and Environment. Unpublished PhD thesis. The University of Western Australia, Perth.

Le Vaillant, M., Barnes, S.J., Fisher, L., Fiorentini, M.L., Caruso, S., 2014. Use and calibration of portable X-Ray fluorescence analysers: application to lithogeochemical exploration for komatiite-hosted nickel sulphide deposits. Geochem. Explor. Environ. Anal. 14, 199−209.

Le Vaillant, M., Barnes, S.J., Fiorentini, M.L., Miller, J., McCuaig, T.C., Muccilli, P., 2015a. A hydrothermal Ni-As-PGE geochemical halo around the miitel komatiite-hosted nickel sulphide deposit, Yilgarn Craton, Western Australia. Econ. Geol. 110, 505−530.

Le Vaillant, M., Saleem, A., Barnes, S., Fiorentini, M., Miller, J., Beresford, S., et al., 2015b. Hydrothermal remobilisation around a deformed and remobilised komatiite-hosted Ni-Cu-(PGE) deposit, Sarah's Find, Agnew Wiluna greenstone belt, Yilgarn Craton, Western Australia. Mineral. Deposita 1−20.

Le Vaillant, M., Barnes, S.J., Fiorentini, M.L., Santaguida, F., Törmänen, T., 2016a. Effects of hydrous alteration on the distribution of base metals and platinum group elements within the Kevitsa magmatic nickel sulphide deposit. Ore Geol. Rev. 729 (Part 1), 128−148.

Le Vaillant, M., Fiorentini, M.L., Barnes, S.J., 2016b. Review of lithogeochemical exploration tools for komatiite-hosted Ni-Cu-(PGE) deposits. J. Geochem. Explor. 168, 1–19.

Lesher, C., 1983. Localization and genesis of komatiite-associated Fe-Ni-Cu sulphide mineralization at Kambalda, Western Australia. The University of Western Australia, Perth.

Lesher, C., 2007. Ni-Cu-(PGE) deposits in the Raglan area, Cape Smith Belt, New Quebec. mineral deposits of Canada: a synthesis of major deposit-types, district metallogeny, the evolution of geological provinces, and exploration. Methods 5, 351–386.

Lesher, C., Barnes, S., 2009. Magmatic Ni-Cu-PGE deposits: Genetic models and exploration.

Lesher, C., Groves, D., 1986a. Controls on the formation of komatiite-associated nickel-copper sulphide deposits. Geology and Metallogeny of Copper Deposits. Springer, pp. 43–62.

Lesher, C., Stone, W., 1996. Exploration geochemistry of komatiites. Igneous trace element geochemistry: applications for massive sulphide exploration. Geol. Assoc. Can., Short Course Notes 12, 153–204.

Lesher, C.M., 1989. Komatiite-associated nickel sulphide deposits, Chapter 5. In: Whitney, J.A., Naldrett, A.J. (Eds.), Ore Deposition Associated with Magmas. Rev. Econ. Geol., 4. pp. 45–100.

Lesher, C.M., Arndt, N.T., 1995. REE and Nd isotope geochemistry, petrogenesis and volcanic evolution of contaminated komatiites at Kambalda, Western Australia. Lithos 34, 127–157.

Lesher, C.M., Campbell, I.H., 1993. Geochemical and fluid dynamic modelling of compositional variations in Archean komatiite-hosted nickel sulphide ores in Western Australia. Econ. Geol. 88, 804–816.

Lesher, C.M., Groves, D.I., 1986b. Controls on the formation of komatiite-associated nickel-copper sulphide deposits. In: Friedrich, G., Genkin, A., Naldrett, A., Ridge, J., Sillitoe, R., Vokes, F. (Eds.), Geology and Metallogeny of Copper Deposits. Springer, Berlin Heidelberg, pp. 43–62.

Lesher, C.M., Keays, R.R., 1984. Metamorphically and hydrothermally mobilized Fe-Ni-Cu sulphides at Kambalda, Western Australia. In: Buchanan, D.L., Jones, M.J. (Eds.), Sulphide Deposits in Mafic and Ultramafic Rocks. Institute of Mineralogy and Metallogeny, London, pp. 62–69.

Lesher, C.M., Keays, R.R., 2002. Komatiite-associated Ni-Cu-(PGE) deposits: mineralogy, geochemistry, and genesis. In: Cabri, L.J. (Ed.), The Geology, Geochemistry, Mineralogy, and Mineral Benefication Of The Platinum-Group Elements. Canadian Institute of Mining, Metallurgy and Petroleum, 54. pp. 579–617.

Lesher, C.M., Arndt, N.T., Groves, D.I., 1984. Genesis of komatiite-associated nickel sulphide deposits at Kambalda, Western Australia: a distal volcanic model. In: Buchanan, D.L., Jones, M.J. (Eds.), Sulphide Deposits in Mafic and Ultramafic Rocks. Institute of Mining and Metallurgy, London, pp. 70–80.

Lesher, C.M., Burnham, O.M., Keays, R.R., Barnes, S.J., Hulbert, L., 2001. Trace-element geochemistry and petrogenesis of barren and ore-associated komatiites. Can. Mineral. 39, 673–696.

Locmelis, M., Pearson, N.J., Barnes, S.J., Fiorentini, M.L., 2011. Ruthenium in komatiitic chromite. Geochim. Cosmochim. Acta 75, 3645–3661.

Locmelis, M., Fiorentini, M.L., Barnes, S.J., Pearson, N.J., 2013. Ruthenium variation in chromite from komatiites and komatiitic basalts—a potential mineralogical indicator for nickel sulphide mineralization. Econ. Geol. 108, 355–364.

Locmelis, M., Fiorentini, M.L., Rushmer, T., Arevalo, R., Adam, J., Denyszyn, S., 2016. Sulphur and metal fertilization of the lower continental crust. Lithos 244, 74–93.

Marston, R.J., 1984. Nickel mineralisation in Western Australia. Geol. Surv.Western Australia Mineral Resour. Bull. 14, 271.

McCuaig, T.C., Hronsky, J.M., 2014. The mineral system concept: the key to exploration targeting. Soc. Econ. Geol. Spec. Publ. 18, 153–175.

McCuaig, T.C., Beresford, S., Hronsky, J., 2010. Translating the mineral systems approach into an effective exploration targeting system. Ore Geol. Rev. 38, 128–138.

McQueen, K.G., 1981. Volcanic-associated nickel deposits from around the Widgiemooltha Dome, Western Australia. Econ. Geol. 76, 1417–1443.

Michailidis, K.M., 1990. Zoned chromites with high Mn-contents in the Fe-Ni-Cr-laterite ore deposits from the Edessa area in northern Greece. Mineral. Deposita 25, 190−197.

Mole, D.R., Fiorentini, M.L., Thebaud, N., McCuaig, T.C., Cassidy, K.F., Kirkland, C.L., et al., 2012. Spatio-temporal constraints on lithospheric development in the southwest-central Yilgarn Craton, Western Australia. Aust. J. Earth Sci. 59, 625−656.

Mole, D.R., Fiorentini, M.L., Cassidy, K.F., Kirkland, C.L., Thebaud, N., McCuaig, T.C., et al., 2013. Crustal Evolution, Intra-Cratonic Architecture and the Metallogeny of an Archaean Craton. Geological Society, London, p. 393. , Special Publications.

Mole, D.R., Fiorentini, M.L., Thebaud, N., Cassidy, K.F., McCuaig, T.C., Kirkland, C.L., et al., 2014. Archean komatiite volcanism controlled by the evolution of early continents. Proc. Natl. Acad. Sci. 111, 10083−10088.

Molnár, F., 2013. Pd-Pt-Au-rich sulphide ores in footwalls of layered mafic-ultramafic igneous complexes. Tutkimusraportti − Geol. Tutkimuskeskus 198 121−125.

Molnar, F., Watkinson, D., 2001. Fluid-inclusion data for Vein-type Cu-Ni-PGE Footwall Ores, sudbury igneous complex and their use in establishing an exploration model for hydrothermal PGE-enrichment around mafic-ultramafic intrusions. Explor. Mining Geol. 10, 125−141.

Molnár, F., Watkinson, D.H., Everest, J.O., 1999. Fluid-inclusion characteristics of hydrothermal Cu-Ni-PGE veins in granitic and metavolcanic rocks at the contact of the Little Stobie deposit, Sudbury, Canada. Chem. Geol. 154, 279−301.

Molnár, F., Watkinson, D.H., Jones, P.C., 2001. Multiple hydrothermal processes in footwall units of the North Range, Sudbury Igneous Complex, Canada, and implications for the genesis of vein-type Cu-Ni-PGE deposits. Econ. Geol. 96, 1645−1670.

Mossman, D., Eigendorf, G., Tokaryk, D., Gauthier-Lafaye, F., Guckert, K.D., Melezhik, V., et al., 2003. Testing for fullerenes in geologic materials: Oklo carbonaceous substances, Karelian shungites. Sudbury Black Tuff. Geology 31, 255−258.

Mudd, G.M., 2010. Global trends and environmental issues in nickel mining: Sulphides versus laterites. Ore Geol. Rev. 38, 9−26.

Naldrett, A., 1966. The role of sulphurization in the fenesis of iron-nickel deposits of the Porcupine District, Ontario. Can. Inst. Mining Metall. Trans 69, 147−155.

Naldrett, A., 2004. Magmatic sulphide deposits: geology, Geochemistry and Exploration. Springer, Heidelberg.

Naldrett, A.J., 1999. World-class Ni-Cu-PGE deposits: key factors in their genesis. Mineral. Deposita 34, 227−240.

Nyman, M.W., Sheets, R.W., Bodnar, R.J., 1990. Fluid-inclusion evidence for the physical and chemical conditions associated with intermediate-temperature PGE mineralization at the New Rambler deposit, Southeastern Wyoming. Can. Mineral. 28, 629−638.

Olivo, G.R., Theyer, P., 2004. Platinum-group minerals from the McBratney PGE-Au prospect in the Flin Flon greenstone belt, Manitoba, Canada. Can. Mineral. 42, 667−681.

Pagé, P., Barnes, S.-J., 2016. The influence of chromite on osmium, iridium, ruthenium and rhodium distribution during early magmatic processes. Chem. Geol. 420, 51−68.

Paterson, H.L., 1984. Nickeliferous sediments and sediment-associated nickel ores at Kambalda, Western Australia. Sulphide deposits in mafic and ultramafic rocks.

Péntek, A., Molnar, F., Watkinson, D., Jones, P.C., 2008. Footwall-type Cu-Ni-PGE mineralization in the Broken Hammer area, Wisnet township, North Range, Sudbury structure. Econ. Geol. 103, 1005−1028.

Péntek, A., Molnár, F., Watkinson, D.H., Jones, P.C., Mogessie, A., 2011. Partial melting and melt segregation in footwall units within the contact aureole of the Sudbury Igneous Complex (North and East Ranges, Sudbury structure), with implications for their relationship to footwall Cu−Ni−PGE mineralization. Int. Geol. Rev. 53, 291−325.

Péntek, A., Molnar, F., Tuba, G., Watkinson, D.H., Jones, P.C., 2013. The significance of partial melting processes in hydrothermal low sulphide Cu-Ni-PGE mineralization within the footwall of the Sudbury Igneous Complex, Ontario, Canada. Econ. Geol. 108, 59−78.

Perring, C.S., 2015a. A 3-D geological and structural synthesis of the leinster area of the Agnew-Wiluna Belt, Yilgarn Craton, Western Australia, with special reference to the volcanological setting of komatiite-associated nickel sulphide deposits. Econ. Geol. 110, 469−503.

Perring, C.S., 2015b. Volcanological and structural controls on mineralization at the mount keith and cliffs komatiite-associated nickel sulphide deposits, Agnew-Wiluna Belt, Western Australia—Implications for Ore genesis and targeting. Econ. Geol. 110, 1669−1695.

Peters, W.S., 2006. Geophysical exploration for nickel sulphide mineralization in the Yilgarn Craton. Econ. Geol. 13 Spec. Publ. 167−193.

Pirajno, F., González-Álvarez, I., 2013. A re-appraisal of the Epoch nickel sulphide deposit, Filabusi Greenstone Belt, Zimbabwe: a hydrothermal nickel mineral system? Ore Geol. Rev. 52, 58−65.

Redman, B.A., Keays, R.R., 1985. Archaean basic volcanism in the eastern goldfields province, yilgarn block, Western Australia. Precambrian Res. 30, 113−152.

Ripley, E.M., Li, C., 2013. Sulphide saturation in mafic magmas: is external Sulphur required for magmatic Ni-Cu-(PGE) ore genesis? Econ. Geol. 108, 45−58.

Robertson, J.C., Barnes, S.J., Le Vaillant, M., 2015. Dynamics of magmatic sulphide droplets during transport in silicate melts and implications for magmatic sulphide ore formation. J. Petrol. 56, 2445−2472.

Rosengren, N.M., Grguric, B.A., Beresford, S.W., Fiorentini, M.L., Cas, R.A.F., 2007. Internal stratigraphic architecture of the komatiitic dunite-hosted MKD5 disseminated nickel sulphide deposit, Mount Keith Domain, Agnew-Wiluna Greenstone Belt, Western Australia. Mineral. Deposita 42, 821−845.

Ross, J.R., Travis, G.A., 1981. The nickel sulphide deposits of Western Australia in global perspective. Econ. Geol. 76, 1291−1329.

Salama, W., Anand, R., Verral, M., 2016. Mineral exploration and basement mapping in areas of deep transported cover using indicator heavy minerals and paleoredox fronts, Yilgarn Craton, Western Australia. Ore Geol. Rev. 72, 485−509.

Seat, Z., Stone, W.E., Mapleson, D.B., Daddow, B.C., 2004. Tenor variation within komatiite-associated nickel sulphide deposits: insights from the Wannaway Deposit, Widgiemooltha Dome, Western Australia. Mineral. Petrol. 82, 317−339.

Seat, Z., Beresford, S.W., Grguric, B.A., Mary Gee, M.A., Grassineau, N.V., 2009. Reevaluation of the role of external sulphur addition in the genesis of Ni-Cu-PGE deposits: evidence from the Nebo-Babel Ni-Cu-PGE deposit. West Musgrave, Western Australia. Available from: http://dx.doi.org/10.2113/gsecongeo.104.4.521.

Sproule, R., Lesher, C., Ayer, J., Thurston, P., Herzberg, C., 2002. Spatial and temporal variations in the geochemistry of komatiites and komatiitic basalts in the Abitibi greenstone belt. Precambrian Res. 115, 153−186.

Summons, T., Green, D., Everard, J., 1981. The occurrence of chromite in the Andersons Creek area, Beaconsfield, Tasmania. Econ. Geol. 76, 505−518.

Sun, S.S., Nesbitt, R.W., 1977. Chemical heterogeneity of the Archean mantle, composition of the Earth and mantle evolution. Earth Planet. Sci. Lett. 35, 429−480.

Trofimovs, J., Tait, M.A., Cas, R.A.F., McArthur, A., Beresford, S.W., 2003. Can the role of thermal erosion in strongly deformed komatiite-Ni-Cu-(PGE) deposits be determined? Perseverance, Agnew-Wiluna Belt Western Australia. Aust. J. Earth Sci. 50, 199−214.

Tuba, G., Molnár, F., Ames, D.E., Péntek, A., Watkinson, D.H., Jones, P.C., 2014. Multi-stage hydrothermal processes involved in "low-sulphide" Cu(-Ni)-PGE mineralization in the footwall of the Sudbury Igneous Complex (Canada): Amy Lake PGE zone, East Range. Mineral. Deposita 49, 7−47.

Willett, G., Eshuys, E., Guy, B., 1978. Ultramafic rocks of the Widgiemooltha-Norseman area, Western Australia: petrological diversity, geochemistry and mineralisation. Precambrian Res. 6, 133−156.

Wood, S.A., 2002. The aqueous geochemistry of the platinum-group elements with applications to ore deposits. In: Cabri, L.J. (Ed.), The Geology, Geochemistry, Mineralogy and Mineral Beneficiation of Platinum-Group Elements. Canadian Institute of Mining and Metallurgy, pp. 211−249.

Woodall, R., Travis, G., 1969. The Kambalda Nickel Deposits, Western Australia. Institution of Mining and Metallurgy.

Woolrich, P., Cowden, A., Giorgetta, N.E., 1981. The chemical and mineralogical variations in the nickel mineralization associated with the Kambalda Dome. Econ. Geol. 76, 1629–1644.

Yuan, M., Etschmann, B., Liu, W., Sherman, D.M., Barnes, S.J., Fiorentini, M.L., et al., 2015. Palladium complexation in chloride- and bisulphide-rich fluids: Insights from ab initio molecular dynamics simulations and X-ray absorption spectroscopy. Geochim. Cosmochim. Acta 161, 128–145.

METALLIC ORE DEPOSITS ASSOCIATED WITH MAFIC TO ULTRAMAFIC IGNEOUS ROCKS

Edward M. Ripley and Chusi Li

Indiana University, Bloomington, IN, United States

CHAPTER OUTLINE

3.1 INTRODUCTION

Ore deposits that are associated with mafic to ultramafic igneous rocks can generally be divided into those where the metals of interest are hosted in oxide minerals and those where the metals of interest are held as sulfides or are strongly associated with sulfides (e.g., PGE alloys, arsenides, bismuthinides). The oxide ore bodies include stratiform, podiform, and breccia-related chromite, magnetite-rich layers (often Ti and V-bearing), and ilmenite-rich layers or discordant bodies. Sulfide-associated ore bodies include massive, net-textured, and disseminated Ni-Cu-PGE (locally Co) occurrences and PGE-rich reefs containing disseminated sulfides. In this review we first briefly describe the deposit types and key geological-lithological features (Table 3.1). Genetic theories for the origin of the deposits are then presented. Finally, we discuss concentration mechanisms of metals where similar processes may be involved for very different types of deposits. Fig. 3.1 shows

Table 3.1 Tectonic Environments of Ore Deposits Hosted by Mafic to Ultramafic Igneous Rocks

Geotectonic Setting	Metals	Examples
Meteorite impact	Ni-Cu-PGE′	Sudbury
Intraplate magmatism; mantle plume	PGE-Cr-Ti-V	Bushveld Complex
		Stillwater Complex
		Great Dyke
Flood basalt provinces; mantle plumes; intracontinental rifting	Ni-Cu-PGE-Ti	Noril'sk
		Duluth Complex
		Eagle
Komatiites; hydrous mantle plume; subduction	Ni	Kambalda
		Raglan
Subduction-related magmatism; backarc extensional systems	Ni-Cu-PGE-Ti	Xiarihamu
		Ural-Alaskan intrusions
		Grenville Province
Arc-continent collisions	Ni-Cu-PGE-Ti	Hongqiling
		Grenville Province
Rifted continental margin	Ni-Cu-PGE	Pechenga

the geographic locations of the deposits referenced in this review. Interested readers are referred to several other reviews on mafic-rock related ore deposits, including Barnes et al. (2017, Ni deposits), Barnes and Ripley (2016, PGE deposits), Charlier et al. (2015, Fe-Ti-V-P deposits), Woodruff et al. (2013, Fe-Ti deposits), Zientek (2012, PGE deposits), Schulte et al. (2012, Cr deposits), Naldrett (2011, Ni-Cu-PGE deposits), Schulz et al. (2010, Ni-Cu deposits), Barnes and Lightfoot (2005, Ni deposits), and Cawthorn et al. (2005a, PGE-Cr-V deposits). Two books dealing with the genesis of Ni-Cu-PGE deposits may be of particular interest; comprehensive treatments of magmatic sulfide deposits (Naldrett, 2004) and Ni-sulfide mineralization associated with the Sudbury Complex (Lightfoot, 2017).

3.2 DEPOSIT TYPES

3.2.1 NI-CU-PGE SULFIDE DEPOSITS

Several types of sulfide-rich Ni-Cu-PGE deposits exist, with virtually all characterized by sulfide assemblages composed principally of pyrrhotite, chalcopyrite, pentlandite, ± cubanite or pyrite. Textures of mineralization include finely-disseminated sulfide minerals, sulfide globules, net-textured assemblages where interstices between silicate minerals are rich in sulfides, and massive sulfide accumulations. Ore tonnage and grade may vary greatly (Fig. 3.2). Komatiite-hosted mineralization is typically mined for Ni only, and is found near the base of lava flows or within spatially associated subvolcanic intrusive rocks (e.g., Lesher, 1989; Lesher and Keays, 2002; Barnes, 2006). Low- to intermediate-tonnage deposits are associated with small intrusions that are part of magma

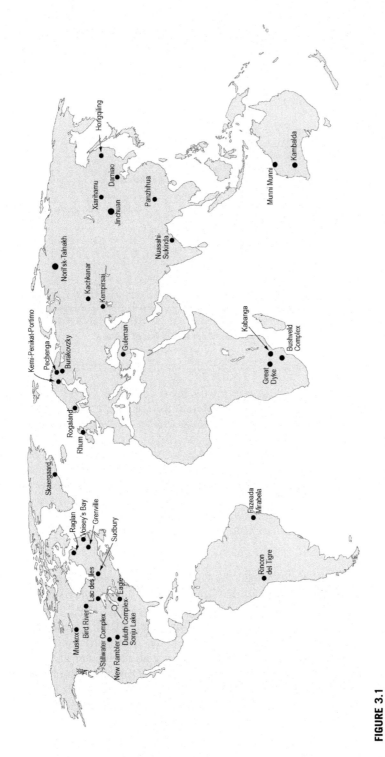

FIGURE 3.1

Locations of the Ni-Cu-PGE, Cr, and Fe-Ti oxide deposits referenced in this review.

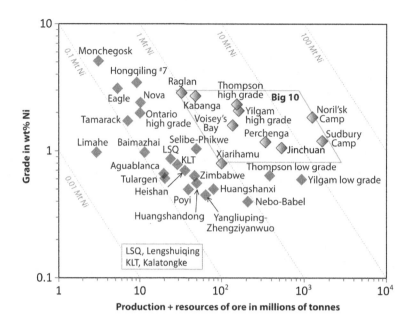

FIGURE 3.2

Grade and tonnage of several Ni deposits.

After Naldrett, A.J., 2011. Fundamentals of magmatic sulfide deposits. In: Li, C., Ripley, E.M. (Eds.), Society of Economic Geologists Reviews in Economic Geology 17, 1–50, with additional data from Wei, C.T., Zhou, M.-F., Zhao,T.P., 2013b. Differentiation of nelsonitic magmas in the formation of the ~ 1.74 Ga Damiao Fe-Ti-P ore deposit, North China. Contrib. Mineral. Petrol. 165, 1341-1362 (Wei et al., 2013b); Xie, W., Song, X.-Y., Chen, L.-M., Deng, Y.-F., Zheng, W.-Q., Wang, Y.-S., et al., 2014. Geochemistry insights on the genesis of the subduction-related Heishan magmatic Ni-Cu-(PGE) deposit, Gansu, northwestern China, at the southern margin of the Central Asian Orogenic Belt. Econ. Geol. 109, 1563–1583; Li, C., Zhang, Z., Li, W., Wang, Y., Sun, T., Ripley, E.M., 2015. Geochronology, petrology and Hf-S isotope geochemistry of the newly-discovered Xiarihamu magmatic Ni-Cu deposit in the Qinghai-Tibet plateau, western China. Lithos, 216–217, 224–240; Xue, S., Qin, K., Li, C., Tang, D., Mao, Y., Ripley, E.M., 2016. Geochronological, Petrological and Geochemical Constraints on Ni-Cu Sulfide Mineralization in the Poyi Ultramafic-troctolitic Intrusion in the NE Rim of Tarim Craton, Western China. Econ. Geol. 111, 1465–1484.

conduit systems. These ore types are exploited for Ni, Cu, Co, and variable amounts of PGEs. Examples include Voisey's Bay, Eagle, Tamarack, Xiarihamu, and several other small occurrences in China (e.g., Ripley and Li, 2011; Lightfoot et al., 2012; Ding et al., 2012; Taranovic et al., 2015; Li et al., 2015; Wei et al., 2013a; Xie et al., 2014; Xue et al., 2016). Mineralized sills in the Noril'sk district of Siberia (e.g., Naldrett et al., 1995) and in the Alxa district of Gansu Province, China (Jinchuan deposit: e.g., Lehmann et al., 2007; Duan et al., 2016) are also interpreted as magma conduits but the tonnage of the deposits is large (Fig. 3.2). Disseminated sulfide mineralization that occurs near the base of composite intrusions includes deposits of the Duluth Complex (Ripley, 2014) and the Platreef of the Bushveld Complex (McDonald and Holwell, 2011). The large-tonnage deposits of the world-famous Sudbury Intrusive Complex (Keays and Lightfoot, 2004; Lightfoot, 2017) represent the sole example of massive sulfide accumulations in embayments

(A)   Km - 100s of km   (B)   10s of m - km

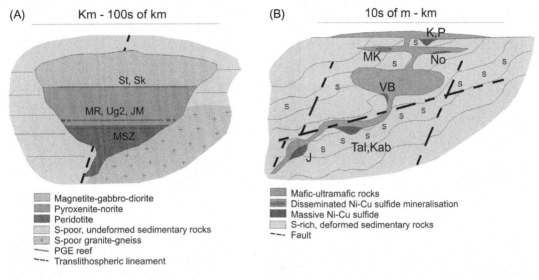

FIGURE 3.3

Schematic representation of PGE-rich reefs in layered intrusions (A) and Ni-Cu-PGE deposits in conduits (B) (after Maier, W.D., Groves, D.I., 2011. Temporal and spatial controls on the formation of magmatic PGE and Ni-Cu deposits. Mineral. Deposita 46, 841—857). *St*, Stella; *Sk*, Skaergaard; *MR*, Merensky Reef; *JM*, JM Reef; *UG2*, UG2 chromitite; *MSZ*, Main Sulfide Zone—Great Dyke; *J*, Jinchuan; *Tal*, Talnakh; *Kab*, Kabanga; *VB*, Voisey's Bay; *No*, Noril'sk; *MK*, Mount Keith; *K*, Kambalda; *P*, Pechenga.

and fractures beneath an impact-produced melt sheet. Fig. 3.3 from Maier (2005) summarizes the relative stratigraphic positions of both Ni-Cu-rich and PGE-rich deposits.

Sulfide mineralization associated with komatiites may be massive or may fill interstices between spinifex olivine plates (e.g., Barnes et al., 2017). Troughs below komatiite flows attest to the elevated temperatures of the Mg-rich magma flows and the process of thermomechanical erosion (Lesher and Keays, 2002; Barnes, 2006). Lesher and Keays (2002) state that large extrusive komatiite-associated Ni-Cu-PGE deposits occur at or near the bases of volcanic cycles in which the lower komatiite flows are thicker, more magnesian, more channelized and commonly intercalated with metasedimentary rocks. Fig. 3.4 from Lesher (2007) illustrates the occurrence of massive sulfides at the base of komatiite flows in contact with pelitic rocks in the Raglan district of Quebec.

Sulfide mineralization in the conduit style deposits may occur as net-textured (Fig. 3.5) or massive sulfide (Fig. 3.6) accumulations where the conduit widens (e.g., Voisey's Bay, Fig. 3.7; Eagle, Fig. 3.8). In some cases sulfide mineralization occurs near the base of sills, or horizontal portions of the conduit system (e.g., Noril'sk, Fig. 3.9). The Ni-Cu-PGE deposits of the Pechenga area occur near the base of differentiated sills that intruded sulfidic and carbonaceous sedimentary rocks. Because the parental magmas are thought to have been ferropicrites and there are no associated flood basalts, the ores have often been assigned to a distinct petro-tectonic class (Hanski et al., 2011). Where magma did extrude, spinifex textures developed that are very similar to those found in komatiites. Disseminated sulfide mineralization also occurs in conduits, but is economically

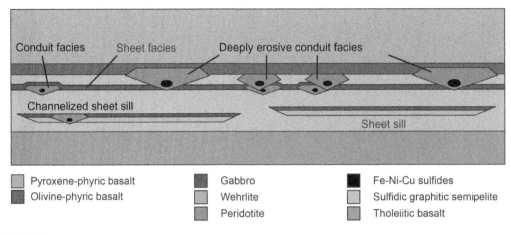

Pyroxene-phyric basalt
Olivine-phyric basalt

Gabbro
Wehrlite
Peridotite

Fe-Ni-Cu sulfides
Sulfidic graphitic semipelite
Tholeiitic basalt

FIGURE 3.4

Schematic illustration of massive sulfides located at the bottom of komatiite flows at the contact with sulfidic semipelites, Raglan area, Quebec.

From Lesher, C.M., 2007. Ni-Cu-(PGE) Deposits in the Raglan Area, Cape Smith Belt, New Québec. In: Goodfellow, W.D. (Ed),
Mineral Resources of Canada: A Synthesis of Major Deposit-types, District Metallogeny, the Evolution of Geological Provinces, and
Exploration Methods. Geological Survey of Canada and Mineral Deposits Division of the Geological Association of Canada Special
Publication 5, 351–386.

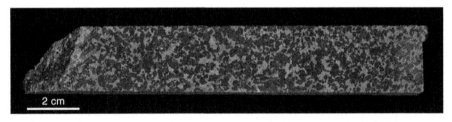

2 cm

FIGURE 3.5

Net-textured sulfide mineralization from the Eagle deposit, Michigan. Sulfide minerals occur interstitially to olivine and pyroxene.

significant only where larger sill-like bodies are present and ore tonnage may be large (e.g., Noril'sk). Disseminated sulfides occur in the large-tonnage, but low-grade occurrences within sheet-style intrusions of the Duluth Complex (Fig. 3.10). These deposits might logically be considered as conduit deposits, since intrusions of the Duluth Complex probably fed overlying flood basalts, but it appears that the deposits are in stacked sills where additional magma did not pass. The disseminated sulfide mineralization in the Platreef of the Bushveld Complex represents particularly PGE-rich occurrences that may be related to the reefs that will be described below, but the magma was concentrated near the contact of the intrusion with surrounding country rocks.

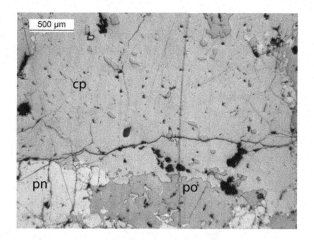

FIGURE 3.6

Photomicrograph of massive sulfide mineralization from the Eagle deposit; *po*, pyrrhotite; *cp*, chalcopyrite; *pn*, pentlandite.

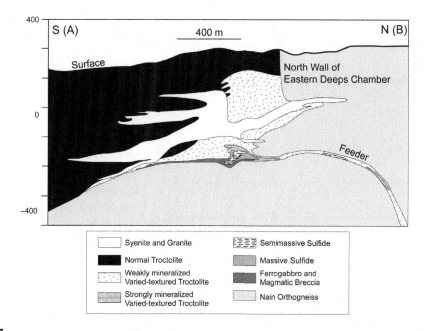

FIGURE 3.7

Longitudinal section of the Voisey's Bay deposit showing the entry of the feeder dike into the Eastern Deeps chamber.

After Naldrett, A.J., Kinnaird, J.A., Wilson, A., Yudovskaya, M., McQuade, S., Chunnett, G., et al., 2009. Chromite composition and PGE content of Bushveld chromitites: Part 1—the Lower and Middle Groups. Appl. Earth Sci. 118, 131–161.

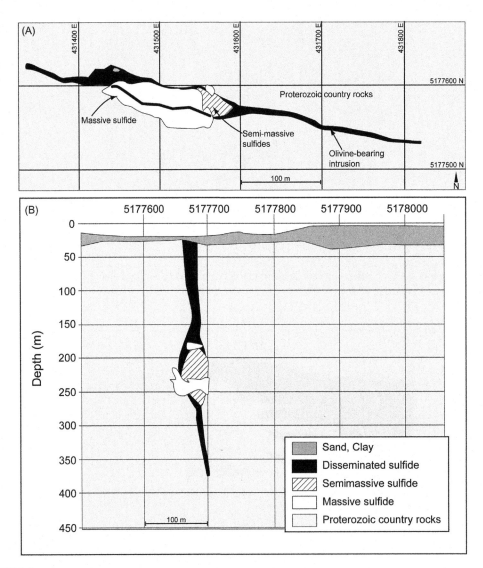

FIGURE 3.8

Geologic map and cross section of the Eagle intrusion, Michigan.

From Ripley, E.M., Li, C., 2011. A review of conduit-related Ni-Cu-(PGE) sulfide mineralization at the Voiseys Bay deposit, Labrador, and the Eagle deposit, northern Michigan. In Li, C., Ripley, E.M. (Eds.), Society of Economic Geologists Reviews in Economic Geology 17, 181–198.

3.2.2 LOW-SULFIDE PGE-RICH DEPOSITS

Maier (2005) classified PGE mineralization into seven types. Virtually all of these types consist of 1−5 vol.% disseminated sulfides hosted in a variety of mafic-rock types. Reefs are layers in

FIGURE 3.9

Disseminated sulfide mineralization overlying massive sulfides at the Kharaelakh deposit, Noril'sk.

intrusive bodies that contain elevated PGE contents. There are only a few deposits that are mined specifically for PGEs in the world (Fig. 3.11); most of the world's supply of PGEs comes from reef-style deposits that are mined in the Bushveld Complex of South Africa and the Great Dyke of Zimbabwe. Maier's classification includes: contact reefs, silicate-hosted reefs in ultramafic lower portions of layered intrusions, PGE-rich chromitite layers, silicate-hosted PGE reefs within the interlayered mafic-ultramafic portions of layered intrusions, PGE reefs in the upper portions of layered intrusions, transgressive iron-rich ultramafic pipes, and a rare type of deposit hosted by

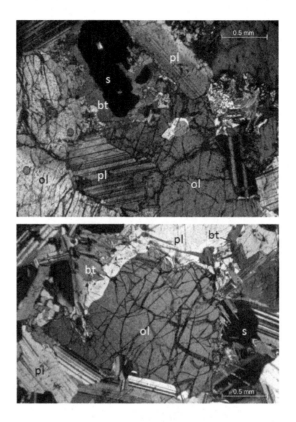

FIGURE 3.10

Sulfide minerals (s) and biotite (bt) in the interstices between olivine (ol) and plagioclase (pl) grains in troctolites from the Duluth Complex.

quartz-monazite veins. We include the vein-type occurrences in the general category of hydrothermal PGE deposits.

Contact reefs occur along the base and sidewalls of layered intrusions. The Platreef of the Bushveld Complex is the type example (Holwell and McDonald, 2006). Host rocks consist of thick sequences of mafic to ultramafic rocks, including gabbronorite and pyroxenite, with only minor peridotite or anorthosite. Silicate-hosted PGE reefs in the lower portions of layered intrusions include the Main Sulfide Zone of the Great Dyke (Wilson and Wilson, 1981) where PGE grades may reach several ppm (e.g., Naldrett and Wilson, 1990; Oberthür, 2011). Grades are typically lower in other examples, including reefs in the Munni Munni intrusion in Australia (Barnes, 1993) and the Fazenda Mirabela intrusion in Brazil (Barnes et al., 2011). PGE-enriched chromitite layers occur in the lower to central portions of many layered intrusions (Naldrett et al., 2011). The thickness of mineralized seams varies from a few millimeters to ~1−5 m. Grades are normally higher in the chromitites than in the reefs in ultramafic rocks. In the UG2 chromitite of the Bushveld Complex total PGE grades average from 7 to 9 ppm (Kinnaird et al., 2002). Not all chromitites are

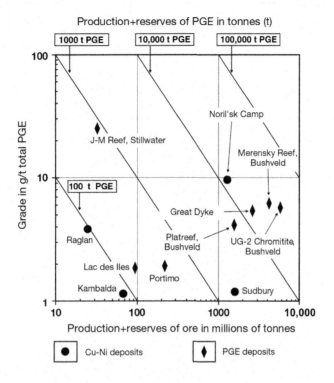

FIGURE 3.11

Grade and tonnage data for PGE deposits.

From Naldrett, A.J., 2011. Fundamentals of magmatic sulfide deposits. In: Li, C., Ripley, E.M. (Eds.), Society of Economic Geologists Reviews in Economic Geology 17, 1–50.

PGE-rich and the origin of PGE enrichment remains controversial. Silicate-hosted PGE reefs within the interlayered portions of layered intrusions include the type example of the Merensky Reef in the Bushveld Complex (e.g., Barnes and Maier, 2002), the J-M Reef of the Stillwater Complex (Barnes and Naldrett, 1985; Boudreau, 1988; Zientek et al., 2002), and the SJ, AP, and PV reefs of the Penikat intrusion in Finland (Halkoaho et al., 1990a,b; Alapieti and Lahtinen, 2002). In the Merensky Reef in the western segment of the Bushveld Complex, sulfides ($\sim 1-3$ volume %) occur in predominantly ultramafic rocks or melanorites at the base of a cyclic unit that overlies anorthosite or leuconorite (Fig. 3.12). In the eastern segment of the Bushveld Complex the mineralized interval is toward the top of a 2–4 m pyroxenite layer. The J-M Reef of the Stillwater is characterized by very high Pd grades that may reach 100 ppm, and average 14 ppm. The Reef occurs in troctolitic to anorthositic rocks of what is known as the Reef Package (Corson et al., 2002) in the first olivine-bearing unit of the Banded Series (Fig. 3.13). The SJ reef of the Penikat intrusion occurs predominantly in ultramafic rocks at the base of a cyclic unit, but feldspathic rocks in the immediate footwall may also be strongly mineralized (Halkoaho et al., 1990b). The PV and AP reefs occur primarily in plagioclase-rich rocks.

Pegmatitic norite

Sulfides

Chromitite

Anorthosite

11 cm

FIGURE 3.12

Merensky Reef sequence from the western portion of the Bushveld Complex.

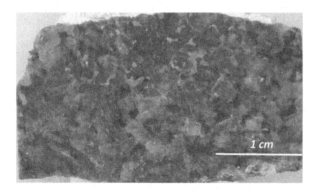

1 cm

FIGURE 3.13

Sulfide mineralization within anorthosite in the J-M Reef of the Stillwater Complex.

PGE reefs in the upper portions of layered intrusions occur in magnetitites, magnetitie-rich gabbros, and gabbronorites. Thicknesses of the PGE-enriched intervals may vary. In the Upper Zone of the Bushveld Complex a one decimeter-wide zone of disseminated sulfides with less than 1 ppm of both Pt and Pd occurs in the immediate footwall of the Main Magnetite Layer (von Gruenewaldt, 1973). The Triple Group of the Skaergarrd Complex hosts the Platinova Reef, which is enriched in Au and Cu as well as PGEs (Andersen et al., 1998; Holwell and Keays, 2014). In the Sonju Lake intrusion of the Midcontinent Rift System in Minnesota a c. 100-m section of ferrogabbro contains up to 1 ppm Pd (Miller, 1999; Park et al., 2004). PGE enrichment also occurs in magnetite-rich rocks in the Stella intrusion (Maier et al., 2003), Rincon del Tigre intrusion (Prendergast et al., 1998), and the Hongge intrusion (Zhong et al., 2002). Transgressive iron-rich ultramafic pipes may host extremely rich PGE mineralization in the Bushveld complex. The mineralization is heterogeneous and the bodies tend to be small so the occurrences tend not to be of economic significance.

The Waterberg deposit occurs in veins overlying the Rustenberg layered Series of the Bushveld Complex (McDonald and Tredoux, 2005). The veins may contain hundreds of ppm Pt and lesser amounts of Pd and Au. The New Rambler deposit of Wyoming (McCallum et al., 1976; Nyman et al., 1990) is another confirmed hydrothermal Pt deposit; several others have been described by Bursztyn and Olivo (2010).

The Lac des Isles deposit in Canada is a PGE mine whose geologic setting does not readily fit into the above classes. Mineralization is hosted in a metamorphosed, adcumulate gabbronorite breccia or within a 15−25 m thick schist containing amphibole, chlorite, and talc (Barnes and Gomwe, 2011). From ∼1 to 5 vol.% sulfide minerals occur in the gabbronorite breccias. The deposit most closely resembles the mineralization of the Platreef in the Bushveld Complex.

3.2.3 CHROMITE DEPOSITS

There are two principal types of chromite deposits (e.g., Stowe, 1994). Chromite layers that occur in large layered intrusions such as the Bushveld Complex (Eales, 1987; Mondal and Mathez, 2007), the Stillwater Complex (Jackson, 1961; Page, 1972), the Muskox Intrusion (Irvine, 1970, 1977, 1980), the Bird River Sill (Scoates, 1983), the Great Dyke (Wilson and Tredoux, 1990), the Kemi Intrusion (Alapieti et al., 1989), the Burakovzky Intrusion (Higgins et al., 1997), the Rhum Intrusion (O'Driscoll et al., 2010), and the Nuasahi and Sukinda Massifs (Mondal et al., 2006; Mondal, 2009) are known as stratiform deposits, and host most of the world's reserves of chromium. Several of the occurrences in the Nuasahi and Sukinda areas are brecciated, but are thought to have been derived from stratiform accumulations (Mondal, 2009). A second type, known as podiform deposits, is found in ophiolite sequences, generally enveloped by dunite (Arai, 1997). Layers of chromite in stratiform deposits may vary from less than a centimeter to 5−8 m, and may be laterally continuous or bifurcate (Fig. 3.14). Podiform chromite deposits consist of irregular pods or veinlets of aggregated chromite, often with nodular or orbicular textures, with limited lateral extent. Approximately 75% of the world's production of chromite comes from stratiform deposits in South Africa and India (Fig. 3.15), and 25% from podiform deposits in Kazakhstan (e.g., Kempirsai district; Melcher et al., 1997) and Turkey (e.g., Guleman ophiolite; Thayer, 1964; Usumezsoy, 1990).

Stratiform chromite occurrences typically are found in the ultramafic portions of layered intrusions, although some occur in higher stratigraphic horizons where plagioclase may be an important mineral. In many layered intrusions, chromite layers occur as parts of cyclic intervals characterized by peridotite-chromite-orthopyroxenite, or chromitite, harzburgite, orthopyroxenite. However many of the chromite occurrences in the Bushveld Complex are bounded by orthopyroxenite. Chromite is the primary host of Cr; layers may be massive (>90% chromite) or may contain disseminated chromite with variable quantities of associated cumulate and interstitial minerals. Most stratiform chromite deposits are Precambrian in age; e.g., Stillwater ∼2.7 Ga, Bushveld ∼2.0 Ga.

Podiform chromite deposits are parts of cumulate sequences that are part of the Moho transition to the upper mantle section of ophiolites (Thayer, 1964; Greenbaum, 1977; Arai, 2010). Massive pods of chromite ore may be accompanied by lateral zones of disseminated chromite that can be traced for hundreds of meters. The major podiform chromite deposits are relatively larger deposits

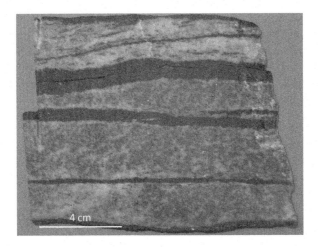

FIGURE 3.14

Thin chromite layers alternating with anorthosite and melanorite from the Bushveld Complex.

Chromium

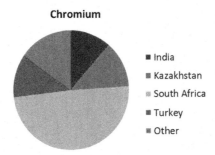

- India
- Kazakhstan
- South Africa
- Turkey
- Other

FIGURE 3.15

World chromium production in 2015. Data from USGS Mineral Resources Program.

(median size is 20,000 metric tons) and occur in Kazakhstan, Turkey, Cuba, Philippines, Iran, and New Caledonia. The minor podiform chromite deposits are smaller deposits (median size is 130 metric tons) that are found in California, Oregon, and Tibet. Few podiform-type deposits occur from the Late Proterozoic, but most deposits are of Mesozoic to Tertiary age.

3.2.4 MAGNETITE-ILMENITE DEPOSITS

Enrichments of magnetite, titanomagnetite, and ilmenite may occur in both ultramafic and mafic-rock types. There are several distinct types of deposits but the simplest classification is based on stratiform versus transgressive types. Nearly massive magnetitite layers occur in the Upper Zone of the Bushveld Complex, where the Main Magnetitite Layer ranges from 1 to 2 m in thickness (von Gruenewaldt, 1973). Over 50% of the world's V is produced from this layer (Crowson, 2001). Magnetitite layer 21 is normally $\sim$10 m thick (Maier et al., 2013). In subzones A and B of the

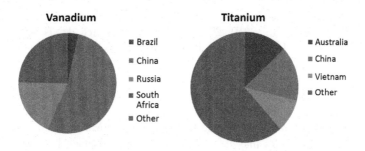

FIGURE 3.16

World vanadium and titanium production in 2015. Data from USGS Mineral Resources Program.

Upper Zone, cyclic layers are composed of magnetitite, gabbronorite and anorthosite (Maier et al., 2013). In the C subzone ilmenite is present in the oxide layers, which are overlain by ferrodiorites (von Gruenewaldt, 1979).

Magnetite, titanomagnetite, and ilmenite are present as disseminated grains in gabbros from the $\sim$ 260 Ma Emeishan Large Igneous Province in China (Pang et al., 2010). Massive oxides also occur locally. The largest deposits occur in the lower portions of the gabbroic layered intrusions, with a few smaller deposits found in higher levels of the intrusive stratigraphy. China is the world's leading producer of both Ti and V (Fig. 3.16), with a substantial proportion of the production coming from the oxide deposits in the Emeishan Large Igneous Province. The four largest Fe-Ti-oxide ore deposits in the province are the Panzhihua, Hongge, Baima, and Taihe deposits (Zhou et al., 2005; Tang et al., 2016).

The bulk of V production from Russia comes from the Kachkanar intrusion in the Ural mountains. The intrusion consists principally of clinopyroxenites and is considered an ultramafic Ural-Alaskan type intrusion (Pushkarev, 2014). Both massive titanomagnetitite and titanomagnetite disseminated (25–30%) in clinopyroxenite are present. Concentrations of TiO_2 and V_2O_5 average 2 and 0.5 wt.%, respectively.

Fe-Ti oxide deposits containing high concentrations of ilmenite also occur associated with massif-type anorthosites (e.g., Woodruff et al., 2013; Charlier et al., 2015, 2006). Apatite enrichment may also occur in P-rich nelsonites (e.g., Dymek and Owens, 2001). The occurrences are characterized by enrichment in ilmenite within gabbroic to noritic rock types. In the Tellnes ilmenite deposit in Norway, ilmenite occurs with plagioclase and orthopyroxene as cumulus minerals in an ilmenite norite that intrudes the central part of the Åna-Sira anorthosite of the Rogaland Anorthosite Province ($\sim$ 920–930 Ma); several other Fe-Ti oxide deposits were mined in the massif in the past (e.g., Duchesne, 1999; Charlier et al., 2015). There are only two other currently producing ilmenite deposits of this type. One is the Lac Tio deposit hosted in the Allard Lake anorthosite ($\sim$ 1061 ma) in the Grenville province of Quebec (Charlier et al., 2015, 2010), and the other is the Damiao deposit, hosted in the Ga Damiao anorthosite complex in the North China Craton (Zhao et al., 2009; Wei et al., 2013a; Wang et al., 2017). The Lac Tio orebody is also an ilmenite-rich norite that intrudes anorthosite. Although the intrusions cut anorthosite, the presence of orthopyroxene and plagioclase attest to oxide accumulation from a silicate magma. At the Damiao deposit several discordant ore bodies are present, hosted by anorthosite and leuconorite. Titaniferous magnetite is the primary ore mineral at Damiao, but an ilmenite $\pm$ apatite assemblage is also common. Layers of apatite, magnetite, and ilmenite-rich gabbro also occur in the Sept Iles intrusion in Quebec (Namur et al., 2010, 2012). Other ilmenite-rich deposits may be essentially massive oxides (e.g., the Big Island massive hemo-ilmenite dike that crosscuts the Allard Lake anorthosite) and their origins are controversial (see below). Contacts between the massive Fe-Ti oxides and anorthosite are sharp, and the transgressive nature of these bodies is clear.

Ilmenite enrichment may also occur in ultramafic rock types, although none of these occurrences has been mined. In the Duluth Complex of Minnesota oxide-rich ultramafic intrusions composed of dunite to pyroxenite may cut layered troctolitic rocks (Miller et al., 2002; Ripley et al., 1998). The oxide ultramafic bodies may be present as stringers to much larger ovoid bodies about $500 \times 150 \times 200$ m. The principal ore minerals are titanomagnetite and ilmenite, which may occur as euhedral primocryst grains or filling interstices between silicate minerals (Fig. 3.17). Large

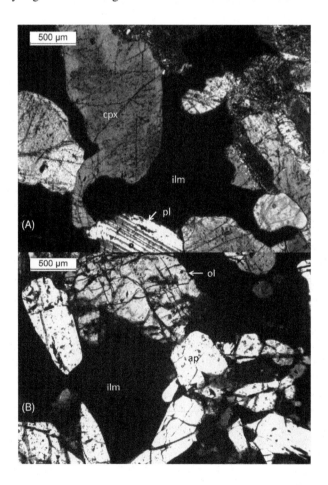

FIGURE 3.17

(A) Ilmenite filling interstices between clinopyroxene and plagioclase in an oxide-rich intrusion from the Duluth Complex. (B) Apatite-rich ultramafic oxide intrusion in the Duluth Complex.

resources of TiO_2 exist in these small intrusions; Hauck et al. (1997) estimated resources of as much as 245 million tons of ore with greater than 10% TiO_2 for the intrusions in the Duluth Complex.

3.3 GENETIC MODELS OF ORE FORMATION

3.3.1 SULFIDE-RICH NI-CU-PGE DEPOSITS

Models for the genesis of Ni-Cu-PGE deposits focus on the importance of sulfide liquid immiscibility in mafic and ultramafic magmas (e.g., Naldrett, 2011). Metals such as Ni, Cu, and PGEs are strongly chalcophile and are thought to be held in the mantle primarily as sulfide minerals (Alard et al., 2000; Lorand et al., 1999, 2013; Fonseca et al., 2012). Experimental studies have verified that these metals are strongly partitioned into immiscible sulfide liquids (e.g., Mungall and Brenan, 2014). Partition or Distribution coefficients (D) for Pt and Pd (concentration of Pt in sulfide liquid/ concentration of Pt in silicate liquid) exceed 5×10^5 (Mungall and Brenan, 2014), whereas the D value for Ni is ~ 500 (Peach et al., 1990; Patten et al., 2013) and for Cu ranges from ~ 500 to 1200 (Ripley et al., 2002). If metal-rich sulfide liquid can accumulate in a magmatic system, the potential for ore formation exists. Because metal-rich sulfide liquids are dense, gravitational accumulation has been proposed as a primary driving force for sulfide liquid accumulation (Naldrett, 2004).

The mantle is thought to contain $\sim 200-250$ ppm S (Palme and O'Neill, 2014). In order to promote sulfide liquid saturation in a magma, and to generate metal-rich sulfide liquids, several processes that may lead to sulfide liquid saturation in a magma have been proposed. Fractional crystallization may lead to sulfide liquid saturation but the amount of sulfide liquid that becomes immiscible is limited by the amount of S that the magma may dissolve. When sulfide liquid saturation is reached via fractional crystallization an amount known as the cotectic sulfide proportion will form as the system crystallizes. Although small volume disseminated sulfide ore bodies could be generated by this process it is unlikely that large ore bodies could be generated in the same way. It is often the case that sulfide saturation is reached via fractional crystallization after considerable crystallization of olivine. Because Ni is sequestered by olivine, Ni-rich sulfide ore bodies are also unlikely products. For many ore bodies studies involving S isotopes have shown that the addition of S from country rocks has been an important process for generating sufficient sulfide mass. In the case of komatiite-hosted deposits the addition of sulfide from underlying country rocks has been shown to be a particularly important process for ore genesis (Lesher and Keays, 2002; Fiorentini et al., 2010). The availability of Ni during high-degree melting of mantle olivine is clear. The relatively early addition of sulfide to komatiitic melts would be expected to lead to reaction with Ni-O complexes in the melt to generate sulfide liquids enriched in Ni.

The models for most other sulfide-rich Ni-Cu-PGE deposits are similar in that the requirements are the emplacement of a metal-bearing magma, addition of crustal sulfide to promote early sulfide saturation, and accumulation of immiscible sulfide liquid, often via gravitationally driven processes. Finely-disseminated sulfides in mafic/ultramafic rocks may represent the attainment of sulfide saturation at depth, and emplacement of sulfide with magmas where surface tension and momentum have acted to aid in the rise of dense sulfide (de Bremond d'ars and Arndt, 2001). A frequently asked question is if externally derived S is a prerequisite for the formation of a magmatic Ni-Cu-PGE sulfide-rich ore body (Keays and Lightfoot, 2010; Ripley and Li, 2013). The answer predominantly lies in the size of the magmatic system. If a system is large enough (i.e., a large mass of mantle-derived magma with ~ 200 ppm S), then with efficient accumulation of sulfide liquid a sulfide-rich ore body could form. Even for systems where the input of magma has been large (Large Igneous Provinces like the Siberian Traps that host the Noril'sk deposits) this seems to have been a very rare process; measurements of S isotopes indicate that the addition of crustal S has been important in these systems as well.

The requirement for a metal-bearing (fertile) magma is perhaps best illustrated using data from the reef-type Sonju Lake intrusion in Minnesota (Miller, 1999; Park et al., 2004). Fig. 3.18 shows that Pd and Pt are enriched in a ferrogabbro layer, and represent the time when sulfide saturation was reached via fractional crystallization in the evolving magma chamber. However, no Ni enrichment is observed because Ni was removed by early crystallization of olivine prior to sulfide

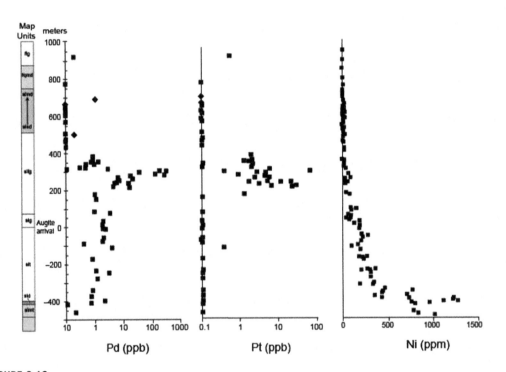

FIGURE 3.18

Stratigraphic column from the Sonju Lake intrusion, showing the depletion of Ni upwards, and the enrichment in Pt and Pd in the ferrogabbro.

Modified from Miller, J.D., 1999, Geochemical evaluation of platinum group element (PGE) mineralization in the Sonju Lake intrusion, Finland, Minnesota. Minnes. Geol. Surv. Inf.n Circ. 44, 31, and Park, Y.-R., Ripley, E.M., Miller, J.D., Li, C., Mariga, J., Shafer, P., 2004. Stable isotopic constraints on fluid-rock interaction and Cu-PGE-S redistribution in the Sonju Lake Intrusion, Minnesota. Econ. Geol. 99, 325–338.

saturation being attained. If metals have been removed by olivine crystallization (in the case of Ni), or via an early separation of sulfide liquid, the formation of an ore body is unlikely, even if assimilation of country rock S does occur.

3.3.2 SULFIDE-POOR PGE REEF-TYPE DEPOSITS

The processes that have been proposed for the formation of PGE reefs are very different from those suggested for sulfide-rich deposits, although they are firmly rooted in the importance of sulfide liquid immiscibility and high D values for the PGEs. For all sulfide deposits the "R-factor" equation of Campbell and Naldrett (1979) may be used to evaluate ore compositions. The equation is based on closed system two-component mass balance and relates the composition, or tenor, of sulfide liquid to the initial composition of a silicate liquid that reaches sulfide liquid saturation, the appropriate D value, and the mass ratio, or R-factor, of silicate to sulfide liquid. The expression

$$C_{\text{metal-sulfide}} = [C_{\text{metal-silicate liquid initial}} D (R + 1)]/ (R + D)$$

illustrates that high R-factors may lead to very high metal tenors but small amounts of sulfide. Analyses of basaltic rocks and sills and dikes that may be parental to several layered intrusions suggest that 10 ppb Pt and Pd are reasonable values for parental melts. One commonly cited model for reef formation was initially proposed by Barnes and Naldrett (1985) and Campbell et al. (1983). They suggested that the addition of a new magma pulse of a different composition from that present in the chamber promotes the attainment of sulfide saturation via magma mixing. Turbulent convection in the chamber allows the small amount of sulfide liquid generated to interact with a large volume of magma and sequester PGEs, again related to very high D values. The metal-rich sulfide droplets then settle to form the reef deposit. In this model the attainment of sulfide saturation via magma mixing and the generation of very high R-factors by turbulence are critical factors.

Another model has been proposed by Boudreau and coworkers (e.g., Boudreau and McCallum, 1992; Boudreau and Meurer, 1999). This is frequently referred to as the hydrodynamic model, and focuses on the role of late-magmatic hydrothermal fluids in dissolving PGEs held in disseminated sulfide minerals and moving them upward in a layered sequence. Chloride-rich fluids are suggested to be responsible for the transport of the PGEs, with deposition related to changes in intensive parameters when a drop in fluid velocity is encountered. The upward-moving-fluid model circumvents some of the problems associated with the attainment of high R-factors in magma bodies, but still requires extraction of PGEs from a thick sequence of rocks beneath the reefs that contain PGEs.

A third model for reef-type PGE deposits centers on the concentration of PGE in the initial magma being much greater than 10 ppb (Keays et al., 2011). In the case of the J-M Reef of the Stillwater Complex, an R-factor on the order of 500,000 would be required to account for the sulfide tenor of ~ 2500 ppm (14 ppm Pd average grade with 0.2 wt.% S). In order to reduce the need for such high R-factors and the amount of in situ silicate liquid required to supply the Pd present in the J-M Reef, Keays et al. (2011) suggested that the initial magma concentrations may have exceeded 10 ppb Pd. They proposed that in staging chambers external crustal S was assimilated and PGEs sequestered from magma that flowed through the chambers holding the sulfide liquid. It remains unclear how PGE-rich a silicate magma could become without crystallization of PGE minerals whose solubility tend to be low. It is also possible that the PGE-enriched reef represents the emplacement of sulfide-saturated and metal-rich melt that originated in the staging chamber.

A recent model for PGE reefs has been proposed by Maier et al. (2013). They suggest that sulfide saturation was attained in a magma chamber as a result of fractional crystallization, and PGE-rich cotectic proportions were produced in cumulates within the chamber. As new pulses of magma entered the chamber, gradual subsidence of the center of the chamber resulted and liquefaction and slumping of semiconsolidated cumulates at the top of the mush column occurred. Sorting of the slurries followed, with sulfides concentrated into reefs, often resulting in subcotectic proportions of sulfide minerals. A difficulty with this model may be that the R-factors necessary for fractional crystallization alone to be important for producing very PGE-rich sulfides cannot be maintained as crystallization proceeds.

3.3.3 CHROMITE DEPOSITS

Two models have emerged for the genesis of stratiform chromite deposits, but both relate in some form to concentration of chromite in magma chambers. We will discuss general metal concentration processes below and discuss only genetic features of the models here. Naldrett et al. (2012) have referred to the two models as those that pertain to the formation of chromite enrichment in (1) a location outside of the present location, or (2) in situ within the host igneous sequence. Eales (2000), Mondal and Mathez (2007), and Voordouw et al. (2009) have all argued that orthopyroxene and chromite have coprecipitated in a staging chamber, with chromite then winnowed out to form a chromite slurry which spread across the floor of the underlying crystal pile. Lesher et al. (2014) suggested that chromite was produced in magma conduits via partial melting of the silicate portion of iron formations and the reaction of residual magnetite with Cr-bearing magma. The fine-grained, upgraded chromite would then have been transported to overlying sills and dispersed as layers. Maier and Barnes (2008) proposed another model for the accumulation of chromite that originated in another location. They suggested that partially consolidated chromitites were thickened by subsidence and slumping, producing a cumulate slurry. Models that focus on in situ crystallization of chromite also include various methods for the accumulation of chromite only. Increase in the fO_2 of the magma was proposed to cause the crystallization of chromite alone by Cameron and Desborough (1964) and Ulmer (1969). Cameron (1980), Lipin (1993), and Cawthorn (2005b) suggested that increases in pressure were responsible for shifting phase relationships such that chromite would be the sole liquidus mineral. Irvine (1977) proposed for the chromite in the Muskox Intrusion that mixing between magma in the chamber and felsic melt at the upper contact of the intrusion promoted saturation in chromite only. Irvine (1977) also proposed that mixing of a newly injected primitive magma and fractionated magma resident in the chamber could lead to chromite crystallization before olivine. Several authors working in the Bushveld have applied this model (Eales, 1987; Scoon and Teigler, 1994; Naldrett et al., 2009). Kruger et al. (2002) also proposed a model for chromite genesis calling upon mixing between a granophyric melt produced at the roof of the Bushveld Complex, resident magma and newly introduced primitive magma.

Models for podiform chromite include selective concentration of chromite in a conduit (Leblanc and Ceuleneer, 1992) and reaction between a primitive melt and mantle harzburgite to form a Si-rich secondary melt. Mixing between the primary and secondary melts produces a melt that is supersaturated in chromite (Arai and Yurimoto, 1994; Zhou et al., 1996). Ballhaus (1998) also proposed that podiform chromite may form when olivine-rich melts mingle with more siliceous melts in low-P chambers. Matveev and Ballhaus (2002) suggested that podiform chromite would collect in a fluid phase that separated from a water-rich olivine-chromite saturated melt during passage through the uppermost mantle.

3.3.4 MAGNETITE-ILMENITE DEPOSITS

Toplis and Carroll (1995) have shown that mantle-derived melts are not saturated with magnetite or ilmenite. Residual melts may show enrichment in FeO and TiO_2 until Fe-Ti oxide saturation is reached. Two processes have been invoked to explain significant concentration of Fe-Ti oxides: liquid immiscibility and fractional crystallization combined with crystal sorting. Immiscibility between oxide liquids has been proposed by Zhou et al. (2005) to account for occurrences of massive titanomagnetitite in the Emeishan large igneous province of China. Although textures are similar to those found in accumulations of immiscible sulfide liquid, experimental evidence (Lindsley, 2003) has indicated that temperatures needed for oxide liquid immiscibility are much too high to be appropriate for basaltic systems. Pang et al. (2008a,b) proposed for the Panzhihua Intrusion that early crystallized Fe-Ti oxides collected in the magma chamber via gravitational segregation. Kolker (1982), Philpotts (1967), and Zhou et al. (2013) have also proposed that immiscible Fe-Ti-P melts may have been important for ore genesis. Zhou et al. (2013) suggested that two stages of magma evolution were important for deposits in the Panzhihua-type system. The first stage involved production of a Fe-Ti-P-rich gabbroic melt and a coexisting syenitic magma. Immiscible Fe-Ti oxide melts are proposed to then have separated from the gabbroic magma. Charlier et al. (2011), Charlier and Grove (2012), and Namur et al. (2012) have shown that liquid immiscibility can produce Fe-rich and Si-rich liquids. Fe-Ti-rich ferrogabbros may form from the Fe-rich liquids, with potential for the generation of oxide-rich layers. Fischer et al. (2016) suggest that the ferrogabbros of the Bushveld Complex and the magnetite layers may have been produced by this process.

To explain Fe-Ti oxide deposits associated with massif-type anorthosites, Charlier et al. (2006) proposed that fractional crystallization of a TiO_2-rich magma led to early ilmenite saturation in the Tellnes Intrusion. Buoyant separation of plagioclase was a key process in producing the ilmenite-rich ores (Charlier at al., 2007). Woodruff et al. (2013) developed a model for Fe-Ti oxide deposits that centered on the crystallization of Fe-Mg silicates and plagioclase that undergo density segregation, while buoyant plagioclase is removed to leave a Fe-Ti enriched crystal-liquid mush. Crustal assimilation and further crystallization could lead to increases in fO_2, the early crystallization of ilmenite and the formation of massive hemo-ilmenite bodies. An alternative path results in the crystallization of titanomagnetite, ilmenite, and apatite bodies in labradorite-type anorthosites or concentrations in layered intrusions with early-formed olivine. Wei et al. (2013a) suggested that both differentiation of a Fe-Ti-rich magma and liquid immiscibility were important processes for the generation of Fe-Ti-P ores.

The origin of oxide-rich ultramafic rocks, such as those present as small intrusive bodies in the Duluth Complex and the Kachkanar intrusion, remains unclear. Hypotheses include partial melting of iron formation and infiltration metasomatism of ultramafic rocks by Fe-Ti-P-rich fluids expelled during the late stages of crystallization of underlying gabbroic cumulates (Severson and Hauck, 1990). Ripley et al. (1998) proposed that related nelsonite occurrences in the Duluth Complex were produced as a result of Fe-Ti-P liquid immiscibility.

3.4 TECTONIC SETTINGS OF MAFIC ROCK-RELATED ORE DEPOSITS

Large PGE deposits such as those in the Bushveld Complex, the Great Dyke, and the Stillwater Complex occur within the central portions of cratons. For this reason no particular plate tectonic control is evident. Maier and Groves (2011) suggest that the buoyant subcontinental lithospheric mantle not only can support the mafic magmas, but also can shield intrusions from tectonic deformation and dismemberment. The relative quiescence permitted the development of continuous reefs in large intrusions. It remains uncertain as to whether any of the large layered intrusions that host PGE deposits fed overlying flood basalts (e.g., Lipin and Zientek, 2002). In the case of the Bushveld Complex, VanTongeren et al. (2010) argue that the felsic volcanics that overlie the Complex represent the last felsic differentiates of the magmatic system, and that extrusion of earlier mafic material did not occur. Thermal plumes impinging on the crust accompanied by decompression melting may have been responsible for the generation of the mafic magma, with extrusion of strongly fractionated felsic magma being the only volcanic expression.

Ni-Cu-PGE deposits occur in a number of plate tectonic settings (Table 3.1). Begg et al. (2010) suggest that most Ni-Cu deposits formed at craton margins. Magma conduit systems appear to provide the dynamic environment necessary for extensive interaction with sulfidic country rocks and potential upgrading as sulfide liquid interacts with passing silicate magma. These environments can be produced at either divergent settings where mantle plumes may be coincident with rifting, or at subduction zones where melting of asthenospheric mantle may be promoted due to dehydration of down-going sediments. Slab roll-back may also initiate back-arc spreading and provide another favorable setting for dynamic magma emplacement and interaction with country rocks. The probable importance of water in the production of komatiitic liquids has led to the suggestion that komatiites form at subduction zones (e.g., Parman et al., 2004), whereas others propose that a hydrous mantle plume was important for komatiite genesis in an extensional environment (Sobolev et al., 2016).

Rivers (1997) has proposed that Proterozoic massif anorthosite complexes that host Fe-Ti oxide deposits were produced in extensional environments. One area is a back-arc basin inboard from an active continental-margin magmatic arc. The second setting is a collisional orogeny during periods of tectonic collapse. The Fe-Ti oxide deposits in the Duluth Complex and the Emeishan province formed in association with flood basalt volcanism. The Duluth Complex is part of the Midcontinent Rift system, which developed in response to intracontinental rifting and the arrival of a mantle plume (Stein et al., 2016).

3.5 METAL CONCENTRATION MECHANISMS

All metallic ore deposits represent anomalous concentrations of metals in the Earth's crust. For some deposits, the mechanism through which the concentration occurs is clear, but in many, the processes involved remain elusive. In the case of Ni-rich deposits melting of olivine in the mantle should generate magmas with sufficient amounts of Ni that could be concentrated, from even low-degree melting of the mantle. For elements like Cu and PGEs low-degree mantle melting may leave behind immiscible sulfide liquids enriched in these elements (e.g., Naldrett, 2011). When mantle melting exceeds $9-12$ wt.% all of the Cu and most of the PGEs initially in the mantle should then reside in the partial melt. Fonseca et al. (2012) and Lorand et al. (2013) suggest that Os, Ir, Ru, and Rh may remain in the mantle as alloys even if they were initially held within base-metal sulfides. Dilution occurs when mantle melting is high-degree; komatiites that may represent $25-30\%$ mantle melting (Arndt, 2003; Lesher and Keays, 2002) contain an abundance of Ni for potential ore formation provided that external S is introduced to the magma. However, relative Cu and PGE concentrations in the primary magma may be low even though the full mantle component of PGEs has been liberated. Parental magmas for most other sulfide-rich Ni-Cu-PGE deposits contain sufficient metal to form economic concentrations as long as sulfide supersaturation is reached early in the crystallization history of the magma. Deposits like the Platinova Reef in the Skaergaard Intrusion (Keays and Tegner, 2015; Andersen et al., 1998) illustrate the situation where Cu and PGE enrichments may occur in gabbroic portions of an intrusion but Ni is low due to incorporation in early-crystallized olivine. Ore-grade enrichments in PGEs and Cu normally require processes in addition to collection by immiscible sulfide liquids. Brügmann et al. (1993), Kerr and Leitch (2005), and Li et al. (2009) have all proposed methods of "upgrading" of sulfide deposits in magma conduit systems by new pulses of pristine magma that may interact with the accumulated sulfide liquid. Because of large D values, metals like the PGEs may become strongly concentrated as magma-sulfide liquid interaction continues. The upgrading process, in addition to physical processes that favor sulfide collection, is one reason why conduits host many sulfide ore deposits.

PGE reefs such as the J-M Reef in the Stillwater Complex and the Merensky Reef in the Bushveld Complex require the extraction of metal from a very large volume of magma. In the case of the J-M Reef average Pd grades are 14 ppm, with 2000 ppm S and a calculated sulfide tenor of 2550 ppm Pd. In order to produce such high metal tenors extremely high R-factors are required (see above), but given a magma with 10 ppb Pd a column of magma of 1400 m would be necessary to provide the mass of Pd present in the J-M Reef. How the extraction of such large amounts of Pd is accomplished is a major question. As described above, researchers have proposed the interaction of sulfide liquid with a large mass of magma at very high R-factors, extraction from a large thickness of underlying cumulates via late stage fluids, and production of PGE-enriched magmas at depth in staging chambers via an upgrading process.

The problem of attaining high metal contents in silicate liquids extends to chromite deposits as well. In the case of chromite in the Bushveld Complex, Naldrett et al. (2012) showed that using the amount of Cr_2O_3 thought to be appropriate for the Bushveld parental magma (~ 0.15 wt.%), chromite would not crystallize before orthopyroxene. In order for chromite to appear on the liquidus before orthopyroxene at least 0.25 wt.% Cr_2O_3 was required, and only 0.04% of this amount could actually be removed in forming chromite. As an example, Naldrett et al. (2012) showed that for a 70 cm thick chromitite seam containing 80 vol.% chromite and 35 wt.% Cr_2O_3, 350 vertical meters of magma would be required if all of the Cr_2O_3 were removed. If only 0.04 wt.% or less were removed, as their calculation indicated, then the column of magma would have been 938 m. Eales (2000) showed that there are ~ 9.5 m of massive chromitite in the Bushveld, requiring about 13 km of magma. Maier and Teigler (1995) and Eales (2000) pointed out that there are only $1.2-1.7$ km of Critical Zone cumulates in the

Bushveld layered sequence. Naldrett et al. (2012) concluded that the Cr_2O_3 content of the parental magma must have been above 0.2 wt.%, and that the "missing" magma needed to supply the required amount of Cr_2O_3 for the massive chromitites escaped and is now represented by satellite intrusions bordering the Bushveld Complex. Experiments by Poustovetov and Roeder (2000) showed that basaltic magmas could dissolve up to 0.275 wt.% Cr_2O_3. High-degree mantle melts that produced komatiites contain approximately 0.22 wt.% Cr_2O_3 (e.g., Barnes and Fiorentini, 2012), so the figure needed for chromite saturation in the Bushveld parental magma appears reasonable. However, the amount of magma from which the Cr must have been extracted is large, and the questions of where the missing magma went and the mechanism of escape remain a challenging problem.

The concentration of Ti in most mantle-derived magmas appears to be sufficient for the generation of ilmenite-rich deposits after a significant degree of fractional crystallization, although high-Ti parental magmas appear to be the source of the deposits in the Emeishan large igneous province (Zhou et al., 2013). Although concentration mechanisms such as fractional crystallization and efficient removal of components like plagioclase (e.g., Charlier et al., 2006) and concentration of immiscible Fe-Ti-P-rich liquids (Zhou et al., 2013; Charlier et al., 2015), can occur simultaneously, a key question is the amount of magma that must be processed to account for the mass of Ti in the deposits. In some cases the amount of Ti present in a cumulate sequence would be sufficient to account for the Ti in the ore deposit. In other cases, the amount of Ti appears to exceed the amount thought to have been available and additional magma would be required, much as is the case for the Bushveld chromitites. In the case of deposits such as those associated with small ultramafic bodies (Urals, Duluth Complex, etc.) the amount of Ti requires addition from sources and processes that are not obvious.

Toplis and Corgne (2002) proposed that oxygen fugacity is a major control on the partitioning of V between melt and magnetite. High partition coefficients and the formation of vanadium-rich magnetite are favored at low-fO_2 conditions. Redox conditions may thus be a strong factor in the formation of V-rich versus V-poor magnetite in intrusions.

3.6 SUMMARY

In summary, magmatic ore deposits associated with mafic-ultramafic rocks occur in a variety of geological settings. The metals of interest occur primarily as sulfide or oxide assemblages. Ni, Cu, and PGE deposits may occur either as sulfide-rich varieties occurring as disseminated, net-textured and massive mineralization, or as PGE-rich, low-sulfide reefs. Oxide deposits include chromite as either stratiform accumulations or irregular podiform deposits. Vanadium-bearing magnetite and ilmenite and Fe-Ti-P deposits may occur as massive layers or disseminated enrichments in gabbroic rocks within large layered intrusions, or as transgressive massive occurrences within gabbroic to anorthositic intrusions. Ilmenite, titanomagnetite, and apatite may also occur in massive or disseminated form within small mafic-ultramafic intrusions. Both chemical and physical processes contribute to the enrichment in ore elements and the ultimate formation of a mineable mass of material. In some cases, e.g., PGE reef-type deposits, the enrichment in metals is particularly notable and unusual processes of enrichment must be responsible for ore formation.

REFERENCES

Alapieti, T.T., Kujanpää, J., Lahtinen, J.J., Papunen, H., 1989. The Kemi stratiform chromitite deposit, northern Finland. Econ. Geol. 84, 1057–1077.

Alapieti, T.T., Lahtinen, J.J., 2002. Platinum-group element mineralization in layered intrusions of northern Finland and the Kola Peninsula, Russia. In: Cabri, L.J. (Ed.), The Geology, Geochemistry, Mineralogy and Mineral Beneficiation of Platinum-Group Elements, Canadian Institute of Mining, Metallurgy and Petroleum Special, 54. pp. 507–546.

Alard, O., Griffin, W.L., Lorand, J.-P., Jackson, S.E., O'Reilly, S.Y., 2000. Non-chondritic distribution of the highly siderophile elements in mantle sulfides. Nature 407, 891–894.

Andersen, J.C.Ø., Rasmussen, H., Nielsen, T.F.D., Ronsbo, J.G., 1998. The triple group and the platinova Au and Pd reefs in the Skaergaard intrusion: stratigraphic and petrographic relations. Econ. Geol. 93, 485–509.

Arai, S., 2010. Chromites: An enigmatic mantle rock type. J. Geogr. 119, 392–410.

Arai, S., 1997. Origin of podiform chromitites. Asian J. Earth Sci. 15, 303–310.

Arai, S., Yurimoto, H., 1994. Podiform chromitites of the Tari-Misaka ultramafic complex, Southwestern Japan, as mantle-melt interaction products. Econ. Geol. 89, 1279–1288.

Arndt, N., 2003. Komatiites, kimberlites and boninites. J. Geophys. Res. 1008 ECV 5-3-5-11.

Ballhaus, C., 1998. Origin of podiform chromite deposits by magma mingling. Earth Planet. Sci. Lett. 156, 185–193.

Barnes, S.-J., Ripley, E.M., 2016. Highly siderophile and strongly chalcophile elements in magmatic ore deposits. In: Harvey, J., Day, J.M. (Eds.), Highly Siderophile and Strongly Chalcophile Elements in High Temperature Geochemistry and Cosmochemistry. Reviews in Mineralogy and Geochemistry, 81. pp. 725–774.

Barnes, S.-J., Gomwe, T.S., 2011. The Pd Deposits of the Lac des Iles Complex, Northwestern Ontario. In: Li, C., Ripley, E.M. (Eds.), Society of Economic Geologists Reviews in Economic Geology, 17. pp. 351–370.

Barnes, S.-J., Lightfoot, P., 2005. Formation of magmatic nickel sulfide ore deposits and processes affecting the Cu and PGE contents. Econ. Geol. 100th Anniver. vol. 179–214.

Barnes, S.-J., Maier, W.D., 2002. Platinum-group element distributionsin the Rustenburg layered suite of the Bushveld Complex, South Africa. In: Cabri, L.J. (Ed.), The Geology, Geochemistry, Mineralogy and Mineral Beneficiation of Platinum-Group Elements, Canadian Institute of Mining, Metallurgy and Petroleum Special, 54. pp. 431–458.

Barnes, S.J., Holwell, D.A., Le Vaillant, M., 2017. Magmatic sulfide ore deposits. Elements 13, 89–95.

Barnes, S.J., Fiorentini, M.L., 2012. Komatiite magmas and sulfide nickel deposits: A comparison of variably endowed Archean terranes. Econ. Geol. 107, 755–780.

Barnes, S.J., Osborne, G., Cook, D., Barnes, L., Maier, W.D., Godel, B., 2011. The Santa Rita Ni sulfide deposit in the Fazenda Mirabela intrusion, Bahia, Brazil: geology, sulfide geochemistry and genesis. Econ. Geol. 106, 1083–1110.

Barnes, S.J., 2006. Komatiite-hosted nickel sulfide deposits: geology, geochemistry, and genesis. Soc. Econ. Geol. Special Publ. 13, 51–118.

Barnes, S.J., 1993. Partitioning of the PGE and gold between silicate and sulphide magmas in the Munni Munni Complex, Western Australia. Geochim. Cosmochim. Acta 57, 1277–1290.

Barnes, S.J., Naldrett, A.J., 1985. Geochemistry of the J-M (Howland) Reef of the Stillwater Complex, Minneapolis Adit Area. I. Sulfide Chemistry and Sulfide-Olivine Equilibrium. Econ. Geol. 80, 627–645.

Begg, G.C., Hronsky, J.A.M., Arndt, N.T., Griffin, W.L., O'Reilly, S.Y., Hayward, N., 2010. Lithospheric, cratonic, and geodynamic setting of Ni-Cu-PGE sulfide deposits. Econ. Geol. 105, 1057–1070.

Boudreau, A.E., Meurer, W.P., 1999. Chromatographic separation of the platinum-group elements, base metals, gold, and sulfur during degassing of a compacting and solidifying igneous crystal pile. Contrib. Mineral. Petrol. 134, 174–185.

Boudreau, A.E., McCallum, I.S., 1992. Concentration of platinum-group elements by magmatic fluids in layered intrusions. Econ. Geol. 87, 1830−1848.

Boudreau, A.E., 1988. Investigations of the Stillwater Complex: Part IV. The role of volatiles in the petrogenesis of the J-M Reef, Minneapolis adit section. Can. Mineral. 26, 193−208.

Brügmann, G.E., Naldrett, A.J., Asif, M., Lightfoot, P.C., Gorbachev, N.S., Fedorenko, V.A., 1993. Siderophile and chalcophile metals as tracers of the evolution of the Siberian trap in the Noril'sk region, Russia. Geochim. Cosmochim. Acta 57, 2001-1018.

Bursztyn, N.E., Olivo, G.R., 2010. PGE rich Ni-Cu sulfide mineralization in the Flin Flon greenstone belt, Manitoba, Canada: implications for hydrothermal remobilization of platinum-group elements in basic-ultrabasic sequences. Econ. Geol. 105, 1469−1490.

Cameron, E.N., 1980. Evolution of the lower critical zone, central sector, eastern Bushveld Complex. Econ. Geol. 75, 845−871.

Cameron, E.N., Desborough, G.A., 1964. Origin of certain magnetite bearing pegmatites in the eastern part of the Bushveld Complex, South Africa. Econ. Geol. 59, 197−225.

Campbell, I.H., Naldrett, A.J., Barnes, S.J., 1983. A model for the origin of the platinum-rich sulfide horizons in the Bushveld and Stillwater Complexes. J. Petrol. 24, 133−165.

Campbell, I.H., Naldrett, A.J., 1979. The influence of silicate:sulphide ratios on the geochemistry of magmatic sulphides. Econ. Geol. 74, 1503−1506.

Cawthorn, R.G., Barnes, S.J., Ballhaus, C., Malitch, K.N., 2005a. Platinum group element, chromium, and vanadium deposits in mafic and ultramafic rocks. Econ. Geol. 100th Anniv. vol. 215−251.

Cawthorn, R.G., 2005b. Pressure fluctuations and the formation of the PGE rich Merensky and chromitite reefs, Bushveld Complex. Mineral. Deposita 40, 231−235.

Charlier, B., Namur, O., Bolle, O., Latypov, R., Duchesne, J.-C., 2015. Fe−Ti−V−P ore deposits associated with Proterozoic massif-type anorthosites and related rocks. Earth Sci. Rev. 141, 56−81.

Charlier, B., Grove, T.L., 2012. Experiments on liquid immiscibility along tholeiitic liquid lines of descent. Contrib. Mineral. Petrol. 164 (1), 27−44.

Charlier, B., Namur, O., Toplis, M.J., Schiano, P., Cluzel, N., Higgins, M.D., et al., 2011. Large-scale silicate liquid immiscibility during differentiation of tholeiitic basalt to granite and the origin of the Daly gap. Geology 39, 907−910.

Charlier, B., Namur, O., Malpas, S., de Marneffe, C., Duchesne, J.C., Vander Auwera, J., et al., 2010. Origin of the giant Allard Lake ilmenite ore deposit (Canada) by fractional crystallization, multiple magma pulses, and mixing. Lithos 117, 119−134.

Charlier, B., Skar, O., Korneliussen, A., Duchesne, J.-C., vabder Auwera, J., 2007. Ilmenite composition in the Tellnes Fe-Ti deposit, SW Norway: fractional crystallization, postcumulus evolution and ilmenite-zircon relation. Contrib. Mineral. Petrol. 154, 119−134.

Charlier, B., Duchesne, J.-C., Vander Auwera, J., 2006. Magma chamber processes in the Tellnes ilmenite deposit (Rogaland Anorthosite Province, SW Norway) and the formation of Fe-Ti ores in massif-type anorthos. Chem. Geol. 234, 264−290.

Corson, S.R., Childs, J.F., Dahy, J.P., Keith, D.W., Koski, M.S., LeRoy, L.W., 2002. The Reef Package Stratigraphy That Contains the J-M Platinium-Palladium Reef of the Stillwater Complex, Montana. 9th International Platinum Symposium. Billings, Montana, pp. 101−102.

Crowson, P., 2001. Mineral handbook 2000-2001. Mining Journal Books, Edenbridge, pp 486.

de Bremond d'ars, J., Arndt, N.T., Hallot, E., 2001. Analog experimental insights into the formation of magmatic sulfide deposits. Earth Planet. Sci. Lett. 186, 371−381.

Ding, X., Ripley, E., Li, C., 2012. PGE geochemistry of the Eagle Ni-Cu-(PGE) deposit, Upper Michigan: Constraints on ore genesis in a dynamic magma conduit. Mineral. Deposita 47, 89−104.

Duan, J., Li, C., Qian, Z., Jiao, J., Ripley, E.M., Feng, Y., 2016. Multiple S isotopes, zircon Hf isotopes, whole-rock Sr-Nd isotopes, and spatial variations of PGE tenors in the Jinchuan Ni-Cu-PGE deposit, NW China. Mineral. Deposita 51, 557−574.

Duchesne, J.C., 1999. Fe-Ti deposits in Rogaland anorthosites (south Norway): geochemical characteristics and problems of interpretation. Mineral. Deposita 34, 182−198.

Dymek, R.F., Owens, B.E., 2001. Petrogenesis of apatite-rich rocks (nelsonites and oxide-apatite gabbronorite) associated with massif anorthosite. Econ. Geol. 96, 797−815.

Eales, H.V., 2000. Implications of the chromium budget of the Western Limb of the Bushveld Complex. SA J. Geol. 103, 141−150.

Eales, H.V., 1987. Upper Critical zone chromitites at R.P.M. Union section mine. In: Stowe, C.W. (Ed.), Evolution of Chromium ore Fields. Van Nostrand Reinhold, New York, NY, pp. 144−168.

Fiorentini, M.L., Beresford, S.W., Rosengren, N.M., Barley, M.E., McCuaig, T.C., 2010. Contrasting komatiite belts, associated Ni-Cu- (PGE) deposit styles and assimilation histories. Aust. J. Earth Sci. 57, 543−566.

Fischer, L.A., Wang, M., Charlier, B., Namur, O., Roberts, R.J., Veskler, I.V., et al., 2016. Immiscible iron- and silica-rich liquids in the Upper Zone of the Bushveld Complex. Earth Planet. Sci. Lett. 443, 108−117.

Fonseca, R., Laurenz, V., Mallmann, G., Lugue, A., Hoehne, N., Jochum, K.P., 2012. New constraints on the genesis and long-term stability of Os-rich alloys in the Earth's mantle. Geochim. Cosmochim. Acta 87, 227−242.

Greenbaum, D., 1977. The chromitiferous rocks of the Troodos ophiolite complex, Cyprus. Econ. Geol. 72, 1175−1192.

Halkoaho, T.A.A., Alapieti, T.T., Lahtinen, J.J., 1990a. The Ala-Penikkat PGE reefs in the Penikat layered intrusion, northern Finland. Mineral. Petrol. 42, 23−38.

Halkoaho, T.A.A., Alapieti, T.T., Lahtinen, J.J., 1990b. The Sompujärvi PGE reef in the Penikat layered intrusion, northern Finland. Mineral. Petrol. 42, 39−55.

Hanski, E.J., Luo, Z.-Y., Oduro, H., Walker, R.J., 2011. The pechenga Ni-Cu sulfide deposits, Northwestern Russia: a review with new constraints from the feeder dikes. In: Li, C., Ripley, E.M. (Eds.), Society of Economic Geologists Reviews in Economic Geology, 17. pp. 145−162.

Hauck, S.A., Severson, M.J., Zanko, L.M., Barnes, S.J., Morton, P., Aliminas, H.V., et al., 1997. An overview of the geology and oxide, sulfide, and platinum group element mineralization along the western and northern contacts of the Duluth Complex. In: Ojakangas, R.W., Dickas, A.B., Green, J.C. (Eds.), Middle Proterozoic to Cambrian Rifting, Central North America. Geological Society of America Special, 312. pp. 137−185.

Higgins, S.J., Snyder, G.A., Mitchell, J.N., Taylor, L.A., Sharkov, E.V., Bogatikov, O.A., et al., 1997. Can. J. Earth Sci. 34, 390−406.

Holwell, D.A., McDonald, I., 2006. Petrology, geochemistry, and the mechanisms determining the distribution of platinum-group element and base metal sulfide mineralization in the Platreef at Overysel, northern Bushveld Complex, South Africa. Mineral. Deposita 41, 575−598.

Holwell, D.A., Keays, R.R., 2014. The formation of low-volume, high-tenor magmatic PGE-Au sulfide mineralization in closed systems: evidence from precious and base metal geochemistry of the platinova reef, skaergaard intrusion, East Greenland. Econ. Geol. 109, 387−406.

Irvine, T.N., 1970. Crystallization sequences in the Muskox intrusion and other layered intrusions. 1. Olivine-pyroxene-plagioclase relations. In: Visser, D.J.L., von Gruenewaldt, G. (Eds.), Symposium on the Bushveld Igneous Complex and Other Layered Intrusions, 1. Geological Society of South Africa Special Publication, pp. 441−476.

Irvine, T.N., 1977. Origin of chromitite layers in the Muskox intrusion and other stratiform intrusions; a new interpretation. Geology 5, 273−277.

Irvine, T.N., 1980. Infiltration metasomatism, adcumulus growth, and double diffusive fractional crystallization in the Muskox intrusion and other layered intrusions. In: Hargaves, R.B. (Ed.), Physics of Magmatic Processes. Princeton University Press, Princeton, pp. 325–383.

Jackson, E.D., 1961. Primary textures and mineral associations in the ultramafic zone of the Stillwater Complex, Montana. U.S. Geol. Surv. Prof. 358, 1–106.

Keays, R.R., Lightfoot, P.C., 2004. Formation of Ni-Cu platinum-group element sulfide mineralization in the Sudbury impact melt sheet. Mineral. Petrol. 82, 217–258.

Keays, R., Lightfoot, P., 2010. Crustal sulfur is required to form magmatic Ni-Cu sulfide deposits: Evidence from chalcophile element signatures of Siberian and Deccan Trap basalts. Mineral. Deposita 45, 241–257.

Keays, R.R., Lightfoot, P.C., Hamlyn, P.R., 2011. Sulfide saturation history of the Stillwater Complex, Montana: chemostratigraphic variations in platinum group elements, Mineral. Deposita, 47. pp. 151–173.

Keays, R.R., Tegner, C., 2015. Magma chamber processes in the formation of the low-sulphide magmatic Au-PGE mineralization of the platinova reef in the skaergaard intrusion, East Greenland. J. Petrol. 56, 2319–2340.

Kerr, A., Leitch, A.M., 2005. Self-destructive sulfide segregation systems and the formation of high-grade magmatic ore deposits. Econ. Geol. 100, 311–332.

Kinnaird, J.A., Kruger, F.J., Nex, P.A.M., Cawthorn, R.G., 2002. Chromitite formation - a key to understanding processes of platinum enrichment. Trans. Inst. Mining Metal. 111, 23–35.

Kolker, A., 1982. Mineralogy and geochemistry of Fe-Ti oxide and apatite (nelsonite) deposits and evaluation of the liquid immiscibility hypothesis. Econ. Geol. 77, 1146–1148.

Kruger, F.J., Kinnaird, J.A., Nex, P.A.M., Cawthorn, R.G., 2002. Chromite Is the Key to PGE (abs). 9th International Platinum Symposium. Billings, Montana, p. 211.

Leblanc, M., Ceuleneer, G., 1992. Chromite crystallization in a multicellular magma flow: evidence from a chromitite dike in the Oman ophiolite. Lithos 27, 231–257.

Lehmann, J., Arndt, N., Windley, B., Zhou, M.-F., Wang, C.Y., Harris, C., 2007. Field relationships and geochemical constraints on the emplacement of the jinchuan intrusion and its Ni-Cu-PGE sulfide deposit, Gansu, China. Econ. Geol. 102, 75–94.

Lesher, C.M., 1989. Komatiite-associated nickel sulfide deposits. Rev. Econ. Geol. 4, 44–101.

Lesher, C.M., Keays, R.R., 2002. Komatiite-associated Ni-Cu-(PGE) deposits. In: Cabri, L. (Ed.), Geology, Mineralogy, and Beneficiation of the Platinum-Group Elements, Canadian Institute of Mining, Metallurgy and petroleum Special, 54. pp. 579–618.

Lesher, C.M., 2007. Ni-Cu-(PGE) deposits in the raglan area, Cape Smith Belt, New Québec. In: Goodfellow, W.D. (Ed.), Mineral Resources of Canada: A Synthesis of Major Deposit-types, District Metallogeny, the Evolution of Geological Provinces, and Exploration Methods. Geological Survey of Canada and Mineral Deposits Division of the Geological Association of Canada Special Publication, 5. pp. 351–386.

Lesher, C.M., Carson, H.J.E., Metsaranta, R.T., Houle, M.G., 2014, genesis of chromite deposits by partial melting, physical transport, and dynamic upgrading of silicate-magnetite facies iron formation (abs). 12th International Pt Symposium, Yekaterinburg, Russia, 36.

Li, C., Ripley, E.M., Naldrett, A.J., 2009. A new genetic model for the giant Ni-Cu-PGE sulfide deposits associated with the Siberian flood basalts. Econ. Geol. 104, 291–301.

Li, C., Zhang, Z., Li, W., Wang, Y., Sun, T., Ripley, E.M., 2015. Geochronology, petrology and Hf-S isotope geochemistry of the newly-discovered Xiarihamu magmatic Ni-Cu deposit in the Qinghai-Tibet plateau, western China. Lithos 216-217, 224–240.

Lightfoot, P.C., Keays, R.R., Evans-Lamswood, D., Wheeler, R., 2012. S saturation history of Nain Plutonic Suite mafic intrusions: origin of the Voisey's Bay Ni-Cu-Co sulfide deposit, Labrador, Canada. Mineral. Deposita 47, 23–50.

Lightfoot, P.C., 2017. Nickel Sulfide Ores and Impact Melts Origin of the Sudbury Igneous Complex. Elsevier, Amsterdam, p. 662.

Lindsley, D.H., 2003. Do Fe-Ti Oxide Magmas Exist? Geology: Yes; Experiments: No!, 9. Norges Geolgiske Undersokelse Special Publication, pp. 34−35.

Lipin, B.R., 1993. Pressure increases, the formation of chromite seams, and the development of the ultramafic series in the Stillwater Complex, Montana. J. Petrol. 34, 955−976.

Lipin, B.R., Zientek, M.L., 2002. The Stillwater Complex, Montana: The Root of a Flood Basalt Province? (abs). 9th International Platinum Symposium. Billings, Montana, p. 265.

Lorand, J.-P., Gros, M., Pattou, L., 1999. Fractionation of platinum-group element in the upper mantle: a detailed study in Pyrenean orogenic peridotites. J. Petrol. 40, 951−987.

Lorand, J.-P., Luguet, A., Alard, O., 2013. Platinum-group element systematics and petrogenetic processing of the continental upper mantle: A review. Lithos 164-167, 2−21.

Maier, W.D., Teigler, B., 1995. A facies model for the Western Bushveld Complex. Econ. Geol. 90, 2343−2349.

Maier, W.D., Barnes, S.-J., Gartz, V., Andrews, G., 2003. Pt-Pd reefs in magnetitites of the Stella layered intrusion, South Africa: A world of new exploration opportunities for platinum group elements, Geology, 31. pp. 885−888.

Maier, W.D., 2005. Platinum-group element (PGE) deposits and occurrences: mineralization styles, genetic concepts, and exploration criteria. J. Afr. Earth Sci. 41, 165−191.

Maier, W.D., Barnes, S.-J., 2008. Platinum-group elements in the UG1 and UG2 chromitites, and the Bastard Reef, at Impala platinum mine, Western Bushveld Complex, South Africa: evidence for late magmatic cumulate instability and reef constitution. SA J. Geol. 111, 159−176.

Maier, W.D., Groves, D.I., 2011. Temporal and spatial controls on the formation of magmatic PGE and Ni-Cu deposits. Mineral. Deposita 46, 841−857.

Maier, W.D., Barnes, S.-J., Groves, D.I., 2013. The Bushveld Complex, South Africa: formation of platinum-palladium, chrome- and vanadium-rich layers via hydrodynamic sorting of a mobilized cumulate slurry in a large, relatively slowly cooling, subsiding magma chamber. Mineral. Deposita 48, 1−56.

Matveev, S., Ballhaus, C., 2002. Role of water in the origin of podiform chromite deposits. Earth Planet. Sci. Lett. 203, 235−243.

McCallum, M.E., Loucks, R.R., Carlson, R.R., Cooley, E.F., Doerge, T.A., 1976. Platinum metals associated with hydrothermal copper ores of the New Rambler mine, Medicine Bow Mountains, Wyoming. Econ. Geol. 71, 1429−1450.

McDonald, I., Tredoux, M., 2005. The history of the Waterbug deposit; Why South Africa's first platinum mine failed. Appl. Earth Sci. 264−272.

McDonald, I., Holwell, D., 2011. Geology of the northern Bushveld complex and the setting and genesis of the Platreef Ni-Cu-PGE deposit. In: Li, C., Ripley, E.M. (Eds.), Society of Economic Geologists Reviews in Economic Geology, 17. pp. 297−327.

Melcher, F., Grum, W., Simon, G., Thalhammer, T.V., Stumpfl, E.F., 1997. Petrogenesis of the ophiolitic giant chromite deposits of Kempirsai, Kazakhstan: a study of solid and fluid inclusions in chromite, J. Petrol, 38. pp. 1419−1458.

Miller, J.D., 1999. Geochemical evaluation of platinum group element (PGE) mineralization in the Sonju Lake intrusion, Finland, Minnesota. Minnes. Geol. Surv. Inf. Circ. 44, 31.

Miller, J.D., Green, J.C., Severson, M.J., Chandler, V.W., Hauck, S.A., Peterson, D.M., et al., 2002. Geology and mineral potential of the Duluth Complex and related rocks of northeastern Minnesota. Minnes. Geol. Surv. Rep. 58, 207.

Mondal, S.K., Ripley, E.M., Li, C., Frei, R., 2006. The genesis of Archean chromitites from the Nuasahi and Sukinda massifs in the Singhbhum craton, India. Precambrian. Res. 148, 45−66.

Mondal, S.K., Mathez, E.A., 2007. Origin of the UG2 chromitite layer, Bushveld Complex. J. Petrol. 48, 495−510.

Mondal, S.K., 2009. Chromite and PGE deposits of Mesoarchean ultramafic-mafic suites within the greenstone belts of the Singhbhum craton, India: implications for mantle heterogeneity and tectonic setting, J. Geol. Soc. India, 73. pp. 36−51.

Mungall, J.E., Brenan, J.M., 2014. Partitioning of platinum-group elements and Au between sulfide liquid and basalt and the origins of mantle-crust fractionation of the chalcophile elements. Geochim. Cosmochim. Acta 125, 265−289.

Naldrett, A.J., Wilson, A.H., 1990. Horizontal and vertical variation in the noble metal distribution in the Great Dyke of Zimbabwe: A model for the origin of the PGE mineralization by fractional segregation of sulfide. Chem. Geol. 88, 279−300.

Naldrett, A.J., Fedorenko, V.A., Lightfoot, P.C., Kunilov, V.E., Gorbachev, N.S., Doherty, W., et al., 1995. Ni-Cu-PGE deposits of the Noril'sk region Siberia: Their formation in conduits for flood basalt volcanism. Trans. Inst. Mining Metal. 104, B18−B36.

Naldrett, A.J., 2004. Magmatic Sulfide Deposits-Geology, Geochemistry and Exploration. Springer, Berlin, New York, Heidelberg, pp 729.

Naldrett, A.J., Kinnaird, J.A., Wilson, A., Yudovskaya, M., McQuade, S., Chunnett, G., et al., 2009. Chromite composition and PGE content of Bushveld chromitites: Part 1—the Lower and Middle Groups. Appl. Earth Sci. 118, 131−161.

Naldrett, A.J., 2011. Fundamentals of magmatic sulfide deposits. In: Li, C., Ripley, E.M. (Eds.), Society of Economic Geologists Reviews in Economic Geology, 17. pp. 1−50.

Naldrett, A.J., Wilson, A., Kinnaird, J., Yudovskaya, M., Chunnett, G., 2012. The origin of chromitites and related PGE mineralization in the Bushveld Complex: new mineralogical and petrological constraints. Mineral. Deposita 47, 209−232.

Namur, O., Charlier, B., Toplis, M.J., Higgins, M.D., Liegeois, J.-P., Vander Auwera, J., 2010. Crystallization sequence and magma chamber processes in the ferrobasaltic Sept Iles layered intrusion, Canada. J. Petrol. 51, 1203−1216.

Namur, O., Charlier, B., Holness, M.B., 2012. Dual origin of Fe-Ti-P gabbros by immiscibility and fractional crystallization of evolved tholeiitic basalts in the Sept Iles layered intrusion. Lithos 154, 100−114.

Nyman, M.W., Sheets, R.W., Bodnar, R.J., 1990. Fluid inclusion evidence for the physical and chemical conditions associated with intermediate-temperature PGE mineralization at the New Rambler deposit, southeastern Wyoming. Can. Mineral. 28, 629−638.

Oberthür, T., 2011. Platinum-group element mineralization of the main sulfide zone, Great Dyke, Zimbabwe. In: Li, C., Ripley, E.M. (Eds.), Society of Economic Geologists Reviews in Economic Geology, 17. pp. 329−349.

O'Driscoll, B., Emeleus, C.H., Donaldson, C.H., Daly, J.S., 2010. Cr-spinel seam petrogenesis in the Rhum layered suite, NW Scotland: cumulate assimilation and in situ crystallization in a deforming crystal mush. J. Petrol. 51, 1171−1201.

Page, N.J., 1972. Stillwater Complex, Montana- Structure, mineralogy, and petrology of the Basal Zone with emphasis on the occurrence of sulfides. US Geol. Surv. Prof. 1038, pp 61.

Palme, H., O'Neill, H.St.C., 2014. Cosmochemical estimates of mantle composition. In: Holland, H., Turekian, K. (Eds.), Treatise on Geochemistry, second ed. pp. 1−39.

Pang, K.N., Li, C., Zhou, M.-F., Ripley, E.M., 2008b. Abundant Fe-Ti oxide inclusions in olivine from the Panzhihua and Hongge layered intrusions, SW China: evidence for early saturation of Fe-Ti oxides in ferrobasaltic magma. Contrib. Mineral. Petrol. 156, 307−321.

Pang, K.N., Zhou, M.-F., Lindsley, D., Zhao, D.G., Malpas, J., 2008a. Origin of Fe-Ti oxides ores in mafic intrusions: evidence from the Panzhihua intrusion, SW China. J. Petrol. 49, 295−313.

Pang, K.-N., Zhou, M.-F., Qi, L., Shellnutt, G., Wang, C.Y., Zhao, D., 2010. Flood basalt-related Fe−Ti oxide deposits in the Emeishan large igneous province, SW China. Lithos 119, 123−136.

Parman, S.W., Grove, T.L., Dann, J.C., de Wit, M.J., 2004. A subduction origin for komatiites and cratonic lithospheric mantle. SA J. Geol. 107, 107−118.

Park, Y.-R., Ripley, E.M., Miller, J.D., Li, C., Mariga, J., Shafer, P., 2004. Stable isotopic constraints on fluid-rock interaction and Cu-PGE-S redistribution in the Sonju Lake Intrusion, Minnesota. Econ. Geol. 99, 325−338.

Patten, C., Barnes, S.-J., Mathez, E.A., Jenner, F.E., 2013. Partition coefficients of chalcophile elements between sulfide and silicate melts and the early crystallization history of sulfide liquid: LA-ICP-MS analysis of MORB sulfide droplets. Chem. Geol. 358, 170−188.

Peach, C.L., Mathez, E.A., Keays, R.R., 1990. Sulphide melt-silicate melt distribution coefficients for the noble metals and other chalcophile metals as deduced from MORB: implications for partial melting. Geochim. Cosmochim. Acta 54, 3379−3389.

Philpotts, A.R., 1967. Origin of certain iron-titanium oxide and apatite rocks. Econ. Geol. 62, 303−315.

Poustovetov, A.A., Roeder, P.L., 2000. The distribution of Cr between basaltic melt and chromian spinel as an oxygen geobarometer. Can. Mineral. 39, 309−317.

Prendergast, M., Bennett, M., Henicke, G., 1998. Platinum exploration in the Rincon del Tigre Complex, eastern Bolivia. Trans. Inst. Mining Metal. 107, B39−B47.

Pushkarev, E., 2014, The Ural platinum Belt: The Kachkanar titanomagnetite deposit in clinopyroxenite. Field Trip 6, 12th International Pt Symposium, Yekaterinburg, Russia.

Ripley, E.M., Severson, M.J., Hauck, S.A., 1998. Evidence for sulfide and Fe-Ti-P -rich liquid immiscibility in the Duluth Complex, Minnesota. Econ. Geol. 93, 1052−1062.

Ripley, E.M., Brophy, J.G., Li, C., 2002. Copper solubility in a basaltic melt and sulfide liquid/silicate melt partition coefficients of Cu and Fe. Geochim. Cosmochim. Acta 66, 2791−2800.

Ripley, E.M., Li, C., 2011. A review of conduit-related Ni-Cu-(PGE) sulfide mineralization at the Voiseys Bay deposit, Labrador, and the Eagle deposit, northern Michigan. In: Li, C., Ripley, E.M. (Eds.), Society of Economic Geologists Reviews in Economic Geology, 17. pp. 181−198.

Ripley, E.M., Li, C., 2013. Sulfide saturation in mafic magma: is external sulfur required for magmatic Ni-Cu-(PGE) ore genesis? Econ. Geol. 108, 45−58.

Ripley, E.M., 2014. Ni-Cu-PGE Mineralization in the Partridge River, South Kawishiwi, and Eagle Intrusions: A Review of Contrasting Styles of Sulfide-Rich Occurrences in the Midcontinent Rift System. Econ. Geol. 109, 309−324.

Rivers, T., 1997. Lithotectonic elements of the Grenville province: Review and tectonic implications. Precambrian. Res. 86, 117−154.

Scoates, R.F.J., 1983. A preliminary stratigraphic examination of the ultramafic zone of the Bird River sill. Manitoba Dep. Energy Mines Rep. Field Activit. 1983, 70−83.

Scoon, R.N., Teigler, B., 1994. Platinum-group element mineralization in the Critical Zone of the Western Bushveld Complex: I. Sulfide poor chromitites below the UG-2. Econ. Geol. 89, 1094−1121.

Schulte, R.F., Taylor, R.D., Piatak, N.M., Seal II, R.R., 2012. Stratiform chromite deposit model. U.S. Geological Survey Scientific Investigations Report 2010−5070−E 131.

Schulz, K.J., Chandler, V.W., Nicholson, S.W., Piatak, N., Seall II, R.R., Woodruff, L.G., et al., 2010. Magmatic sulfide nickel-copper deposits related to picrite and (or) tholeiitic basalt dike-sill complexes − A preliminary deposit model. U.S. Geological Survey Open-File Report 2010-1179, pp 25.

Severson, M.J., Hauck, S.A., 1990. Geology, geochemistry, and stratigraphy of a portion of the Partridge River Intrusion: Duluth. Natural Resources Research Institute Technical Report NRRI/GMIN-TR-89-11, 235pp.

Sobolev, A.V., Asafov, E.V., Gurenko, A.A., Arndt, N.T., Batanova, V.G., 2016. Komatiites reveal an Archean hydrous deep mantle reservoir. Nature 531, 628−632.

Stein, S., Stein, C., Kley, J., Keller, R., Merino, M., Wolin, E., et al., 2016. New insights into North America's Midcontinent Rift. EOS. 97, 10−16.

Stowe, C.W., 1994. Compositions and tectonic settings of chromite deposits through time. Econ. Geol. 89, 528−546.

Tang, Q., Zhang, Z., Li, C., Wang, Y., Ripley, E.M., 2016. Neoproterozoic subduction-related basaltic magmatism in the northern margin of the Tarim Craton: Implications for Rodinia reconstruction. Precambrian. Res. 286, 370−378.

Taranovic, V., Ripley, E.M., Li, C., Rossell, D., 2015. Petrogenesis of the Ni-Cu-PGE sulfide-bearing Tamarack Intrusive Complex, Midcontinent Rift System, Minnesota. Lithos 212−215, 16−31.

Thayer, T.P., 1964. Principal features and origin of podiform chromite deposits and some observations on the Guleman-Soridag district, Turkey. Econ. Geol. 59, 1497−1524.

Toplis, M.J., Carroll, M.R., 1995. An experimental study of the influence of oxygen fugacity on Fe-Ti- oxide stability, phase relations, and mineral-melt equilibria in ferro-basaltic systems. J. Petrol. 36, 1137−1170.

Toplis, M., Corgne, A., 2002. An experimental study of element partitioning between magnetite, clinopyroxene and iron-bearing silicate liquids with particular emphasis on vanadium. Contrib. Mineral. Petrol. 144, 22−37.

Ulmer, G.C., 1969. Experimental investigations of chromite spinels. Econ. Geol., Monograph 4, 114−181.

Usumezsoy, S., 1990. On the formation of mode of the Guleman chromite deposits, Turkey. Mineral. Deposita 25, 89−95.

Van Tongeren, J.A., Mathez, E.A., Kelemen, P.B., 2010. A Felsic End to Bushveld Differentiation. J. Petrol. 51, 1891−1912.

von Gruenewaldt, G., 1973. The main and upper zones of the Bushveld Complex in the Roossenekal area, eastern Transvaal. Trans. Geol. Soc. SA 76, 207−227.

von Gruenewaldt, G., 1979. A review of some recent concepts of the Bushveld Complex, with particular reference to sulfide mineralization. Can. Mineral. 17, 233−256.

Voordouw, R., Gutzmer, J., Beukes, N.J., 2009. Intrusive origin for upper group (UG1, UG2) stratiform chromitite seams in the Dwars River area, Bushveld Complex, South Africa. Mineral. Petrol. 97, 75−94.

Wang, M., Veksler, I., Zhang, Z., Hou, T., Keiding, J.T., 2017. The origin of nelsonite constrained by melting experiment and melt inclusions in apatite: the Damiao anorthosite complex, North China Craton. Gondwanna Res. 42, 163−176.

Wei, B., Wang, C.Y., Li, C., Sun, Y., 2013a. Origin of PGE-depleted Ni-Cu sulfide mineralization in the Triassic Hongqiling No.7 orthopyroxenite intrusion, Central Asian Orogenic Belt, NE China. Econ. Geol. 108, 1813−1831.

Wei, C.T., Zhou, M.-F., Zhao, T.P., 2013b. Differentiation of nelsonitic magmas in the formation of the ∼1.74 Ga Damiao Fe-Ti-P ore deposit, North China. Contrib. Mineral. Petrol. 165, 1341−1362.

Wilson, A.H., Tredoux, M., 1990. Lateral and vertical distribution of the platinum-group elements and petrogenetic controls on the sulfide mineralization in the PI pyroxenite layer of the Darwendale Subchamber of the Great Dyke, Zimbabwe. Econ. Geol. 85, 556−584.

Wilson, J.A., Wilson, J.F., 1981. The Great 'Dyke'. In: Hunter, D.R. (Ed.), Precambrian of the Southern Hemisphere. Elsevier, Amsterdam, pp. 572−578.

Woodruff, L.G., Nicholson, S.W., Fey, D.L., 2013. A deposit model for magmatic iron-titanium oxide deposits related to Proterozoic massif anorthosite plutonic suites. U.S. Geological Survey Scientific Investigations Report 2013-5091, 47.

Xie, W., Song, X.-Y., Chen, L.-M., Deng, Y.-F., Zheng, W.-Q., Wang, Y.-S., et al., 2014. Geochemistry insights on the genesis of the subduction-related Heishan magmatic Ni-Cu-(PGE) deposit, Gansu, northwestern China, at the southern margin of the Central Asian Orogenic Belt. Econ. Geol. 109, 1563−1583.

Xue, S., Qin, K., Li, C., Tang, D., Mao, Y., Ripley, E.M., 2016. Geochronological, Petrological and Geochemical Constraints on Ni-Cu Sulfide Mineralization in the Poyi Ultramafic-troctolitic Intrusion in the NE Rim of Tarim Craton, Western China. Econ. Geol. 111, 1465−1484.

Zhao, T.P., Chen, W., Zhou, M.F., 2009. Geochemical and Nd-Hf isotopic constraints on the origin of the ∼1.74−Ga Damiao anorthosite complex. North China Craton. Lithos 113, 673−690.

Zhong, H., Zhou, X.H., Zhou, M.-F., Sun, M., Liu, B.-G., 2002. Platinum-group element geochemistry of the Hongge Fe-V-Ti deposit in the Pan-Xi area, southwestern China. Mineral. Deposita 37, 226−239.

Zhou, M.-F., Robinson, P.T., Lesher, C.M., Keays, R.R., Zhang, C.J., Malpas, J., 2005. Geochemistry, petrogenesis and metallogenesis of the Panzhihua Gabbroic Layered Intrusion and associated Fe-Ti-V oxide deposits, Sichuan Province, SW China. J. Petrol. 46, 2253−2280.

Zhou, M.-F., Robinson, P.T., Malpas, J., Li, Z., 1996. Podiform chromitites in the Luobusa Ophiolite (Southern Tibet): Implications for melt-rock interaction and chromite segregation in the upper mantle. J. Petrol. 37, 3−21.

Zhou, M.-F., Wei, T.C., Wang, C.Y., Prevec, S.P., Liu, P.P., Howarth, J.W., 2013. Two stages of immiscible liquid separation in the formation of Panzhihua-type Fe-Ti-V oxide deposits, SW China. Geosci. Front. 4, 481−502.

Zientek, M.L., 2012. Magmatic ore deposits in layered intrusions − Descriptive model for reef-type PGE and contact-type Cu-Ni_PGE deposits. U.S. Geol. Surv. Open File Rep. 2012-1010, 48.

Zientek, M.L., Cooper, R.W., Corson, S.R., Geraghty, E.P., 2002. Platinum-group element mineralization in the Stillwater Complex, Montana. In: Cabri, L. (Ed.), The Geology, Geochemistry, Mineralogy and Mineral Beneficiation of Platinum-Group Elements, Canadian Institute of Mining, Metallurgy and Petroleum Special, 54. pp. 459−481.

MIXING AND UNMIXING IN THE BUSHVELD COMPLEX MAGMA CHAMBER

4

Jill A. VanTongeren

Rutgers University, New Brunswick, NJ, United States

CHAPTER OUTLINE

4.1 INTRODUCTION

The Rustenberg Layered Suite (RLS) of the Bushveld Complex (hereafter referred to as simply "the Bushveld Complex") is a >8 km thick sequence of mafic-ultramafic rocks emplaced at approximately 2.06 Ga (Scoates and Friedman, 2008; Zeh et al., 2015). The Bushveld Complex is exposed or inferred in 4 limbs: the Northern, Eastern, Western, and Southern (or Bethal) limbs (Fig. 4.1). The Eastern and Western limbs are best known due to greater outcrop exposure and drilling activity. The stratigraphy of the Eastern and Western limbs is generally similar (Eales and Cawthorn, 1996a,b) in terms of thicknesses and occurrences of layering. For this reason, initial researchers of the Bushveld Complex considered the intrusion to be laterally continuous at depth. Connectivity between the Eastern and Western limbs has also been inferred from geophysical data (Cawthorn and Webb, 2001; Webb et al., 2004). The Northern limb has a slightly different stratigraphy from the Eastern and Western limbs, and the Southern limb (or Bethal lobe) is not

Processes and Ore Deposits of Ultramafic-Mafic Magmas through Space and Time. DOI: http://dx.doi.org/10.1016/B978-0-12-811159-8.00005-6

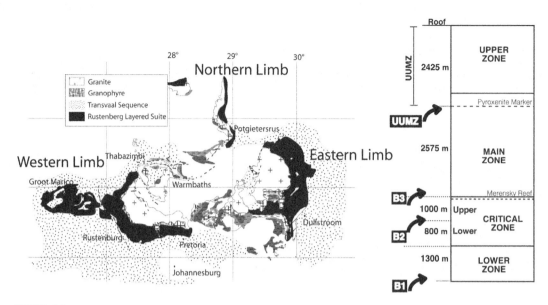

FIGURE 4.1

Map of the Bushveld Complex outcrop and stratigraphy, adapted from VanTongeren, J.A., & Mathez, E.A. (2013). Incoming Magma Composition and Style of Recharge below the Pyroxenite Marker, Eastern Bushveld Complex, South Africa. J. Petrol. 54, 1585−1605. B1, B2, B3, and UUMZ refer to new magma compositions thought to be parental to each portion of the RLS stratigraphy (see text for details). Thicknesses of each unit cited for the Eastern Limb according to Molyneux (2008).

well-characterized due to its limited exposure. Assuming connectivity at depth between all limbs, the Bushveld Complex stretches over 150 km from N to S, and over 450 km from E to W.

The enormous size of the Bushveld Complex, nearly 1 million km^3 of magma (Cawthorn and Walraven, 1998), its highly layered stratigraphy and lateral continuity of layering make it ideal for economic exploitation. The Bushveld Complex contains over 80% of the world's proven reserves of platinum group elements (PGEs). PGEs in the Bushveld Complex are concentrated into two primary layers, the "Merensky Reef" named for Hans Merensky, who discovered the deposit in 1924, and the "UG2 Chromitite" layer, named because it is the second layer of the "upper group" chromitites; several other chromitite layers also contain significant, but noneconomic concentrations of PGE. Several theories about the formation these layers and the concentration of PGE within them have been proposed; many of them involve the input of new magma and progressive magma mixing within the chamber.

This chapter will review the evidence for new magma inputs and mixing within the magma chamber stratigraphy, as well as the present state of understanding regarding the source magmas to the Bushveld Complex. Additionally, the formation of valuable PGE sulfides within the lower portions of the magma chamber, and Fe-Ti ores in the upper portions, is probably linked to unmixing of the magmas in the form of sulfide immiscibility (PGEs) and silicate liquid immiscibility (Fe-Ti). I will review the evidence for immiscibility and the current debate over the scale of immiscibility at the top of the magma chamber.

4.2 GEOLOGICAL SETTING

The magmas contributing to the Bushveld Complex magma chamber intruded into the Transvaal basin at approximately 2054 Ma (Scoates and Friedman, 2008; Zeh et al., 2015). Prior to the emplacement of the Bushveld Complex, the Transvaal basin underwent postrifting thermal subsidence, resulting in the deposition of a marine sequence of mudstones and mature sandstones (Eriksson and Reczko, 1995a,b); minor andesitic volcanism punctuated the dominantly sedimentary sequences. The floor of the Bushveld Complex in the eastern Bushveld, where the contact is well-exposed, consists of metasedimentary units in the north, such as the Magaliesburg Quartzite, and a metavolcanic layer further to the south of the Dullstroom formation. The roof of the intrusion in the eastern Bushveld is laterally variable, with metasediments in the north, a mixture of metasediments and metavolcanic rocks in the central segment, and metavolcanic and granophyric rocks in the southern segment (VanTongeren and Mathez, 2015). The progression in roof lithology from the north to the south in the eastern Bushveld led VanTongeren and Mathez (2015) to conclude that the magmas of the Bushveld Complex intruded at progressively shallower levels as the magma chamber grew (Fig. 4.2).

In many regions the Rooiberg Group felsic volcanics immediately overlie the layered rocks of the Bushveld Complex. The Rooiberg lavas are coeval with the Bushveld Complex (Walraven, 1997) and possess nearly identical Sr, Nd, and O isotopic compositions (Buchanan et al., 2002, 2004; Fourie and Harris, 2011). Hatton and Schweitzer (1995) proposed that the Rooiberg and Bushveld magmas were emplaced at the same time, with the Rooiberg being the result of crustal partial melting in contact with the rising mafic magmas of the Bushveld. Buchanan et al. (2002)

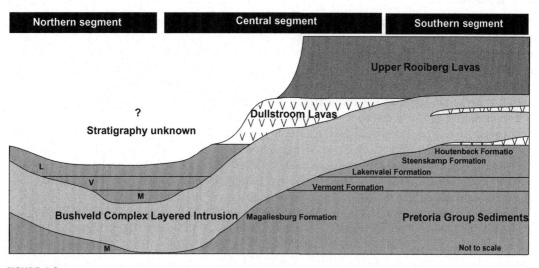

FIGURE 4.2

Emplacement of the Bushveld Complex magmas into the crust in the Eastern Limb.

From VanTongeren, J.A., & Mathez, E.A. (2015). On the relationship between the Bushveld Complex and its felsic roof rocks, part 2: the immediate roof. Contrib. Mineral. Petrol. 170, 56, reproduced with permissions.

followed up on the study of Hatton and Schweitzer (1995) and modeled the evolution of the Rooiberg lavas from a pre-Bushveld staging chamber. Buchanan et al. suggested that the evolution in lava compositions was due to progressive crystallization in a lower crustal magma chamber, which may have also housed the mafic Bushveld magmas. However, VanTongeren et al. (2010) used the nearly identical timing and isotopic composition of the upper Rooiberg Group to argue that the lavas, which are highly enriched in incompatible trace elements, represent the missing melt after fractionation of the Bushveld magma chamber itself. In their view, the Bushveld Complex represents a subvolcanic magma chamber that differentiated to produce the Rooiberg lavas. This suggestion is supported by the overlap in Hf isotopic compositions in zircon crystals from the Bushveld Complex and the Rooiberg Group lavas (VanTongeren et al., 2016).

In addition to the voluminous magmatism represented by the Bushveld Complex layered mafic intrusion (LMI) and associated volcanics, there are several additional mafic-ultramafic bodies emplaced in the Kaapvaal Craton around 2.06 Ga. Over 10 different intrusive bodies make up the Bushveld Complex Magmatic Province (Rajesh et al., 2013), which includes some well-known mafic-ultramafic intrusions such as the Malopo Farms Complex in the NW and Uitkomst Complex to the SE, as well as some lesser-known and compositionally unique intrusions like the alkaline-carbonatite Schiel and Phalaborwa Complexes.

Regardless of how all of these bodies are related, it is clear that a significant amount of volcanic and magmatic activity of various types was taking place in the region during this time period.

4.3 SOURCE

There are two significant, and as yet unresolved, aspects of the mineral deposits found within the Bushveld Complex: (1) the vast size and metal concentration of the deposits; and (2) the uniquely high Pt/Pd ratio. Both aspects are probably related to the source magmas that contributed to the intrusion.

The question of why the Bushveld is so massive has perplexed researchers since its original discovery. It is not the intent of this manuscript to determine why the Bushveld is so large; however, its enormous size is likely due to the tectonic setting in which it was emplaced and the nature of the magmas that were generated. New research suggests that the Bushveld was emplaced and cooled within 1 million years (Zeh et al., 2015), and previous theoretical calculations estimated that the entire magma chamber was emplaced within 75,000 years (Cawthorn and Walraven, 1998). What type of magmatic setting allows for nearly 1 million km^3 of magma to be generated and emplaced in the shallow crust within such a geologically short time period?

Everything from a plume head (e.g., Hatton, 1995; Hatton and Schweitzer, 1995), to melting of subducted Archean oceanic lithosphere (e.g., Clarke et al, 2009; Cawthorn, 2011; Wilson, 2012), to a meteorite impact (e.g., Rhodes, 1975; Elston and Twist, 1989; Elston, 1992) has been postulated to trigger the magmatism associated with the Bushveld Complex.

Proponents of a meteorite-impact origin for the Bushveld Complex note its overall circular shape and the presence of glassy spherulitic textures in some of the Rooiberg Group lavas, suggesting they are superheated, possibly due to formation by impact melting (Hamilton, 1970; Rhodes, 1975; Elston and Twist, 1989; Elston, 1992). However, Buchanan and Reimold (1998) noted the presence of several undeformed sedimentary layers between layers of the Rooiberg lavas, which precludes their formation as impact melt sheets. No shocked quartz or other unequivocal evidence of impact has been recovered from the Bushveld region.

The overall consensus among many researchers today is that the Bushveld Complex is a result of a large asthenospheric plume intersecting the base of the lithosphere, where it interacted significantly with a previously metasomatized subcontinental lithospheric mantle (SCLM, see below) or assimilated significant amounts of continental crust.

Many of the largest LMIs have strikingly similar parent-magma composition (Table 4.1) to that of the Bushveld Complex. This is particularly surprising because the composition is so unique among the range of igneous compositions produced today. The parent-magma compositions deduced for the Bushveld Complex (2.05 Ga), the Stillwater Complex (2.7 Ga), and the Great Dyke (2.57 Ga), are all characterized by high SiO_2 (54−56 wt%), high MgO (13−15 wt%), and low CaO (6−9 wt%). Several other late Archean-early Proterozoic bodies also share this trait (Sun et al., 1989). These so-called siliceous high-magnesium basalts are found globally and include the Munni Munni and the Jimberlana intrusions of Western Australia, the Fennoscandian intrusions of northern Europe and parts of Russia, the Vestfold Hills of Antarctica, and the Dongargarh Supergroup of India. Today, such a composition is typically reserved for boninites resulting from shallow hydrous melting of previously-depleted mantle in a subduction zone setting. However, the three LMIs listed above formed in a stable craton setting and are thought to have been dry melts, as evidenced by the distinct lack of hydrous cumulate phases and paucity of intercumulus hydrous phases.

Many researchers have concluded that a komatiite with 30%−50% crustal contamination is the most viable option. However, there are several issues with this hypothesis. (1) The compositions of both the crust and the primary komatiite melt are unlikely to have been similar in each of the different cratons at different times. (2) The isenthalpic assimilation of such large volumes of continental crust should have a dramatic effect on the composition, stability, and density of the residual

Table 4.1 Comparison of Estimated LMI Parent Magmas

	Bushveld Estimates			Stillwater Estimates		Great Dyke
	Davies et al. (1980)	Harmer and Sharpe (1985)	Sharpe and Hulbert (1985)	Longhi et al. (1983)	McCallum (1988)	Wilson (1982)
SiO_2	55.7	55.28	56.07	52.27	54.1	53.09
TiO_2	0.36	0.4	0.34	0.8	0.6	0.55
Al_2O_3	12.74	10.42	11.47	11.27	12.7	11.06
FeOt	8.79	9.97	9.53	10.4	9.5	9.49
MnO	0.09	0.2	0.18	0.21	−	0.14
MgO	12.44	15.25	12.96	13.08	14.5	15.51
CaO	6.96	6.25	6.68	9.33	7.6	7.68
Na_2O	2.02	1.7	1.68	1.96	0.6	1.79
K_2O	1.03	1.01	0.8	0.67	0.2	0.69
P_2O_5	−	0.08	0.07	−	−	−
Mg#	71.61	73.17	70.80	69.15	73.12	74.45

crust. (3) The isotopic composition of the Great Dyke, in particular, clearly rules out the influence of *any* amount of continental crust—let alone 30%−50%.

Longhi et al. (1983) conducted some of the first experiments on the liquids thought to be parental to the Stillwater Complex of Montana. By adding incremental amounts of a granodioritic composition (assumed to be representative of a partial melt of continental crust) to a primitive komatiite composition, they showed that 30%−50% crustal assimilation would produce a liquid in which orthopyroxene—the dominant liquidus phase in the Stillwater, Bushveld, and Great Dyke—was the first phase to crystallize upon cooling. A komatiite is required for its high temperature, which presumably would allow it to assimilate more crust, and high MgO content.

New evidence in support of an initial komatiite parental magma for the Bushveld Complex comes from a new borehole through the base of the Lower Zone in the Eastern Bushveld, which showed a previously unknown extension of the intrusion, now named the Basal Ultramafic Zone. Wilson (2012) showed that the most likely parental magma to this section is komatiitic ($SiO_2 = \sim 53$ wt%; MgO = ~ 19 wt%).

However, there are several caveats that arise with the komatiite + crustal assimilation hypothesis for the Bushveld and other layered intrusions. The first is the differing crustal compositions available for assimilation. It is unlikely that the crust being assimilated from the Wyoming craton at 2.7 Ga, the Zimbabwe craton at 2.57 Ga and the Kaapvaal craton at 2.0 Ga all had the same composition. Additionally, Herzberg et al. (2010) showed that the secular cooling of the Archean and early Proterozoic mantle has had a large impact on the composition of magmas generated by partial melting through time (particularly MgO). Thus, the compositions of komatiites at 2.7, 2.57, and 2.0 Ga probably would have varied in a significant way.

The second issue with the komatiite + crust hypothesis involves the isenthalpic assimilation of large volumes of crust. Longhi et al. (1983) modeled the isenthalpic assimilation using a ratio of approximately 1 mol of crystallization for every 1 mol of assimilation. Such extensive crystallization would result in the formation of a very large amount of peridotite in the lower crust (the 50% crystallization required in their model for the Bushveld is equivalent to an olivine layer approximately 350 km wide and *at least* 0.5 km thick—assuming a Bushveld Lower Zone 1 km thick). Additionally, Longhi et al. (1983) modeled crustal assimilation by adding incremental amounts of a single granodiorite composition to the komatiite magma. While granodiorite is a reasonable approximation of a lower crust melt, partial melting progresses in incremental degrees. Small degrees of partial melting will produce a distinctly different composition than large amounts. If the change in assimilant composition with melting is not taken into account, the calculated derivative liquids and the quantities required may be misleading.

The third and perhaps most fundamental problem with the komatiite + crust hypothesis is the isotopic composition of the parent magmas. With the exception of the Bushveld Complex, none of these intrusions has an isotopic signature indicative of any significant crustal contribution (Fig. 4.3). The isotopic compositions of the Great Dyke, in particular, overlap almost entirely the field of mantle composition at 2.57 Ga, yet the parental magma Great Dyke had the same high-Si, high-Mg composition as the Bushveld and the Stillwater. Finally, it was recently shown that the isotopic character of the Bushveld may be derived from the SCLM and *not* from the continental crust (Richardson and Shirey, 2008; Zirakparvar et al., 2014; Laurent and Zeh, 2015) (Fig. 4.3).

High-temperature oxygen-isotope values are good recorders of crustal assimilation in mafic magmas. The $\delta^{18}O$ composition of the mantle is approximately 5.7−5.8‰ (Gregory and Taylor,

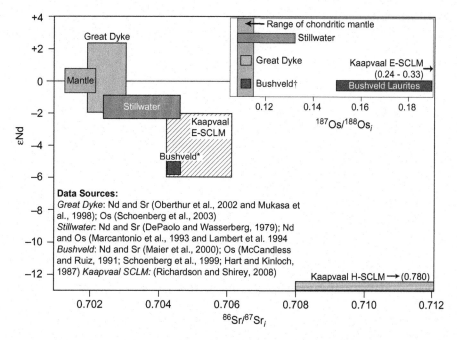

FIGURE 4.3

Isotopic compositions of the Bushveld, Stillwater, and Great Dyke compared with that of the chondritic mantle. Mantle range is from 2.7 to 2.0 Ga. Bushveld* Nd and Sr are most primitive values from Lower Zone samples. Bushveld† Os is from erlichmanite (OsS₂) from the UG1 chromitite layer. The deviation from chondritic mantle for the Bushveld may be the result of mixing with an eclogitic SCLM (E-SCLMS; Richardson and Shirey, 2008). H-SCLM refers to the harzburgitic SCLM. Isotopic compositions shown here for each of the LMIs are only for the most primitive cumulates known to be from a U-type (ultramafic) magma with high Mg and high Si. Other pulses of magma into the Bushveld, Stillwater, and Great Dyke display a larger range of isotopic compositions, but are not considered here because their estimated parent-magma compositions have significantly lower SiO_2 and MgO than those studied here.

1981), but surface alteration and crustal assimilation drive these values significantly higher. The Stillwater Complex has a uniform $\delta^{18}O$ very near that of the mantle, at 5.9‰ (Dunn, 1986). Samples from the Bushveld Complex Lower Zone have highly variable $\delta^{18}O$ from 5.7‰ to 7.6‰ (Harris et al., 2005; Schiffries and Rye, 1989, 1990), with an average of 6.9‰ that is constant throughout the Bushveld. The Great Dyke lacks significant oxygen-isotope data from the lowermost portions of the magma chambers; however, data from the leucocratic section of the intrusion (the Mafic Sequence) have $\delta^{18}O = 6.65‰$ (Chaumba and Wilson, 1997). Schiffries and Rye (1989, 1990) attribute the higher values for the Bushveld to in situ contamination with sedimentary floor rocks, and Chaumba and Wilson (1997) showed that the Great Dyke $\delta^{18}O$ increases with stratigraphic height and could be generated from interaction with crustal xenoliths found within the magma chamber. Another possible explanation, particularly given the uniform values of the Bushveld, may involve some component of recycled crust in the mantle source, as has been proposed for some OIBS (Woodhead et al., 1993).

Many researchers have struggled to reconcile the isotopic data with the major-element compositions. Some have suggested alternative hypotheses such as melting of a heterogeneous mantle source with some component of recycled crust (DePaolo and Wasserburg, 1979; Hart and Kinloch, 1989); others have suggested some interaction and equilibration with the SCLM (Richardson and Shirey, 2008). Yet others have suggested that these magmas may originate entirely within a Si-enriched SCLM (Lambert et al., 1994; Wooden and Mueller, 1988).

A role for the SCLM in the production of the Bushveld is also suggested by the Re-Os isotopic composition of sulfide minerals found in rocks from the Critical Zone (Richardson and Shirey, 2008). Richardson and Shirey (2008) showed that the unique Re-Os isotopic signature of the Bushveld sulfides is also observed in sulfide inclusions in coeval diamonds in eclogitic and harzburgitic xenoliths from the nearby Premier and Venetia kimberlites. They concluded that the Re-Os signature could only be explained by mixing between an asthenospheric melt (presumably plume-sourced) and an eclogitic or harzburgitic SCLM prior to emplacement in the Bushveld Complex magma chamber.

Zirakparvar et al. (2014) also calculated that the Hf-isotope composition and major- element composition of the Bushveld parent magmas requires a significant contribution from the SCLM of the Kaapvaal craton. They showed that the Hf-isotope signatures of zircons from throughout the Bushveld Complex stratigraphy are remarkably homogeneous, and importantly, are also distinct from depleted-mantle compositions at 2.05 Ga. They also showed that no amount of mixing between nearby crustal compositions and an asthenospheric melt could produce the observed Hf isotopic composition. They concluded that the Bushveld magmas must have been derived from partial melting of a previously enriched SCLM.

While the SCLM signature in the cumulates suggests that the Busvheld source magmas spend a considerable amount of time reacting with the SCLM prior to ascent through the thick Kaapvaal lithosphere, this is seemingly at odds with the extraordinarily large volumes of magma erupted over a very short time interval. Silver et al. (2006) suggested that the Bushveld magmas may be generated several million years prior to emplacement by asthenospheric melting. In their model, the magmas were unable to break through the lithosphere and resided above the solidus within the SCLM until a lithospheric fracture zone developed. The Silver et al. (2006) model is able to satisfy both the significant SCLM signature of the magmas and the extremely short duration (<1 Myr) of such large volumes (>1 million km^3) of magma.

The Pt/Pd ratio of the Bushveld Complex magmas is significantly higher (>1.5; Barnes et al., 2010) than most other LMIs and other magmatic deposits around the world (Pt/Pd $= <1$). This unique characteristic results in very high concentrations of Pt in the Bushveld relative to all other producers globally. Barnes et al. (2010) calculated that the high Pt/Pd ratio of the Bushveld Complex cannot be due to crustal assimilation in a mantle plume-sourced magma, but rather, the unique compositions must be generated by interaction with the SCLM.

4.4 EMPLACEMENT AND MIXING

Difficulties in determining the source reservoirs that contributed to the Bushveld Complex magmas arise due to the wide range in parent-magma compositions that are thought to have contributed to its buildup. The Bushveld Complex was intruded in four main pulses of magma (Eales, 2002), each loosely corresponding to a new zone within the stratigraphy—the Lower Zone, Critical Zone, Main Zone, and Upper Zone (Fig. 4.1).

The extensive crystallization of orthopyroxene relative to olivine throughout the Lower Zone suggests that the first pulse of magma, the so-called "B1 magma," was a high-silica (~ 56 wt%) and high-magnesium ($\sim > 13$ wt%) basaltic composition similar to that of a boninite (e.g., Barnes, 1989). Feeder dikes (Sharpe, 1981) surrounding the Bushveld region, and marginal rocks from the Lower and lower Critical Zones, have been found with a similar composition (Barnes et al., 2010). However, the B1 magma compositions found in the feeder dikes are cannot crystallize olivine at shallow pressures (Eales, 2002), and do not have the correct Fe-Mg composition to produce the observed high-Mg# orthopyroxene (e.g., Wilson, 2012; Yudovskaya et al., 2013). The chilled margin of the Basal Ultramafic Sequence identified by Wilson (2012) has the correct compositional characteristics (~ 53 wt% SiO_2 and 19 wt% MgO) to produce the Basal Ultramafics, the noritic Marginal Zone, as well as portions of the Lower Zone cumulates (e.g., Wilson, 2015).

The B2 and B3 magmas are identified by chilled margins found adjacent to the upper Critical Zone and Main Zone respectively (Barnes et al., 2010) where they are interpreted to have been emplaced. They are of tholeiitic basaltic composition and are inferred to have mixed with the resident magma at each level to produce the upper Critical Zone and Main Zone. However, some Main Zone cumulates have more radiogenic Sr-isotope compositions than the B3 magma, leading Kruger (2005) to propose that the Main Zone magmas assimilated some degree of granophyre formed by partial melting at the roof of the intrusion.

Gabbros and diorites from the combined Upper Zone have a uniform $^{87}Sr/^{86}Sr_i = 0.706$. The uniformity of isotopic compositions above the Pyroxenite Marker led Kruger (1987) to suggest that the geochemical Upper Zone-Main Zone boundary was at the Pyroxenite Marker in the Upper Main Zone. Cawthorn (1991) showed a marked change in the major-element composition of minerals and in the whole-rock isotopic composition of the rocks above the Pyroxenite Marker. He suggested that this was evidence for a final large pulse of magma into the system to generate the Upper Zone. VanTongeren and Mathez (2013) further supported this hypothesis and calculated the composition and proportion of new and resident magma at this level.

It is clear that layered intrusions are large crustal magma chambers, the majority of which have received multiple repeated injections of magma during their buildup. In layered intrusions, the emplacement of new magma, and subsequent mixing with resident magma, is commonly used to explain the formation of everything from chromite or magnetite layers to sulfide saturation. Given the importance of magma mixing and emplacement in layered intrusions, it is important to outline the evidence for, and caveats associated with, identifying a new input of magma into the system.

4.4.1 IDENTIFYING MAGMA INPUTS: ISOTOPES

By far the most widely used tool for identifying a new pulse of magma into the magma chamber is the isotopic composition of the cumulates. A change in whole-rock isotopic composition is a clear indicator of a new component in the magma chamber. This component may be a new pulse of magma with a different isotopic signature, or assimilated wall rock.

For example, Kruger (2005) compiled the Sr-isotope composition of whole rocks through the entire Bushveld Complex stratigraphy (Fig. 4.4). He noted that the cumulates in the Lower Zone,

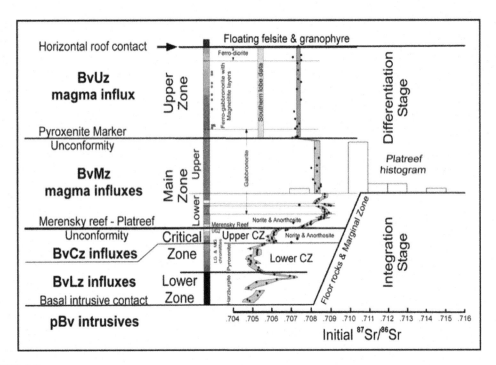

FIGURE 4.4

Whole rock initial Sr-isotope composition of the Bushveld Complex layered rocks from Kruger, F.J. (2005). Filling the Bushveld Complex magma chamber: lateral expansion, roof and floor interaction, magmatic unconformities, and the formation of giant chromitite, PGE and Ti-V-magnetitite deposits. Mineral. Deposita 40, 451–472, reproduced with permissions.

Critical Zone, and lower Main Zone show a highly isotopic stratigraphy and suggested that these rocks were formed during a time when the evolving magma chamber was an open system—what he termed the "Integration Stage" of the Bushveld. The middle to upper Main Zone and Upper Zone, in contrast, show very uniform, homogeneous Sr-isotope compositions, albeit with one large new pulse of magma at the Pyroxenite Marker. Kruger (2005) proposed that these sections must have evolved during a closed-system "Differentiation Stage" of the magma chamber.

The observations of Kruger (2005) are highly useful for determining where a new component must have been introduced; however, there are two caveats associated with this approach. (1) Isotopes cannot easily distinguish between the assimilant and a new magma pulse. For example, the Rooiberg Group lavas and Rashoop Granophyres, which immediately overlay the Bushveld Complex magmas, have Sr-isotope compositions that overlap those of the mafic cumulates (Buchanan et al., 2002, 2004). Thus, if the Bushveld magmas assimilated the overlying roof rock, there would be no distinctive change in the isotopic composition of the resulting cumulates. (2) New magma pulses may have the same isotopic composition as the resident magma, particularly if they are from the same source. In this case, the addition of any amount—whether large or small—of new magma into the system would not be recorded in the isotopic composition of the cumulates.

Recent work looking in detail at the isotopic compositions of individual mineral phases also shows considerable variability at the small scale. In nearly every isotopic system (Sr, Nd, Pb) isotopic disequilibrium has been identified between different mineral phases and even between different grains of the same mineral throughout the Bushveld Complex (Prevec et al., 2005; Chutas et al., 2012; Roelofse et al., 2015). Various models have been put forth to explain the mineral-scale disequilibrium: intrusions of different mineral slurries; mixing of different crystal populations between resident and incoming magma; even late-stage metasomatism affecting different populations. What is clear is that the whole-rock isotopic composition is simply an aggregate of a much more complex system, and future work investigating in situ mineral variations is necessary to shed light on these dynamic and complex magma systems.

4.4.2 IDENTIFYING MAGMA INPUTS: MAJOR ELEMENTS

A more rigorous petrological tool for identifying a new pulse of magma is the changing major-element composition of the phases as well as any changes in the phase assemblage. A new pulse of magma into a LMI is almost always sure to be hotter, and probably more mafic, than the resident fractionating magma. A more mafic and/or hotter pulse of magma can be recognized by a change in the major-element composition of the primary mineral phases, such as an increase in the An content of plagioclase or Mg# of pyroxenes. For example, in the eastern Bushveld the approximately 350 m of stratigraphy below the Pyroxenite Marker show a marked progression from plagioclase An_{57} to An_{75} and orthopyroxene Mg# from 63 to 72 (Fig. 4.5; VanTongeren et al., 2013). This reversal to more primitive compositions is also seen below the Pyroxenite Marker in the western limb (Cawthorn et al., 1991), and clearly indicates a chamber-wide recharge event.

However, shifts in the major-element composition of the mineral phases can be misleading in some cases. For example, Cawthorn (1996) suggested that the increase in orthopyroxene Mg# at the top of the Upper Critical Zone was due to reequilibration with trapped liquid, e.g., the trapped-liquid shift (Barnes, 1986), rather than magma recharge.

An additional indicator of new pulses of magma may be the removal or addition of a phase in the primary phase assemblage. In the case of the Pyroxenite Marker recharge event the addition of a new pulse of magma resulted in the crystallization of orthopyroxene rather than the pigeonite that was previously on the liquidus. In the case of the upper Critical Zone, the gradual appearance of plagioclase on the liquidus, and the resorbed plagioclase found in the lower Critical Zone (e.g., Eales et al., 1990), is attributed to the introduction of the B2 magmas and their mixing with the resident, fractionated B1 magma.

However, there also are several caveats associated with this method of identifying a new magma pulse. First, if the new pulse of magma is large enough, or hot enough, it could overwhelm the crystallization sequence and halt crystallization of the magma until it is fully mixed and has been allowed to cool sufficiently. This would essentially amount to a magmatic disconformity. The resulting stratigraphy would show a large jump in the mineral compositions over a very short vertical interval.

Magmatic "unconformities" also exist, wherein the new magma pulse scours the existing cumulate pile and erodes away the previously crystallized mush zone. A similarly large jump in the plagioclase An content or pyroxene Mg# would result, and in some cases the scouring features are seen in outcrop. Examples of this are found in the Bushveld Complex at the boundary between the

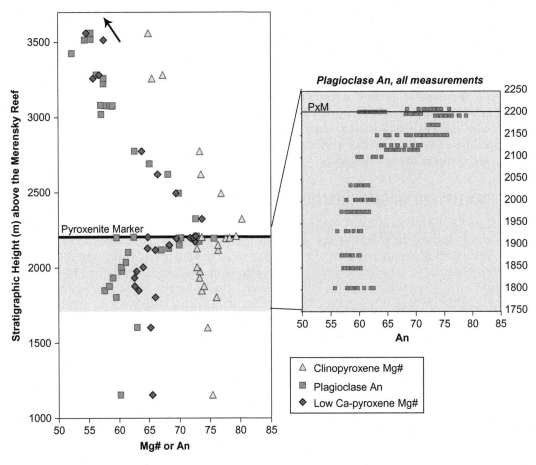

FIGURE 4.5

Reversal in plagioclase and pyroxene compositions associated with the final major pulse of new magma into the Bushveld Complex magma chamber below the Pyroxenite Marker.

Reproduced with permissions from VanTongeren, J.A., & Mathez, E.A. (2013). Incoming Magma Composition and Style of Recharge below the Pyroxenite Marker, Eastern Bushveld Complex, South Africa. J. Petrol. 54, 1585–1605.

Lower Critical Zone and Upper Critical Zone (e.g., Eales et al., 1990; Kruger, 2005), and in the pothole structures that punctuate the Merensky Reef (e.g., Viljoen, 1999).

An additional caveat is in over-estimating the significance of the appearance of new phases, or particularly new layers of mineral phases, to a new pulse of magma into the magma chamber. Early thinking on the formation of chromitite and magnetite layers in the Bushveld Complex proposed that monomineralic layers were directly linked to new pulses of magma entering the system. It was considered that layers were formed by one of two ways. The first is that mixing of new magma with resident magma caused the mixed magma to move into the primary phase volume of a single mineral. This is potentially the case for the formation of chromitite layers due to the topology of the chromite-plagioclase-silica liquidus (e.g., Irvine, 1977). The second mechanism is that the new magma contained an entrained crystal cargo that was deposited upon entering the magma chamber. For example, Mondal and Mathez (2007) showed that the pyroxenites immediately above and below the UG2 chromitite layer are petrographically distinct, but return to the original composition ~ 17 m higher in the stratigraphy. They argue that the monomineralic UG2 chromitite must have been formed, not by magma mixing, but by magma emplacement and deposition of an entrained slurry of chromite onto the floor of the magma chamber.

Attributing the formation of each monomineralic layer in the stratigraphy, however, is unnecessary and cumbersome, as layers of anorthosite, pyroxenite, magnetite, and chromite are laterally continuous for hundreds of meters and in some cases are less than 10 cm thick throughout. A new pulse of magma entering the system and mixing in the same proportions throughout the 450 km + of the Bushveld magma chamber, to produce a 10 cm thick layer, is not a likely scenario. Rather, typical magma chamber dynamics, including convection of the resident magma, compaction, and double-diffusive convection, can cause magmatic layering that is unrelated to the emplacement of new magma. This final point leads to another important consideration with respect to the buildup of the Bushveld magma chamber: the dynamics of magma emplacement and mixing.

4.4.3 DYNAMICS OF MAGMA EMPLACEMENT AND MIXING

It is commonly assumed that new pulses of magma enter the chamber near the interface between the cumulate pile and resident magma, and then instantly mix completely with the resident magma. However, numerous factors can influence where in the magma chamber the new magma will be emplaced. These include the geometry of the chamber, the porosity of the mush zone at the chamber floor, the density and temperature of the incoming magma, the density and temperature stratification of the resident magma, and whether the incoming magma carries an entrained crystal cargo. If a new pulse of magma is significantly less dense than the resident magma, then it can be rise buoyantly to pond near the top of the magma chamber (Huppert et al., 1986). Alternatively, if the new magma is denser than the resident magma, ponding will occur at the base of the chamber (Huppert and Sparks, 1980). Similarly, if the new pulse of magma contains a significant crystal load, then the entrained crystal load will be rapidly deposited on the floor of the intrusion, and mixing will occur near this level due to the turbulence induced by the sinking crystal load.

Huppert and Sparks (1980) investigated several different scenarios for the mixing of hot, dense new magma within an actively convecting magma chamber. Using analogue models, they showed that a new pulse of hotter denser magma into a resident magma chamber will pond at the floor of the intrusion and form a separate layer. Only after cooling, during which diffusion of heat and mass occurs between the two layers, will the layers become able to mix with one another. They suggested that this process may be responsible for the origin of cyclic layering within layered intrusions such as the Bushveld Complex. As a follow-up, Huppert et al. (1986) investigated the input of a more buoyant new magma into a chamber with a dense resident magma. Using laboratory models, they showed that the behavior of the new magma depends on the height/width proportions of the magma chamber and the Reynolds number. For magma chambers that have a greater height than width, the new magma will easily mix with the resident magma, but for magma chambers like the Bushveld Complex, which are significantly wider than they are high, the new magma will form a conduit within the resident magma and rise to near the roof of the intrusion. However, mixing becomes more efficient between new and resident magma, even in the Bushveld case, if the resident magma is turbulently convecting.

The implications of the dynamics of magma emplacement are critical to the interpretation of the layered stratigraphy and the economic deposits in the Bushveld. For example, Campbell et al. (1983) investigated the emplacement of buoyant new magma into the Bushveld magma chamber as a trigger for sulfide saturation and the formation of PGE-rich layers within the stratigraphy. They noted that in order to form the massive quantities of PGE-rich material found in the Bushveld Complex, the sulfide liquid that it formed from must have interacted with the entire magma chamber of silicate melt prior to its deposition. As they write: "If the new pulse enters the chamber with sufficient momentum it will initially jet upwards into the overlying magma then fall back on itself when negative buoyancy overcome the jet's initial momentum. During jetting the new pulse mixes with the host magma to form a hybrid melt which collects at the floor of the chamber, producing a gravity-stratified layer with warm primitive magma overlain by cooler, more fractionated magma" (Campbell et al., 1983, p. 133).

One further consideration to the dynamics of magma mixing is the viscosity barrier. Field observations of basaltic enclaves within basaltic andesite matrices show that even magmas with very similar compositions do not mix readily. An evolved resident magma will have a significantly different viscosity to a hotter, more mafic new pulse of magma and mixing will be dynamically impeded.

For example, Cawthorn et al. (1991) postulated that the new magma that is emplaced the MZ-UZ boundary does not actually mix with the resident magma, but is emplaced and kept as its own separate body due to the difference in viscosity between the resident and new magmas. This hypothesis is not consistent with the gradual increase in all geochemical indices (major-element, trace-element, and isotopic compositions) over >300 m of stratigraphy below the Pyroxenite Marker, followed by a uniform isotopic composition and fractional crystallization sequence from the base of the Upper Zone all the way to the roof. These observations, coupled with the lack of step-wise change in compositions, led VanTongeren and Mathez (2013) to conclude that the new magma that intruded below the Pyroxenite Marker was emplaced incrementally in small amounts where it was able to mix rapidly with the resident magma.

4.5 UNMIXING

4.5.1 SULFIDE LIQUID IMMISCIBILITY

In addition to the emplacement and mixing of new magmas within the Bushveld Complex magma chamber, there are several features of the stratigraphy that can only be explained by a process of unmixing, or liquid immiscibility. It has long been accepted that magmas that reach sulfur saturation will develop an immiscible sulfide liquid (Naldrett, 1969; Skinner and Peck, 1969), which strongly concentrates chalcophile elements such as PGE. Sulfide saturation within the Eastern and Western limbs of the Bushveld is evident within the Critical Zone where vast quantities of sulfide minerals are associated with chromitite layers such as the PGE-rich UG2 chromites, and found within the pyroxenite pegmatite of the Merensky Reef. The PGEs are concentrated within the sulfides crystallized from the immiscible sulfide liquid.

Sulfide saturation in *S*-undersaturated magmas can occur due to either contamination or fractional crystallization. In the eastern and western limbs of the Bushveld Complex, the stratiform, yet interstitial, nature of the sulfides in the UG chromitites and the Merensky Reef suggests that the magmas did not reach sulfide saturation by fractional crystallization (Seabrook et al., 2005). Rather, it is likely that contamination of the resident magma, whether by country- rock assimilation or emplacement of new magma, caused *S*-saturation.

Early studies suggested that magma emplacement and mixing between a hotter, more buoyant, new magma and the resident magma triggered the sulfide saturation in the Bushveld Complex (Campbell et al., 1983; Naldrett and von Gruenewaldt, 1989). The numerous proponents of this model for *S*-saturation call upon immiscible sulfide liquid to physically separate from the significantly larger volume of silicate liquid on the scale of several kilometers. This is because, as noted above, the uniquely high concentrations of PGE in the sulfides of the Bushveld Complex require the sulfide liquid to have scavenged PGE from a very large volume of silicate melt. Contamination by a new magma pulse was also proposed by Cawthorn (2005), although in his model, the *S*-saturation of the resident magma was triggered by the increased pressure associated with the pulse of new magma, and not necessarily the mixing between the magmas.

Contamination by assimilation of the country rock has also been proposed as the main cause of *S*-saturation in the Bushveld Complex (Cawthorn et al., 1985). In the Northern limb, sulfide saturation of the magma occurs almost immediately at the contact between the Bushveld magma and the floor rocks, resulting in the Platreef (Van der Merwe, 1976; Gain and Mostert, 1982). In the Platreef, Rb-Sr isotopes show clear contamination from both dolomites (mainly of the Chuniespoort Group) and Archean granite which make up the floor rocks in the Northern Limb (Barton et al., 1986). On the basis of *O*-isotopes, Harris and Chaumba (2001) proposed an additional role for late-stage hydrothermal fluids in the formation of the Platreef. Perhaps the most unequivocal evidence of country-rock assimilation and contamination in the Platreef comes from the presence of numerous xenoliths of calc-silicate floor material within the pyroxenite layer.

Regardless of the exact mechanism that produced the sulfides, there is currently no debate that once a magma reaches *S*-saturation, an immiscible sulfide liquid forms, which can physically separate and sink through the larger volume of silicate liquid.

4.5.2 SILICATE LIQUID IMMISCIBILITY

In contrast to the consensus over the occurrence of sulfide liquid immiscibility, there is considerable debate over the viability of silicate liquid immiscibility. Silicate liquid immiscibility is thought to occur in highly evolved tholeiitic magmas, particularly in shallow magma chambers (e.g., undergoing crystallization at low P; Roedder, 1951; Veksler et al., 2009; Charlier and Grove, 2012). During silicate liquid immiscibility the magma separates into a Si-rich melt with SiO_2 >60 wt% and an Fe-rich melt with SiO_2 <45 wt% and FeO = 15−30 wt%). According to experiments, the Si-rich melt will concentrate elements such as K, Na; while the Fe-rich melt partitions the REE and HFSE (Watson and Green, 1981; Veksler et al., 2006).

The debate over silicate liquid immiscibility started with its initial proposal by Daly in 1914. Bowen (1928) strongly opposed the idea of silicate liquid immiscibility on the basis of crystallization experiments showing that tholeiitic liquids evolve to a Si-rich composition during fractionation. In one of the most seminal papers on the subject, Philpotts (1982) documented the presence of both Fe-rich and Si-rich melt inclusions in hundreds of samples worldwide from the Cambridge University collection. He showed that basalts from nearly every tectonic location, contained Fe and Si-rich melt inclusion populations and suggested that this was a ubiquitous process during fractional crystallization. Several more studies documenting the presence of both sets of melt inclusions from volcanic rocks in various locations also appeared around that time (see Veksler et al., 2007). More recently, in the ∼3 km-thick Skaergaard Intrusion, both Fe-rich and Si-rich melt inclusions have been found within single crystals of apatite (Jakobsen et al., 2005) and plagioclase (Jakobsen et al., 2011).

The evidence of Fe-rich and Si-rich melt inclusions in single minerals and on the hand-sample scale in gabbroic and dioritic rocks then leads to the question of whether silicate liquid immiscibility occurs only on a small scale (e.g., millimeters to centimeters) or on a larger scale (e.g., meters to hundreds of meters). The question of scale is critical for understanding what role silicate liquid immiscibility might play in the fractionation of magmas and residual liquids.

Charlier et al. (2011) observed both Fe-rich and Si-rich inclusions in apatites from the ∼5 km thick Sept Isles intrusion in Canada. They showed that there were more Fe-rich inclusions in the magnetite-rich layers and more Si-rich inclusions in the overlying diorite layers and suggested that this pattern was indicative of liquid−liquid separation on the order of meters, and proposed that silicate liquid immiscibility was responsible for the layering in the highly evolved portions of the Sept Isles intrusion. Silicate liquid immiscibility on the order of meters was also suggested by Reynolds (1985) as the probable cause of the 21 near monomineralic magnetite layers, and some nelsonite (magnetite-apatite) layers, throughout the Upper Zone of the Bushveld Complex. VanTongeren and Mathez (2012) detailed evidence to support even larger-scale liquid−liquid separation due to silicate liquid immiscibility on the order of hundreds of meters in the Upper Zone. The uppermost 600 m of stratigraphy in this zone contains the most highly evolved compositions in the Bushveld Complex crystallization sequence. Approximately 600 m below the roof of the intrusion, apatite becomes a cumulus phase. From 600 to 300 m below the roof, the rocks are fayalite-diorites with approximately 5−7 modal% apatite and 5−15 modal% magnetite, though due to the layering some sections of stratigraphy contain more apatite than others (e.g., the nelsonite layers described in the Western Limb; Kolker, 1982; Reynolds, 1985; von Gruenewaldt, 1993). This lower 300 m also contains four magnetitite layers (Upper Seams 17−21 as described by Molyneux (1974) in the eastern limb), the uppermost layer, Upper Seam 21, is locally up to 60 m thick and is currently being actively mined for Fe-Ti-P. Directly above magnetitite Upper Seam 21, from ∼300 m below the roof to the roof-contact, the modal mineralogy of the rocks changes drastically. The rocks are still classified as fayalite diorite, but they contain only 1%−3% apatite, 2%−5% oxide (mostly ilmenite over magnetite), and cumulus orthoclase (1%−5%) and quartz (1%−5%) join the liquidus assemblage. There are no magnetitite layers present from 300 m below the roof to the roof of the intrusion.

In addition to these changes in modal mineralogy, the geochemistry of the individual mineral phases also changes above US21. In the lowermost 300 m, plagioclase An, olivine Fo, and clinopyroxene Mg# do not undergo significant fractionation, whereas in the uppermost 300 m, the compositions of these phases evolve from ∼An_{50}−An_{37}, Fo_{35} to Fo_5, and cpx Mg# from 60 to 30 (VanTongeren and Mathez, 2014). The biggest change, however, is observed in the composition of the REEs in apatite above and below US21. In the 300 m below magnetite seam 21, apatite has a REE profile with no Eu anomaly and low concentrations; above the seam, the apatite has approximately $3 \times$ more REE and a distinctive negative Eu anomaly (Fig. 4.6). VanTongeren and Mathez (2012) showed that this change in apatite REE concentration cannot be explained by fractional crystallization or a change in the oxygen fugacity of the system.

The best explanation for the change in REE patterns and concentrations is that the apatite in the lower section crystallized from a segregated Fe-rich melt, while the upper section crystallized from a segregated Si-rich melt. If the two liquids had remained physically in contact with one another, they would remain in chemical equilibrium, and would produce phases with identical concentrations, but different modal abundances. Thus, the observation of two different apatite populations requires that the Bushveld magma underwent large-scale (i.e., hundreds of meters) silicate liquid immiscibility, resulting in physically separated Fe-rich and Si-rich melts at the top of the Bushveld Complex, which then were able to diverge from equilibrium during crystallization (Fig. 4.7). Large-scale silicate liquid immiscibility was previously hypothesized in the Bushveld Complex to explain the occurrence of large discordant bodies of Iron-Rich Ultramafic Pegmatites (IRUPs) observed in the lower portions of the Bushveld stratigraphy at some locations in the western limb (e.g., Reid and Basson, 2002; Scoon and Mitchell, 1994).

The conclusions of VanTongeren and Mathez (2012) were questioned by Cawthorn (2013), who suggested that the contrast in apatite REE compositions may not be due to silicate liquid immiscibility, but rather are the result of variable amounts of trapped liquid (e.g., Barnes, 1986) resetting

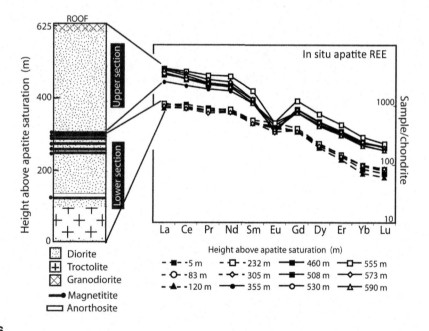

FIGURE 4.6

Evidence for large-scale liquid immiscibility at the top of the Bushveld Complex comes from changes in geochemistry and mineralogy between the lower and upper sections defined by VanTongeren and Mathez (2012).

Adapted from VanTongeren, J.A., & Mathez, E.A. (2012). Large-scale liquid immiscibility at the top of the Bushveld Complex, South Africa. Geology 40, 491–494.

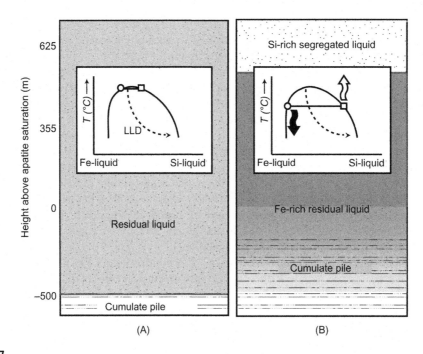

FIGURE 4.7

Schematic illustration of the process of large-scale liquid immiscibility at the top of the Bushveld Complex, resulting in the formation of an Fe-rich lower section and a Si-rich upper section that were physically separated from one another and allowed to chemically diverge from equilibrium.

Adapted from VanTongeren, J.A., & Mathez, E.A. (2012). Large-scale liquid immiscibility at the top of the Bushveld Complex, South Africa. Geology 40, 491–494.

the compositions of the phases within the cumulate pile. Cawthorn (2013) used the apatite REE data of VanTongeren and Mathez (2012) and calculated that the apatites in the lowermost section were reset from their liquidus compositions by reequilibration with more than 40% trapped liquid fraction. However, the calculations of Cawthorn (2013) were based on several unsupported assumptions or incorrect values such as the composition of the starting magma, the modal compositions of the rocks in question, and the partition coefficients of REE in apatite (VanTongeren and Mathez, 2014).

One important test of the large-scale liquid immiscibility theory remains. If the Bushveld Upper Zone magma underwent large-scale liquid immiscibility, then this should be reflected in the populations of melt inclusions trapped within the two different sections of stratigraphy. Fischer et al. (2016) published a dataset of melt inclusion compositions from an Upper Zone drillcore in the Western Limb. They showed that there were both Fe-rich and Si-rich melt inclusions present in apatite grains from two samples spaced roughly 8 m apart (Fig. 4.9).

In both samples analyzed, Fischer et al. (2016) found Fe-rich and Si-rich melt inclusions, which led them to conclude that large-scale separation (hundreds of meters) of the two liquids did not occur, but that perhaps small scale (tens of meters) separation may have occurred.

There are several important caveats associated with the conclusions reached by Fischer et al. (2016). The REE content of the apatites hosting the melt inclusions were not measured, so it is not possible to pinpoint precisely how the samples compare stratigraphically with those of the Eastern limb. This is important because, based on their cumulus mineral assemblage and abundance, both of the samples chosen appear to come from the Fe-rich "lower section" of the stratigraphy identified by VanTongeren and Mathez (2012) in the eastern limb. If the apatites formed from a segregated Fe-rich melt, then we would expect to see a majority proportion of Fe-rich melt inclusions in the apatites, which is exactly what is observed (Fig. 4.8). Only ~ 10 melt compositions out of >100 analyzed by Fischer et al. (2016) have $SiO_2 > 60$ wt%.

In fact, the very small proportion of Si-rich melt inclusions measured in the Fischer et al. (2016) study supports the original hypothesis of VanTongeren and Mathez (2012) of large-scale liquid—liquid separation. If the Fe-rich and Si-rich melts physically separated from one another at high temperature, then both magmas would be expected to continue undergoing silicate liquid immiscibility as they cooled (e.g., Fig. 4.9).

As illustrated in Fig. 4.9, in the case of the Fe-rich magma, this would result in some Si-rich melt segregations within the Fe-rich section; in the case of the Si-rich magma, it would produce a small number of Fe-rich segregations.

In order to determine whether large-scale liquid immiscibility resulting in the formation of two distinctive melts occurred at the top of the Bushveld Complex, it is critical to compare the

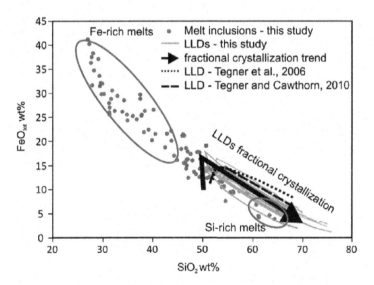

FIGURE 4.8

Melt inclusions analyzed from two samples from the top of the Bushveld Complex show a range of Fe-rich compositions not predicted by models of the liquid line of descent. Si-poor, Fe-rich melt inclusions circled in red must be derived from Fe-rich melt formed due to liquid immiscibility.

Adapted from Fischer, L.A., Wang, M., Charlier, B., Namur, O., Roberts, R.J., Veksler, I.V., et al., 2016. Immiscible iron- and silica-rich liquids in the Upper Zone of the Bushveld Complex. Earth Planet. Sci. Lett. 443, 108–117.

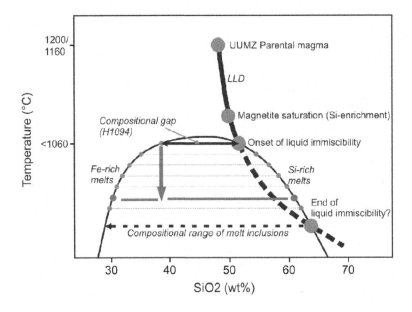

FIGURE 4.9

Solvus between Fe-rich and Si-rich immiscible melts. At high temperature, the two melts are more similar in composition, but as the melts cool, they continue to diverge. Thus, if an Fe-rich melt segregated from a Si-rich melt at high temperature, the Fe-rich composition would continue to undergo some degree of immiscibility during cooling resulting in a fraction of Si-rich melt within the Fe-rich magma in a volume proportion according to the lever rule.

Adapted from Fischer, L.A., Wang, M., Charlier, B., Namur, O., Roberts, R.J., Veksler, I.V., et al., 2016. Immiscible iron- and silica-rich liquids in the Upper Zone of the Bushveld Complex. Earth Planet. Sci. Lett. 443, 108–117.

proportions of Fe-rich and Si-rich melt inclusions in both the postulated Fe-rich section and the Si-rich section of the intrusion. The study of Fischer et al. (2016) is a first step in doing this, but it is necessary to compare their results with similar results from the Si-rich section in the future.

4.6 CONCLUSIONS

The enormous size and compositional structure of the Bushveld Complex LMI suggests that it probably was built up by numerous pulses of magma with variable compositions. The source of the Bushveld magmas is likely to have been a large mantle plume, which may have been initially komatiitic, and which interacted to varying degrees with the SCLM. The plume and SCLM signatures are present in the major- element, isotopic and PGE contents of the Bushveld cumulates. The layered stratigraphy and valuable economic deposits formed by processes of magma mixing and dynamics associated with the emplacement of the magma at various levels. Ore horizons of both PGE and Fe-Ti-V probably formed from large-scale liquid immiscibility, though debate still remains.

REFERENCES

Barnes, S.J., 1986. The effect of trapped liquid crystallization on cumulus mineral compositions in layered intrusions. Contrib. Mineral. Petrol. 93, 524−531.

Barnes, S.J., 1989. Are bushveld U-type parent magmas boninites or contaminated komatiites. Contrib. Mineral. Petrol. 101, 447−457.

Barnes, S.J., Maier, W.D., Curl, E.A., 2010. Composition of the marginal rocks and sills of the rustenburg layered suite, bushveld complex, South Africa: implications for the formation of the platinum-group element deposits. Econ. Geol. 105, 1491−1511.

Barton, J., Cawthorn, R.G., White, J., 1986. The role of contamination in the evolution of the Platreef of the Bushveld Complex. Econ. Geol. 81, 1096−1104.

Bowen, N.L., 1928. The Evolution of Igneous Rocks. Princeton University Press, Princeton.

Buchanan, P.C., Reimold, W.U., 1998. Studies of the rooiberg group, bushveld complex, South Africa: no evidence for an impact origin. Earth. Planet. Sci. Lett. 155, 149−165.

Buchanan, P.C., Reimold, W.U., Koeberl, C., Kruger, F.J., 2002. Geochemistry of intermediate to siliceous volcanic rocks of the Rooiberg Group, Bushveld Magmatic Province, South Africa. Contrib. Mineral. Petrol. 144, 131−143.

Buchanan, P.C., Reimold, W.U., Koeberl, C., Kruger, F.J., 2004. Rb-Sr and Sm-Nd isotopic compositions of the Rooiberg Group, South Africa: early Bushveld-related volcanism. Lithos 75, 373−388.

Campbell, I.H., Naldrett, A.J., Barnes, S.J., 1983. A model for the origin of the platinum-rich sulfide horizons in the Bushveld and Stillwater Complexes. J. Petrol. 24, 133−165.

Cawthorn, R.G., 1996. Re-evaluation of magma compositions and processes in the uppermost Critical Zone of the Bushveld Complex. Mineral. Mag. 60, 131−148.

Cawthorn, R.G., 2005. Pressure fluctuations and the formation of the PGE-rich Merensky and chromitite reefs, Bushveld Complex. Mineral. Deposita 40, 231−235.

Cawthorn, R.G., 2011. Geological interpretations from the PGE distribution in the Bushveld Merensky and UG2 chromitite reefs. J S Afr I. Min. Metall. 111, 67−79.

Cawthorn, R.G., 2013. Rare earth element abundances in apatite in the Bushveld Complex-A consequence of the trapped liquid shift effect. Geology. 41, 603−606.

Cawthorn, R.G., Barton, J., Viljoen, M.J., 1985. Interaction of floor rocks with the Platreef on Overysel, Potgietersrus, northern Transvaal. Econ. Geol. 80, 988−1006.

Cawthorn, R.G., Meyer, P.S., Kruger, F.J., 1991. Major addition of magma at the pyroxenite marker in the western bushveld complex, South-Africa. J. Petrol. 32, 739−763.

Cawthorn, R.G., Walraven, F., 1998. Emplacement and crystallization time for the Bushveld Complex. J. Petrol. 39, 1669−1687.

Cawthorn, R.G., Webb, S.J., 2001. Connectivity between the western and eastern limbs of the Bushveld Complex. Tectonophysics 330, 195−209.

Charlier, B., Grove, T.L., 2012. Experiments on liquid immiscibility along tholeiitic liquid lines of descent. Contrib. Mineral. Petrol. 164, 27−44.

Charlier, B., Namur, O., Toplis, M.J., Schiano, P., Cluzel, N., Higgins, M.D., et al., 2011. Large-scale silicate liquid immiscibility during differentiation of tholeiitic basalt to granite and the origin of the Daly gap. Geology 39, 907−910.

Chaumba, J.B., Wilson, A.H., 1997. An oxygen isotope study of the Lower Mafic Succession of the Darwendale Subchamber of the Great Dyke, Zimbabwe. Chem. Geol. 135, 293−305.

Chutas, N.I., Bates, E., Prevec, S.A., Coleman, D.S., Boudreau, A.E., 2012. Sr and Pb isotopic disequilibrium between coexisting plagioclase and orthopyroxene in the Bushveld Complex, South Africa: microdrilling and progressive leaching evidence for sub-liquidus contamination within a crystal mush. Contrib. Mineral. Petrol. 163, 653−668.

Clarke, B., Uken, R., Reinhardt, J., 2009. Structural and compositional constraints on the emplacement of the Bushveld Complex, South Africa. Lithos 111, 21−36.

Daly, R.A., 1914. Igneous Rocks and Their Origin. McGraw-Hill Book Company, Inc, New York, NY.

Davies, G., Cawthorn, R.G., Barton, J.M., Morton, M., 1980. Parental magma to the bushveld complex. Nature 287, 33−35.

DePaolo, D.J., Wasserburg, G.J., 1979. Sm-Nd age of the stillwater complex and the mantle evolution curve for neodymium. Geochim. Cosmochim. Acta. 43, 999−1008.

Dunn, T., 1986. An investigation of the oxygen isotope geochemistry of the stillwater complex. J. Petrol. 27, 987−997.

Eales, H., Cawthorn, R., 1996a. The bushveld complex. Dev. Petrol. 15, 181−229.

Eales, H.V., 2002. Caveats in defining the magmas parental to the mafic rocks of the Bushveld Complex, and the manner of their emplacement: review and commentary. Mineral. Mag. 66, 815−832.

Eales, H.V., Cawthorn, R.G., 1996b. The bushveld complex. In: Cawthorn, R.G. (Ed.), Layered Intrusions. Elsevier Science, Amsterdam, Netherlands, pp. 181−229.

Eales, H.V., Deklerk, W.J., Teigler, B., 1990. Evidence for magma mixing processes within the critical and lower zones of the northwestern bushveld complex, South-Africa. Chem. Geol. 88, 261−278.

Elston, W.E., 1992. Does the Bushveld−Vredefort system South Africa record the largest known terrestrial impact catastrophe? Lunar Planet. Inst. Contrib. 790, 23−24.

Elston, W.E., Twist, D., 1989. Vredefort−Bushveld enigma of South Africa and the recognition of large terrestrial impact structures: mental leaps and mental obstacles. Abs. Inter. Geol. Cong. 28, 449.

Eriksson, P.G., Hattingh, P.J., Altermann, W., 1995a. An overview of the geology of the transvaal sequence and bushveld complex, South-Africa. Mineral. Deposita 30, 98−111.

Eriksson, P.G., Reczko, B.F.F., 1995b. The sedimentary and tectonic setting of the Transvaal Supergroup floor rocks to the Bushveld complex. J. Afr. Earth Sci. 21, 487−504.

Fischer, L.A., Wang, M., Charlier, B., Namur, O., Roberts, R.J., Veksler, I.V., et al., 2016. Immiscible iron- and silica-rich liquids in the Upper Zone of the Bushveld Complex. Earth Planet. Sci. Lett. 443, 108−117.

Fourie, D.S., Harris, C., 2011. O-isotope study of the bushveld complex granites and granophyres: constraints on source composition, and assimilation. J. Petrol. 52, 2221−2242.

Gain, S.B., Mostert, A., 1982. The geological setting of the platinoid and base metal sulfide mineralization in the Platreef of the Bushveld Complex in Drenthe, north of Potgietersrus. Econ. Geol. 77, 1395−1404.

Gregory, R.T., Taylor, H.P., 1981. An oxygen isotope profile in a section of the Cretaceous oceanic crust, Samail Ophiolite, Oman: evidence for delta18O buffering of the oceans by deep (>5 km) seawater-hydrothermal circulation at mid-ocean ridges, J. Geophys. Res. Solid Earth, 86. pp. 2737−2755.

Hamilton, W., 1970. Bushveld complex - product of impacts? Geol. Soc. South Afr. Special Publication 1, 367−379.

Harmer, R.E., Sharpe, M.R., 1985. Field relations and strontium isotope systematics of the marginal rocks of the eastern bushveld complex. Econ. Geol. 80, 813−837.

Harris, C., Chaumba, J.B., 2001. Crustal contamination and fluid-rock interaction during the formation of the Platreef northern limb of the Bushveld complex, South Africa. J. Petrol. 42, 1321−1347.

Harris, C., Pronost, J.J.M., Ashwal, L.D., Cawthorn, R.G., 2005. Oxygen and hydrogen isotope stratigraphy of the Rustenburg Layered Suite, Bushveld Complex: constraints on crustal contamination. J. Petrol. 46, 579−601.

Hart, S.R., Kinloch, E.D., 1989. Osmium isotope systematics in witwatersrand and bushveld ore-deposits. Econ. Geol. 84, 1651−1655.

Hatton, C.J., 1995. Mantle plume origin for the Bushveld and Ventersdorp magmatic provinces. J. Afr. Earth Sci. 21, 571−577.

Hatton, C.J., Schweitzer, J.K., 1995. Evidence for synchronous extrusive and intrusive Bushveld magmatism. J. Afr. Earth Sci. 21, 579−594.

Herzberg, C., Condie, K., Korenaga, J., 2010. Thermal history of the Earth and its petrological expression. Earth Planet. Sci. Lett. 292, 79–88.

Huppert, H.E., Sparks, R.S.J., 1980. The fluid-dynamics of a basaltic magma chamber replenished by influx of hot, dense ultrabasic magma. Contrib. Mineral. Petrol. 75, 279–289.

Huppert, H.E., Sparks, R.S.J., Whitehead, J.A., Hallworth, M.A., 1986. Replenishment of magma chambers by light inputs. J. Geophys. Res.-Solid 91, 6113–6122.

Irvine, T.N., 1977. Origin of chromitite layers in muskox intrusion and other stratiform intrusions - new interpretation. Geology 5, 273–277.

Jakobsen, J.K., Veksler, I.V., Tegner, C., Brooks, C.K., 2005. Immiscible iron- and silica-rich melts in basalt petrogenesis documented in the Skaergaard intrusion. Geology 33, 885–888.

Jakobsen, J.K., Veksler, I.V., Tegner, C., Brooks, C.K., 2011. Crystallization of the skaergaard intrusion from an emulsion of immiscible iron- and silica-rich liquids: evidence from melt inclusions in plagioclase. J. Petrol. 52, 345–373.

Kolker, A., 1982. Mineralogy and geochemistry of Fe-Ti Oxide and Apatite (Nelsonite) deposits and evaluation of the liquid immiscibility hypothesis. Econ. Geol. 77, 1146–1158.

Kruger, F.J., 2005. Filling the Bushveld Complex magma chamber: lateral expansion, roof and floor interaction, magmatic unconformities, and the formation of giant chromitite, PGE and Ti-V-magnetitite deposits. Mineral. Deposita 40, 451–472.

Kruger, F.J., Cawthorn, R.G., Walsh, K.L., 1987. Strontium isotopic evidence against magma addition in the upper zone of the bushveld complex. Earth. Planet. Sci. Lett. 84, 51–58.

Laurent, O., Zeh, A., 2015. A linear Hf isotope-age array despite different granitoid sources and complex Archean geodynamics: example from the Pietersburg block (South Africa). Earth Planet. Sci. Lett. 430, 326–338.

Lambert, D.D., Walker, R.J., Morgan, J.W., Shirey, S.B., Carlson, R.W., Zientek, M.L., et al., 1994. Re-Os and Sm-Nd isotope geochemistry of the stillwater complex, Montana - implications for the petrogenesis of the J-M Reef. J. Petrol. 35, 1717–1753.

Longhi, J., Wooden, J.L., Coppinger, K.D., 1983. The petrology of high-Mg dikes from the beartooth mountains, Montana - a search for the parent magma of the stillwater complex. J. Geophys. Res. 88, B53–B69.

McCallum, I.S., 1988. Evidence for crustal recycling during the Archean: the parental magmas of the Stillwater Complex. Lunar Planet. Inst. Tech. Rep 88, 92–94.

Molyneux, T.G., 1974. A geological investigation of the Bushveld Complex in Sekhukhuneland and part of the Steelpoort Valley. Trans. Geol. Soc. South Afr. 77, 329–338.

Molyneux, T.G., 2008. Compilation on a scale of 1: 50000 of the geology of the eastern compartment of the Bushveld Complex. Special Publication 1.

Mondal, S.K., Mathez, E.A., 2007. Origin of the UG2 chromitite layer, Bushveld Complex. J. Petrol. 48, 495–510.

Naldrett, A., 1969. A portion of the system Fe–S–O between 900 and 1080 C and its application to sulfide ore magmas. J. Petrol. 10, 171–201.

Naldrett, A., Von Gruenewaldt, G., 1989. Association of platinum-group elements with chromitite in layered intrusions and ophiolite complexes. Econ. Geol. 84, 180–187.

Philpotts, A.R., 1982. Compositions of Immiscible Liquids in Volcanic-Rocks. Contrib. Mineral. Petrol. 80, 201–218.

Prevec, S.A., Ashwal, L.D., Mkaza, M.S., 2005. Mineral disequilibrium in the Merensky Reef, western Bushveld Complex, South Africa: new Sm–Nd isotopic evidence. Contrib. Mineral. Petrol. 149 (3), 306–315.

Rajesh, H.M., Chisonga, B.C., Shindo, K., Beukes, N.J., Armstrong, R.A., 2013. Petrographic, geochemical and SHRIMP U-Pb titanite age characterization of the Thabazimbi mafic sills: extended time frame and a unifying petrogenetic model for the Bushveld Large Igneous Province. Precambrian. Res. 230, 79–102.

Reid, D.L., Basson, I.J., 2002. Iron-rich ultramafic pegmatite replacement bodies within the Upper Critical Zone, Rustenburg Layered Suite, Northam Platinum Mine, South Africa. Mineral. Mag. 66, 895−914.

Reynolds, I.M., 1985. The nature and origin of titaniferous magnetite-rich layers in the upper zone of the bushveld complex − a review and synthesis. Econ. Geol. 80, 1089−1108.

Rhodes, R.C., 1975. New evidence for impact origin of the Bushveld Complex. Geology 3, 549−554.

Richardson, S.H., Shirey, S.B., 2008. Continental mantle signature of Bushveld magmas and coeval diamonds. Nature 453, 910−913.

Roedder, E., 1951. Low Temperature Liquid Immiscibility in the System K2o-Feo-Al2o3-Sio2. Am. Mineral. 36, 282−286.

Roelofse, F., Ashwal, L.D., Romer, R.L., 2015. Multiple, isotopically heterogeneous plagioclase populations in the Bushveld Complex suggest mush intrusion. Chemie der Erde-Geochemistry 75 (3), 357−364.

Schiffries, C.M., Rye, D.M., 1989. Stable isotopic systematics of the bushveld complex. 1. Constraints of magmatic processes in layered intrusions. Am. J. Sci. 289, 841−873.

Schiffries, C.M., Rye, D.M., 1990. Stable isotopic systematics of the bushveld complex. 2. Constraints on hydrothermal processes in layered intrusions. Am. J. Sci. 290, 209−245.

Scoates, J.S., Friedman, R.M., 2008. Precise age of the platiniferous Merensky reef, Bushveld Complex, South Africa, by the U-Pb zircon chemical abrasion ID-TIMS technique. Econ. Geol. 103, 465−471.

Scoon, R.N., Mitchell, A.A., 1994. Discordant iron-rich ultramafic pegmatites in the Bushveld Complex and their relationship to iron-rich intercumulus and residual liquids. J. Petrol. 35, 881−917.

Seabrook, C.L., Cawthorn, R.G., Kruger, F.J., 2005. The Merensky reef, Bushveld Complex: mixing of minerals not mixing of magmas. Econ. Geol. 100, 1191−1206.

Sharpe, M.R., 1981. The chronology of magma influxes to the eastern compartment of the bushveld complex as exemplified by its marginal border groups. J. Geol. Soc. London 138, 307−326.

Sharpe, M.R., Hulbert, L.J., 1985. Ultramafic sills beneath the eastern bushveld complex − mobilized suspensions of early lower zone cumulates in a parental magma with boninitic affinities. Econ. Geol. 80, 849−871.

Silver, P.G., Behn, M.D., Kelley, K., Schmitz, M., Savage, B., 2006. Understanding cratonic flood basalts. Earth Planet. Sci. Lett. 245, 190−201.

Skinner, B.J., Peck, D.L., 1969. An immiscible sulfide melt from Hawaii. Econ. Geol. Monogr. 4, 10−32.

Sun, S.S., McCulloch, M.T., Nesbitt, R.W., 1989. Geochemistry and petrogenesis of Archaean and early Proterozoic siliceous high magnesian basalts. In: Crawford, A.J. (Ed.), Boninites. Unwin Hyman, Winchester, MA, pp. 149−173.

van der Merwe, M., 1976. The layered sequence of the Potgietersrus limb of the Bushveld Complex. Econ. Geol. 71, 1337−1351.

VanTongeren, J.A., Mathez, E.A., 2012. Large-scale liquid immiscibility at the top of the Bushveld Complex, South Africa. Geology 40, 491−494.

VanTongeren, J.A., Mathez, E.A., 2013. Incoming magma composition and style of recharge below the pyroxenite marker, eastern bushveld complex, South Africa. J. Petrol. 54, 1585−1605.

VanTongeren, J.A., Mathez, E.A., 2014. Rare earth element abundances in apatite in the Bushveld Complex-A consequence of the trapped liquid shift effect Comment. Geology 42, E318-E318.

VanTongeren, J.A., Mathez, E.A., 2015. On the relationship between the Bushveld Complex and its felsic roof rocks, part 2: the immediate roof. Contrib. Mineral. Petrol. 170, 56.

VanTongeren, J.A., Mathez, E.A., Kelemen, P.B., 2010. A felsic end to bushveld differentiation. J. Petrol. 51, 1891−1912.

VanTongeren, J.A., Zirakparvar, N.A., Mathez, E.A., 2016. Hf isotopic evidence for a cogenetic magma source for the Bushveld Complex and associated felsic magmas. Lithos 248, 469−477.

Veksler, I.V., 2009. Extreme iron enrichment and liquid immiscibility in mafic intrusions: experimental evidence revisited. Lithos 111, 72−82.

Veksler, I.V., Dorfman, A.M., Borisov, A.A., Wirth, R., Dingwell, D.B., 2007. Liquid immiscibility and the evolution of basaltic magma. J. Petrol. 48, 2187–2210.

Veksler, I.V., Dorfman, A.M., Danyushevsky, L.V., Jakobsen, J.K., Dingwell, D.B., 2006. Immiscible silicate liquid partition coefficients: implications for crystal-melt element partitioning and basalt petrogenesis. Contrib. Mineral. Petrol. 152, 685–702.

Viljoen, M.J., 1999. The nature and origin of the Merensky Reef of the western Bushveld Complex based on geological facies and geophysical data. South Afr. J. Geol. 102, 221–239.

von Gruenewaldt, G., 1993. Ilmenite-apatite enrichments in the Upper Zone of the Bushveld Complex: a major titanium-rock phosphate resource. Int. Geol. Rev. 35, 987–1000.

Walraven, F., 1997. Geochronology of the Rooiberg Group, Transvaal Supergroup, South Africa. Econ. Geol. Res. Unit, Univers. Witwatersrand Information Circular 316.

Watson, E.B., Green, T.H., 1981. Apatite liquid partition-coefficients for the rare-earth elements and strontium. Earth Planet. Sci. Lett. 56, 405–421.

Webb, S.J., Cawthorn, R.G., Nguuri, T., James, D., 2004. Gravity modeling of Bushveld Complex connectivity supported by Southern African Seismic Experiment results. South Afr. J. Geol. 107, 207–218.

Wilson, A.H., 1982. The geology of the great dyke, Zimbabwe – the ultramafic rocks. J. Petrol. 23, 240–292.

Wilson, A.H., 2012. A chill sequence to the bushveld complex: insight into the first stage of emplacement and implications for the parental magmas. J. Petrol. 53, 1123–1168.

Wilson, A.H., 2015. The earliest stages of emplacement of the eastern bushveld complex: development of the lower zone, marginal zone and basal ultramafic sequence. J. Petrol. 56, 347–388.

Wooden, J.L., Mueller, P.A., 1988. Pb, Sr, and Nd isotopic compositions of a suite of late archean, igneous rocks, eastern beartooth mountains – implications for crust-mantle evolution. Earth. Planet. Sci. Lett. 87, 59–72.

Woodhead, J.D., Greenwood, P., Harmon, R.S., Stoffers, P., 1993. Oxygen isotope evidence for recycled crust in the source of Em-Type Ocean Island Basalts. Nature 362, 809–813.

Yudovskaya, M.A., Kinnaird, J.A., Sobolev, A.V., Kuzmin, D.V., McDonald, L., Wilson, A.H., 2013. Petrogenesis of the lower zone olivine-rich cumulates beneath the platreef and their correlation with recognized occurrences in the bushveld complex. Econ. Geol. 108, 1923–1952.

Zeh, A., Ovtcharova, M., Wilson, A.H., Schaltegger, U., 2015. The Bushveld Complex was emplaced and cooled in less than one million years - results of zirconology, and geotectonic implications. Earth Planet. Sci. Lett. 418, 103–114.

Zirakparvar, N.A., Mathez, E.A., Scoates, J.S., Wall, C.J., 2014. Zircon Hf isotope evidence for an enriched mantle source for the Bushveld Igneous Complex. Contrib. Mineral. Petrol. 168.

FURTHER READING

Maier, W.D., Arndt, N.T., Curl, E.A., 2000. Progressive crustal contamination of the Bushveld Complex: evidence from Nd isotopic analyses of the cumulate rocks. Contrib. Mineral. Petrol. 140, 316–327.

Marcantonio, F., Zindler, A., Reisberg, L., Mathez, E.A., 1993. Re-Os isotopic systematics in chromitites from the stillwater complex, Montana, USA. Geochim. Cosmochim. Acta. 57, 4029–4037.

McCandless, T.E., Ruiz, J., 1991. Osmium Isotopes and Crustal Sources for Platinum-Group Mineralization in the Bushveld Complex, South-Africa. Geology 19, 1225–1228.

Mukasa, S.B., Wilson, A.H., Carlson, R.W., 1998. A multielement geochronologic study of the Great Dyke, Zimbabwe: significance of the robust and reset ages. Earth Planet. Sci. Lett. 164, 353–369.

Oberthur, T., Davis, D.W., Blenkinsop, T.G., Hohndorf, A., 2002. Precise U-Pb mineral ages, Rb-Sr and Sm-Nd systematics for the Great Dyke, Zimbabwe — constraints on late Archean events in the Zimbabwe craton and Limpopo belt. Precambrian. Res. 113, 293—305.

Schoenberg, R., Kruger, F.J., Nagler, T.F., Meisel, T., Kramers, J.D., 1999. PGE enrichment in chromitite layers and the Merensky Reef of the western Bushveld Complex; a Re-Os and Rb-Sr isotope study. Earth. Planet. Sci. Lett. 172, 49—64.

Schoenberg, R., Nagler, T.F., Gnos, E., Kramers, J.D., Kamber, B.S., 2003. The source of the Great Dyke, Zimbabwe, and its tectonic significance: evidence from Re-Os isotopes. J. Geol. 111, 565—578.

SECULAR CHANGE OF CHROMITE CONCENTRATION PROCESSES FROM THE ARCHEAN TO THE PHANEROZOIC

5

Shoji Arai[1] and Ahmed H. Ahmed[2,3]

[1]Kanazawa University, Kanazawa, Japan [2]Helwan University, Cairo, Egypt
[3]King Abdulaziz University, Jedda, Saudi Arabia

CHAPTER OUTLINE

5.1 INTRODUCTION

Earth has gradually changed its magmatic, metamorphic, and hydrothermal products with time (Condie, 1985; Martin and Moyen, 2002; Brown, 2007). Stowe (1994) and Hutchinson (2000) summarized secular changes in ore deposits through Earth's history. Stowe (1994) noted that podiform chromitites have been widely produced in the Phanerozoic mantle, in contrast to the stratiform chromitites, which were preferentially formed during the Archean to Early Proterozoic time, but did not refer to the mechanism responsible for this change (Stowe, 1994). In this chapter we revisit and reinterpret this issue based on new data and ideas on the generation of podiform chromitites.

Chromitites (chromite- or chromian spinel-dominant rocks) are widely distributed from the crust to the mantle, and provide us with information on deep-seated magmatic processes. Crustal chromitites are mainly represented by chromite concentrations in stratiform complexes, such as the Great Dyke, Bushveld, and Stillwater intrusions. Podiform chromitites are found in mantle peridotites,

Processes and Ore Deposits of Ultramafic-Mafic Magmas through Space and Time. DOI: http://dx.doi.org/10.1016/B978-0-12-811159-8.00006-8

mainly from ophiolites. The distinction between the two types is sometimes difficult due to intense tectonic disturbance, metamorphism, and weathering, especially in the oldest rocks (e.g., Ferreira Filho et al., 1992). Gornostayev et al. (2004) reported possible 3.0 Ga ophiolitic chromitites from the Ukrainian Shield, but they are closely associated not with mantle harzburgite but with gabbroic rocks. They show no features characteristic of ordinary podiform chromitites. The stratiform chromitites form thin but continuous layers intercalated with silicate-rich layers, such as peridotite, pyroxenite, and anorthosite (Schulte et al., 2012). The podiform chromitites are usually enveloped by dunite within harzburgites in the mantle section to the Moho transition zone of ophiolites, as well as in alpine-type peridotite complexes. They show various shapes, such as irregular pods, lenses, dikes, or bands, and their lateral continuity is limited.

The stratiform chromitites were formed as crystal cumulates from chromite-oversaturated magmas in crustal magma chambers or in large-scale sills (e.g., Irvine, 1977), more or less affected by selective sinking of chromite crystals from cotectic olivine-chromite liquidus phase assemblages. The podiform chromitites fill mantle magma conduits, and possibly represent cumulates from chromite-oversaturated magmas, which were formed by mixing of a primitive magma saturated in olivine + chromite and a more silicic magma resulting from magma-mantle orthopyroxene reactions (Arai and Yurimoto, 1994; Arai and Miura, 2016a). Irrespective of the origin of chromitite, the production of the Mg-rich primary magmas that involved in chromitite production is primarily dependent on the thermal state of the mantle. We reexamine the secular change of these two styles of chromitite production during Earth's history to evaluate a change of magmatic processes in the mantle, which in turn has possibly controlled the Cr distribution throughout the upper mantle to the crust.

5.2 GENERATION OF CHROMITITES

5.2.1 STRATIFORM CHROMITITES

Major layered intrusions such as the Bushveld (e.g., Cameron, 1980, 1982; Mondal and Mathez, 2007), Stillwater (Jackson, 1968; Campbell and Murck, 1993), Great Dyke (Wilson, 1982; Wilson, 1996), Muskox (Irvine and Smith, 1967), and Kemi (Alapieti et al., 1989) intrusions are of Proterozoic age. The layered intrusions commonly contain chromitite layers (=== = stratiform chromitites), which are characterized by a remarkable lateral continuity. Even thin layers can be traceable over several tens of kilometers (Jackson and Thayer, 1972). The host rocks to the chromitites are orthopyroxenites and anorthosites in the Bushveld and Stillwater complexes, but peridotites, including dunites, in the Great Dyke (Wilson, 1982). The magmas involved in the generation of the layered intrusions are mainly boninites or komatiites (e.g., Rollinson, 1997; Mondal et al., 2006; Prendergast, 2008; Mukherjee et al., 2010). The oldest (>4.2 Ga) chromitites are found in anorthosites in western Greenland, and are also part of a large layered intrusion (Ghisler, 1970; Chadwick and Crewe, 1986; Appel et al., 2002).

Sill-like intrusions have been found in Archean greenstone belts (e.g., Prendergast, 2008; Wilson and Nutt, 1990). They show good layered structure more or less similar to that in the typical Proterozoic layered intrusions such as Bushveld and Stillwater, and their chromitites also occur as layers together with ultramafic cumulates. The mid-Archean (c. 3.0 Ga) Shurugwi chromitites in the Zimbabwe Craton (Prendergast, 2008) are a good example. All these layered chromitites can be grouped as the "stratiform chromitites" (cf. Schulte et al., 2012).

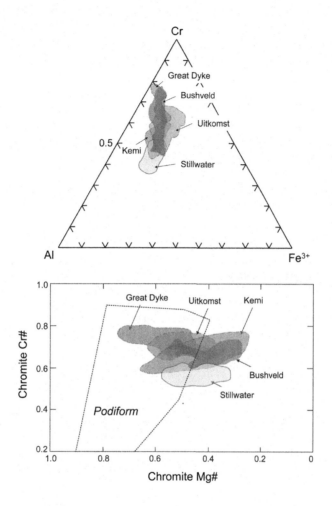

FIGURE 5.1

Chemical characteristics of chromites in stratiform chromitites.

Wilson, A.H., 1982. The geology of the Great "Dyke", Zimbabwe: the ultramafic rocks. J. Petrol. 23, 240–292 for the Great Dyke Complex, Zimbabwe; Naldrett, AJ., Kinnaird, JA., Wilson, A., Yudovskaya, M., McQuade, S., Chunnett G., et al., 2009. Chromite composition and PGE content of Bushveld chromitites: Part 1—the Lower and Middle Groups. Appl. Earth Sci. 118, 131–161 (Naldrett et al., 2009) for the Bushveld Complex, South Africa; Campbell, I.H., Murck, B.W., 1993. Petrology of the G and H chromitite zones in the Mountain View Area of the Stillwater Complex, Montana. J. Petrol. 34, 291–316 for the Stillwater Complex, USA; Alapieti, T.T., Kujanpää, J., Lahtinen, J.J., Papunen, H., 1989. The Kemi stratiform chromitite deposit, northern Finland. Econ. Geol. 84, 1057–1077 for Kemi Complex, Finland; and Yudovskaya M.A., Naldrett A.J., Woolfe J.A.S., Costin G., Kinnaird J.A., 2015. Reverse compositional zoning in the Uitkomst chromitites as an indication of crystallization in a magmatic conduit J. Petrol. 56, 2373–2394 (Yudovskaya et al., 2015) for the Uitkomst Complex, South Africa.

The mechanism for the generation of chromite-oversaturated magmas or the stratiform chromitites has been a matter of debate. Increases in the total pressure (Lipin, 1993) or in the oxygen fugacity (Roeder and Reynolds, 1991) may form them, but the mixing of a primitive magma (e.g., an olivine-chromite cotectic magma) with a felsic magma formed by partial melting of wall rocks (Irvine, 1975; Spandler et al., 2005) or with more fractionated magmas (Irvine, 1977; Campbell and Murck, 1993) has been supported by many studies to date. Supplies of chromite-laden magmas from deeper levels were suggested based on the chromium budget (Mondal and Mathez, 2007; Eales, 2000). The magmas that formed the layered or sill-like intrusions hosting stratiform chromitites successfully transported a large amount of chromium from the mantle to the crust, regardless of the mechanism of chromitite formation.

Chromites from the stratiform chromitite show as a whole a differentiation trend and a limited range of the Cr# (== = Cr/[Cr + Al] atomic ratio), from 0.5 to 0.8, which is positively correlated with the Fe^{3+} ratio (Fig. 5.1). The least evolved chromite, with the highest-Cr# ($\sim$0.8) and lowest Fe^{3+}-chromite, has been found from the Great Dyke (Fig. 5.1). The Mg# (== = Mg/[Mg + Fe^{2+}] atomic ratio) of chromites varies from 0.7 to 0.2, depending both on the chromite/mafic mineral ratio and on the equilibrium temperature during the postmagmatic cooling (Arai, 1980).

5.3 PODIFORM CHROMITITES

Lago et al. (1981) found that the podiform chromitites basically formed as cumulates which filled mantle magma conduits. The chromite-oversaturated magma involved was produced by the mixing of an olivine-chromite cotectic magma with relatively silica-rich magma formed by the magma-peridotite (harzburgite) reaction within the conduit in the mantle (Arai and Yurimoto, 1994; Zhou et al., 1994) (Fig. 5.2). In contrast to the stratiform chromitites in the crust, the formation of

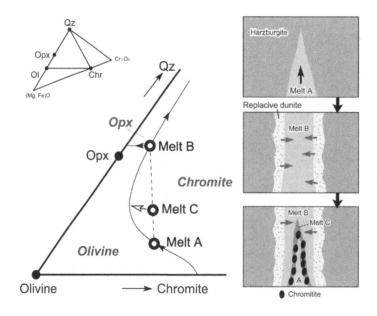

FIGURE 5.2

A model for the podiform chromitites in the mantle after (Arai and Yurimoto, 1994).

From Figure 2 of Arai, S., Miura, M., 2016a. Formation and modification of chromitites in the mantle. Lithos 264, 277–295.

podiform chromitites strongly suggests that a large amount of chromium has been preserved within the upper mantle.

The Cr# of chromites from the podiform chromitites shows a wide range mainly from 0.4 to 0.9 (Fig. 5.3). The $Fe^{3+}/(Cr + Al + Fe^{3+})$ ratio is generally lower than 0.1 (Fig. 5.3). The Mg# of chromites is not correlated with the Cr#, possibly due to the variable chromite/olivine ratios in chromitites (Arai, 1980) (Fig. 5.3).

Some of the pairs of the chromitite pod and enclosing dunite (== = the dunite envelope) are concordant with the fabrics of the ambient mantle peridotite, and others are discordant (Lago et al., 1981; Cassard et al., 1981; Miura et al., 2012). The chromitite-dunite pods were initially discordant, i.e., cutting the deformation structure of the mantle peridotite (Lago et al., 1981), but were rotated into the foliation via deformation (horizontal mantle flow) (Cassard et al., 1981).

The origin of podiform chromitites has been discussed by many people (e.g., Arai and Yurimoto, 1994; Arai and Miura, 2016a; Zhou et al., 1994; Uysal et al., 2009; González-Jiménez et al., 2013; Ahmed and Habtoor, 2015). The magma-harzburgite reaction is included in most models as an elementary step of the chromitite generation processes. The model of magma-harzburgite reaction combined with magma mixing stated above well explains the mode of occurrence of the podiform chromitites. The dunite was left after the reaction between peridotite and invading magma, resulting in the decomposition of mantle orthopyroxene and the precipitation of new olivine (Arai and Yurimoto, 1994) (Fig. 5.2). The dunite thus formed encloses chromitite and replaces

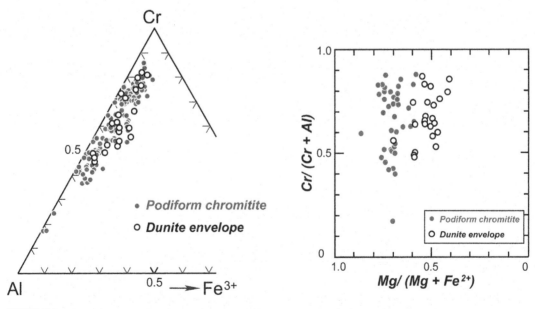

FIGURE 5.3

Chemical characteristics of chromites in podiform chromitites and enclosing dunites.

Data compilation of Arai, S., 1980. Dunite-harzburgite-chromitite complexes as refractory residue in the Sangun-Yamaguchi zone, western Japan. J. Petrol. 21, 141–165 and Arai, S., Miura, M., 2016a. Formation and modification of chromitites in the mantle. Lithos 264, 277–295.

harzburgite along the contact. This means that the chromitite formation is more or less controlled by the chemistry of the wall-rock peridotite (Arai and Abe, 1995; Arai, 1997). This is consistent with the observation that podiform chromitites are most commonly found in moderately depleted harzburgites, which contain small amounts of clinopyroxene and chromite (or chromian spinel) with intermediate values of Cr# 0.4−0.6 (Arai and Miura, 2016a,b; Arai and Abe, 1995; Arai, 1997). Neither fertile lherzolites nor depleted harzburgites are a good host for podiform chromitite (Arai and Miura, 2016a,b; Arai and Abe, 1995; Arai, 1997). It is noteworthy that the chromitite-forming system is far larger than the scales of outcrops or mines, where the budget of Cr cannot be maintained (Arai and Miura, 2016a; Arai and Miura, 2015). We consider the secondary magma (== = melt B in Fig. 5.2) collects Cr and leaves barren dunites (Arai and Miura, 2016a). The secondary magma, possibly with suspended chromite crystals, moves to another place and precipitates the podiform chromitite by mixing with subsequently supplied primitive magmas.

The moderately depleted harzburgites that host podiform chromitites formed as refractory residues after moderate degrees of partial melting (Jaques and Green, 1980; Arai, 1994). They are typically found at the current ocean floor of fast-spreading ridges. This kind of harzburgite, in which

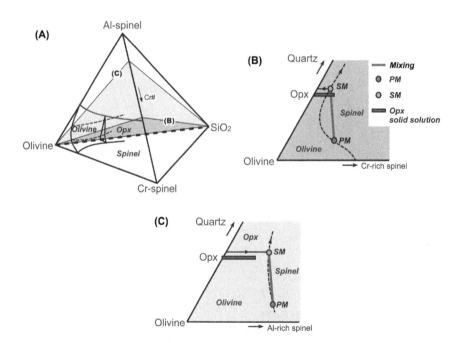

FIGURE 5.4

Chemical control of chromitite formation in a hypothetical quaternary system of olivine-(Cr, Al) spinels-quartz (A). In a high-Cr system, the mixed magma between a primitive magma and a silica-rich one is well in the chromite (spinel) primary field (B) and could precipitate a large amount of chromites. In a low-Cr (high-Al) system, in contrast, the mixed magma is close to the chromite (spinel)-olivine cotectic boundary and could not precipitate a large amount of chromite (spinel) (C) (Miura et al., 2012).

From Figure 10 of Arai, S., Miura, M., 2016a. Formation and modification of chromitites in the mantle. Lithos 264, 277−295.

spinel shows moderate Cr#, 0.4—0.6, may show degrees of partial melting to 15%—20% (Jaques and Green, 1980; Arai, 1994; Hellebrand et al., 2001). These degrees of partial melting basically are attained at lower temperatures than higher degrees (>20%) of partial melting (Jaques and Green, 1980; Takahashi and Kushiro, 1983). Generally speaking, if the ambient temperature of the mantle is higher, a higher degree of partial melting of the mantle peridotite is possible. So the chemistry of residual mantle peridotite is strongly dependent on the thermal state of the mantle.

We can conclude that the formation of podiform chromitites means the preservation of chromium or chromite within the mantle irrespective of the origin. The degree of partial melting of the mantle peridotite is important for the podiform chromitite production: a moderately depleted peridotite, i.e., the harzburgite in which chromite shows a Cr# around 0.4—0.6, provides an optimum condition for the chromitite formation in the mantle (Figs. 5.4 and 5.5).

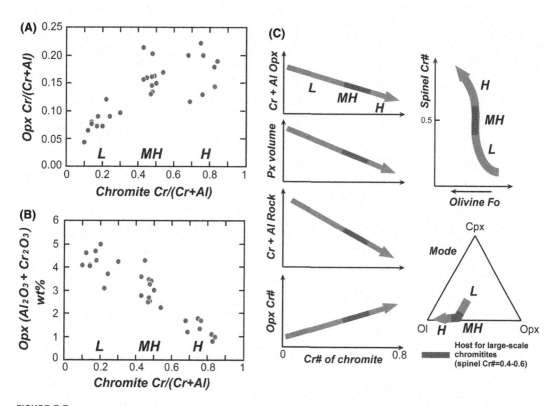

FIGURE 5.5

Diagrams to show the moderately depleted harzburgite, which provides the optimum condition for chromitite formation. Combined with Fig. 5.4, only the moderately depleted harzburgite (MH) has enough high Cr#s and (Cr + Al) contents (A and B). The chemical characteristics of the moderately depleted harzburgite (MH) as compared with lherzolites (L) and highly depleted harzburgites (H) (C).

From Figure 11 of Arai, S., Miura, M., 2016a. Formation and modification of chromitites in the mantle. Lithos 264, 277–295.

5.4 POSSIBLE SECULAR VARIATION OF THE ABUNDANCE OF STRATIFORM CHROMITITE IN THE CRUST

The dimensions of stratiform chromitites appear to decrease in size with time. The Precambrian stratiform chromitites, e.g., those from the Great Dyke, Stillwater, and Bushveld, are large in dimension and have been continuously mined to supply chromium to our society. In contrast, smaller Paleogene layered intrusions formed in relation to the opening of the Northern Atlantic, such as Rum (e.g., Emeleus and Troll, 2014) and Skaergaard (Wager and Brown, 1968; Nielsen, 2004), contain very small-scale chromitites. Chromitites occur as very thin (<5 mm) seams associated with peridotites or troctolites in the Rum complex (Henderson and Suddaby, 1971; O'Driscoll et al., 2009; O'Driscoll et al., 2010). The chromite phase itself is very rare in the Skaergaard intrusion (Wager and Brown, 1968; Nielsen, 2004).

Other Phanerozoic plutonic complexes contain no stratiform chromitites. No chromitites have been found in gabbroic complexes (Paleogene), associated with peridotites, from the Hidaka belt, in the axial zone of Hokkaido, Japan (Komatsu et al., 1983). The Miocene Murotomisaki gabbroic complex of Shikoku Island, Japan, has no chromite concentrations (Yajima, 1972; Hoshide and Obata, 2012).

Chromitites are also rare in the crustal sections of ophiolite (Arai et al., 2004). Chromitite seams or layers have not been found in the layered gabbros (Adachi and Miyashita, 2003) nor in the gabbroic part of the Moho transition zone in the Cretaceous Oman ophiolites (Akizawa and Arai, 2009; Negishi et al., 2013), even though they show good exposures. The MORB magmatism appears to produce no chromitite seams or layers in the current suboceanic lower crust (Dick et al., 2000; Gillis et al., 2014).

5.5 POSSIBLE SECULAR VARIATION OF CHROMITE COMPOSITION IN THE MANTLE PERIDOTITE

The mantle peridotite from ophiolites seems to have changed, on average, in composition, inferred from the Cr# of chromite, with time, although the data from old ophiolites are very scanty (Fig. 5.6). For example, there have been no reports of lherzolites from the Precambrian ophiolites (Ahmed and Habtoor, 2015; Ahmed, 2013; Liipo et al., 1995), although they are common in Phanerozoic ophiolites (Menzies and Allen, 1974; Khedr et al., 2013). Chromite compositions in Precambrian rocks should be treated with care because chromites may have been modified in chemistry during metamorphism (e.g., Ahmed et al., 2001). The Cr# of chromite easily increases with or without an increase in Fe^{3+} via chlorite formation in metamorphosed peridotites or chromitites (Ahmed et al., 2001; Arai et al., 2006; Merlini et al., 2009).

The harzburgites from the Late Archean (2.5 Ga) Zunhua and Dongwanzi ophiolites, North China Craton, are highly depleted and contain high-Cr# (0.7–0.9) chromite (Li et al., 2002; Huang et al., 2004) (Fig. 5.6). Harzburgites in the Early Proterozoic ophiolites from Finland (Jormua and Outokumpu; c. 2.0 Ga) (Vuollo et al., 1995; Kontinen, 1987) also show a seemingly depleted character; the Cr# of chromite ranges from 0.4 to 0.9 (Liipo et al., 1995; Peltonen and Kontinen, 2004). However, the high-Cr# (>0.6) chromites may be alteration products because of the intense

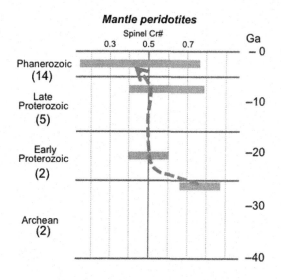

FIGURE 5.6

Possible secular change of the composition of mantle peridotites from ophiolites in terms of the Cr# of chromites. Parenthesized are numbers of ophiolitic complex examined. For the Phanerozic ophiolites see (González-Jiménez et al., 2013); (Jaques and Green, 1980; Arai, 1994; Hellebrand et al., 2001) for the Late Proterozoic ophiolites (== = Pan-African ophiolites); (Liipo et al., 1995; Vuollo et al., 1995) for the Early Proterozoic ophiolites; (Li et al., 2002; Huang et al., 2004) for the Archean ophiolites.

serpentinization or metamorphism of the host peridotite (Liipo et al., 1995; Li et al., 2002). The high Zn in some chromites (Liipo et al., 1995) is due to metamorphism (Gahlan and Arai, 2007; Gahlan et al., 2006). The Zn was possibly supplied from decomposing olivine together with Mn and Ni (see Table 2 of Liipo et al., 1995) upon serpentinization (Gahlan and Arai, 2007; Gahlan et al., 2006). The primary chromite may have had intermediate Cr# around 0.4−0.6 (Fig. 5 of Liipo et al., 1995) (Fig. 5.6). The Finnish Early Proterozoic ophiolitic peridoties are not very depleted but similar in chromite Cr# to modern abyssal harzburgites (e.g., Arai, 1994; Dick and Bullen, 1984).

Peridotites from the mantle section of Late Proterozoic ophiolites (e.g., the Pan-African ophiolites) are also depleted; their chromites mainly show high Cr#, 0.6−0.8 (Ahmed and Habtoor, 2015; Khedr and Arai, 2013; Ahmed, 2013; Farahat et al., 2011) (Fig. 5.7). They are more depleted than the abyssal harzburgites, in which chromites show lower Cr# (<0.6) (Arai and Matsukage, 1996; Dick and Natland, 1996) (Fig. 5.7). Khedr and Arai (Khedr and Arai, 2016) recently found moderately depleted harzburgites that contain chromites with intermediate Cr#s (0.4−0.5) from the southern Eastern Desert, Egypt (Figs. 5.6 and 5.7).

Most of the mantle peridotites from the Phanerozoic ophiolites are less depleted than the Precambrian ones. In the former, the Cr# of the chromites is lower than 0.7 (Arai, 1997; Arai, 1994), although some others contain higher-Cr# (up to 0.9) chromites (e.g., England and Davies, 1973; Rammlmair et al., 1987) (Fig. 5.6).

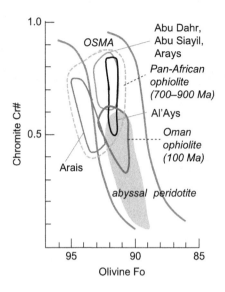

FIGURE 5.7

Differences in the Fo (olivine) and the Cr# (chromite) from peridotites between the Proterozoic ophiolites (Pan-African ophiolites) and the younger Oman ophiolite and the abyssal peridotites; (Ahmed and Habtoor, 2015; Khedr and Arai, 2013; Ahmed, 2013) for the Pan-African ophiolites; Arai (1997) for the abyssal peridotites; Arai (2006) for the main cluster of peridotites from the Oman ophiolite.

In summary, *on average*, the Archean mantle peridotites may have been more depleted than the Proterozoic ones, which were in turn more depleted than the Phanerozoic ones (Fig. 5.6).

5.6 POSSIBLE SECULAR VARIATION OF CHROMITE COMPOSITION IN PODIFORM CHROMITITES

The Cr# of chromites in podiform chromitites may have changed with time in accordance with that of the mantle peridotite described above (cf. Arai, 1997) (Fig. 5.8). The Late Archean chromitites from the Zunhua and Dongwanzi ophiolites, China, are composed of high-Cr# (0.7–0.9) chromites (Li et al., 2002; Huang et al., 2004). The Early Proterozoic Finnish chromitites contain high-Cr# (0.6–0.7) chromites, but they are less Cr-rich than the Archean Zunhua and Dongwanzi chromitites (Cr#, 0.7–0.9). The Late Proterozoic Pan-African chromitites are also composed of high-Cr# (0.6–0.9) chromites (Ahmed and Habtoor, 2015; Ahmed, 2013; Menzies and Allen, 1974; Farahat et al., 2011; Khedr and Arai, 2016; Khudeir et al., 1992; Ahmed et al., 2009; El Ela and Farahat, 2009).

Chromitites from the Phanerozoic ophiolites show a wide range in chromite Cr#, from 0.3 to 0.9 (compiled by Arai, 1997) (Fig. 5.3), possibly depending on the tectonic setting of formation (cf. Arai and Miura, 2016a,b; Arai and Miura, 2015). The low Cr# (<0.5) of chromites particularly characterizes the Phanerozoic chromitites (e.g., Arai, 1980; Thayer, 1964; Henares et al., 2010).

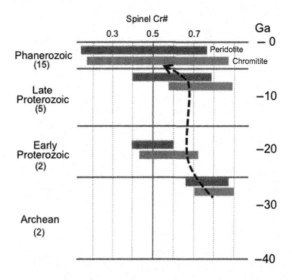

FIGURE 5.8

Possible secular change of the Cr# of chromite in podiform chromitites; Data sources are the same as in Fig. 5.6.

For example, some chromitite pods from Camagüey, Cuba (of the Upper Jurassic to Lower Cretaceous), are composed of very low-Cr# (0.3−0.4) chromite (e.g., Ahmed et al., 2009). Thayer (Thayer, 1964) especially reported a remarkably low-Cr# (0.177) chromite (chromian spinel) in a chromitite from North Carolina, USA.

5.7 POSSIBLE SECULAR VARIATION ON THE ABUNDANCE OF CHROMITITE PODS IN THE MANTLE

The abundance of chromitite in the upper mantle section has seemingly increased with time. As far as we know, there have been no large mines working Precambrian podiform chromitites. The size of individual chromitite pods also appears to have changed with time (Fig. 5.9). This apparent variation may be consistent with an increase in the abundance of podiform chromitites, although there is no clear assurance that the pod size is a proxy for the chromitite abundance. We should treat this issue with care because the old chromitite pods are subject to change in shape and size due to deformation. Although not clearly showing a secular change, chromitite pods from the Precambrian ophiolites are apparently smaller in size than those from Phanerozoic ophiolites. The known Archean chromitite pods (Zunhua and Dongwanzi ophiolites) are mostly less than 20 m across (rarely up to 100 m in a longer dimension) (Li et al., 2002; Huang et al., 2004; Kusky, 2013). The Early Proterozoic chromitite pods from the Jormua and Outokumpu ophiolites are also very small, mostly around 5 × 0.7 m in size (Peltonen and Kontinen, 2004; Vuollo, 2013). The Late Proterozoic Pan-African chromitite pods usually show thin disc-like forms up to 30 m in length (Khedr and Arai, 2013; Ahmed, 2013; Menzies and Allen, 1974; Khedr and Arai, 2016) (Fig. 5.10).

Age	Examples	Size	References
Phanerozoic	Kempirsai Tari-Misaka	< 1500 x 150 m < 200 x 30 m	Distler et al. (2008) Matsumoto et al. (2002)
Late Proterozoic	Pan-African	< 30 m long	Our observations
Early Proterozoic	Outokumpu Jormua	< 5 x 0.7 m?	Vuollo (pers. com.) Peltonen et al. (2004)
Archean	Zunhua Dongwanz	Rarely > 100 m long (mostly < 20 m)	Kusky (pers. com.) Huang et al. (2004)

Increasing?

FIGURE 5.9

A summary of the approximate sizes of individual chromitite pods. A possibly secular variation can be observed.

FIGURE 5.10

Examples of the size of chromitite pod from the Pan-African ophiolites. (A): The largest pod from the Al'Ays ophiolite, Saudi Arabia (see Ahmed and Habtoor, 2015). (B), (C): Hollows formed by complete mining of chromitites from the Bou-Azzer ophiolite, Morocco (see Ahmed et al., 2009). The size and shape of pod are well observed. (B) shows the largest pod in Bou-Azzer.

In contrast, large chromitite pods are rather common in Phanerozoic ophiolites. The largest pod that has ever been reported is the "40 Years of the KazSSR" deposit (Distler et al., 2008) from the Early Paleozoic Kempirsai ophiolite, in the southern Urals (Melcher et al., 1999). The shape of the "40 Years of the KazSSR" pod is very complex but shows a dimension of around 1.5 km × 150 m (Fig. 2 of Distler et al., 2008). The largest pod composed of low-Cr# (around 0.5) chromite may be the "Nana-Go" (== = seventh) deposit from the Wakamatsu mine in the Tari-Misaka complex (part of the Ordovician Oeyama ophiolite), SW Japan (Arai, 1980), with a dimension of 200 × 30 m (Matsumoto et al., 2002). The Cretaceous Guleman ophiolite, Turkey, is famous for its rich chromitite production (e.g., Thayer, 1964; Üşümezsoy, 1990); the Guleman chromitite pods have been mined since the 1930s.

5.8 DISCUSSION

The oldest ophiolite (chromitite) is just 2.5 billion years old, but according to the proposed ophiolitic pulse, older complexes equivalent to ophiolites have been cyclically formed at intervals of around 1 billion years (e.g., Vaughan and Scarrow, 2003). They are the so-called greenstone belts (e.g., de Wit, 2004), but we do not find the mantle peridotites that contain podiform chromitites there. It is noteworthy that both the ophiolites and the greenstone belts were possibly representative of the oceanic lithosphere at that time (Vaughan and Scarrow, 2003; de Wit, 2004). In summary, the Archean mafic-ultramafic sills with clearly layered structures host the stratiform chromitites instead of podiform chromitites (e.g., Prendergast, 2008; Wilson and Nutt, 1990).

As stated above, the layered intrusions that host stratiform chromitites were prevalent in the Archean to the Early Proterozoic, and the ophiolites that host podiform chromitites were commonly produced from the Late Proterozoic to the Phanerozoic. This means that Cr was preferentially transported to the crust via relatively Cr-rich magmas, formed by high-degree partial melting, instead of precipitation in the mantle as the podiform chromitites, in Archean to Early Proterozoic time (Fig. 5.11). This in turn indicates that the mechanisms that generate chromite-oversaturated magmas were more easily available within the crust before the Early Proterozoic, but within the mantle after the Proterozoic in Earth's history (Fig. 5.11). This is possibly due to the highly depleted character of the upper mantle in Archean to the Early Proterozoic due to the earlier production of larger amounts of mafic to ultramafic magma (cf. Stowe, 1994). Based on the model of podiform chromitite formation in the mantle (Arai and Yurimoto, 1994; Arai and Miura, 2016a; Miura et al., 2012; Arai and Abe, 1995), the highly depleted harzburgite-magma system is not optimum for the production of large-scale podiform chromitites (Figs. 5.4 and 5.5). Orthopyroxenes in the highly depleted harzburgite are low in both modal amount and in (Cr + Al), and could not release a large amount of (Cr + Al) to form chromite (or chromitite) on interaction between the harzburgite and magmas (Fig. 5.5).

In the Late Proterozoic to the Phanerozoic, less-depleted peridotites were prevalent in the upper mantle through an ascent of the fertile peridotites from deeper mantle and lower degrees of partial melting, which has left the moderately depleted residual harzburgites in the upper mantle. They provide an optimum condition for the production of large volumes of podiform chromitite as discussed by Arai and Abe (1995) and Arai (1997).

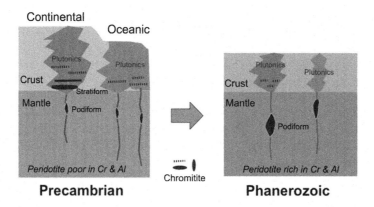

FIGURE 5.11

Cartoon to indicate a switch of the main way of chromitite production in Earth's history. In the Precambrian, especially the Archean, large amounts of Mg-, Cr-rich magmas were produced in the mantle by high degrees of partial melting and transported to the surface to form stratiform chromities in layered intrusions. This gave rise to the formation of highly depleted mantle peridotite, which is not a good host for a large amount of podifrom chromitite. In the Phanerozoic, moderate degrees of partial melting left moderately depleted harzburgite in the mantle, and supplied a good host for podiform chromitite formation through the peridotite-magma interaction and related processes. This mechanism left large amounts of chromitite (or Cr) preferentially in the mantle. The crustal stratiform chromitite is small in amount—if any—as compared with that in the old time.

In conclusion, in the Archean to the Early Proterozoic, chromitites were mainly produced as stratiform bodies in the crust because the mechanism for production of a large amount of chromitite was restricted by the highly depleted character of the upper mantle peridotite (Fig. 5.11). In the Late Proterozoic to the Phanerozoic the highly depleted upper mantle was replaced with less depleted peridotite (mainly mildly depleted harzburgite), which was a good host for chromitite pods (see Arai, 1997) (Fig. 5.5). Cr has been preserved in the upper mantle as podiform chromitites, rather than being transported upward to form large-scale stratiform chromitites in the crust (Fig. 5.11).

ACKNOWLEDGEMENTS

We are grateful to Sisir K. Mondal for inviting us to contribute to this book. We also thank T.M. Kusky and J. Vuollo for giving us information on podiform chromitites from China and Finland, respectively. T.M. Kusky and an anonymous reviewer gave us critical comments, which were helpful in revision. M. Miura helped us to prepare some figures. Bill Griffin kindly made English edition.

REFERENCES

Adachi, Y., Miyashita, S., 2003. Geology and petrology of the plutonic complexes in the Wadi fizh area: multiple magmatic events and segment structure in the northern Oman ophiolite. Geochem. Geophys. Geosyst. 4, 8619. Available from: http://dx.doi.org/10.1029/2001GC000272.

Ahmed, A.H., 2013. Highly depleted harzburgite-dunite-chromitite complexes from the Neoproterozoic ophiolite, south Eastern Desert, Egypt: a possible recycled upper mantle lithosphere. Precamb. Res. 233, 173−192.

Ahmed, A.H., Arai, S., Attia, A.K., 2001. Petrological characteristics of the Pan African podiform chromitites and associated peridotites of the Proterozoic ophiolite complexes, Egypt. Mineral. Deposita 36, 72−84.

Ahmed, A.H., Habtoor, A., 2015. Heterogeneously depleted Precambrian lithosphere deduced from mantle peridotites and associated chromitite deposits of Al'Ays ophiolite, Northwestern Arabian Shield, Saudi Arabia. Ore Geol. Rev. 67, 279−296.

Ahmed, H.A., Arai, S., Abdel-Aziz, Y.M., Ikenne, M., Rahimi, A., 2009. Platinum-group elements distribution and spinel composition in podiform chromitites and associated rocks from the upper mantle section of the Neoproterozoic Bou Azzer ophiolite, Anti-Atlas, Morocco. J. Afr. Earth Sci. 55, 92−104.

Akizawa, N., Arai, S., 2009. Petrologic profile of peridotite layers under a possible Moho in the northern Oman ophiolite: an example from Wadi Fizh. J. Mineral. Petrol. Sci. 104, 389−394.

Alapieti, T.T., Kujanpää, J., Lahtinen, J.J., Papunen, H., 1989. The Kemi stratiform chromitite deposit, northern Finland. Econ. Geol. 84, 1057−1077.

Appel, C.C., Appel, P.W.U., Rollinson, H.R., 2002. Complex chromite textures reveal the history of an early Archean layered ultramafic body in Western Greenland. Mineral. Mag. 66, 1029−1041.

Arai, S., 1980. Dunite-harzburgite-chromitite complexes as refractory residue in the Sangun-Yamaguchi zone, western Japan. J. Petrol. 21, 141−165.

Arai, S., 1994. Characterization of spinel peridotites by olivine-spinel compositional relationships: review and interpretation. Chem. Geol. 113, 191−204.

Arai, S., 1997. Control of wall-rock composition on the formation of podiform chromitites as a result of magma/peridotite interaction. Resour. Geol. 47, 177−187.

Arai, S., Abe, N., 1995. Reaction of orthopyroxene in peridotite xenoliths with alkali basalt melt and its implication for genesis of alpine-type chromitite. Am. Mineral. 80, 1041−1047.

Arai, S. and Kadoshima, K., 2006. Unpublised data.

Arai, S., Matsukage, K., 1996. Petrology of the gabbro-troctolite-peridotite complex from Hess Deep, equatorial Pacific: implications for mantle-melt interaction within the oceanic lithosphere. Proc. Ocean Drilling Prog. Sci. Results 147, 135−155.

Arai, S., Miura, M., 2015. Podiform chromitites do form beneath mid-ocean ridges. Lithos 232, 143−149.

Arai, S., Miura, M., 2016a. Formation and modification of chromitites in the mantle. Lithos 264, 277−295.

Arai, S., Miura, M., 2016b. Reply to the comment of rollinson and adetunji "podiform chromitites do form beneath mid-ocean ridges" by Arai, S. and Miura, M. Lithos 254−255, 134−136.

Arai, S., Shimizu, Y., Ismail, S.A., Ahmed, A.H., 2006. Low-T formation of high-Cr spinel with apparently primary chemical characteristics within podiform chromitite from Rayat, northeastern Iraq. Mineral. Mag. 70, 499−508.

Arai, S., Uesugi, J., Ahmed, A.H., 2004. The upper crustal podiform chromitite from the northern Oman ophiolite as the stratigraphically shallowest chromitite in ophiolite and its implication for Cr concentration. Contrib. Mineral. Petrol. 147, 145−154.

Arai, S., Yurimoto, H., 1994. Podiform chromitites of the Tari-Misaka ultramafic complex, southwestern Japan, as mantle-melt interaction products. Econ. Geol. 89, 1279−1288.

Brown, M., 2007. Metamorphic conditions in orogenic belts: a record of secular change. Inter. Geol. Rev. 49, 193−234.

Cameron, E.N., 1980. Evolution of the Lower Critical Zone, central sector, eastern Bushveld Complex, and its chromite deposits. Econ. Geol. 75, 845−871.

Cameron, E.N., 1982. The Upper Critical Zone of the Eastern Bushveld Complex − Precursor of the Merensky Reef. Econ. Geol. 77, 1307−1327.

Campbell, I.H., Murck, B.W., 1993. Petrology of the G and H chromitite zones in the Mountain View Area of the Stillwater Complex, Montana. J. Petrol. 34, 291–316.

Cassard, D., Nicolas, A., Rabinovitch, M., Moutte, J., Leblanc, M., Prinzhofer, A., 1981. Structural classification of chromite pods in southern New Caledonia. Econ. Geol. 76, 805–831.

Chadwick, B., Crewe, M.A., 1986. Chromite in the Early Archean Akilia association (ca. 3,800 m.y.), Ivisârtoq region, Inner Godthâbsfjord, southern West Greenland. Econ. Geol. 81, 184–191.

Condie, K.C., 1985. Secular variation in the composition of basalts: an index to mantle evolution. J. Petrol. 24, 545–563.

De Wit, M.J., 2004. Archean greenstone belts do contain fragments of ophiolites. In: Kusky, T.M. (Ed.), Precambrian Ophiolites and Related Rocks. Elsevier, Amsterdam, pp. 599–614.

Dick, H.J.B., Bullen, T., 1984. Chromian spinel as a petrogenetic indicator in abyssal and alpine type peridotites and spatially associated lavas. Contrib. Mineral. Petrol. 86, 54–76.

Dick, H.J.B., Natland, J.H., 1996. Late-stage melt evolution and transport in the shallow mantle beneath the East Pacific Rise. Proc. Ocean Drilling Program, Sci. Results 147, 103–134.

Dick, H.J.B., Natland, J.H., Alt, J.C., Bach, W., Bideau, D., Gee, J.S., et al., 2000. A long in situ section of the lower ocean crust: results of ODP Leg 176 drilling at the Southwest Indian Ridge. Earth Planet. Sci. Lett. 179, 31–51.

Distler, V.V., Kryachko, V.V., Yudovskaya, M.A., 2008. Ore petrology of chromite-PGE mineralization in the Kempirsai ophiolite complex. Mineral. Petrol. 92, 31–58.

Eales, H.V., 2000. Implications of the chromium budget of the Western Limb of the Bushveld Complex. SA J. Geol. 103, 141–150.

El Ela, F.F.A., Farahat, E.S., 2009. Neoproterozoic podiform chromitites in serpentinites of the Abu Meriewa-Hagar Dungash district, Eastern Desert, Egypt: geotectonic implications and metamorphism, Island Arc, 19. pp. 151–164.

Emeleus, C.H., Troll, V.R., 2014. The Rum Igneous Centre, Scotland. Mineral. Mag. 78, 805–839.

England, R.N., Davies, H.L., 1973. Mineralogy of ultramafic cumulates and tectonites from Eastern Papua. Earth Planet. Sci. Lett. 17, 416–425.

Farahat, E.S., Hoinkes, G., Mogessie, A., 2011. Petrogenetic and geotectonic significance of Neoproterozoic suprasubduction mantle as revealed by the Wizer ophiolite complex, Central Eastern Desert, Egypt. Inter. J. Earth Sci. 100, 1433–1450.

Ferreira Filho, C.F., Nilson, A.A., Naldrett, A.J., 1992. The Niquelândia mafic-ultramafic complex, Goias, Brazil: a contribution to the ophiolite x stratiform controversy based on new geological and structural data. Precamb. Res. 59, 125–143.

Gahlan, H.A., Arai, S., 2007. Genesis of peculiarly zoned Co, Zn and Mn-rich chromian spinel in serpentinite of Bou-Azzer ophiolite, anti-Atlas, Morocco. J. Mineral. Petrol. Sci. 102, 69–85.

Gahlan, H.A., Arai, S., Ahmed, A.H., Ishida, Y., Abdel-Aziz, Y.M., Rahim, A., 2006. Origin of magnetite veins in serpentinite from the late Proterozoic Bou-Azzer ophiolite, Anti-Atlas, Morocco: an implication for mobility of iron during serpentinization, J. Afr. Earth Sci., 46. pp. 318–330.

Ghisler, M., 1970. Pre-metamorphic folded chromite deposits of stratiform type in the Early Precambrian of West Greenland. Mineral. Deposita 5, 223–236.

Gillis, K.M., Snow, J.E., Klaus, A., Abe, N., Adrião, Á.B., Akizawa, N., et al., 2014. Primitive layered gabbros from fast-spreading lower oceanic crust. Nature 505, 204–207.

González-Jiménez, J.-M., Griffin, W.L., Proenza, J.A., Gervilla, F., O'Reilly, S.Y., Akbulut, M., et al., 2013. Chromitites in ophiolites: how, where, when, why?, Part II. A review and new ideas on the crystallization of chromitites. Lithos 189, 140–158.

Gornostayev, S.S., Walker, R.J., Hanski, E.J., Popovchenko, S.E., 2004. Evidence for the emplacement of ca. 3.0 Ga mantle-derived mafic ultramafic bodies in the Ukrainian Shield. Precamb. Res. 132, 349–362.

Hellebrand, E., Snow, J.E., Dick, H.J.B., Hofmann, A.W., 2001. Coupled major and trace elements as indicators of the extent of melting in mid-ocean-ridge peridotites. Nature 410, 677−681.

Henares, S., González-Jiménez, J.M., Gervilla, F., Proenza, J.A., Rodríguez, González-Pontón, R.B., 2010. Las cromititas del Complejo Ofíolitico de Camagüey, Cuba: un ejemplo de cromitas ricas en Al. Bol. Soc. Geol. Mexicana 62, 173−185 (in Spanish with English abstract)

Henderson, P., Suddaby, P., 1971. Nature and origin of the chrome-spinel of the Rhum layered intrusion. Contrib. Mineral. Petrol. 33, 211−231.

Hoshide, T., Obata, M., 2012. Amphibole-bearing multiphase solid inclusions in olivine and plagioclase from a layered gabbro: origin of the trapped melts. J. Petrol. 53, 419−440.

Huang, X., Li, J., Kusky, T.M., Chen, Z., 2004. Microstructures of the Zunhua 2.50 GA podiform chromite, North China Craton and implications for the deformation and rheology of the Archean oceanic lithospheric mantle. In: Kusky, T.M. (Ed.), Precambrian Ophiolites and Related Rocks. Elsevier, Amsterdam, pp. 321−337.

Hutchinson, W.R., 2000. Mineral deposits as guides to supracrustal evolution. In: O'Connell, R.J., Fyfe, W.S. (Eds.), Evolution of the Earth. American Geophysical Union, Washington, DC, pp. 120−140.

Irvine, T.N., 1975. Crystallization sequences in the Muskox intrusion and other layered intrusions. 2. Origin of chromitite layers and similar deposits of other magmatic ores. Geochim. Cosmochim. Acta 39, 911−1020.

Irvine, T.N., 1977. Origin of chromitite layers in the Muskox intrusion and other stratiform intrusions: a new interpretation. Geology 5, 273−277.

Irvine, T.N., Smith, C.H., 1967. The ultramafic rocks of the Muskox Intrusion, Northwest Territories, Canada. In: Wyllie, P.J. (Ed.), Ultramafic and Related Rocks. John Wiley, New York, pp. 39−49.

Jackson, E.D., 1968. The chromite deposits of the Stillwater Complex, Montana. In: Ridge, J.D. (Ed.), Ore Deposits of the United States, 1933-1967, The Graton-Sales Volume, 2. The American Institute of Mining, Metallurgical, and Petroleum Engineers, Inc., New York, pp. 1495−1510.

Jackson, E.D., Thayer, T.P., 1972. Some criteria for distinguishing between stratiform, and alpine peridotite-gabbro complexes. Proc. 24th Geol. Cong. Sec. 2, 289−296.

Jaques, A.L., Green, D.H., 1980. Anhydrous melting of peridotite at 0-15 kb pressure and the genesis of tholeiitic basalts. Contrib. Mineral. Petrol. 73, 287−310.

Khedr, M.Z., Arai, S., 2013. Origin of Neoproterozoic ophiolitic peridotites in south Eastern Desert, Egypt, constrained from primary mantle mineral chemistry. Mineral. Petrol. 107, 807−828.

Khedr, M.Z., Arai, S., 2016. Chemical variations of mineral inclusions in Neoproterozoic high-Cr chromitites from Egypt: evidence of fluids during chromitite genesis. Lithos 240-243, 309−326.

Khedr, M.Z., Arai, S., Python, M., 2013. Petrology and chemistry of basal lherzolites above the metamorphic sole from Wadi Sarami, central Oman ophiolite. J. Mineral. Petrol. Sci. 108, 13−24.

Khudeir, A.A., El Haddod, M.A., Leake, B.E., 1992. Compositional variation in chromite from the Eastern Desert, Egypt. Mineral. Mag. 56, 567−574.

Komatsu, M., Miyashita, S., Maeda, J., Osanai, Y., Toyoshima, T., 1983. Disclosing of a deepest section of continental-type crust up-thrust as the final event of collision of arcs in Hokkaido, north Japan. In: Hashimoto, M., Uyeda, S. (Eds.), Accretion Tectonics in the Circum-Pacific Regions. Terra Scientific Publishing Co., Tokyo, pp. 149−165.

Kontinen, A., 1987. An Early Proterozoic ophiolite − the Jormua mafic-ultramafic complex, northeastern Finland. Precamb. Res. 35, 313−341.

Kusky, T.K., 2013. Personal communications.

Lago, B.L., Rabinowicz, M., Nicolas, A., 1981. Podiform chromite ore deposits: a genetic model. J. Petrol. 23, 103−125.

Li, J., Kusky, T.M., Huang, X., 2002. Archean podiform chromitites and mantle tectonites in ophiolitic mélange, North China Craton: a record of early oceanic mantle processes. GSA Today 2002, 4−11.

Liipo, J., Vuollo, J., Nykänen, V., Pirainen, T., Pekkarinen, T., Tuokko, I., 1995. Chromites from the early Proterozoic Outokumpu-Jormua Ophiolitic Belt: a comparison with chromitites from Mesozoic ophiolites. Lithos 36, 15−27.

Lipin, B.R., 1993. Pressure increases, the formation of chromite seams, and the development of the ultramafic series in the Stillwater Complex, Montana. J. Petrol. 34, 955−976.

Martin, H., Moyen, J.-F., 2002. Secular changes in tonalite-trondhjermite-granodiorite composition as markers of the progressive cooling of Earth. Geology 30, 319−322.

Matsumoto, I., Arai, S., Yamane, T., 2002. Significance of magma/peridotite reaction for size of chromitite: example for Wakamatsu chromite mine of the Tari-Misaka ultramafic complex, southwestern Japan. Shigen-Chishitu 52, 135−146 (in Japanese with English abstract)

Melcher, F., Grum, W., Thalhammer, T.V., Thalhammer, O.A.R., 1999. The giant chromite deposits at Kempirsai, Urals: constrains from trace element (PGE, REE) and isotope data. Mineral. Petrol. 34, 250−272.

Menzies, M., Allen, C., 1974. Plagioclase lherzolite-residual mantle relationships within two eastern Mediterranean ophiolites. Contrib. Mineral. Petrol. 45, 197−213.

Merlini, A., Grieco, G., Deilla, V., 2009. Ferritchromite and chromian-chlorite formation in mélange-hosted Kalkan chromitite (Southern Urals, Russia). Am. Mineral. 94, 1459−1467.

Miura, M., Arai, S., Ahmed, A.H., Mizukami, T., Okuno, M., Yamamoto, S., 2012. Podiform chromitite classification revisited: a comparison of discordant and concordant chromitite pods from Wadi Hilti, northern Oman ophiolite. J. Asian Earth Sci. 59, 52−61.

Mondal, S.K., Mathez, E.A., 2007. Origin of the UG2 chromitite layer, Bushveld Complex. J. Petrol. 48, 495−510.

Mondal, S.K., Ripley, Li,C., Frei, R., 2006. The genesis of Archean chromititites from the Nuashi and Sukinda massifs in the Singhbhum Craton, India. Precamb. Res. 148, 45−66.

Mukherjee, R., Mondal, S.K., Rosing, M.T., Frei, R., 2010. Compositional variations in the Mesoarchean chromites of the Nuggihalli schist belt, Western Dharwar Craton (India): potential parental melts and implications for tectonic setting. Contrib. Mineral. Petrol. 160, 865−885.

Naldrett, A.J., Kinnaird, J.A., Wilson, A., Yudovskaya, M., McQuade, S., Chunnett, G., et al., 2009. Chromite composition and PGE content of Bushveld chromitites: Part 1—the Lower and Middle Groups. Appl. Earth Sci. 118, 131−161.

Negishi, H., Arai, S., Yurimoto, H., Ito, S., Ishimaru, S., Tamura, A., et al., 2013. Sulfide-rich dunite within a thick Moho transition zone of the northern Oman ophiolite: implications for the origin of Cyprus-type sulfide deposits. Lithos 164-167, 22−35.

Nielsen, T.F.D., 2004. The shape and volume of the Skaergaard Intrusion, Greenland: implications for mass balance and bulk composition, J. Petrol., 45. pp. 507−530.

O'Driscoll, B., Donaldson, C.H., Daly, J.S., Emeleus, C.H., 2009. The roles of melt infiltration and cumulated assimilation in the formation of anorthosite and a Cr-spinel seam in the Rum Layered itrusions, NW Scotland. Lithos 111, 6−20.

O'Driscoll, B., Emeleus, C.H., Donaldson, C.H., Daly, J.S., 2010. Cr-spinel seam petrogenesis in the Rum Layered Suite, NW Scotland: cumulate assimilation and *in situ* crystallization in a deforming crystal mush. J. Petrol. 51, 1171−1201.

Peltonen, P., Kontinen, A., 2004. The Jormua Ophiolite: a mafic-ultramafic complex from an ancient ocean-continent transition zone. In: Kusky, T.M. (Ed.), Precambrian Ophiolites and Related Rocks. Elsevier, Amsterdam, pp. 35−71.

Prendergast, M.D., 2008. Archean komatiitic sill-hosted chromite deposits in the Zimbabwe Craton. Econ. Geol. 103, 981−1004.

Rammlmair, D., Rashck, H., Steiner, H.R.L., 1987. Systematics of chromitite occurrences in Central Palawan, Philippines. Mineral. Deposita 22, 190−197.

Roeder, P.L., Reynolds, I., 1991. Crystallization of chromite and chromium solubility in basaltic melts. J. Petrol. 32, 909−934.

Rollinson, H., 1997. The Archean komatiite-related Inyala chromitite, southern Zimbabwe. Econ. Geol. 92, 95−107.

Schulte, R.F., Taylor, R.D., Piatak, N.M., Seal II, R.R., 2012. Stratiform Chromite Deposit Model, Chapter E of Mineral Deposit Model for Resource Assessment. U.S. Geological Survey Scientific Investigation Report 2010-5070-E, U.S. Geological Survery, Reston, Virginia, 131p.

Spandler, C., Mavrogenes, J., Arculus, R., 2005. Origin of chromitites in layered intrusions: evidence from chromite-hosted melt inclusions from the Stillwater Complex. Geology 33, 893−896.

Stowe, C.W., 1994. Compositions and tectonic settings of chromite deposits through time. Econ. Geol. 89, 528−546.

Takahashi, E., Kushiro, I., 1983. Melting of a dry peridotite at high pressures and basalt magma genesis. Am. Mineral. 68, 859−879.

Thayer, T.P., 1964. Principal features and origin of podiform chromitite deposits, and some observations on the Guleman-Soridag district, Turkey. Econ. Geol. 59, 1497−1524.

Üşümezsoy, Ş., 1990. On the formation mode of the Guleman chromitite deposits, Turkey. Mineral. Deposita 25, 89−95.

Uysal, I., Tarkian, M., Sadiklar, M.B., Zaccariani, F., Meisel, T., Gruti, G., et al., 2009. Petrology of Al- and Cr-rich ophiolitic chromitites from the Mugla, SW Turkey: implications from composition of chromite, solid inclusions of platinum-group mineral, silicate, and base-metal mineral, and Os-isotope geochemistry. Contrib. Mineral. Petrol. 158, 659−674.

Vaughan, A.P.M., Scarrow, J.H., 2003. Ophiolite obduction pulsed as a proxy indicator of superplume events? Earth Planet. Sci. Lett. 213, 407−416.

Vuollo, J., 2013. Personal communications.

Vuollo, J., Liipo, J., Nykänen, V., Pirainen, T., Oekkarinen, L., Tuokko, I., et al., 1995. An Early Proterozoic podiform chromitite in the Outokump Ophiolite, Finland. Econ. Geol. 90, 445−452.

Wager, L.R., Brown, G.M., 1968. Layered igneous rocks. Oliver and Boyd, Edinburgh, 588 pp.

Wilson, A.H., 1982. The geology of the Great "Dyke", Zimbabwe: the ultramafic rocks, J. Petrol., 23. pp. 240−292.

Wilson, A.H., 1996. The Great Dyke of Zimbabwe. In: Cawthorn, R.G. (Ed.), Layered Intrusions. Elsevier, Amsterdam, pp. 365−402.

Wilson, J.F., Nutt, T.H.C., 1990. The nature and occurrence of mineralization in the Early Precambrian crust of Zimbabwe. In: Naqvi, S.M. (Ed.), Precambrian Continental Crust and Its Economic Resources. Elsevier, Amsterdam, pp. 555−591.

Yajima, T., 1972. Petrology of the Murotomisaki gabbroic complex. J. Jap. Assoc. Mineral. Petrol. Econ. Geol. 67, 218−241.

Yudovskaya, M.A., Naldrett, A.J., Woolfe, J.A.S., Costin, G., Kinnaird, J.A., 2015. Reverse compositional zoning in the Uitkomst chromitites as an indication of crystallization in a magmatic conduit. J. Petrol. 56, 2373−2394.

Zhou, M.F., Robinson, P.T., Bai, W.J., 1994. Formation of podiform chromitites by melt/rock interaction in the upper mantle. Mineral. Deposita 29, 98−101.

PETROGENETIC EVOLUTION OF CHROMITE DEPOSITS IN THE ARCHEAN GREENSTONE BELTS OF INDIA

Ria Mukherjee[1] and Sisir K. Mondal[2]

[1]*University of the Witwatersrand, Johannesburg, South Africa* [2]*Jadavpur University, Kolkata, West Bengal, India*

6.1 INTRODUCTION

Archean greenstone belts represent some of the earliest records of the Earth's lithospheric history (DeWit and Ashwal, 1995); hence, their study is crucial for understanding the compositional character of the mantle, mantle processes, and the tectonic settings that were prevalent in the early Earth. The greenstone belts represent a complex geological environment, where rocks belonging to multiple tectonic regimes occur in close association with one another. The greenstone belts are mainly composed of interlayered volcanic ultramafic-mafic rocks that comprise 10%—50% of the belt (compositionally komatiites to komatiitic basalts and tholeiites; DeWit and Ashwal, 1995; Hunter and Stowe, 1997), and sediments (turbidites, graywackes, meta-argillites, quartzites, cherts, and barites, banded iron formations (BIFs); Hunter and Stowe, 1997), which are surrounded by the tonalite-trondhjemite granodiorite

Processes and Ore Deposits of Ultramafic-Mafic Magmas through Space and Time. DOI: http://dx.doi.org/10.1016/B978-0-12-811159-8.00007-X

gneisses (TTG); all rocks in a greenstone belt are metamorphosed to low-grade greenschist to amphibo-lite facies, and are strongly sheared and deformed. Sill-like plutonic ultramafic-mafic rocks that are genetically related to high-Mg magmas like komatiites, boninites, and high-Mg siliceous basalts, also constitute the Archean greenstone belts (Rollinson, 1997; Mondal et al., 2001; Prendergast, 2008).

The study of the plutonic ultramafic-mafic rocks in greenstone belts is often underemphasized, but they are significant because they provide important clues about the magmatic processes that operated in the Archean, and because these rocks host economically important deposits of chromite, Ti-V bearing magnetite, Ni-sulfide, and platinum-group elements (PGEs) (Rollinson, 1997; Mondal et al., 2006; Arndt et al., 2008; Mukherjee et al., 2010; Fiorentini et al., 2012; Barnes et al., 2013). Chromite deposits are common in the Archean greenstone belts of Shurugwi in the Zimbabwe craton (Prendergast, 1984; Stowe, 1987), the Sukinda-Nuasahi-Jojohatu Massifs within the Iron-Ore Group (IOG) greenstone belts in the Singhbhum craton (eastern India; Mondal, 2009), the Nuggihalli-Holenarsipur-Krishnarajpet greenstone belts in the Western Dharwar craton (WDC, southern India; Mukherjee et al., 2010), the Jamestown igneous complex from the Barberton green-stone belt in the Kaapvaal craton (South Africa; DeWit et al., 1987), the Bird River Sill within the Bird River greenstone belt in the Superior craton (Canada; Ohnenstetter et al., 1986; Scoates and Scoates, 2013), the chromite-bearing Ring of Fire intrusions in the McFaulds Lake greenstone belt in Northern Ontario (Mungall and Staff, 2008; Laarman, 2013), and the Coobina deposit from the Jimblebar greenstone belt in Western Australia (Pilbara craton; Barnes and Jones, 2013).

Chromite is an efficient petrogenetic tool as its composition is characteristically distinct and diverse in different deposits (Irvine, 1965; Dick and Bullen, 1984; Roeder, 1994; Stowe, 1994; Rollinson, 1995a, 2008; Kamenetsky et al., 2001; Mondal et al., 2006; Mukherjee et al., 2010, 2015). This is largely because the chemical composition of chromite is controlled by mantle-melting processes that are typical of a particular tectonic setting (e.g., Dick and Bullen, 1984; Barnes and Roeder, 2001; Kamenetsky et al., 2001 and references therein). Therefore, the study of chromite deposits in Archean greenstone belts can help to reveal the nature of the mantle and the tectonic processes prevalent in the Archean. Also, the host ultramafic-mafic rocks to the chromitites in Archean greenstone belts are typi-cally metamorphosed and hydrothermally altered and are not reliable indicators, thus making chromites indispensable in Archean petrogenetic studies. Furthermore, studying the secular changes in chromite composition in different types of globally occurring chromite deposits can be very useful in tracing the evolution of the Earth's upper mantle through geological time (Stowe, 1994; Arai and Ahmed, 2013).

Despite chromite being an indispensable petrogenetic tool, the composition of this mineral is not always robust, and can be susceptible to variations, which may arise due to subsolidus reequilibration with the surrounding silicate minerals and interstitial melts (e.g., Jackson, 1969; Hamlyn and Keays, 1979; Scowen et al., 1991; Rollinson, 1995b). Oxidation and alteration of chromites to ferritchromit and magnetite rims is commonly seen in Alpine-type ultramafic complexes (e.g., Burkhard, 1993; Gervilla et al., 2012; González-Jiménez et al. 2014), and in plutonic ultramafic-mafic rocks and koma-tiites from Archean greenstone belts (e.g., Barnes, 2000; Mondal et al., 2006; Mukherjee et al., 2010, 2015). This alteration is considered to be related to serpentinization of the host rock (e.g., Burkhard, 1993; Mukherjee et al., 2010, 2015) or its subsequent prograde (Evans and Frost, 1975; Bliss and Maclean, 1975; Loferski and Lipin, 1983; Abzalov, 1998) or retrograde metamorphism (Gervilla et al., 2012; González-Jiménez et al., 2014). Accessory chromites ($\approx 5-10$ modal % chromite; Irvine, 1965, 1967; Jackson, 1969; Roeder et al., 1979) that are disseminated in the host silicate rocks, or the chro-mites in chromitites containing high silicate modes (≈ 60 modal % chromite; Mukherjee et al., 2010,

2015) are the most affected by such compositional variabilities. Chromites in massive chromitites (>75 modal % chromite) commonly retain primary compositions (Mondal et al., 2006; Mukherjee et al., 2010, 2015), although metamorphosed chromites from massive chromitites within ophiolites may be altered in terms of their minor and trace elements (e.g., González-Jiménez et al., 2014; Colás et al., 2014). Therefore, liquidus chromite compositions should be recovered before using chromites as petrogenetic indicators (Rollinson, 1995a).

In this chapter we have described the chromite deposits hosted in Archean greenstone belts from the Indian Shield (Fig. 6.1). Significant chromite deposits that account for $\approx 96\%-98\%$ of the total resources of Cr in India are hosted within plutonic ultramafic-mafic rocks of the Nuasahi-Sukinda Massifs, which are associated with volcano-sedimentary supracrustal rocks of the IOG greenstone belts in the Singhbhum craton (eastern India; Mondal et al., 2006; Mondal, 2009). These rocks also host significant PGE mineralization (Mondal and Baidya, 1997; Mondal et al., 2001; Mondal and Zhou, 2010; Khatun et al., 2014), which is the only PGE deposit that has been reported from India. The remaining $\approx 2\%-4\%$ of the chromite deposits in India are concentrated in the Nuggihalli-Holenarsipur-Krishnarajpet greenstone belts situated in the WDC (southern India; Mukherjee et al., 2010, 2012, 2014). This chapter essentially provides recent updates on the petrology, geochemistry (bulk-rock and isotope), and geochronology of the ultramafic-mafic rocks and their chromite deposits from the Nuasahi-Sukinda Massifs and the Nuggihalli greenstone belt. The broad aim of this review is to discuss the petrogenesis of these Indian greenstone belt-hosted chromite deposits, and compare them with similar deposits in globally distributed greenstone belts, so as to understand their bearing on Earth's evolution in the Archean, and to explore whether the ultramafic-mafic magmatism responsible for ore mineralization was linked to any large-scale global event, such as the amalgamation or break-up stage of a supercontinent cycle.

6.2 **TYPES OF CHROMITE DEPOSITS**

The chromite deposits were originally classified by Thayer (1960) as stratiform and podiform types, based on the morphological outline of the ore body. According to this classification, stratiform chromitite refers to conformable layers usually of great lateral extent, which are hosted within ultramafic-mafic cumulate rocks in large layered intrusions that occur within stable continental crust. Examples include the Bushveld Complex (South Africa), the Stillwater Complex (USA), the Muskox intrusion (Canada), and the Great Dyke (Zimbabwe). Other smaller stratiform chromitite bodies occur in the Kemi Complex (Finland), Skaergaard intrusion (Greenland), and the Campo Formoso Complex and Ipueira-Medrados sill (Brazil) (Misra, 2000). Archean chromitite deposits hosted in low-grade greenstone belts, such as the Bird River sill (Canada), and in high-grade metamorphic terrains where they are interlayered with anorthosite, such as the Fisknæsset Complex (southwest Greenland), are also examples of stratiform chromitites.

Podiform chromitites are pod-shaped or lenticular ore bodies that are discordant to the host rocks and are usually observed in ophiolites (Thayer, 1960), which represent obducted oceanic crust. Examples include ophiolites in the United States (Oregon), Canada (Thetford Complex and Bay of islands), Albania, Greece, Turkey, Cyprus (Troodos), Oman (Semail), Iran, Pakistan, USSR, Philippines, New Guinea, New Caledonia, Madagascar, Sudan and Cuba (Misra, 2000). Most chromitite

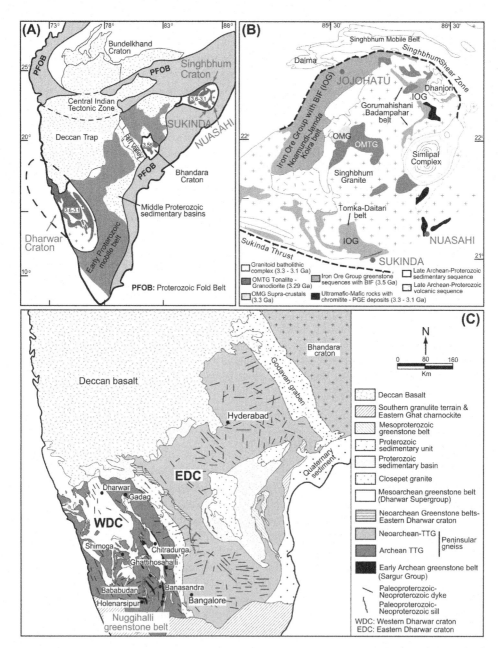

FIGURE 6.1

(A) Geology of the Indian Shield showing the location of the Singhbhum and Dharwar cratons (see Mondal, 2009; modified after Radhakrishna, B.P., Naqvi, S.M., 1986. Precambrian continental crust of India and its evolution. J. Geol. 94, 145–166 (Radhakrishna and Naqvi, 1986) and Leelananadam, C., Burke, K., Ashwal, L.D., Webb, S.J., 2006. Proterozoic mountain building in Peninsular India: an analysis based primarily on alkaline rock distribution.

(Continued)

ore bodies in Archean greenstone belts (e.g., Selukwe in Zimbabwe, and Nuasahi-Sukinda in India) were considered as podiform-type, because of their shape (pod-shaped), intensely deformed character and the peridotitic nature of the host rock. However, the former can be well distinguished from the ophiolite-hosted deposits due to their cumulate nature (Stowe, 1994). In addition, ophiolites not only host discordant pod-shaped chromitite layers in the mantle-section, but also include stratiform chromitite layers that are hosted within ultramafic cumulate rocks in the crustal section.

Hence, Stowe (1994) further revised the classification of chromite deposits and based it on the tectonic setting of the deposit, rather than the shape of the ore body. According to this classification, Stowe (1994) suggested a "Bushveld-type" chromite deposit, which refers to conformable layers of chromitite that usually form in regions of crustal extension such as in large mafic layered intrusions, and are associated with a peak in continental growth rates, and the "Ophiolite-hosted" chromite deposit, which includes both podiform and stratiform chromitites hosted in ophiolites.

The other types of chromite deposits are the Archean deposits that are found in different geological environments (e.g., Stowe, 1994), and are compositionally distinctive, suggesting derivation of the chromites from various types of parental magmas (Rollinson, 1995a). Stowe (1994) classified the Archean deposits into two types: (1) the 3.5−2.7 Ga greenstone belt-hosted deposits, where chromitite ore bodies are part of large sill-like ultramafic complexes, and occur as deformed lenses or pods that are enclosed in serpentinized dunites and (2) the 3.8−2.9 Ga Archean deposits where extremely deformed stratiform chromitite layers are interlayered with anorthosite and surrounded by gneisses in a high-grade metamorphic terrain (upper-amphibolite to granulite facies).

The Archean greenstone belt-hosted chromite deposits are usually associated with high-Mg metavolcanics and sediments such as cherts, because of which these deposits were considered to represent the ancient oceanic lithosphere (Stowe, 1994). The relationship of the sill-like chromitite-bearing ultramafic-mafic rocks and the associated metavolcanic suite remains ambiguous (DeWit and Ashwal, 1995; Mondal, 2009). They have been considered to represent sill-like intrusions (Mondal et al., 2006; Prendergast, 2008; Mukherjee et al., 2012), or subvolcanic sill-like feeders to the overlying volcanic rocks that were tectonically emplaced alongside the volcanics during later deformation events (e.g., Naldrett and Turner, 1977; Lesher and Groves, 1986). Alternatively, they are also considered as the lower cumulate portions of thick komatiitic extrusions (e.g., Barnes, 2006 and references therein). Resolving these questions requires precise geochronological and geochemical studies. The Archean greenstone belt-hosted chromitite bodies occur in stable cratonic areas around the world such as in the

---

◀ Geol. Mag. 143, 1−18 (Leelanandam et al., 2006)). The location of the Nuasahi-Sukinda Massifs in Singhbhum craton is also highlighted. (B) Geology of the Singhbhum craton (see Mondal, 2009; modified after Saha, A.K., 1994. Crustal Evolution of Singhbhum North Orissa, Eastern India. Mem. Geol. Soc. India no. 27, p.341 (Saha, 1994) and Sengupta, S., Acharya, S.K., De Smeth, J.B., 1997. Geochemistry of Archean volcanic rocks from Iron Ore supergroup, Singhbhum, eastern India. Proc. Indian Acad. Sci. (Earth Planet. Sci.) 106, 327−342 (Sengupta et al., 1997)). (C) Geology of the Dharwar craton showing location of the Nuggihalli greenstone belt (see Mukherjee et al., 2015; modified after Murthy, N.G.K., 1987. Mafic Dyke Swarms of the Indian Shield, Mafic Swarms. Geological Association of Canada Special Paper 34, pp. 393−400 (Murthy, 1987); cited in Devaraju, T.C., Viljoen, R.P., Sawkar, R.H., Sudhakara, T.L., 2009. Mafic and ultramafic magmatism and associated mineralization in the Dharwar craton, southern India. J. Geol. Soc. India 73, 73−100 (Devaraju et al., 2009)).

Table 6.1 Global Occurrences of Archean Greenstone Belt-Hosted Chromite Deposits

Name of Deposit	Location	Age (Ga)	Dimension of Chromitite Ore Body	Lithological Association
Selukwe greenstone belt	Zimbabwe craton	3.3[a]	Length: 150–1700 m (down plunge); width: 15–25 m	Talc-carbonate schist hosted chromitite lenses in a highly deformed nappe-structured terrain
Barberton greenstone belt (Jamestown Igneous Complex)	Kaapvaal craton (South Africa)	3.5[b,c,d]	Stratiform chromitite layer—width: 20–30 cm Chromitite pods—width: 10–50 cm	Metapyroxenite hosts chromitite layers; small chromitite pods hosted within deformed talc-carbonate rocks
Coobina sill	Jimblebar greenstone belt in Pilbara craton (Western Australia)	2.9[e,f]	Length: 350 m; width: 2 cm–2 m	Dunite-hosted chromitite lenses overlain by peridotite-leucogabbro/anorthositic metagabbro rock association
Bird River sill	Bird River greenstone belt in Superior craton (Canada)	2.7[g]	Length: 600 m; width: 1 mm–1 m	Layered sequence of chromitite-bearing dunite, peridotite, pyroxenite- (ultramafic series) followed successively by gabbro and anorthosite rocks (mafic series)
Ring of Fire intrusion: (e.g., Blackbird, Big Daddy, Black Thor, and Black Label deposits	McFaulds Lake greenstone belt in Superior craton (Northern Ontario, Canada)	2.7[h]	Big Daddy: length: 1.4 km; width: 60 m Black Label: length: 2.2 km Black Thor: length: 2.6 km; width: tens of meter to 100 m	Serpentinized dunite or peridotite-hosted chromitite lenses overlain by pyroxenite and leucogabbros
Inyala chromite	North Marginal Zone (southern Zimbabwe)	2.7[i]	Length: 800 m; width: 170 m	Dunite-hosted chromitite lenses enclosed by orthopyroxenite and amphibolites

[a]*Stowe (1987).*
[b]*DeWit et al. (1987).*
[c]*Kröner et al. (1996).*
[d]*Van Kranendonk et al. (2009).*
[e]*Witt et al. (1998).*
[f]*Barnes and Jones (2013).*
[g]*Scoates and Scoates (2013).*
[h]*Laarman (2013) and references therein.*
[i]*Rollinson (1997).*

Zimbabwe craton (e.g., Selukwe), the North Marginal zone in southern Zimbabwe (e.g., Inyala), the Kaapvaal craton in South Africa (e.g., Barberton), the Singhbhum (e.g., Nuasahi-Sukinda-Jojohatu) and Western Dharwar (e.g., Nuggihalli-Holenarsipur-Krishnarajpet) cratons in India, the Superior craton in Canada (e.g., Bird River sill, Ring of Fire intrusions-Blackbird, Big Daddy, Black Thor and Black Label), and the Pilbara craton (e.g., Coobina sill) in Western Australia (Table 6.1).

The association of highly calcic anorthosites (e.g., Ashwal, 1993) and Fe-rich chromitites occurs exclusively in the Archean, where they form part of layered igneous complexes that are characterized by ultramafic sequences that are thin relative to the thick gabbro-anorthosite association (Rollinson et al., 2010). These layered complexes are associated with volcano-sedimentary supracrustal rocks that occur within TTG's in highly deformed terrains that have undergone high-grade metamorphism, usually reaching the granulite grade ($>600°C$; $6-13$ kb; Subramanium, 1956; Windley et al., 1981; Polat et al., 2010; Rollinson et al., 2010). The stratiform chromitites occurring in these high-grade terrains are thought to represent "Bushveld-type" chromitite layers that were emplaced onto an unstable crust (Stowe, 1994). Although these complexes make up a minor proportion of the early crust, they still provide crucial clues about petrogenetic and geodynamic processes that were operative in the early Earth (Windley et al., 1981; Polat et al., 2010; Rollinson et al., 2010). Such anorthosite-chromitite layered complexes are found in Greenland (Fisknæsset Complex), southern India (Sittampundi Complex), South Africa (Messina layered intrusion), and Mauritania (Guelb el Azib Complex).

6.3 CHROMITE DEPOSITS IN THE INDIAN ARCHEAN GREENSTONE BELTS

6.3.1 NUASAHI-SUKINDA MASSIFS, EASTERN INDIA

The Nuasahi-Sukinda Massifs occur in the Singhbhum craton in eastern India (Figs. 6.1A and B and 6.2A and B), and comprise a layered sequence of plutonic ultramafic-mafic rocks that are associated with volcano-sedimentary supracrustal rocks of the IOG greenstone belts (Mondal et al., 2006; Mondal, 2009). The Singhbhum craton consists of Paleoarchean to Mesoarchean (3.3–3.1 Ga) granite/gneiss "basement" (e.g., Singhbhum granite, Nilgiri granite, Bonai granite, Mayurbhanj granite), which are interleaved with the IOG rocks of the greenstone belts (Fig. 6.1B; Mondal, 2009). The Nuasahi and Sukinda Massifs in the Singhbhum craton host the largest chromite deposits in India. In addition, a gabbro-breccia hosted PGE mineralization occurs in both these massifs (Mondal and Baidya, 1997; Mondal et al., 2001; Khatun et al., 2009; Mondal and Zhou, 2010).

The volcano-sedimentary rocks of the IOG are mainly BIFs, phyllites, tuffaceous shales, quartz-arenite, argillite, ferruginous quartzite, dolomite, acid to intermediate volcanics, and mafic volcanics (Saha, 1994). Komatiites, spinifex textured komatiitic basalts, and high-Mg basalts also occur in the IOG (Sahu and Mukherjee, 2001; Mondal et al., 2006). The rocks in the IOG have all been metamorphosed to greenschist and lower amphibolites facies, and are extensively hydrothermally altered. However, the layered ultramafic-mafic plutonic rocks still preserve primary cumulate textures and mineralogy (Mondal et al., 2006). These layered rocks are closely associated with gabbroic intrusions (3.2–3.1 Ga; Augé et al., 2003) that host Ti-V bearing magnetitite layers. The intrusions occur in the Nuasahi-Nilgiri-Gorumahishani-Badampahar areas, which hereafter will be termed as the NNGB-gabbroic suite (Mondal, 2009; Fig. 6.1B).

The IOG rocks are exposed in three greenstone belts: (1) the Gorumahishani-Badampahar greenstone belt in the east; (2) the Noamundi-Jamda-Koira greenstone belt in the west; and (3) the Tomka-Daitari greenstone belt in the south (Mondal, 2009; Fig. 6.1B). The Sukinda Massif occurs in the southern greenstone belt, while the Nuasahi Massif occurs to the northeast of Sukinda

(Fig. 6.1B). The age of sedimentation in these greenstone belts was constrained between 3.3 and 3.1 Ga (Saha et al., 1988) based on field observations. However, U-Pb dating of zircons from dacitic rocks in the Tomka-Daitari IOG greenstone belt has provided more precise constraints of 3506.8 ± 2.3 Ma (Mukhopadhyay et al., 2008). The chromitite-bearing ultramafic-mafic plutonic rocks have not been dated yet, but their age may range between 3.3 and 3.1 Ga, based on field relations and age of the associated rocks (Mondal et al., 2006; Mondal, 2009). Mondal et al. (2007) have shown that the Sm-Nd isotopic data of the major ultramafic-mafic igneous suites in the Singhbhum craton such as the NNGB, high-Mg metavolcanic rocks of the IOG, and the amphibolites of the Older Metamorphic Group (OMG), plot close to the 3.3 Ga reference isochron (e.g., Sharma et al., 1994), indicating a contemporaneous formation history for these rock suites (Fig. 6.1B).

The chromitite-bearing plutonic ultramafic-mafic rocks in the Nuasahi-Sukinda Massifs were emplaced within the IOG rocks as sills, with which they were later cofolded (Mondal et al., 2006 and references therein). The chromitite layers were thought by previous workers to represent stratiform chromitites in large layered intrusions such as the Bushveld and the Stillwater Complexes (Mukherjee, 1966; Haldar, 1967; Chakraborty and Chakraborty, 1984). However, extensive shearing, strike-slip faulting and fractures occur in the ultramafic rocks of the massifs, and the chromitite bodies occur as discontinuous pods that are hosted within highly deformed and serpentinized dunites (Mondal et al., 2006). Based on these observations some workers considered these chromitites to represent Archean ophiolites (e.g., Varma, 1986).

The Nuasahi Massif (≈ 10 km^2 area) occurs in the southeastern part of the Singhbhum craton and extends for nearly 5 km with widths of ≈ 400 m (Fig. 6.2A). The layered ultramafic-mafic massif was divided into three rock suites (Mondal et al., 2001) that have a NW-SE to NNW-SSE trend, and the igneous layering dips 60−80 degrees (Mondal et al., 2006). Suite 1 is represented by a layered sequence of steeply dipping rocks that from west to east are exemplified by orthopyroxenite ($\approx 80-120$ m) and harzburgites, followed by serpentinized dunites ($\approx 130-150$ m thick) hosting chromitite layers, which in turn are followed by another orthopyroxenite layer (Fig. 6.2A); the western orthopyroxenite layer has a coarser grain size than its counterpart in the east. Three distinct dismembered chromitite seams, from the west to the east side of the massif, are named by the local mining authorities as: Durga ($\approx 3-4$ m thick), Laxmi 2 ($\approx 2-3$ m thick), and Laxmi 1 ($\approx 1.6-2$ m thick, Fig. 6.2A). The chromitites share sheared and faulted contacts with the host dunite, but locally they may show gradational contacts (Fig. 6.3A). The dunite on the other hand shares sharp, but not intrusive contacts, with the orthopyroxenite layers flanking it (Mondal et al., 2006). The ultramafic rocks are followed by gabbroic rocks in both west and the east, which constitute the Suite 2 rocks (Fig. 6.2A). The eastern gabbro is hosting Ti-V rich magnetitite bands at the upper part. The gabbros grade into dioritic rocks in the northeastern part of the massif. The PGE and base metal sulfide (BMS) rich chromitite and ultramafic breccias occur at the contact between the eastern orthopyroxenite layer in Suite 1 and the eastern gabbroic rocks in Suite 2 (Fig. 6.2A; Mondal et al., 2001; Mondal and Zhou, 2010). Suite 3 is represented by the late intrusive rocks such as dolerite, pegmatitic gabbro and quartz-diorite. Among these the pegmatitic gabbro constitutes the groundmass of the breccias, which is also locally rich in sulfide and PGE (Mondal and Zhou, 2010).

The Sukinda Massif occurs in the southern part of the Singhbhum craton (Fig. 6.2B). It is the largest ultramafic body (≈ 25 km $\times 400$ m) in the craton, but the rocks in this massif have undergone deep and extensive lateritic weathering (Mondal et al., 2006). The ultramafic belt appears as a

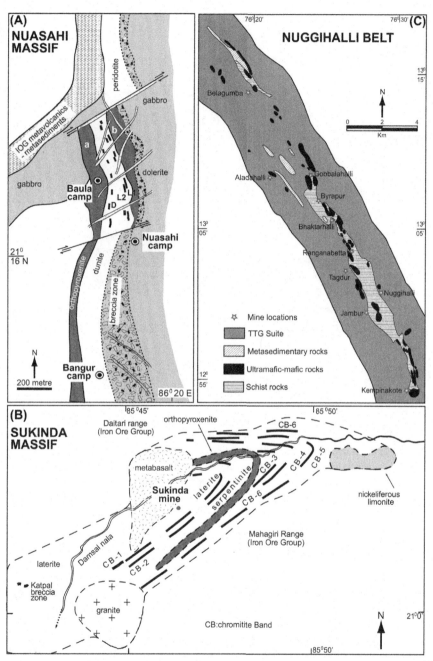

(Continued)

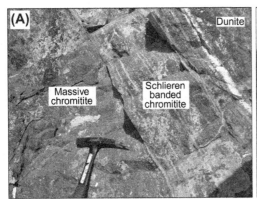

FIGURE 6.3

Field photographs showing: (A) fine-scale (mm) layering within chromitite seam and contact between dunite and the chromitite seam in the Nuasahi Massif, Singhbhum craton; (B) Lens-shaped chromitite seam enclosed within serpentinized dunite (now lateritized) in Sukinda Massif, Singhbhum craton; (C) Lens-shaped chromitite seam (mined out) enclosed within serpentinized dunite in Nuggihalli chromite deposits, Western Dharwar craton.

◄ **FIGURE 6.2**

(A) Geology of the Nuasahi Massif showing occurrences of chromitite seams and PGE-rich breccia zone (see Mondal, 2009). (B) Geology of the Sukinda Massif showing occurrences of chromitite seams (reproduced from Mondal, 2009). (C) Geology of the Nuggihalli schist belt showing locations of the chromite mining districts (see Mukherjee et al., 2010).

Modified after Jafri, S.H., Khan, N., Ahmed, S.M., Saxena, R., 1983. Geology and geochemistry of Nuggihalli schist belt, Dharwar craton, Karnataka, India. In: Naqvi, S.M., Rogers, J.J.W. (Eds.), Precambrian of South India. Memoir. Geol. Soc. India No 4, pp. 110–120 (Jafri et al., 1983); cited in Devaraju, T.C., Viljoen, R.P., Sawkar, R.H., Sudhakara, T.L., 2009. Mafic and ultramafic magmatism and associated mineralization in the Dharwar craton, southern India. J. Geol. Soc. India 73, 73–100 (Devaraju et al., 2009).

southwesterly plunging body with a subvertically dipping south limb, and a moderately dipping (40−50 degrees) north limb (Basu et al., 1997). The ultramafic lithologies in Sukinda are comparable with those found in the Nuasahi Massif, although for the former the exposures are poor as the rocks are covered by thick laterites ($\approx$30 m; Mondal et al., 2006). Chromitite layers in Sukinda occur within serpentinized dunites (Fig. 6.2B), and are very fragmented and deformed in the western part of the massif, where they are impregnated by pegmatitic gabbro and dioritic rocks; a PGE-mineralized breccia zone was reported from this area (e.g., Sarkar et al., 2001), which is very similar to the one observed in Nuasahi. The contact between the orthopyroxenite and the dunite is again sharp and the former rocks are quite unaltered, as in the Nuasahi Massif.

The chromitite seams in the Nuasahi-Sukinda Massifs occur as lens-shaped bodies, but are remarkably continuous along strike (Figs. 6.2A and B and 6.3A and B). The ore body is nearly subvertical and can be quite sheared, rotated and offset by several NE and NNW trending faults (Mondal et al., 2006). In Sukinda, the chromitites are relatively more folded and faulted, and there are at least six dissected bands with lengths ranging from 200 m to 3 km (Mondal et al., 2006). The chromitites are characterized by various textures, depending on the mode of interstitial silicate minerals, such as schlieren bands, spotted bands, clot-textured, net-textured, nodular, and antinodular textured chromitites (Mondal et al., 2006; Khatun et al., 2014; Figs. 6.3A and 6.4). The schlieren and spotted chromitites may grade vertically into massive chromitite within the same seam, and may show rhythmic banding with olivine-rich layers (Fig. 6.3A). In clot-textured chromitites, the chromite grains within clots are much finer, compared to the surrounding grains comprising the rock, and occur embedded within a serpentinized matrix (Mondal et al., 2006). The Nuasahi chromitite seams show gravity-controlled features such as slumping, slump-folds, pseudo-current bedding and convolute structures (Halder, 1967). The Sukinda massive chromitites show thin alternate laminae (0.5−1.5 cm thick) of

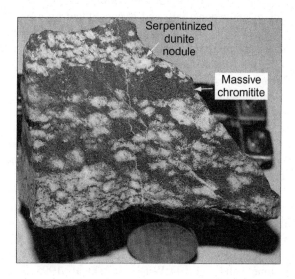

FIGURE 6.4

Typical antinodular chromitite (Durga seam) from the Nuasahi Massif, Singhbhum craton; nodules represent dunite now pseudomorphed by serpentinite. Sample overall shows schlieren-banded character of the chromitite.

chromite and olivine within massive chromitite layers (e.g., Chakraborty and Chakraborty, 1984). The massive chromitites (90—95 modal % chromite) in the Nuasahi-Sukinda Massifs contain unaltered chromites with inclusions of clinopyroxene and phlogopite (Mondal et al., 2013); the interstitial phases are serpentine, chlorite, talc, magnesite and minor base metal sulfides (BMSs).

6.3.2 **NUGGIHALLI BELT, SOUTHERN INDIA**

The Dharwar craton has been subdivided into an eastern and a western component by a 500 km long N-S trending body of alkali-feldspar rich granite, known as the Closepet Granite (Fig. 6.1C; Mukherjee et al., 2010, 2012).The Eastern Dharwar craton is younger (2.7—2.5 Ga; Dey, 2012) than the WDC, which is comprised of several early Archean greenstone belts (3.4—3.0 Ga) that are also known as the Sargur Group (Swami Nath and Ramakrishnan, 1981). The Sargur Group of rocks occurs as discontinuous linear belts ($\approx 30-60$ km $\times \approx 2-6$ km), which can be traced along a ≈ 250 km long N-S zone in the craton (Mukherjee et al., 2012; Fig. 6.1C).

The rocks of the Sargur Group are comparable to the IOG rocks in the Singhbhum craton (eastern India), and are represented by conformable volcanic-sedimentary rock assemblages that are surrounded by the TTG (Fig. 6.1B and C). The Sargur Group of rocks are deformed and metamorphosed to lower greenschist to lower amphibolite facies, despite which primary igneous textures are preserved and the protoliths can still be identified. The igneous lithologies occur as both intrusive and extrusive ultramafic-mafic rocks with a compositional range from komatiites to komatiitic basalts and tholeiites (Ramakrishnan, 2009), now represented by tremolite-actinolite-chlorite-hornblende-quartz bearing schists. The sedimentary rocks are fuchsite-quartzites, BIFs, bedded barites, and kyanite-garnet bearing quartzites and mica schists (Ramakrishnan, 1981). The type area of the Sargur Group is the high-grade (amphibolite-granulite) Sargur schist belt that occurs south of the Nuggihalli greenstone belt, and is constituted by a cluster of mafic-ultramafic enclaves that occur as synformal keels and antiformal hinges floating in the TTGs. These enclaves in all probability represent the deeper crustal roots of the linear low-grade ultramafic-mafic Sargur Group greenstone belts to the north (Ramakrishnan, 2009).

Sill-like chromitite-bearing layered ultramafic-mafic rocks occur commonly within the Sargur Group, as for example in the Nuggihalli-Holenarsipur-Krishnarajpet-Nagamangala greenstone belts (Fig. 6.1C). Amongst these greenstone belts, the Nuggihalli belt has been studied in the greatest detail, and will therefore be discussed further. The chromitite-bearing layered ultramafic-mafic rocks in Nuggihalli occur as *en echelon*, lenticular fragments (Figs. 6.2C and 6.3C) due to superposed folding common in the greenstone belts (Mukherjee et al., 2012). The plutonic rocks in the Nuggihalli greenstone belt were dated at 3125 ± 120 Ma by whole rock Sm-Nd methods (Mukherjee et al., 2012), while Jayananda et al. (2008) dated the komatiitic rocks from the WDC by bulk-rock Sm-Nd methods at 3352 ± 110 Ma, which corresponds with the age of the plutonics within the error limits. Thus, the Sargur Group supracrustals formed over a protracted time period between 3.4 and 3.0 Ga (Mukherjee et al., 2012), which is consistent with the observation that very short but distinct pulses of magmatism spanning up to 50 Ma, generate the komatiitic-tholeiitic rock sequences in greenstone belts (e.g., DeWit and Ashwal, 1995; Bryan and Ernst, 2008). Stratigraphically the Sargur Group in the WDC is followed by the late Archean (2.9—2.6 Ga; Taylor et al., 1984; Bhaskar Rao et al., 1992; Kumar et al., 1996) greenstone belts or the Dharwar Supergroup of rocks, following an unconformity, with TTG as the basement for the younger rocks.

The 3.1 Ga Nuggihalli greenstone belt ($\approx 60 \times 2$ km) appears as a linear NNW-SSE trending belt in the WDC (Mukherjee et al., 2010, 2012, 2015; Figs. 6.1C and 6.2C). The greenstone belt is comprised of a plutonic chromitite-bearing sill-like layered ultramafic-mafic unit that is surrounded by schistose metavolcanic rocks, minor metasediments and TTG. The chromiferous ultramafic rocks occur as dismembered lenticular bodies in the greenstone belt (Figs. 6.2C and 6.3C). The layered ultramafic-mafic plutonic rock sequence has been divided into lower and upper ultramafic units, which are separated by a gabbro unit. The lower ultramafic unit is comprised of serpentinized dunite and peridotite (now tremolite-talc-chlorite-actinolite schist) that hosts chromitite ore bodies, which are then followed by pyroxenite and gabbro.

The gabbro shows layering, and contains two conformable bands of titaniferous-vanadiferous magnetite at its base and top. The upper ultramafic unit is comprised of chromitite-bearing serpentinized dunite and peridotite (now tremolite-talc-chlorite-actinolite schists) that are further surrounded by the metavolcanic schists and TTG (Fig. 6.2C). Anorthosite has been reported to occur in the greenstone belt in association with serpentinites, but is not consistently exposed in all outcrops. The contact between the metavolcanic schists and TTG is obscured by soil cover but a concordant relationship can be deciphered. The metavolcanic schists are represented by metagabbro, amphibolite schist, chlorite-quartz-actinolite schist, amphibole-chlorite schist, and talc-chlorite schists. The rocks show strong penetrative deformation fabrics, and exhibit lens-shaped geometry that conforms to the general appearance of the sill-like ultramafic-mafic unit in the greenstone belt (Fig. 6.2C).

The chromitite ore bodies in Nuggihalli range in length from ≈ 50 to 500 m with a width of ≈ 15 m, and exhibit various shapes like sack-shaped, sigmoidal, lenticular, pod-shaped, and folded bodies (≈ 0.5 m length; width ≈ 0.3 m), owing to the superposed folding common in the greenstone belts (Fig. 6.3C) (Mukherjee et al., 2010, 2012, 2015); chromitites also preserve primary igneous layering with dunite (now serpentinite) in which the contact between the two rock units is gradational. Contacts between the chromitite body and the host serpentinite are highly sheared in this greenstone belt. The strikes of the ore bodies vary but are predominantly N-S and NW-SE with a nearly vertical dip (75−80 degrees) toward the east; locally the strike is E-W with dip toward the north. The upper chromitite occurring above the gabbro unit appear as an elongate, lenticular body (length ≈ 100 m; width ≈ 15 m; N-S trend, dipping west), which is more altered with a higher mode of carbonate (mainly magnesite).

The chromitite ($\approx 60-85$ modal % chromite) in Nuggihalli exhibits primary igneous layering where individual seams contain layers of massive, spotted, clot-textured, schlieren-banded and net-textured chromitite, depending on the modal proportion of the intercumulus silicate mineral (now mainly serpentine (Mukherjee et al., 2010). The chromites in massive chromitite are coarse-grained, euhedral, polygonal and less deformed, which is also seen for the Nuasahi-Sukinda chromitites; Nuggihalli massive chromitites may contain relics of interstitial olivine and pyroxene grains in an otherwise serpentinized and chloritized matrix (Mondal et al., 2006; Mukherjee et al., 2010, 2015). Chromites may also contain primary inclusions of orthopyroxene, olivine and clinopyroxene that are spherical to euhedral in shape (Mukherjee et al., 2010). Accessory chromite grains in serpentinized dunites and peridotites ($\approx 5-10$ modal % chromite), as well as those in silicate-rich chromitites (≈ 50 modal % chromite), are compositionally zoned with a moderately to completely altered cores and a rim of ferritchromit and rarely magnetite (Mukherjee et al., 2010, 2015); ferritchromit may also occur along fractures, and in some extensively altered chromite grains, the grains are homogeneously modified to ferritchromit (Mukherjee et al., 2010, 2015). Compositional zoning is also observed in accessory chromites hosted within serpentinized dunite in the Nuasahi-Sukinda Massifs (Mondal et al., 2006). The altered grains in Nuggihalli may exhibit a corroded and skeletal texture, and are usually surrounded by flakes of chlorite (Mukherjee et al., 2010). Usually the chromite grains host disseminated sulfides (e.g., millerite-heazlewoodite with minor niccolite; $20-50\,\mu m$) in the interstitial spaces, and as inclusions ($0.5-1.0\,\mu m$) along with minor platinum-group minerals (e.g., laurite; Mukherjee et al., 2015).

6.4 COMPOSITIONAL CHARACTERISTICS OF INDIAN CHROMITES FROM ARCHEAN GREENSTONE BELTS

6.4.1 MAJOR ELEMENT COMPOSITION OF CHROMITES

The Archean greenstone belt-hosted chromites from massive chromitites (>70 modal % chromite) are usually characterized by high Cr# ($100 \times Cr/(Cr + Al)$ molar ratio) and moderate Mg# ($100 \times Mg/(Fe^{2+} + Mg)$ molar ratio). The chromites from the Nuasahi Massif in the Singhbhum craton (eastern India) have Cr# of $78-87$ and Mg# of $66-82$, while the chromites from the Sukinda Massif have Cr# of $75-81$ and Mg# of $62-73$ (Fig. 6.5A; Mondal et al., 2006). In comparison, the Nuggihalli chromites have slightly higher Cr# of $85-86$ than the Sukinda chromites, and a lower Mg# of $38-58$ compared to the chromites from the Nuasahi-Sukinda Massifs (Fig. 6.5A; Mukherjee et al., 2010). Interestingly, chromites from the chromitite ore bodies in the Nuasahi-Sukinda Massifs do not display compositional variations across the stratigraphy, and there seem to be no significant compositional differences between chromites from massive chromitites and those in silicate-rich chromitites from the same ore body (e.g., net-textured, schlieren-banded and clot-textured chromitites) (Mondal et al., 2006). However, the chromites from the Nuggihalli chromitite ore bodies show compositional variations with respect to location, and even within the same sample (Mukherjee et al., 2010). The chromites from the Byrapur mining district are both optically and compositionally homogeneous and massive and silicate-rich chromitites have similar Cr# values. The cores of these chromite grains show a similar range of values and represent primitive compositions (Mukherjee et al., 2010). The primary compositional variation in the Byrapur chromites along the Mg# axis is indicative of fractional crystallization (Fig. 6.5A), which is commonly observed in layered intrusions (e.g., Rollinson, 1995b; Barnes and Roeder, 2001). Chromites from the Bhaktarhalli mining district also are chemically and optically homogeneous and unzoned, like the ones from Byrapur. However, the Bhaktarhalli chromites show higher Mg# ($51-55$) and a lower Cr# ($78-79$), due to the higher concentrations of Al_2O_3 and MgO (Mukherjee et al., 2010).

Different fields in Fig. 6.5A and B show the compositional range of natural spinels from different geological environments as identified by Barnes and Roeder (2001). Primary chromites in chromitites from the Nuasahi and Sukinda Massifs plot within the field expected for chromites from Archean greenstone belts (Fig. 6.5A and B; Mondal et al., 2006). The chromites from massive chromitites in Nuggihalli plot in the same field (Fig. 6.5A and B; Mukherjee et al., 2010).

In the Cr-Al-Fe^{3+} ternary diagram (Fig. 6d of Mondal et al., 2006), the Nuasahi-Sukinda chromites show two distinct compositional trends where: (1) relatively less-altered accessory chromite grains have constant Cr/Fe^{3+} ratios and variable Cr/Al ratios while (2) altered (e.g.,

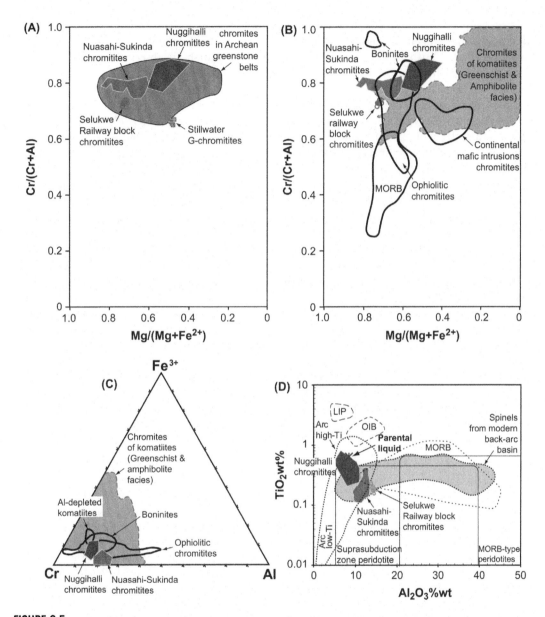

FIGURE 6.5

Tectonic discrimination diagrams: (A) Variation of Cr-ratio and Mg-ratio of primary chromite compositions from the Nuasahi-Sukinda Massifs and the Nuggihalli greenstone belt compared with composition of chromites from globally occurring Archean greenstone belts (shaded field from Rollinson, H., 1995b. The relationship between chromite chemistry and the tectonic setting of Archean ultramafic rocks. In: Blenkinsop, T.G., Tromps, P. (Eds.),

(Continued)

compositionally zoned) chromites in serpentinites show variation in Fe^{3+} with a moderate range in Cr/Al ratios. The Nuggihalli chromites also show similar trends in this diagram (Fig. 7e, f of Mukherjee et al., 2010); however, restricted chemical variation is observed in unaltered and unzoned chromites from massive chromitites from both the Nuasahi and Sukinda Massifs as well as from the Nuggihalli belt (Figs. 6.5C and 6.6A; Mondal et al., 2006; Mukherjee et al., 2010).

Accessory chromites (≈ 5 modal % chromite) hosted within the serpentinized dunites show very low Mg# (≈ 15), variable and high Cr# (≈ 95), and enrichment in TiO_2 (1.5 wt%) and MnO (1.4 wt%) (Mondal et al., 2006). Some of these grains are compositionally zoned, and the rims plot in the fields of metamorphic magnetite and ferritchromit (altered chromite), along with chromites from the IOG high-Mg metavolcanic rocks (Mondal et al., 2006). This trend is not unique to the chromites from the Nuasahi-Sukinda Massifs, but is also seen in the Nuggihalli chromites that are hosted within serpentinized dunite (Cr# = 2−99; Mg# = 1−38), and in silicate-rich chromitites (Cr# = 6−99; Mg# = 6−48; Mukherjee et al., 2010). Some of the Nuggihalli chromites are compositionally zoned with a dark core that is surrounded by a bright rim of ferritchromit (Fig. 6.6B). The contact between the core and the surrounding ferritchromit rim is always sharp (Mukherjee et al., 2010, 2015). The cores are modified and show a range of Cr# from 64 to 79 and Mg# from 8 to 15, which falls within the field of chromites from komatiites that have been modified by greenschist facies metamorphism (Barnes and Roeder, 2001). The rims show lower Mg# (2−6), and higher Cr# (85−99) and Fe^{3+}# ($100 \times Fe^{3+}/Cr + Al + Fe^{3+}$) (23−77) (Mukherjee et al., 2010).

The zoned grains in Nuggihalli show enrichment of Cr_2O_3, Al_2O_3 and MgO in the core, and a progressive loss of these elements toward the rim, which also show strong enrichment of Fe_2O_3 (Fig. 6.6B; Mukherjee et al., 2010, 2015). FeO shows negligible to minor enrichment in the rim (Fig. 6.6B). TiO_2 also follows the trend of Fe_2O_3 and shows strong enrichment in the rim (≈ 1.3 wt %) (Mukherjee et al., 2010). The compositional zoning and formation of ferritchromit and magnetite is related to hydrothermal alteration and oxidation of the chromite grains during the serpentinization event; alteration in the Nuggihalli chromites was pervasive, as shown by the altered

◀ Sub-Saharan Economic Geology. Amsterdam, Balkema, pp. 7−23), the Seluwe greenstone belt and the Stillwater Complex (data from Mondal, S.K., Ripley, E.M., Li, C., Frei, R., 2006. The genesis of Archean chromitites from the Nuasahi and Sukinda massifs in the Singhbhum craton, India. Precambrian Res. 148, 45−66); (B) Cr-ratio and Mg-ratio variations of primary chromite compositions from the Nuasahi-Sukinda Massifs and the Nuggihalli greenstone belt compared with chromite compositions from different types of magmas, ophiolitic chromitites and chromitites from continental mafic intrusions. Also shown is the field of chromites from komatiites that have experienced greenschist to amphibolite facies metamorphism. All fields are from Barnes, S.J., Roeder, P.L., 2001. The range of spinel compositions in terrestrial mafic and ultramafic rocks. J. Petrol. 42, 2279−2302; (C) Trivalent cation plot of primary chromite compositions from the Nuasahi-Sukinda Massifs and the Nuggihalli greenstone belt compared with fields of chromite composition from boninite, Al-depleted komatiite, ophiolite and altered (greenschist-amphibolite facies metamorphosed) komatiites. All fields are from Barnes, S.J., Roeder, P.L., 2001. The range of spinel compositions in terrestrial mafic and ultramafic rocks. J. Petrol. 42, 2279−2302; (D) TiO_2-Al_2O_3 variation in Cr-spinels from Nuasahi-Sukinda Massifs and the Nuggihalli greenstone belt compared with composition of Cr-spinels from modern-day tectonic settings. Fields are from Kamenetsky, V.S., Crawford, A.J., Meffre, S., 2001. Factors controlling chemistry of magmatic spinel: an empirical study of associated olivine, Cr−spinel and melt inclusions from primitive rocks. J. Petrol. 42, 655−671.

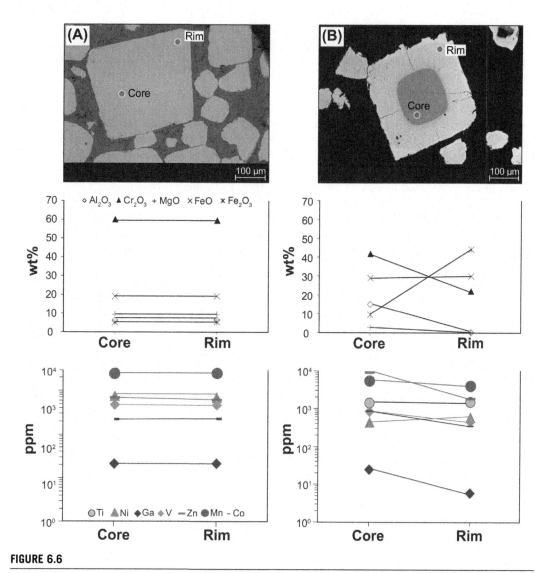

FIGURE 6.6

Variation of major element (analyzed by EPMA) and trace element (analyzed by LA-ICPMS) distributions in Nuggihalli chromites (from Mukherjee, R., Mondal, S.K., González-Jiménez, J.M., Griffin, W.L., Pearson, N.J., O' Reilly, S.Y., 2015. Trace-element fingerprints of chromite, magnetite and sulphides from the 3.1 Ga ultramafic-mafic rocks of the Nuggihalli greenstone belt, Western Dharwar craton (India). Contrib. Mineral. Petrol. 169, 59. doi:10.1007/s00410-015 1148-1): (A) Unaltered chromite from massive chromitite in Byrapur and (B) Compositionally zoned chromite grain from silicate-rich chromitite in Tagdur.

composition of the cores (Mukherjee et al., 2010). Ferritchromit usually occurs along chromite grain boundaries and fractures, which are the pathways for fluids to interact with the grain. The loss of Cr, Al, and Mg, and a simultaneous infiltration of Fe^{3+} and Ti into the chromite grains were brought about by fluids, as they are efficient agents of transportation and diffusion (Mukherjee et al., 2010). The slower rates of diffusion of the trivalent cations relative to the divalent cations in chromites caused a compositional gradient to develop across the grains (e.g., Rollinson, 1995b), which is observed as zoning. Metamorphism can cause significant increases in the rates of diffusion of cations, which perhaps facilitated the formation of optically unzoned but completely ferritchromitized grains in Nuggihalli (Figs. 5c and 6b of Mukherjee et al., 2010).

The interstitial silicates in massive chromitites from the Nuasahi-Sukinda Massifs and Nuggihalli are represented by olivine and orthopyroxene, and some of these minerals also occur as inclusions within chromites (Mondal et al., 2006; Mukherjee et al., 2010). The compositions of these silicates are much more extreme (Fo_{96-97}, En_{96-97}; Mondal et al., 2006; Mukherjee et al., 2010) and variable when they occur as interstitial phases in chromitites, compared to their compositions in dunites (Nuasahi-Sukinda = Fo_{90-92}; Nuggihalli = Fo_{83-95}; Bidyananda and Mitra, 2005; Mondal et al., 2006) or orthopyroxenites (En_{90-93}) where they are modally abundant. The most extreme compositions are in fact shown by inclusions of olivine (Fo_{96}–Fo_{98}) and orthopyroxene (En_{94-99}) within chromite (Mondal et al., 2006; Mukherjee et al., 2010). Such compositional variations are attributed to subsolidus reequilibration processes (Mondal et al., 2006; Mukherjee et al., 2010), which however seem to be insignificant where olivine and chromite are modally abundant (i.e., dunites and chromitites). For this reason the compositions of chromite from massive chromitites were utilized to compute liquidus compositions for the parental melt (Mondal et al., 2006; Mukherjee et al., 2010).

In Nuggihalli, chromite-olivine (Fo_{96}) pairs from massive chromitites of Byrapur were used by Mukherjee et al. (2010) to calculate the reequilibration temperature, using the equation of Roeder et al. (1979) where T (K) $= ((\alpha \times 3480) + (\beta \times 1018) + (\gamma \times 1720) + 2400)/((\alpha \times 2.23) + (\beta \times 2.56) - (\gamma \times 3.08) - 1.47 + 1.987 \ln K_D)$; $\alpha = Cr/R^{3+}$, $\beta = Al/R^{3+}$, $\gamma = Fe^{3+}/R^{3+}$, R^{3+} refers to the trivalent cations in chromite structure and K_D refers to $(MgO/FeO)_{olivine}/(MgO/FeO)_{chromite}$. Very low temperatures of around 464–521°C were obtained, which indicates that the chromite-olivine pair in Nuggihalli reequilibrated until the serpentinization event destroyed the olivine. Similar calculations for the Archean Inyala chromite deposit in Zimbabwe (e.g., Rollinson, 1997) and from Nuasahi, India (e.g., Mondal et al., 2006) yielded equilibration temperatures that were slightly higher, around 885–1020°C.

6.4.2 TRACE ELEMENT COMPOSITION OF CHROMITES

Chromites from massive chromitites of the Nuggihalli greenstone belt were studied for their trace elements using laser ablation ICPMS (LA-ICPMS) by Mukherjee et al. (2015). This was the first study that characterized trace elements in chromites from Archean greenstone belts (Figs. 6.6–6.8), as most studies have focused on chromites from Phanerozoic ophiolites or oceanic peridotites (Dare et al., 2009; Pagé and Barnes, 2009; González-Jiménez et al., 2014; Pagé et al., 2012; Colás et al., 2014). Compared to major elements in chromites, the trace elements are more sensitive to parameters like oxygen fugacity, and they show larger variations during partial melting and fractional crystallization (Dare et al., 2009; González-Jiménez et al., 2014). The trace elements that enter the

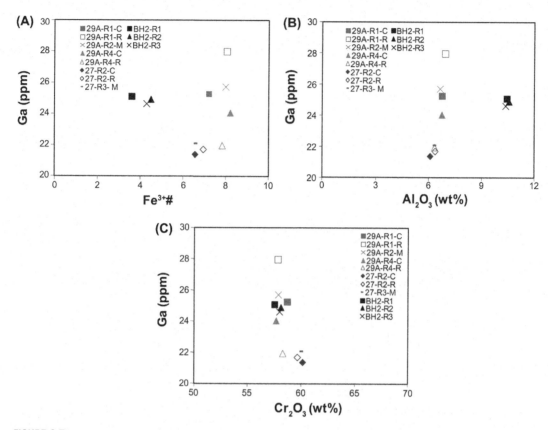

FIGURE 6.7

(A) Fe^{3+}# versus Ga, (B) Al_2O_3 versus Ga, and (C) Cr_2O_3 versus Ga in unaltered chromites from massive chromitites of Byrapur (samples 29A, 27) and Bhaktarhalli (sample BH2). The abbreviations C, R, and M refer to the core, rim, and middle part of the chromite grains.

From Mukherjee, R., Mondal, S.K., González-Jiménez, J.M., Griffin, W.L., Pearson, N.J., O' Reilly, S.Y., 2015. Trace-element fingerprints of chromite, magnetite and sulphides from the 3.1 Ga ultramafic-mafic rocks of the Nuggihalli greenstone belt, Western Dharwar craton (India). Contrib. Mineral. Petrol. 169, 59. doi:10.1007/s00410-015 1148-1.

trivalent (Cr, Al, Fe^{3+}) octahedral site of the spinel structure, e.g., Ti, Ga, V, and Co, are much more sensitive to these processes. Thus, trace elements in chromites are good for fingerprinting the parental melt compositions and tectonic settings of the chromite deposits (Figs. 6.6–6.8; Dare et al., 2009; Page and Barnes, 2009; Mukherjee et al., 2015).

The trace elements were studied in massive chromitites from the Tagdur, Byrapur and Bhaktarhalli mines. The unaltered chromite grains from Nuggihalli (e.g., Byrapur and Bhaktarhalli chromites) did not show any variations in trace element concentration, which is consistent with the uniformity observed in terms of their major elements (Fig. 6.6A; Mukherjee et al., 2010, 2015); therefore, they are ideal for petrogenetic studies. Amongst the unaltered chromites, trace-element

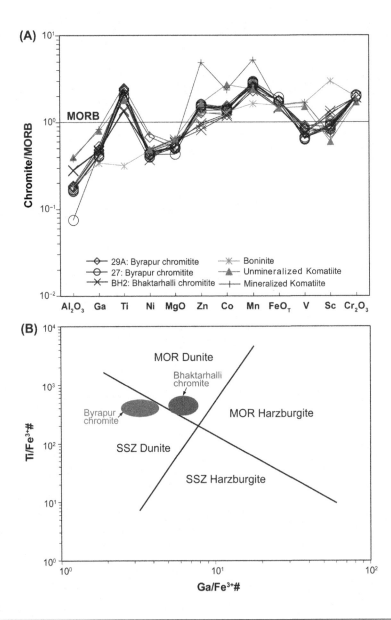

FIGURE 6.8

(A) MORB-normalized trace and major elements in unaltered chromites from massive chromitites of Byrapur and Bhaktarhalli from the Nuggihalli greenstone belt (from Mukherjee, R., Mondal, S.K., González-Jiménez, J.M., Griffin, W.L., Pearson, N.J., O' Reilly, S.Y., 2015. Trace-element fingerprints of chromite, magnetite and sulphides from the 3.1 Ga ultramafic-mafic rocks of the Nuggihalli greenstone belt, Western Dharwar craton (India). Contrib. Mineral. Petrol. 169, 59. doi:10.1007/s00410-015 1148-1) compared with chromite compositions

(Continued)

concentrations are slightly higher in the Byrapur chromites compared to the Bhaktarhalli ones, which is indicative of a different nature for the parental melt or the tectonic setting of the former (Mukherjee et al., 2015). In contrast, trace element concentrations in altered chromites (e.g., Tagdur chromites) show considerable variation across single grains, where Zn, Ga, Mn, and V usually decrease from the core to the ferritchromit rim, while Ti and Ni show an increase in the rim (Fig. 6.6B; Mukherjee et al., 2015). The cores of the altered grains show higher concentrations of Zn, Co, Mn, and V, and a moderate enrichment in Ga compared to the unaltered chromite grains (Fig. 6.6A and B). This indicates that the cores of the altered grains were enriched in trace element concentrations during the alteration process (Mukherjee et al., 2015).

The chromite grains in chromitites may have experienced heterogeneous degrees of alteration and ferritchromit formation, despite which, all the altered grains show similar distributions (except for Mn and V) and variations in trace element concentration in the sample (Fig. 6.6B). During alteration, the trivalent cations Ga and V decrease concomitantly with Cr from the core, while Fe^{3+} and Ti^{4+} are enriched in the rim, indicating that Ga and V behave similarly to Cr, and are being substituted by Fe^{3+} and Ti^{4+} during alteration (Mukherjee et al., 2015). Ga also behaves similarly to Fe^{3+} due to their similar valence and ionic radii (e.g., Dare et al., 2009), and hence these elements can substitute for one another. Such primary substitution is observed in the unaltered Byrapur chromite grains from Nuggihalli (Fig. 6.7A); insufficient data for the Bhaktarhalli chromites show no distinct trends for these grains. Ga also shows positive correlations with Al_2O_3 in the unaltered chromite grains, indicating that it can substitute for Al^{3+} but not Cr^{3+}, with which it shows a negative correlation (Fig. 6.7B and C; Mukherjee et al., 2015). In the altered grains, the divalent cations Zn and Mn behave similarly to Mg during alteration; they show lower concentrations in the rims and are substituted by Ni, which is enriched in the rim (2362–3584 ppm; Fig. 6.6B; Mukherjee et al., 2015).

Intergrain and intersample variations in trace-element distribution are common in Nuggihalli chromites, but are not commonly observed for the major elements (Figs. 6.6 and 6.7). The alteration of chromite to ferritchromit and magnetite was heterogeneous, and related to hydrothermal activity during serpentinization (Mukherjee et al., 2010, 2015). The fluids responsible for serpentinization were responsible for the addition of Ni (from altering olivine), Co and Ti in the altered grains. The local presence of carbonates was responsible for the enrichment of Zn and Mn in some grains (e.g., Barnes, 2000), and these grains also show uniform enrichment in all trace elements, which indicates that fluids of different nature (during serpentinization) have the effect of changing the trace-element distribution patterns in chromite grains; carbonate-rich fluids are reactive enough to alter chromite (Mukherjee et al., 2015). In the Nuasahi Massif (Singhbhum craton), low-temperature carbonate-rich fluids have completely altered chromite grains to form the hydrous

---

◀ in boninites (Bonin Islands; Pagé and Barnes, 2009), nickel-sulfide unmineralized (McAuliffe Well, Round Hill, Wiluna, Yunndage; Yilgarn craton, Western Australia; Yao, 1999) and mineralized komatiites (Honeymoon Well, Snake Hill; Yilgarn craton, Western Australia; Yao, 1999); (B) Plot of unaltered chromite compositions from the Byrapur and Bhaktarhalli mines in the Ga/Fe^{3+}# versus Ti/Fe^{3+}# tectonic discrimination diagram of Dare et al. (2009). The Byrapur chromites plot in the SSZ field and the Bhaktarhalli chromites plot in the MOR field (from Mukherjee, R., Mondal, S.K., González-Jiménez, J.M., Griffin, W.L., Pearson, N.J., O' Reilly, S.Y., 2015. Trace-element fingerprints of chromite, magnetite and sulphides from the 3.1 Ga ultramafic-mafic rocks of the Nuggihalli greenstone belt, Western Dharwar craton (India). Contrib. Mineral. Petrol. 169, 59. doi:10.1007/s00410-015 1148-1).

chrome carbonate, stichtite [$Mg_6Cr_2(OH)16CO_3 \cdot 4H_2O$], which commonly occurs in chromitites and in serpentinized dunite and peridotite. Stichtite occurs predominantly as veinlets or along fractures in chromite grains, but in dunites it may completely pseudomorph chromite grains (Mondal and Baidya, 1996).

6.5 CONSTRAINTS ON PARENTAL MAGMA COMPOSITION AND PROBABLE TECTONIC SETTINGS

6.5.1 NUASAHI-SUKINDA MASSIFS, SINGHBHUM CRATON

As already discussed, chromites are efficient petrogenetic tools as their composition is reflective of the nature of the parental melt and the tectonic settings in which they crystallize (e.g., Maurel and Maurel, 1982; Dick and Bullen, 1984; Kamenetsky et al., 2001). The aluminum and titanium concentrations in chromite are a function of the concentration of these elements in the parental melts from which the chromites crystallize (e.g., Maurel and Maurel, 1982; Kamenetsky et al., 2001). The Al_2O_3 (wt%) content of the parental liquid for chromites from the Nuasahi-Sukinda Massifs were calculated by Mondal et al. (2006) using the equation proposed by Maurel and Maurel (1982) for spinel-liquid equilibrium at 1 bar, where: $(Al_2O_3)_{spinel} = 0.035(Al_2O_3)^{2.42}_{melt}$. The Al_2O_3 of the calculated parental melt was in the range of $10-12$ wt%. The FeO/MgO ratio of the parental melt for the chromites from the Nuasahi-Sukinda Massifs was determined by Mondal et al. (2006) using the formula proposed by Maurel and Maurel (1982) where: $\ln(FeO/MgO)_{spinel} = 0.47 - 1.07Y^{Al}_{spinel} + 0.64Y^{Fe3+}_{spinel} + \ln(FeO/MgO)_{liquid}$, where FeO and MgO are in wt%, $Y^{Al}_{spinel} = Al/(Cr + Al + Fe^{3+})$ and $Y^{Fe3+}_{spinel} = Fe^{3+}/(Cr + Al + Fe^{3+})$. The FeO/MgO (wt%) of the melt was also calculated by using the K_D of 0.3 for olivine (dunite)-liquid equilibrium and by using $K_D = 0.23$ for calcium-free pyroxene (orthopyroxenite)-liquid equilibrium (Roeder and Emslie, 1970). The values obtained from the above calculations yield a range of $0.3-1.0$, and the ratio is higher for the orthopyroxenites from the eastern part of Nuasahi than the ones from the western part of the massif, indicating a fractionated nature of the parental melt for the former (Mondal et al., 2006).

Based on these calculations and tectonic discrimination diagrams (Fig. 6.5), Mondal et al. (2006) have found that the compositions of chromites from the Singhbhum craton show similarities with those from other Archean greenstone belts, such as Selukwe (Zimbabwe craton). The chromites from the Nuasahi-Sukinda Massifs were inferred by the authors to have crystallized from a high-Mg, low-Al, suprasubduction zone (SSZ)-related boninitic magma, or a low Al-komatiitic magma common in greenstone belts, which intruded into the volcano-sedimentary rocks of the IOG. In addition, Mondal et al. (2006) found that some of the chromite-bearing IOG high-Mg metavolcanic rocks from the Singhbhum craton are compositionally similar to boninites from subduction zones and back-arc rift settings. Petrological and experimental studies (e.g., Allégre, 1982; Grove et al., 1999; Wilson et al., 2003) have shown that komatiites and komatiitic basalts within Archean greenstone belts in some cases can form by hydrous mantle melting, at relatively low temperatures in suprasubduction settings. The parental magma of the Nuasahi-Sukinda chromites may thus have been generated in response to the dehydration of a subducting slab (Fig. 6.9A).

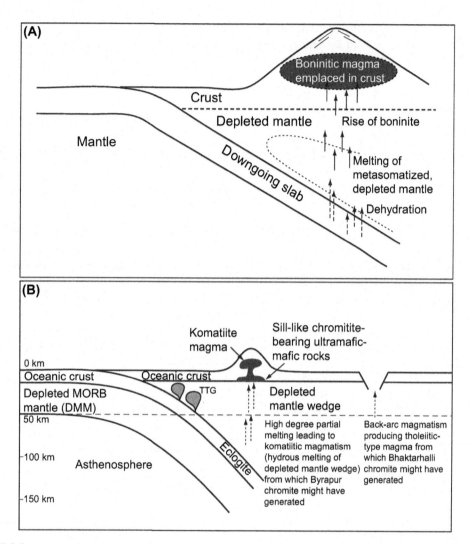

FIGURE 6.9

(A) Schematic representation of possible suprasubduction zone (SSZ) setting for the formation of ultramafic-mafic rocks in the Nuasahi and Sukinda Massifs (reproduced from Mondal, S.K., Ripley, E.M., Li, C., Frei, R., 2006. The genesis of Archean chromitites from the Nuasahi and Sukinda massifs in the Singhbhum craton, India. Precambrian Res. 148, 45−66); (B) Model showing the probable geodynamic setting of the chromitite-bearing sill-like ultramafic−mafic rocks, surrounding metavolcanic schists and the TTG in the Western Dharwar craton (after Mukherjee, R., Mondal, S.K., González-Jiménez, J.M., Griffin, W.L., Pearson, N.J., O' Reilly, S.Y., 2015. Trace-element fingerprints of chromite, magnetite and sulphides from the 3.1 Ga ultramafic-mafic rocks of the Nuggihalli greenstone belt, Western Dharwar craton (India). Contrib. Mineral. Petrol. 169, 59. doi:10.1007/s00410-015 1148-1).

Based on oxygen- and Os-isotopic studies of chromites from massive chromitites, Mondal et al. (2003, 2007) found no evidence for crustal contamination in the parental melts of these chromitites. The primitive mineral compositions of chromite, olivine and orthopyroxene from massive chromitites, dunites and orthopyroxenites, were taken as evidence that the parental melts for the Nuasahi-Sukinda chromites were formed by high degrees of mantle melting that characterized Archean magmatism (Mondal et al., 2006). A suprasubduction-zone setting was also suggested for the origin of the gneissic OMG and the OMTG rocks (e.g., Fig. 6.1B) in the Singhbhum craton by Saha et al. (2004).

The Os-isotope study by Mondal et al. (2007) found very unradiogenic $^{187}Os/^{188}Os$ ratios for chromites from massive chromitites of the Nuasahi-Sukinda Massifs ($^{187}Os/^{188}Os_{Nuasahi} = 0.10410 - 0.10739$ and $^{187}Os/^{188}Os_{Sukinda} = 0.10377 - 0.10536$). The low Os isotopic ratios are comparable to ultramafic rocks and xenoliths that are derived from the subcontinental lithospheric mantle (SCLM), and to chromitites and peridotites belonging to the Archean cratonic shields, particularly those of pre-Great Dyke occurrences in the Archean greenstone belts within the Zimbabwe craton (2.7–3.5 Ga; Nägler et al., 1997). Initial γ_{Os} values (at 3.2 Ga) of unaltered chromites from massive chromitite layers are consistently negative (range from -0.79 to -2.67), indicating a depleted nature of the mantle source; the average initial $^{187}Os/^{188}Os$ values for the Nuasahi and Sukinda chromites are statistically indistinguishable (Nuasahi = 0.10314 ± 00069; Sukinda = 0.10287 ± 0.00045) (Mondal et al., 2007). The Os isotopic studies on chromites date the initial melt extraction from the mantle at 3.7 Ga, and the isotopic values indicate that a subchondritic component of Os was derived from the SCLM beneath the Singhbhum craton (Mondal et al., 2007). An alternative possibility could be that the early Archean depleted mantle remained within the convecting mantle for a few hundred million years, before being added to the SCLM beneath the Singhbhum craton (Mondal et al., 2007a). The high Mg# of olivine and orthopyroxene, along with the high Cr# and Mg# of chromites, their unradiogenic Os-isotope compositions and old Re-Os melt depletion ages indicated that large-scale melt extraction occurred during early to mid-Archean (Mondal et al., 2007). Petrochemical characteristics indicate a strong similarity between the Archean chromitite deposits from the Singhbhum and Zimbabwe cratons (Mondal et al., 2007 and references therein). Re-Os isotopic studies (Nägler et al., 1997; Mondal et al., 2007) on chromites from both the cratons strongly suggest that depleted mantle components were recycled into the SCLM by early subduction processes, and that stabilization of the lithospheric mantle began as early as 3.7 Ga. Limited Sm-Nd data for the IOG metavolcanic rocks also support a long-term depleted source within the Singhbhum craton ($\varepsilon_{Nd} = 2.63$; Mondal et al., 2007). The depleted mantle source signature for the Nuasahi-Sukinda chromitites was further corroborated by PGE geochemical studies of these rocks (e.g., Mondal, 2007; Khatun et al., 2014; Coggon et al., 2015).

6.5.2 NUGGIHALLI GREENSTONE BELT, WESTERN DHARWAR CRATON

Similar calculations of parental melts using the relations of Maurel and Maurel (1982) and Kamenetsky et al. (2001) for the unaltered Nuggihalli chromites yielded values of $Al_2O_3 \approx 8-9$ wt% and FeO/MgO of 1.2–1.6 (Mukherjee et al., 2010). The Al_2O_3 value of the parental melt obtained from the Bhaktarhalli massive chromitite was slightly higher (10.5–10.6 wt%) than the ones computed from Byrapur. The TiO_2 content of the parental melt ($\approx 0.5-0.9$ wt%) was obtained empirically from the graphical relation shown by Kamenetsky et al. (2001), which illustrates a linear relationship between the TiO_2 concentration in chromites and in the parental liquid. The parental melt for the Nuggihalli chromites showed closest resemblance with the high-Mg komatiitic basalts that occur commonly in Archean greenstone belts, and with komatiitic rocks reported from contemporaneous greenstone belts in the WDC (Mukherjee et al., 2010). The compositions of Bhaktarhalli chromites from Nuggihalli show similarities with the compositional range of chromites from the Selukwe and Inyala greenstone belts in the Zimbabwe craton, and the Nuasahi-Sukinda chromites from the Singhbhum craton (Mukherjee et al., 2010). The Byrapur chromites show slightly higher variation in Cr# (Fig. 7a of Mukherjee et al., 2010). Although the calculations for the Nuggihalli chromites indicate a high-Mg komatiitic basalt to be the possible parental melt for the Nuggihalli chromites, the tectonic discrimination diagrams suggest derivation from boninitic magmas that are generated in modern day back-arc settings or in subduction zones (high TiO_2 area; Fig. 6.5; Mukherjee et al., 2010). This discrepancy was explained by Mukherjee et al. (2010) by the fact that komatiitic basalts within the Archean greenstone belts are petrographically, geochemically and genetically similar to boninites, such as in the Barberton greenstone belt in South Africa (e.g., Parman et al., 2001, 2004).

The trace-element data for the Nuggihalli chromites were also used to evaluate the composition of the parental magma (Mukherjee et al., 2015). The trace elements and selected major elements of unaltered Nuggihalli chromites closely resemble the composition of chromites in komatiites that lack nickel-sulfide mineralization (Fig. 6.8A; Mukherjee et al., 2015), which is in accordance with the result obtained for parental melts calculated from major-element data of the Nuggihalli chromites (Mukherjee et al., 2010). Mukherjee et al. (2015) used the tectonic discrimination diagram of Ga/Fe^{3+}# versus Ti/Fe^{3+} that was introduced by Dare et al. (2009) particularly for chromites in mantle peridotites from oceanic and ophiolite settings, to differentiate between chromites that formed in SSZs and those that formed in a midocean ridge (MOR) setting. Dare et al. (2009) mention in their work that the tectonic discrimination diagram should not be used for chromitite and lava spinels, as their history of subsolidus exchange would be different to that of the peridotites, and that future work should determine the boundaries for chromitite and lava spinels. Despite the constraints and limitations, as a very small database exists for trace elements in chromite, Mukherjee et al. (2015) used the tectonic discrimination diagram of Dare et al. (2009), which showed the Byrapur chromites to plot in the SSZ field while those from Bhaktarhalli plot in the MOR field (Fig. 6.8B; Mukherjee et al., 2015). These results again are consistent with previous results obtained from unaltered chromite compositions in Nuggihalli (e.g., Fig. 6.5; Mukherjee et al., 2010, 2015).

Mukherjee et al. (2015) put forward a model (Fig.6.9B) to explain the close spatial relation between the TTGs, the sill-like ultramafic-mafic rocks, and the surrounding metavolcanics in the Nuggihalli greenstone belt. The model shows the probable geodynamic setting in the WDC, where the TTGs formed by melting of subducting oceanic crust, with simultaneous formation of the komatiite by hydrous melting of the depleted mantle wedge. The chromitite-bearing ultramafic-mafic rocks represented the plutonic part of this magmatism (e.g., Byrapur chromitites), while the surrounding metavolcanic rocks represented the extrusive phase of the same magmatism (Fig. 6.9B; Mukherjee et al., 2012). The isotopic (Sm-Nd) and geochemical study conducted by Mukherjee et al. (2012) on the plutonic chromitite-bearing ultramafic-mafic rocks and the metavolcanics from Nuggihalli, indicated that the two rock suites are comagmatic, and that the mantle source for these rocks seem to be depleted (average ε_{Nd} at 3.1 Ga = +2.7, Mukherjee et al., 2012; ε_{Nd} at 3.4 Ga = +0.5 to +6.1 for komatiites, Jayananda et al., 2008). The relatively Al-rich parental magma obtained through calculation from the Bhaktarhalli chromites is indicative of a tholeiite-type melt that might have formed in a back-arc setting (Fig. 6.9B). The tholeiitic magma perhaps mixed with komatiitic basalts in subduction settings to form the relatively Al-rich chromite deposits of the Nuggihalli greenstone belt (such as the Bhaktarhalli chromite deposits), as well as the magnetite ore-bearing gabbroic suite of the plutonic complex (Mukherjee et al., 2012, 2015). Subsequent accretion of the back-arc setting with the arc environment was considered by Mukherjee and coworkers (2015) as an explanation for the close spatial association observed between the back-arc and arc-derived chromite deposits in the Nuggihalli greenstone belt.

Mukherjee et al. (2015) attempted to provide a quantitative estimate of partition coefficients for trace elements between the Nuggihalli chromites and their inferred parental melt, which was taken to be a komatiitic melt from Archean greenstone belts; partition coefficients were also calculated for a boninitic parental melt. Any partition coefficient calculation is based on the assumption that the crystal and melt were in equilibrium. However, for the Nuggihalli greenstone belt, calculated partition coefficients were not representative of real values, as the actual melt composition is not preserved any more (Mukherjee et al., 2015). The partition coefficients were calculated using the equation $D = C^i_{chromite}/C^i_{melt}$, where $C^i_{chromite}$ is the concentration of trace element "i" in chromite from the Nuggihalli massive chromitites, and C^i_{melt} is the concentration of trace elements in komatiites and boninites (Mukherjee et al., 2015). In this case, the authors considered the mineral partition coefficient to be similar to the bulk distribution coefficient, since chromite was the only phase crystallizing from the melt that accumulated to form the massive chromitites (e.g., Irvine, 1977; Mondal and Mathez, 2007). The partition coefficient calculations for a komatiitic melt showed that Ga ($D = 4.4-5.2$), V ($D = 2.4-6.6$), Co ($D = 1.5-3.0$), and Zn ($D = 3.6-11$) were compatible in chromite, while Mn ($D = 0.9-1.7$) and Ti ($D = 0.9-1.3$) were less strongly so, and Ni and Sc are incompatible (Mukherjee et al., 2015). The partition coefficients calculated with late Archean boninitic melts, and boninites from Cape Vogel, showed similar results where V ($D = 1.9-3.9$) and Co ($D = 3.2-4.6$) were compatible in chromite, while Mn ($D = 0.7-2.2$) and Ti ($D = 0.4-1.7$) were less compatible. The only contrast in the case of boninites was that Ni was observed to be compatible ($D = 2.8-5.3$) in chromite (Mukherjee et al., 2015). Klemme et al. (2013) conducted experiments to constrain the partition coefficients of trace elements between spinel and basaltic melt at 1 atm, 1200−1430°C and fO_2 between log − 12 and −0.7. They observed that trivalent cations in spinel structure controlled the partitioning of Ti, Sc, and the high-field-strength elements (HFSEs), such that their concentration will be highest in Fe^{3+}-enriched spinel, followed by the Cr- and Al-enriched spinel species. The data from the Nuggihalli chromites showed that Ti is moderately compatible in the chromite structure, and becomes more compatible with alteration and enrichment of Fe^{3+} in the chromite structure; Sc is always incompatible except in case of chromites with ambient interstitial carbonates, and the concentrations of most HFSEs (e.g., Zr, Nb, Hf, Ta) were below the detection limits for the Nuggihalli chromites (Mukherjee et al., 2015).

6.6 FORMATION OF CHROMITITES IN THE INDIAN ARCHEAN GREENSTONE BELTS

The formation of a chromitite layer is in itself a very complex problem, and the authors therefore refer the readers to the work of Mondal and Mathez (2007) for a comprehensive review of the different theories of chromitite genesis and their pitfalls. However, to summarize, formation of monomineralic chromitite layers may occur via various processes such as an increase in total pressure within the magma chamber (Cameron, 1977; Lipin, 1993; Cawthorn, 2005), or an increase in pressure due to shock waves associated with earthquakes (Cawthorn, 2015), assimilation by the parental magma of siliceous country rocks or roof rocks (Irvine, 1975; Kinnaird et al., 2002; Spandler et al., 2005), mixing of a primitive (ultramafic U-type) and an evolved (anorthositic A-type) magma (e.g., Sharpe and Irvine, 1983), or mixing of magmas of differing temperatures and fO_2 (Cameron and Desborough, 1969; Ulmer, 1969; Murck and Campbell, 1986), replenishment of the magma chamber by chromite-crystal bearing mush (e.g., Eales, 2000; Mondal and Mathez, 2007; Eales and Costin, 2012), and lastly, the addition of water (e.g., Nicholson and Mathez, 1991; Sisson and Grove, 1993; Gaetani et al., 1994). All these models have partially been able to solve the problem, but none can single-handedly explain the formation of chromitite layers. However, the most widely accepted theory for chromitite formation has been that of magma mixing (Irvine, 1977), which suggests the mixing of primary olivine-saturated melts with somewhat evolved residual melts in the magma chamber, to generate a hybrid melt that is saturated in chromite, owing to the curved nature of the olivine-chromite cotectic in the MgO-Cr_2O_3-SiO_2 system. The chromitite layers in the Nuasahi-Sukinda Massifs have been explained by Mondal et al. (2006) as formed either in response to mixing of magmas, or due to the suppression of silicate mineral crystallization in response to elevated H_2O concentrations in the parent melts. The chromitites from the Nuggihalli greenstone belt are considered to also have formed by magma-mixing processes, where the Byrapur chromitites perhaps formed by mixing of komatiites with slightly evolved komatiitic basalts, while mixing between a tholeiitic magma with komatiitic basalts generated the slightly more evolved Bhaktarhalli chromitites (Mukherjee et al., 2012, 2015).

6.7 BROADER IMPLICATIONS

The original depositional environment of greenstone belt lithologies is always a topic of dispute. Suggestions are plentiful, ranging from entirely intraoceanic (oceanic crust or oceanic plateau) settings to oceanic ridge-fore arc basin-, marginal basin-, island arc-, and active continental rift environments (e.g., Kusky and Kidd, 1992; DeWit, 1998). Geochemical and experimental work on komatiitic rocks from greenstone belts suggests their formation from a diapirically rising plume of mantle rocks generated at great depths (e.g., Arndt, 2003), or via hydrous mantle-melting processes in subduction zone-like settings in the Archean (e.g., Grove et al., 1999; Parman et al., 2004). Geochemical studies of high-Mg low-Ti siliceous volcanic rocks, or boninites, in greenstone belts support the involvement of early subduction-accretion processes in continent formation and crustal evolution (e.g., Polat et al., 2002; Frei and Jensen, 2003). The combination of Re-Os, U-Pb, and O isotopes, along with trace element studies, suggests that major portions of early formed cratons (e.g., Kaapvaal) grew and stabilized by $3.1 - 2.9$ Ga (Shirey et al., 2004) through subduction processes. Os-isotope study of sulfide inclusions in diamond within peridotites and eclogites derived from the SCLM, has been utilized by some researchers (e.g., Pearson et al., 1995; Shirey et al., 2004) as evidence for Archean subduction-related crustal recycling processes in the middle Archean. The research results presented by Mondal and coworkers (2006, 2007) and by Saha et al. (2004) for the ultramafic-mafic igneous suites from the Singhbhum craton, as well as by Mukherjee et al. (2010, 2012, 2015) for the Nuggihalli ultramafic-mafic rocks from the WDC, also suggest that plate-tectonic processes analogous to those of the present day occurred even in the early Archean.

Furnes et al. (2007) suggested that the ancient rocks of southwest Greenland (≈ 3.8 Ga) represented a vestige of the Earth's oldest ophiolite. The authors provided evidences such as the presence of a sheeted-dyke complex, cogenetic pillow lavas, rocks of boninitic affinity and oxygen isotopic character of the pillow lavas and dykes indicating seafloor hydrothermal metamorphism, to represent an Isua ophiolitic complex, which signify a vestige of the Archean SSZ oceanic crust. In addition, there are geochemical and field evidences which show that some of the ultramafic-mafic bodies in Archean cratons (e.g., 3.5 Ga Jamestown ophiolites, in Kaapvaal craton, DeWit et al., 1987; 2.5 Ga Dongwanzi ophiolite complex in North China craton, Kusky et al., 2001) originated as part of the oceanic lithosphere, which again suggests that modern day plate-tectonic processes operated in the Archean. Recent studies by Majumdar et al. (2016) show that serpentinization of dunites in the Nuasahi Massif occurred during interaction of the dunite with a boron-rich solution (based on boron-enrichment observed in lizardites), which was taken as inference for oceanic setting for the chromitite-bearing dunites in the massif. Past oxygen isotope studies of these rocks (Mondal et al., 2003) also indicated varying degrees of isotopic reequilibration to have occurred due to the interaction of the ultramafic-mafic rocks with evolved seawater.

There is an increased abundance in the occurrence of ultramafic-mafic rocks within greenstone belts from the early to the late Archean (i.e., from 3500 to 2700 Ma), indicating a peak in high-Mg ultramafic-mafic magmatism that was concentrated in the Archean (e.g., DeWit and Ashwal, 1995; Grove and Parman, 2004). This period of high-Mg ultramafic-mafic magmatism spanned almost 800 Ma, and may have been associated with a supercontinent cycle; also such large-scale magmatism may have been responsible for the depleted nature of the Archean mantle (Mukherjee et al., 2012 and references therein). Supercontinent cycles involve repetitive amalgamation and break up of large blocks of the Earth's continent with a periodicity of $\approx 800-900$ Ma (Rogers and Santosh, 2003). In the Archean, there seems to have been large hypothesized supercratons like Vaalbaara (3.5 Ga; break up ≈ 2.7 Ga), Superia (2.7 Ga; break up ≈ 2.4 Ga), and Slavia (2.6 Ga; break up $\approx 2.2-2.0$ Ga) from which the Kaapvaal, Superior, and Slave cratons were derived, respectively (Bleeker, 2003). A supercontinent cycle is responsible for the formation of Large Igneous Provinces (LIPs) that represent emplacement of large volumes of ultramafic-mafic magmatism in pulses of shorter duration (1–5 Ma, maximum up to 50 Ma), in every 10–30 million years of the Earth's history (e.g., Ernst et al., 2008 and references therein). Emplacement ages of LIPs are found to correlate with periods of supercontinent break up, with the break up being initiated by mantle plume activity (Bryan and Ernst, 2008). The komatiite-tholeiite associations in Archean greenstone belts and their plutonic equivalents are also considered to be manifestations of such LIPs in the Archean (Bryan and Ernst, 2008).

In the Indian Shield, although most of the craton formation events are recorded to have started as early as 3.7–3.5 Ga, the peak crust building events are rather clustered around 3.3–3.1 Ga (such as in Singhbhum craton), when the smaller cratonic nuclei (Dharwar-Bhandara Singhbhum-Bundelkhand) may have started to assemble to form a bigger continent (Mondal, 2009). The narrow range of Re-Os and Sm-Nd melt depletion ages for all the Archean rock suites so far studied from the Singhbhum craton, and corresponding available crystallization dates, strongly suggest large degree melt extraction events and concomitant major crust building processes during middle to late

Archean in the Singhbhum craton (Mondal et al., 2007). Other Archean cratons such as the Kaapvaal and Zimbabwe cratons of southern Africa show similar kinds of broadly contemporaneous events of amalgamation, which may also suggest the formation of supercontinent(s) in the late Archean (e.g., Myers, 1995). This bigger continent may have experienced a break-up due to major tectonic disturbances that led to the development of the early Proterozoic mobile belts such as in the Singhbhum craton (e.g., Singhbhum Mobile Belt; Fig. 6.1B; Mondal, 2009). The "contact type" metasomatic PGE mineralization in the Nuasahi and Sukinda Massifs of the Singhbhum craton occur in brecciated rocks, which may be linked to deep-penetrating shear zones that were developed mostly in the late Archean during formation of this big continent (Mondal, 2009). Large volumes of fluids passed through the fault zones and were responsible for base metal and gold mineralization (e.g., Kolar schist belt, Dharwar craton; Singhbhum Shear Zone, Singhbhum craton; Mondal, 2009). The similarity in SCLM depletion history between parts of the Indian Shield and the South African Shield suggests that India was part of the African continent even in the early Archean, and that amalgamation of a supercontinent involved cratons from both the Indian Shield and South African Shield during the late Archean (e.g., Bleeker, 2003; Rogers and Santosh, 2003; Srivastava, 2008).

An uneven temporal variation exists with respect to the distribution of metal deposits in the Earth's geological past (Barley and Groves, 1992). For example, chromite deposits are predominant in the early to mid-Archean greenstone belts (3500−2900 Ma). Slightly younger (≈ 2.7 Ga) chromite deposits within greenstone belts occur in the sill-like ultramafic complex of the Bird River sill deposit in the Superior craton, Canada (Ohnenstetter et al., 1986), and in the Ring of Fire intrusions in Ontario (Laarman, 2013). The distribution of chromite deposits has been shown to occur over discrete periods in the Earth's history, which may be considered to be a reflection of secular changes in the Earth's changing tectonic regimes and global heat flow (Stowe, 1994). The greenstone belt chromites are characterized by very high Mg# (60−80) and Cr# (68−78), which indicates high degrees of melting and derivation of the parental melts from a depleted mantle source (Stowe, 1994; Rollinson, 1997; Mondal et al., 2006; Mukherjee et al., 2010). Similar contemporaneous early to mid-Archean stratiform chromite deposits occur in high-grade gneissic terrains like the Fiskenæsset and Akilia anorthositic complexes (West Greenland; 2973 ± 28 Ma; Polat et al., 2010), and the Sittampundi complex (south India; 2935 ± 60 Ma; Bhaskar Rao et al., 1996). The initial ε_{Nd} of the high-grade chromitite deposits also indicates a depleted nature of the mantle source (Fiskenæsset $+3.3 \pm 0.7$, Polat et al., 2010; Sittampundi $+1.85 \pm 0.16$, Bhaskar Rao et al., 1996). The observed discrete temporal distribution of different styles of metal deposits was explained by Barley and Groves (1992) to be an outcome of plate-tectonic cycles, rather than to progressive changes in the tectonic processes of the Earth.

The preponderance of ore deposits of chromite from ≈ 3.5 to 2.7 Ga, and Ni-sulfide and PGE from ≈ 3.0 to 2.7 Ga are thus related to large-scale high-Mg ultramafic-mafic plutonic and volcanic events in the Earth that are connected to supercontinent cycles (Mukherjee et al., 2012 and references therein). There could possibly have been two supercontinents at 3.5 and 2.7 Ga, with an overlap between the break up of the older hypothesized supercontinent, and the formation of the younger one; such overlaps between break up and assembly have been reported for Rodinia (Ernst et al., 2008). The SSZ setting for the 3.1 Ga chromite deposits of the Nuasahi-Sukinda Massifs and the Nuggihalli greenstone belt perhaps indicates that these deposits were related to the amalgamation phase of a supercontinent which culminated at 2.7 Ga (Mondal, 2009; Mukherjee et al., 2012).

REFERENCES

Abzalov, M.Z., 1998. Chrome spinels in gabbro-wehrlite intrusions of the Pechenga area, Kola Peninsula, Russia: emphasis on alteration features. Lithos 43, 109–134.

Allègre, C.J., 1982. Genesis of Archean komatiites in a wet ultramafic subducted plate. In: Arndt, N.T., Nisbet, E.G. (Eds.), Komatiites. George Allen and Unwin, London, pp. 495–500.

Arai, S., Ahmed, A.H., 2013. Secular change of the chromite concentration process from the Archean to Phanerozoic. Goldschmidt abstract. Mineralogical Mag. 606 p.

Arndt, N.T., 2003. Komatiites, kimberlites and boninites. J. Geophys. Res. 108, 2293 p.

Arndt, N.T., Lesher, C.M., Barnes, S.J., 2008. Komatiite. Cambridge University Press, UK, 467 p.

Ashwal, L.D., 1993. Anorthosites. Springer-Verlag, Berlin, 422 p.

Augé, T., Cocherie, A., Genna, A., Armstrong, R., Guerrot, C., Mukherjee, M.M., et al., 2003. Age of the Baula PGE mineralization (Orissa, India) and its implications concerning the evolution of the Singhbhum Archean nucleus. Precambrian Res. 121, 85–101.

Barley, M.E., Groves, D.I., 1992. Supercontinent cycles and the distribution of metal deposits through time. Geology 20, 291–294.

Barnes, S.J., 2000. Chromite in Komatiites. II. Modification during greenschist to mid amphibolite facies metamorphism. J. Petrol. 41, 387–409.

Barnes, S.J., 2006. Komatiites: Petrology, Volcanology, Metamorphism, and Geochemistry. Society of Economic Geologists Special Publication 13, pp. 13–49.

Barnes, S.J., Jones, S., 2013. Deformed Chromitite Layers in the Coobina Intrusion, Pilbara Craton, Western Australia. Econ. Geol. 108, 337–354.

Barnes, S.J., Roeder, P.L., 2001. The range of spinel compositions in terrestrial mafic and ultramafic rocks. J. Petrol. 42, 2279–2302.

Barnes, S.J., Heggie, G.J., Fiorentini, M.L., 2013. Spatial variation in platinum group element concentrations in ore-bearing komatiite at the Long-Victor deposit, Kambalda Dome, Western Australia: enlarging the footprint of nickel sulfide orebodies. Econ. Geol. 108, 913–933.

Basu, A., Maitra, M., Roy, P.K., 1997. Petrology of mafic ultramafic complex of Sukinda valley, Orissa. Indian Miner. 50, 271–290.

Bhaskar Rao, Y.J., Sivaraman, T.V., Pantulu, C.V.C., Gopalan, K., Naqvi, S.M., 1992. Rb–Sr ages of late Archean metavolcanics and granites, Dharwar craton, south India and evidence for early Proterozoic thermotectonic event(s). Precambrian Res. 59, 145–170.

Bhaskar Rao, Y.J., Chetty, T.R.K., Janardhan, A.S., Gopalan, K., 1996. Sm–Nd and Rb–Sr ages and P–T history of the Archean Sittampundi and Bhavani layered meta anorthosite complexes in Cauvery shear zone, South India: evidence for Neoproterozoic reworking of Archean crust. Contrib. Mineral. Petrol. 125, 237–250.

Bidyananda, M., Mitra, S., 2005. Chromitites from komatiitic affinity from the Archean Nuggihalli greenstone belt in South India. Mineral. Petrol. 84, 169–187.

Bleeker, W., 2003. The late Archean record: a puzzle in ca. 35 pieces. Lithos 91, 99–134.

Bliss, N.W., Maclean, W.H., 1975. The paragenesis of zoned chromite from central Manitoba. Geochim. Cosmochim. Acta 39, 973–990.

Bryan, S.E., Ernst, R.E., 2008. Revised definition of large igneous provinces (LIPs). Earth Sci. Rev. 86, 175–202.

Burkhard, D.J.M., 1993. Accessory chromium spinels: their coexistence and alteration in serpentinites. Geochim. Cosmochim. Acta 57, 1297–1306.

Cameron, E.N., 1977. Chromite in the central sector of the eastern Bushveld Complex, South Africa. Am. Mineral. 62, 1082–1096.

Cameron, E.N., Desborough, G.A., 1969. Occurrence and characteristics of chromite deposits-eastern Bushveld Complex. Econ. Geol. Monogr. 4, 23–40.

Cawthorn, R.G., 2005. Pressure fluctuations and the formation of the PGE-rich Merensky and chromitite reefs, Bushveld Complex. Mineral. Dep. 40, 231–235.

Cawthorn, R.G., 2015. Bushveld Complex. In: Charlier, B., Namur, O., Latypov, R., Tegner, C. (Eds.), Layered Intrusions. Springer International Publishing, New York City, pp. 517–587.

Chakraborty, K.L., Chakraborty, T.L., 1984. Geological features and origin of the chromite deposits of Sukinda valley, Orissa, India. Miner. Dep. 19, 256–265.

Coggon, J.A., Luguet, A., Lorand, J.P., Fonseca, R., Appel, P., Mondal, S.K., et al., 2015. Early mantle evolution and the late veneer new perspectives from highly siderophile elements. In: Abstract DI31A-2575 Presented at 2015 Fall Meeting, AGU, San Francisco, Calif., 14-18 Dec.

Colás, V., González-Jiménez, J.M., Griffin, W.L., Fanlo, I., Gervilla, F., O'Reilly, S.Y., et al., 2014. Fingerprints of metamorphism in chromite: new insights from minor and trace elements. Chem. Geol. 389, 137–152.

Dare, S.A.S., Pearce, J.A., McDonald, I., Styles, M.T., 2009. Tectonic discrimination of peridotites using fO_2–Cr# and Ga–Ti–Fe^{III} systematics in chrome-spinel. Chem. Geol. 261, 199–216.

Devaraju, T.C., Viljoen, R.P., Sawkar, R.H., Sudhakara, T.L., 2009. Mafic and ultramafic magmatism and associated mineralization in the Dharwar craton, southern India. J. Geol. Soc. India 73, 73–100.

DeWit, M.J., 1998. On Archean granites, greenstones, cratons and tectonics: does the evidence demand a verdict?. Precambrian Res. 91, 181–226.

DeWit, M.J., Ashwal, L.D., 1995. Greenstone belts: what are they? S. Afr. J. Geol. 98, 505–520.

DeWit, M.J., Hart, R.A., Hart, R.J., 1987. The Jamestown ophiolite complex, Barberton mountain belt: a section through 3.5 Ga oceanic crust. J. Afr. Earth Sci. 6, 681–730.

Dey, S., 2012. Evolution of Archaean crust in the Dharwar craton: the Nd isotope record. Precambrian Res. 227, 227–246.

Dick, H.J.B., Bullen, T., 1984. Chromian spinel as a petrogenetic indicator in abyssal and alpine-type peridotites and spatially associated lavas. Contrib. Mineral. Petrol. 86, 54–76.

Eales, H.V., 2000. Implications for the chromium budget of the western limb of the Bushveld Complex. S. Afr. J. Geol. 103, 141–150.

Eales, H.V., Costin, G., 2012. Crustally contaminated komatiite: primary source of the chromitites and Marginal, Lower, and Critical Zone magmas in a staging chamber beneath the Bushveld Complex. Econ. Geol. 107, 645–665.

Ernst, R.E., Wingate, M.T.D., Buchan, K.L., Li, Z.X., 2008. Global record of 1600–700 Ma Large Igneous Provinces (LIPs): implications for the reconstruction of the proposed Nuna (Columbia) and Rodinia supercontinents. Precambrian Res. 160, 159–178.

Evans, B.W., Frost, B.R., 1975. Chrome-spinel in progressive metamorphism-a preliminary analysis. Geochim. Cosmochim. Acta 39, 959–972.

Fiorentini, M.L., Beresford, S.W., Barley, M.E., Duuring, P., Bekker, A., Rosengren, N., et al., 2012. District to camp controls on the genesis of komatiite-hosted nickel sulfide deposits, Agnew-Wiluna greenstone belt, Western Australia: insights from the multiple sulfur isotopes. Econ. Geol. 107, 781–796.

Frei, R., Jensen, B.K., 2003. Re–Os, Sm–Nd isotope– and REE systematics on komatiites and pillow basalts from the Earth's oldest oceanic crustal fragments (Isua Supracrustal Belt, W Greenland). Chem. Geol. 196, 163–191.

Furnes, H., DeWit, M.J., Staudigel, H., Rosing, M., Muehlenbachs, K., 2007. A vestige of Earth's Oldest Ophiolite. Science 315, 1704–1707.

Gaetani, G.A., Grove, T.L., Bryan, W.B., 1994. Experimental phase relations of basaltic andesite from hole 839B under hydrous and anhydrous conditions. In: Hawkins, J., Parson, L., Allan, J. (Eds.), Proceedings of the Ocean Drilling Program: Scientific Results, 135. Ocean Drilling Program, College Station, TX, pp. 557–563.

Gervilla, F., Padrón-Navarta, J., Kerestedjian, T., Sergeeva, I., González-Jiménez, J.M., Fanlo, I., 2012. Formation of ferrian chromite in podiform chromitites from the Golyamo Kamenyane serpentinite, Eastern Rhodopes, SE Bulgaria: a two-stage process. Contrib. Mineral. Petrol. 164, 1–15.

González-Jiménez, J.M., Griffin, W.L., Gervilla, F., Proenza, J.A., O'Reilly, S.Y., Pearson, N.J., 2014. Chromitites in ophiolites: how, where, when, why? Part II. The crystallization of chromitites. Lithos 189, 140–158.

Grove, T.L., Parman, S.W., 2004. Thermal evolution of the earth as recorded by komatiites. Earth Planet. Sci. Lett. 219, 173–187.

Grove, T.L., Parman, S.W., Dann, J.C., 1999. Conditions of magma generation for Archean komatiites from the Barberton Mountainland, South Africa. In: Fei, Y., Bertka, C.M., Mysen, B.O. (Eds.), Mantle Petrology; Field Observations and High-Pressure Experimentation; A Tribute to Francis R (Joe) Boyd. Geochem. Soc. Spl. Publ. 6, pp. 155–167.

Haldar, D., 1967. Some observations on the chromiferous ultramafic and the associated rocks around Nuasahi, Keonjhar district, Orissa. Indian Miner. 21, 196–204.

Hamlyn, P.R., Keays, R.R., 1979. Origin of chromite compositional variation in the Panton Sill, Western Australia. Contrib. Mineral. Petrol. 69, 75–82.

Hunter, D.R., Stowe, C.W., 1997. A historical review of the origin, composition, and setting of Archean greenstone belts (Pre-1980). In: DeWit, M.J., Ashwal, L.D. (Eds.), Greenstone Belts. Clarendon Press, Oxford, pp. 5–29.

Irvine, T.N., 1965. Chromian spinel as a petrogenetic indicator. Part I, Theory. Can. J. Earth Sci. 2, 648–671.

Irvine, T.N., 1967. Chromian spinel as a petrogenetic indicator. Part 2. Petrologic applications. Can. J. Earth Sci. 4, 71–103.

Irvine, T.N., 1975. Crystallization sequences in the Muskox intrusion and other layered intrusions II. Origin of chromitite layers and similar deposits of other magmatic ores. Geochim. Cosmochim. Acta 39, 991–1020.

Irvine, T.N., 1977. Origin of chromitite layers in the Muskox intrusion and other layered intrusions: a new interpretation. Geology 5, 273–277.

Jackson, E.D., 1969. Chemical variation in coexisting chromite and olivine in chromitite zones of the Stillwater complex. Econ. Geol. Monogr. 4, 41–71.

Jafri, S.H., Khan, N., Ahmed, S.M., Saxena, R., 1983. Geology and geochemistry of Nuggihalli schist belt, Dharwar craton, Karnataka, India. In: Naqvi, S.M., Rogers, J.J.W. (Eds.), Precambrian of South India. Memoir. Geol. Soc. India No 4, pp. 110–120.

Jayananda, M., Kano, T., Peucat, J.J., Channabasappa, S., 2008. 3.35 Ga komatiite volcanism in the Western Dharwar craton, southern India: constraints from Nd isotopes and whole-rock geochemistry. Precambrian Res. 162, 160–179.

Kamenetsky, V.S., Crawford, A.J., Meffre, S., 2001. Factors controlling chemistry of magmatic spinel: an empirical study of associated olivine, Cr–spinel and melt inclusions from primitive rocks. J. Petrol. 42, 655–671.

Khatun, S., Mondal, S.K., Balaram, V., Rosing, M.T., Frei, R., 2009. Geochemistry of Mesoarchean Sukinda chromite deposits (India): Implications for gabbro-breccia hosted PGE mineralization. Geochim. Cosmochim. Acta 73 (13 Suppl.), 647.

Khatun, S., Mondal, S.K., Zhou, M.-F., Balaram, V., Prichard, H.M., 2014. Platinum-group element (PGE) geochemistry of Mesoarchean ultramafic–mafic cumulate rocks and chromitites from the Nuasahi Massif, Singhbhum Craton (India). Lithos 205, 322–340.

Kinnaird, J.A., Kruger, F.J., Nex, P.A.M., Cawthorn, R.G., 2002. Chromite formation—a key to understanding processes of platinum enrichment. Trans. Inst. Mining Metal. 111, B23–B35.

Klemme, S., Wijbrans, C.H., Vollmer, C., Menneken, M., Berndt, J., 2013. Experimental study of trace element partitioning between spinel and silicate melts: Effects of oxygen fugacity and spinel composition. Goldschmidt Abstract. Mineral. Mag. 1476 p.

Kröner, A., Hegner, E., Wendt, J.I., Byerly, G.R., 1996. The oldest part of the Barberton granitoid-greenstone terrain, South Africa: evidence for crust formation between 3.5 and 3.7 Ga. Precambrian Res. 78, 105—124.

Kumar, A., Bhaskar Rao, Y.J., Sivaraman, T.V., Gopalan, K., 1996. Sm—Nd ages of Archean metavolcanics of the Dharwar craton, South India. Precambrian Res. 80, 205—216.

Kusky, T.M., Kidd, W.S.F., 1992. Remnants of an Archean oceanic plateau, Belingwe greenstone belt, Zimbabwe. Geology 20, 43—46.

Kusky, T.M., Li, J.H., Tucker, R.D., 2001. The Archean Dongwanzi Ophiolite Complex, North China Craton: 2.505-billion-year-old Oceanic Crust and Mantle. Science 292, 1142—1145.

Laarman, J.E., 2013. A Detailed Metallogenic Study of the McFaulds Lake Chromite Deposits, Northern Ontario (Ph.D. thesis). University of Western Ontario.

Leelananadam, C., Burke, K., Ashwal, L.D., Webb, S.J., 2006. Proterozoic mountain building in Peninsular India: an analysis based primarily on alkaline rock distribution. Geol. Mag. 143, 1—18.

Lesher, C.M., Groves, D.I., 1986. Controls on the formation of komatiite—associated nickel copper sulfide deposits. In: Friedrich, G.H. (Ed.), Geology and Metallogeny of Copper Deposits. Springer, Berlin, pp. 63—90.

Lipin, B.R., 1993. Pressure increases in the formation of chromite seams and the development of the ultramafic series in the Stillwater complex, Montana. J. Petrol. 34, 955—976.

Loferski, P.J., Lipin, B.R., 1983. Exsolution in metamorphosed chromite from the Red Lodge district, Montana. Am. Mineral. 68, 777—789.

Majumdar, A.S., Hövelmann, J., Vollmer, C., Berndt, J., Mondal, S.K., Putnis, A., 2016. Formation of Mg-rich olivine pseudomorphs in serpentinized dunite from the Mesoarchean Nuasahi Massif, eastern India: insights into the evolution of fluid composition at the mineral-fluid interface. J. Petrol. 57, 3—26.

Maurel, C., Maurel, P., 1982. Etude expérimentale de la solubilité du chrome dans les bains silicatés basiques et sa distribution entre liquide et minéraux coexistants: conditions d'existence du spinelle chromifére. Bull. Minéral. 105, 197—202.

Misra, K.C., 2000. Understanding Mineral Deposits. Springer, Netherlands, 845 p.

Mondal, S.K., 2007. PGE distributions in Mesoarchean chromitites and mafic-ultramafic rocks in the Singhbhum Craton (India): evidence for presence of a subchondritic source mantle domain. In: Abstract Published in Geochim. Cosmochim. Acta; Goldschmidt 2007 (Cologne).

Mondal, S.K., 2009. Chromite and PGE deposits of Mesoarchaean ultramafic-mafic suites within the greenstone belts of the Singhbhum craton, India: implications for mantle heterogeneity and tectonic setting. J. Geol. Soc. India 73, 36—51.

Mondal, S.K., Baidya, T.K., 1996. Stichtite [$Mg_6Cr_2(OH)_{16}CO_3.4H_2O$] in the Nuasahi ultramafites, Orissa, India—its transformation at elevated temperatures. Mineral. Mag. 60, 836—840.

Mondal, S.K., Baidya, T.K., 1997. Platinum-group minerals from the Nuasahi ultramafic mafic complex, Orissa, India. Mineral. Mag. 61, 902—906.

Mondal, S.K., Mathez, E.A., 2007. Origin of the UG2 chromitite layer, Bushveld complex. J. Petrol. 48, 495—510.

Mondal, S.K., Zhou, M.-F., 2010. Enrichment of PGE through interaction of evolved boninitic magmas with early formed cumulates in a gabbro—breccia zone of the Mesoarchean Nuasahi massif (eastern India). Mineral. Dep. 45, 69—91.

Mondal, S.K., Baidya, T.K., Rao, K.N.G., Glascock, M.D., 2001. PGE and Ag mineralization in a breccia zone of the Precambrian Nuasahi ultramafic-mafic complex, Orissa, India. Can. Mineral. 39, 979—996.

Mondal, S.K., Ripley, E.M., Li, C., Ahmed, A.H., Arai, S., Liipo, J., et al., 2003. Oxygen isotopic compositions of Cr-spinels from Archean to Phanerozoic chromite deposits. In: Abstract published in Geochim. Cosmochim. Acta, 18S, A30: Goldschmidt 2003, (Japan).

Mondal, S.K., Ripley, E.M., Li, C., Frei, R., 2006. The genesis of Archean chromitites from the Nuasahi and Sukinda massifs in the Singhbhum craton, India. Precambrian Res. 148, 45−66.

Mondal, S.K., Frei, R., Ripley, E.M., 2007. Os isotope systematics of meso-Archean chromitite-PGE deposits in the Singhbhum Craton (India): Implications for the evolution of lithospheric mantle. Chem. Geol. 244, 391−408.

Mondal, S.K., Arai, S., Payot, B.D., Tamura, A., 2013. Lithospheric mantle connection of clinopyroxene inclusions in chromites from the Archean Nuasahi ultramafic-mafic complex (India). Mineral. Mag. 77 (5), 1779 p.

Mukherjee, S., 1966. The Nuasahi−Nilgiri igneous complex. Bull. Geol. Soc. India 1, 34−37.

Mukherjee, R., Mondal, S.K., Rosing, M.T., Frei, R., 2010. Compositional variations in the Mesoarchean chromites of the Nuggihalli schist belt, Western Dharwar craton (India): potential parental melts and implications for tectonic setting. Contrib. Mineral. Petrol. 160, 865−885.

Mukherjee, R., Mondal, S.K., Frei, R., Rosing, M.T., Waight, T.E., Zhong, H., et al., 2012. The 3.1 Ga Nuggihalli chromite deposits, Western Dharwar craton (India): Geochemical and isotopic constraints on mantle sources, crustal evolution and implications for supercontinent formation and ore mineralization. Lithos 155, 392−409.

Mukherjee, R., Mondal, S.K., Zhong, H., Bai, Z.-J., Balaram, V., Ravindra Kumar, G.R., 2014. Platinum group element geochemistry of komatiite-derived 3.1 Ga ultramafic−mafic rocks and chromitites from the Nuggihalli greenstone belt, Western Dharwar craton (India). Chem. Geol. 386, 190−208.

Mukherjee, R., Mondal, S.K., González-Jiménez, J.M., Griffin, W.L., Pearson, N.J., O' Reilly, S.Y., 2015. Trace-element fingerprints of chromite, magnetite and sulphides from the 3.1 Ga ultramafic-mafic rocks of the Nuggihalli greenstone belt, Western Dharwar craton (India). Contrib. Mineral. Petrol. 169 (59), . Available from: http://dx.doi.org/10.1007/s00410-0151148-1.

Mukhopahyay, J., Beukes, N.J., Armstrong, R.A., Zimmermann, U., Ghosh, G., Medda, R.A., 2008. Dating the oldest greenstone in India: a 3.51-Ga precise U-Pb SHRIMP zircon age for dacitic lava of the southern Iron Ore Group, Singhbhum craton. J. Geol. 116, 449−461.

Mungall, J.E., Staff, N.G., 2008. Formation of massive chromitite by assimilation of iron formation in the Blackbird deposit, Ontario, Canada. In: Eos Transaction AGU 89 (53), Fall Meeting Supplement, Abstract V11A−2014.

Murck, B.N., Campbell, I.H., 1986. The effects of temperature, oxygen fugacity and composition on the behavior of chromium in basic and ultramafic melts. Geochim. Cosmochim. Acta 50, 1871−1887.

Murthy, N.G.K., 1987. Mafic Dyke Swarms of the Indian Shield, Mafic Swarms. Geological Association of Canada Special Paper 34, pp. 393−400.

Myers, J.S., 1995. The generation and assembly of an Archean supercontinent: evidence from the Yilgarn Craton, Western Australia. In: Coward, M.P., Ries, A.C. (Eds.), Early Precambrian Processes. Geol. Soc. London, pp. 143−154.

Nägler, T.F., Kramers, J.D., Kamber, B.S., Frei, R., Prendergast, M.D.A., 1997. Growth of subcontinental lithospheric mantle beneath Zimbabwe started at or before 3.8 Ga: Re Os study on chromites. Geology 25, 983−986.

Naldrett, A.J., Turner, A.R., 1977. The geology and petrogenesis of a greenstone belt and related nickel-sulfide mineralization at Yakabindie, Western Australia. Precambrian Res. 5, 43−103.

Nicholson, D.M., Mathez, E.A., 1991. Petrogenesis of the Merensky Reef in the Rustenburg section of the Bushveld Complex. Contrib. Mineral. Petrol. 107, 293−309.

Ohnenstetter, D., Watkinson, D.H., Jones, P.C., Talkington, R., 1986. Cryptic compositional variation in laurite and enclosing chromite from the Bird River Sill, Manitoba. Econ. Geol. 81, 1159–1168.

Pagé, P., Barnes, S.-J., 2009. Using trace elements in chromites to constrain the origin of podiform chromitites in the Thetford Mines ophiolite, Québec, Canada. Econ. Geol. 104, 997–1018.

Pagé, P., Barnes, S.-J., Bédard, J.H., Zientek, M.L., 2012. In situ determination of Os, Ir, and Ru in chromites formed from komatiite, tholeiite and boninite magmas: implications for chromite control of Os, Ir and Ru during partial melting and crystal fractionation. Chem. Geol. 302 (303), 3–15.

Parman, S.W., Grove, T.L., Dann, J.C., 2001. The production of Barberton komatiites in an Archean subduction zone. Geophys. Res. Lett. 28, 2513–2516.

Parman, S.W., Grove, T.L., Dann, J.C., 2004. A subduction model for komatiites and cratonic lithosphere mantle. S. Afr. J. Geol. 107, 107–118.

Pearson, D.G., Shirey, S.B., Carlson, R.W., Boyd, F.R., Pokhilenko, N.P., Shimizu, N., 1995. Re–Os, Sm–Nd and Rb–Sr isotope evidence for thick Archean lithospheric mantle beneath the Siberian craton modified by multistage metasomatism. Geochim. Cosmochim. Acta 59, 959–977.

Polat, A., Hofmann, A.W., Rosing, M.T., 2002. Boninite-like volcanic rocks in the 3.7 – 3.8 Ga Isua greenstone belt, West Greenland: geochemical evidence for intra-oceanic subduction zone processes in the early earth. Chem. Geol. 184, 231–254.

Polat, A., Frei, R., Scherstén, A., Appel, P.W.U., 2010. New age (ca. 2970 Ma), mantle source composition and geodynamic constraints on the Archean Fiskenæsset anorthosite complex, SW Greenland. Chem. Geol. 277, 1–20.

Prendergast, M.D., 1984. Chromium deposits of Zimbabwe. Chromium Rev. 2, 5–9.

Prendergast, M.D., 2008. Archean komatiitic sill–hosted chromite deposits in the Zimbabwe craton. Econ. Geol. 103, 981–1004.

Radhakrishna, B.P., Naqvi, S.M., 1986. Precambrian continental crust of India and its evolution. J. Geol. 94, 145–166.

Ramakrishnan, M., 1981. Nuggihalli and Krishnarajpet belts. In: Swami Nath, J., Ramakrishnan, M. (Eds.), Early Precambrian Supracrustals of Southern Karnataka. Geol. Surv. India Mem. No 112, pp. 61–70.

Ramakrishnan, M., 2009. Precambrian mafic magmatism in the western Dharwar craton, Southern India. J. Geol. Soc. India 73, 101–116.

Roeder, P.L., 1994. Chromite: from the fiery rain of chondrules to the Kilauea Iki lava lake. Can. Mineral. 32, 729–746.

Roeder, P.L., Emslie, R.F., 1970. Olivine-liquid equilibria. Contrib. Mineral. Petrol. 29, 275–289.

Roeder, P.L., Campbell, I.H., Jamieson, H.E., 1979. A re-evaluation of the olivine-spinel geothermometer. Contrib. Mineral. Petrol. 68, 325–334.

Rogers, J.J.W., Santosh, M., 2003. Supercontinents in Earth history. Gond. Res. 6, 357–368.

Rollinson, H., 1995a. Composition and tectonic settings of chromite deposits through time. Econ. Geol. 90, 2091–2092.

Rollinson, H., 1995b. The relationship between chromite chemistry and the tectonic setting of Archean ultramafic rocks. In: Blenkinsop, T.G., Tromps, P. (Eds.), Sub-Saharan Economic Geology. Balkema, Amsterdam, pp. 7–23.

Rollinson, H., 1997. The Archaean komatiite–related Inyala chromitite, Southern Zimbabwe. Econ. Geol. 92, 98–107.

Rollinson, H., 2008. The geochemistry of mantle chromitites from the northern part of the Oman ophiolite: inferred parental melt compositions. Contrib. Mineral. Petrol. 156, 273–288.

Rollinson, H., Reid, C., Windley, B., 2010. Chromitites from the Fiskenæsset Anorthositic Complex, West Greenland: Clues to Late Archaean Mantle Processes. Geol. Soc. London Spl. Publ. 338, pp. 197–212.

Saha, A.K., 1994. Crustal Evolution of Singhbhum North Orissa, Eastern India. Mem. Geol. Soc. India No. 27, p. 341.

Saha, A.K., Ray, S.L., Sarkar, S.N., 1988. Early history of the Earth: evidence from the eastern Indian Shield. In: Mukhopadhyay, D. (Ed.), Precambrian of the Eastern Indian Shield. Mem. Geol. Soc. India, No. 8, pp. 13–37.

Saha, A.K., Basu, A.R., Garzione, C.N., Bandyopadhyay, P.K., Chakraborti, A., 2004. Geochemical and petrological evidence for subduction-accretion processes in the Archean Eastern Indian Craton. Earth Planet. Sci. Lett. 220, 91–106.

Sahu, N.K., Mukherjee, M.M., 2001. Spinifex textured komatiite from Badampahar-Gorumahishani schist belt, Mayurbhanj district, Orissa. J. Geol. Soc. India 57, 529–534.

Sarkar, N.K., Mallik, A.K., Panigrahi, D., Ghosh, S.N., 2001. A note on the occurrence of breccia zone in the Katpal chromite lode, Dhenkanal district, Orissa. Indian Mineral. 55 (3–4), 247–250.

Scoates, J.S., Scoates, R.F.J., 2013. Age of the Bird River Sill, Southeastern Manitoba, Canada, with implications for the secular variation of layered intrusion-hosted stratiform chromite mineralization. Econ. Geol. 108, 895–907.

Scowen, P.A.H., Roeder, P.L., Helz, R.T., 1991. Reequilibration of chromite within Kilauea Iki lava lake, Hawaii. Contrib. Mineral. Petrol. 107, 8–20.

Sengupta, S., Acharya, S.K., De Smeth, J.B., 1997. Geochemistry of Archean volcanic rocks from Iron Ore supergroup, Singhbhum, eastern India. Proc. Indian Acad. Sci. (Earth Planet. Sci.) 106, 327–342.

Sharma, M., Basu, A.R., Ray, S.L., 1994. Sm–Nd isotopic and geochemical study of the Archean tonalite-amphibolite association from the eastern Indian Craton. Contrib. Mineral. Petrol. 117, 45–55.

Sharpe, M.R., Irvine, T.N., 1983. Melting relations of two Bushveld chilled margin rocks and implications for the origin of chromitite, Carnegie Institution Geophysical Laboratory Yearbook, vol. 82. pp. 295–300.

Shirey, S.B., Richardson, S.H., Harris, J.W., 2004. Age, paragenesis, and composition of diamonds and evolution of the Precambrian mantle lithosphere of southern Africa. S. Afr. J. Geol. 107, 91–106.

Sisson, T.W., Grove, T.L., 1993. Experimental investigations of the role of H2O in calc alkaline differentiation and subduction zone magmatism. Contrib. Mineral. Petrol. 113, 143–166.

Spandler, C., Mavrogenes, J., Arculus, R., 2005. Origin of chromitites in layered intrusions: evidence from chromite-hosted melt inclusions from the Stillwater Complex. Geology 33, 893–896.

Srivasatva, R.K., 2008. Global intracratonic Boninite-Norite magmatism during the Neoarchean–Paleoproterozoic: evidence from the Central Indian Bastar Craton. Int. Geol. Rev. 50, 61–74.

Stowe, C.W., 1987. Chromite deposits of the Shurugwi greenstone belt, Zimbabwe. In: Stowe, C.W. (Ed.), Evolution of Chromium Ore Fields. Van Nostrand Reinhold Company, New York, NY, pp. 71–88.

Stowe, C.W., 1994. Compositions and tectonic settings of chromite deposits though time. Econ. Geol. 89, 528–546.

Subramaniam, A.P., 1956. Mineralogy and petrology of the Sittampundi complex, Salem district, Madras State, India. Geol. Soc. Am. Bull. 67, 317–390.

Swami Nath, J., Ramakrishnan, M., 1981. Early Precambrian Supracrustals of Southern Karnataka. Geol. Surv. India Mem. 112, 363 p.

Taylor, P.N., Chadwick, B., Moorbath, S., Ramakrishnan, M., Viswanatha, M.N., 1984. Petrography, chemistry and isotopic ages of peninsular gneisses, Dharwar acid volcanic rocks and the Chitradurga granite with special reference to the late Archaean evolution of the Karnataka craton. Precambrian Res. 23, 349–375.

Thayer, T.P., 1960. Some critical differences between alpine-type and stratiform peridotite gabbro complexes. In: Sorgenfrei, T. (Ed.), International Geol. Congress Rept., 21st session, Copenhagen, vol. 13, Det Berligske Bogtrykkeri, Copenhagen, pp. 247–259.

Ulmer, G.C., 1969. Experimental investigations of chromite spinels. Econ. Geol. Monogr. 4, 114–131.

Van Kranendonk, M.J., Kröner, A., Hegner, E., Connelly, J., 2009. Age, lithology and structural evolution of the c. 3.53 Ga Theespruit Formation in the Tjakastad area, southwestern Barberton Greenstone Belt, South Africa, with implications for Archaean tectonics. Chem. Geol. 261, 115–139.

Varma, O.P., 1986. Some aspects of ultramafic and ultrabasic rocks and related chromite metallogenesis with examples from eastern region of India. Proceedings of the 73rd Session. Indian Science Congress Association, Delhi, pp. 1–72.

Wilson, A.H., Shirey, S.B., Carlson, R.W., 2003. Archean ultradepleted komatiites formed by hydrous melting of cratonic mantle. Nature 423, 858–861.

Windley, B.F., Bishop, F.C., Smith, J.S., 1981. Metamorphosed layered igneous complexes in Archean Granulite-Gneiss Belts. Ann. Rev. Earth Planet. Sci. 9, 175–198.

Witt, W.K., Hickman, A.H., Townsend, D., Preston, W.A., 1998. Mineral potential of the Archaean Pilbara and Yilgarn Cratons, Western Australia. J. Austr. Geol. Geophys. 17 (3), 201–221.

Yao, S., 1999. Chemical Composition of Chromites from Ultramafic Rocks: Application to Mineral Exploration and Petrogenesis (Ph.D. thesis). Macquarie University, Sydney, 174 p.

CHAPTER

7

NEW INSIGHTS ON THE ORIGIN OF ULTRAMAFIC-MAFIC INTRUSIONS AND ASSOCIATED NI-CU-PGE SULFIDE DEPOSITS OF THE NORIL'SK AND TAIMYR PROVINCES, RUSSIA: EVIDENCE FROM RADIOGENIC- AND STABLE-ISOTOPE DATA

Kreshimir N. Malitch[1,2], Elena A. Belousova[2], William L. Griffin[2], Inna Yu Badanina[1], Rais M. Latypov[3] and Sergey F. Sluzhenikin[4]

[1]*Institute of Geology and Geochemistry, Ural Branch of Russian Academy of Sciences (UB RAS), Ekaterinburg, Russia* [2]*Macquarie University, Sydney, NSW, Australia* [3]*University of the Witwatersrand, Wits, South Africa* [4]*Institute of Geology of Ore Deposits, Petrography, Mineralogy, and Geochemistry, Russian Academy of Sciences (IGEM RAS), Moscow, Russia*

CHAPTER OUTLINE

Processes and Ore Deposits of Ultramafic-Mafic Magmas through Space and Time. DOI: http://dx.doi.org/10.1016/B978-0-12-811159-8.00008-1

7.1 INTRODUCTION

World-class Ni-Cu-platinum-group-element (PGE) sulfide deposits (e.g., Noril'sk-1, Talnakh and Kharaelakh) occur within the Noril'sk-Talnakh region of northern Siberia, Russia. Despite the long-term study of the ore-bearing ultramafic-mafic Noril'sk-type intrusions, they remain a subject of ongoing debate related to their origin, spanning different, sometime contradictory ideas. One of the burning questions concerns the involvement of different sources of silicate and ore material parental to individual lithological units and associated Ni-Cu-PGE sulfide ores. The origin of Noril'sk type intrusions has been explained by several mechanisms: the differentiation of a single magma (Zen'ko, 1983; Krivolutskaya, 2014a); emplacement of multiple distinct magmas (Likhachev, 1978; Tuganova, 1991; Czamanske et al., 1995); a lava-conduit model (Rad'ko, 1991; Naldrett et al., 1995); a crust-mantle interaction model (Likhachev, 1994; Pushkarev, 1997, 1999), assimilation (Li et al., 2003) and metasomatic models for the ores (Zolotukhin et al., 1975) and intrusions (Zotov, 1979 among others). None of these ideas predominate and frequently even coauthors contradict each other (see Arndt et al., 2003; Malitch et al., 2014; Sluzhenikin et al., 2014). The disagreement on the origin of the Noril'sk-type intrusions and associated ores is partly due to the unresolved issues regarding (1) the relationships between ore-bearing intrusions and lavas, (2) the composition of parental magmas and the extent of interaction between these magmas and the continental crust and subcontinental lithospheric mantle (SCLM), (3) sources of the ore metals, (4) the role of assimilation in the ore formation.

For example, it is commonly assumed that ultramafic-mafic intrusions and associated Ni-Cu-PGE deposits of the Noril'sk area genetically linked to the ~ 250 Ma Siberian flood-basalt volcanism (e.g., Campbell et al., 1992; Naldrett et al., 1995). However, based on petrography, bulk chemical analyses and phase equilibria, it was shown that there is no compositional overlap between the gabbroic intrusions and the volcanic rocks (Latypov, 2002, 2007), implying that intrusions and basalts are products of crystallization of two distinct magma types. Growing evidence for temporal differences between the chalcophile element-depleted basalts and the sulfide-rich Noril'sk-type intrusions has come from recent U-Pb studies of zircon and baddeleyite (Malitch et al., 2010a,b, 2012, 2014a). These studies advocate an extended period of mafic-ultramafic activity, consistent with multiple magmatic events during the protracted evolution of the economic ultramafic-mafic intrusions of the Noril'sk area (e.g., Noril'sk-1, Talnakh, and Kharaelakh), implying that their temporal and spatial relationships to volcanic rocks could be coincidental. This is probably one of the reasons why the search for similar deposits within Siberia and similar volcanic provinces (e.g., Karoo, Deccan, Greenland) has not produced any positive results.

Mafic-ultramafic intrusions and Ni-Cu-PGE deposits in the Noril'sk-Talnakh region (Russia) are considered to be closely linked, indicating that primitive mantle-derived materials are intrinsic to their petrogenesis (e.g., Tuganova, 2000; Arndt et al., 2005; Malitch et al., 2013). Despite general agreement that the primitive melts were derived from a plume source, the extent of interaction between these melts and the continental crust and SCLM has been debated

(Sharma et al., 1992; Lightfoot et al., 1993; Wooden et al., 1993; Czamanske et al., 1994; Ivanov, 2007; Dobretsov et al., 2008; Zhang et al., 2008; Malitch et al., 2010b, 2013). Hf-isotope data on zircon and baddeleyite, which reflect their source region, provided a new set of constraints for the origin of the Noril'sk-type intrusions. Hf-Nd isotope studies at Kharaelakh (Malitch et al., 2010b) and Noril'sk-1 (Malitch et al., 2013) have shown a significant range in initial $^{176}Hf/^{177}Hf$ values of zircon along with a less pronounced range in the initial $^{143}Nd/^{144}Hf$ values of rocks, which favored a model of mixing between the mantle and lithospheric sources. Thus, the in situ Hf-isotope composition of zircons records the evolution of the magma chamber, with input of magmas from different sources, and/or progressive contamination by country rocks. A prolonged period for the concentration of the ore components in deep-seated staging chambers during this interaction might be a key factor for formation of economic deposits.

The stable-isotope compositions of metals such as Cu can be used to trace their sources in high-temperature magmatic Ni-Cu-PGE sulfide deposits (Zhu et al., 2000; Larson et al., 2003; Malitch et al., 2014b; Ripley et al., 2015). Combined Cu-S stable-isotope studies (e.g., Malitch et al., 2014b) have provided new insights into the coupling of copper- and sulfur-isotope compositions in high-temperature massive and disseminated sulfide ores typical of world-class economic deposits, and several subeconomic and noneconomic accumulations related to various ultramafic-mafic intrusions of the Noril'sk Province. The observed $\delta^{65}Cu$ variability has been interpreted to represent a primary signature of the ores at Talnakh and Noril'sk-1, whereas a magmatic fractionation of copper isotopes and/or assimilation of the ore material from external sources have not been ruled out in the case of the Kharaelakh ores. Sulfur-isotope studies based solely on elevated $\delta^{34}S$ values of sulfides (Godlevsky and Grinenko, 1963; Grinenko, 1966; Li et al., 2003 among others) strongly argued in favor of contamination models during the formation of Noril'sk-type deposits, though the extent to which sulfates (with $\delta^{34}S$ values between 18‰ and 22‰, Ripley et al., 2010; Malitch and Latypov, 2011) would melt/dissolve and contribute to the magma remains unresolved. These contamination models are in the obvious conflict with low $\delta^{34}S$ values (i.e., a mean of 0.39‰ with a standard deviation of 1.55‰), characteristic of disseminated sulfide ores from the Zub-Marksheider intrusion (Malitch et al., 2014b) and radiogenic Pb-Os isotope data (Wooden et al., 1992; Walker et al., 1994; Arndt et al., 2003; Malitch and Latypov, 2011), which recorded mantle-like S-Os-Pb isotope values. The reason for the isotopically-heavy sulfur in the Ni-Cu-PGE sulfide ores of the Noril'sk-type economic deposits has been a matter of considerable debate since the early 1960s (e.g., Godlevsky and Grinenko, 1963; Vinogradov and Grinenko, 1966; Grinenko, 1966, 1984, 1985; Kuz'min and Tuganova, 1977; Godlevsky and Likhachev, 1986; Tuganova, 2000; Li et al., 2003; Ripley et al., 2003, 2010; Ripley and Li, 2003, 2013; Petrov et al., 2009; Malitch and Petrov, 2010; Malitch and Latypov, 2011; Krivolutskaya, 2014a,b, 2016; Malitch et al., 2014b).

Understanding the processes behind the formation of economically important ore deposits is crucial for mineral exploration in underexplored regions. It was considered that the promising targets for the Ni-Cu-PGE prospecting within the Taimyr Province (Fig. 7.1) are linked to the Dyumtaley and Binyuda ultramafic-mafic intrusions located in the Binyuda-Tarey and Luktakh ore districts, respectively (Dyuzhikov et al., 1995; Komarova et al., 1999; Malitch et al., 2002; Romanov et al., 2011). Recent U-Pb and Hf-Nd-Sr-Cu-S isotope studies of the ore-bearing Dyumtaley and Binyuda intrusions provided a set of baseline isotope signatures for lithologies and ores (Badanina et al., 2014; Krivolutskaya, 2014b; Malich et al., 2014; Malitch et al., 2016). These data would allow us to test the existing models, which have been developed for variously mineralized intrusions from the Noril'sk Province.

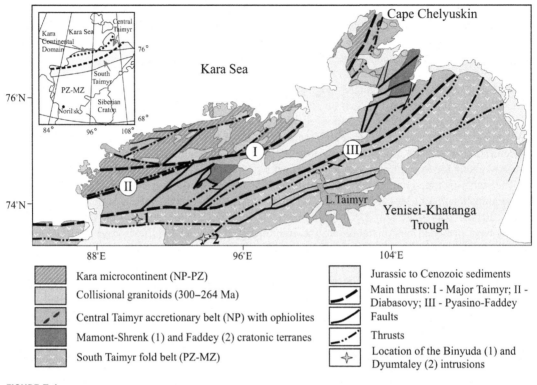

FIGURE 7.1

Geological sketch map of the Taimyr Peninsula.

Modified after Vernikovsky, V.A., Vernikovskaya, A.E., 2001. Central Taimyr accretionary belt (Arctic Asia): Meso-Neoproterozoic tectonic evolution and Rodinia breakup. Precamb. Res. 110, 127–141 (Vernikovsky and Vernikovskaya, 2001); Proskurnin, V.F., Vernikovsky, V.A., Metelkin, D.V., Petrushkov, B.S., Vernikovskaya, A.E., Gavrish, A.V., et al., 2014. Rhyolite-granite association in the Central Taimyr zone: evidence of accretionary-collisional events in the Neoproterozoic. Russian Geology and Geophysics 55, 18-32 (Proskurnin et al., 2014).

In order to place geochronological and isotope-geochemical constraints on timing and sources of materials involved in the generation of variously mineralized intrusions within the Noril'sk and Taimyr Provinces we aim to: (1) compare available U-Pb and Re-Os isotope data for intrusions and sulfide ores, and (2) discuss their Hf-Nd-Sr radiogenic and Cu-S stable isotope systematics. Finally, we attempt to highlight the key geophysical and isotopic-geochemical indicators that can be used in exploration for Ni-Cu-PGE sulfide ores.

7.2 GEOLOGICAL BACKGROUND, MINERALIZATION, AND SAMPLE LOCATIONS

Ni-Cu-PGE sulfide deposits related to the Noril'sk-1, Talhakh, and Kharaelakh (also known as Oktyabr'skoe) mafic-ultramafic intrusions are located in the northwestern part of the Siberian

Craton (Russia). The intrusions are controlled by a long-lived intracontinental paleorift (Tuganova, 2000) and have thicknesses up to 360 m and lengths up to 25 km. The intrusions usually exhibit the following upward succession: Zone 1—fine-grained gabbro along the contact between the country rocks and intrusions; Zone 2—taxitic-textured rocks composed of olivine gabbro and troctolite together with "patches" of ultramafic rocks; Zone 3—ultramafic rocks composed of sulfide-bearing werhlite, plagioclase-bearing dunite and troctolite; Zone 4—olivine gabbro and troctolite gabbro; Zone 5—olivine-bearing gabbro and gabbro; Zone 6—leucogabbro with lenses of ultramafic and taxitic-textured rocks; and Zone 7—gabbro and diorite with lenses of coarse-grained diorite.

Up to three distinct compositional groups of olivine, Fo_{92-95}, Fo_{72-82}, and Fo_{65}, have been identified in the taxitic-textured rocks which represent an extremely heterogeneous variety of ultramafic and mafic rocks (Zolotukhin et al., 1975). Li et al. (2003) and Arndt et al. (2003) also identified two distinct groups of olivine, which could be derived from distinct magmas. However, the most magnesian compositions have not been identified because of the large extent of serpentinization.

Sulfide ores are present as (1) massive or vein-like ores at the lower contacts of the intrusions, (2) vein-like and disseminated ores in host rocks, and (3) disseminated ores in ultramafic and taxitic-textured rocks in Zones 2, 3, and 6. Different ore types have different total resources of base metals and PGE (Table 7.1), and distinct PGE concentrations and modes of platinum-group minerals (PGMs).

Sheet-like massive orebodies can extend for several kilometers with thicknesses up to 45 m and they also penetrate the host rocks as thin sheets and stringers. Massive ores are mainly composed of pyrrhotite (90%) and minor chalcopyrite, pentlandite, and PGMs. Subordinate chalcopyrite-rich ores usually have a zoned structure from pyrrhotite, through chalcopyrite and cubanite, to mooihoekite ($Cu_9Fe_9S_{16}$). Chalcopyrite-rich ores are more Pt- and Pd-enriched than pyrrhotite-rich ores (Tuganova and Malich, 1990). Vein-like and disseminated ores occur as haloes around massive ores with thicknesses ranging from 1 to 16 m. Disseminated ores are the main repository of PGE

Table 7.1 Proportions of Total Reserves of Various Ore Types in the Norilsk Area

Ore Type	Reserves (%)				Average Metal Contents in Ores		
	Total Ore	Ni	Cu	PGE	Ni (%)	Cu (%)	PGE (g/t)
Massive	10.3	41.7	31.8	20.6	3.2	4.6	10.8
Vein/disseminated	7.4	8.1	16.4	13.3	0.9	3.3	9.8
Cu-dominant, Disseminated	82.3	50.2	50.2	66.1	0.5	0.9	4.3
PGE-dominant, Low sulfide horizon					0.3	0.2	3–6

Data compiled from Lyul'ko, V.A., Amosov, Y.N., Kozyrev, S.M., Komarova, M.Z., Ryabikin, V.A., Rad'ko, V.A., et al., 2002. The state of the ore base of non-ferrous and noble metals in the Norilsk region, with guidelines of top-priority geological and exploration works. Rudy i Metally 5, 66–82 (in Russian) (Lyul'ko et al., 2002); ore from the low sulfide horizon has not been adequately evaluated to include in reserves but 90 and 120 Moz of PGE may occur in this setting at Norilsk and Talnakh and Oktvabr'skoe (combined); After Malitch, K.N., Latypov, R.M., Badanina, I.Yu., Sluzhenikin, S.F., 2014b. Insights into ore genesis of Ni-Cu-PGE sulfide deposits of the Noril'sk Province (Russia): evidence from copper and sulfur isotopes. Lithos 204, 172–187.

(Table 7.1); they occur in taxitic-textured and ultramafic horizons (zones 2 and 3), and are 30−90 m thick.

The "low-sulfide" horizon in the upper part of the intrusions is a relatively new economic aspect of the Noril'sk-type deposits (Sluzhenikin et al., 1994). The low-sulfide horizon is 0.1−30 m thick, and occurs exclusively in the taxitic-textured rocks, with lenses of chromitite and chromite-olivine segregations (e.g., dunite) within zone 6. The low-sulfide horizon contains up to 3 vol.% of sulfides with Cu and Ni ranging from 0.2 to 0.3 wt.%, and PGE concentrations (up to 60 ppm) comparable with or overcoming that of the Ni-Cu-PGE disseminated ores (Table 7.1).

The geologic setting, morphological, lithologic, mineralogical, and isotope-geochemical features of ultramafic-mafic intrusions and closely linked economic PGE-Cu-Ni deposits are described in numerous publications summarized in Table 7.2.

One of the fundamental publications on this topic is Naldrett et al. (1992). They distinguished five groups of intrusions in the Noril'sk region (Table 7.4 on page 988): (1) alkaline and subalkaline plutons, (2) high-Ti dolerite dikes in the northeastern part of the Noril'sk region, (3) dolerite sills in the Noril'sk region, (4) differentiated mafic-ultramafic intrusions that are not related to the ore junctions, (5) differentiated mafic-ultramafic intrusions in the vicinity of the ore junctions and (6) unclassified intrusions. Group 5 is subdivided into the Talnakh, Chernogorsk, Zubovsky, Dvugorbinsky, Kruglogorsky, and Lower Talnakh subgroups, and unclassified intrusions. According to Naldrett et al. (1992, Fig. 12) the later include Tangaralakh, Tomulakh, Pyasino-Vologochan, South Noril'sk, and several other unnamed intrusions.

Table 7.2 Contributions on the Noril'sk-Type Intrusions and Associated Ni-Cu-PGE Sulfide Deposits

Geology

Urvantsev (1967), Godlevsky (1959), Dodin and Batuev (1971), Fedorenko (1991), Duzhikov et al. (1992), Likhachev (1994), Lyul'ko et al. (1994), Simonov et al. (1994), Zen'ko and Czamanske (1994a,b), Naldrett (2004), Ryabov et al. (2000), Turovtsev (2002), Sluzhenikin et al. (2014)

Geochronology

Campbell et al. (1992), Walker et al. (1994), Dalrymple et al. (1995), Kamo et al. (1996), Malitch et al. (2010a, 2010b, 2012); Malitch and Latypov (2011); Burgess and Bowring (2015)

Mineralogy and geochemistry

Grinenko (1966, 1984, 1985, 1990), Kuz'min and Tuganova (1977), Distler et al. (1979, 1999), Genkin et al. (1981), Genkin and Evstigneeva (1986), Godlevsky and Likhachev (1986), Tuganova and Malich (1990), Wooden et al. (1992), Czamanske et al. (1994), Distler (1994), Torgashin (1994), Zen'ko and Czamanske (1994); Naldrett et al. (1996), Dodin et al. (2001), Komarova et al. (2000, 2002), Tuganova (2000), Cabri et al. (2003), Latypov (2002, 2007), Arndt et al. (2003), Li et al. (2003); Ripley et al. (2003, 2010), Pokrovskii et al. (2005), Petrov et al. (2009); Malitch et al. (2010b); Krivolutskaya et al. (2011, 2012), Sluzhenikin and Krivolutskaya (2015), Sluzhenikin and Mokhov (2015); Krivolutskaya (2014a)

Origin

Kotulsky (1946), Zolotukhin et al. (1975), Likhachev (1978), Zen'ko (1983), Distler et al. (1988); Rad'ko (1991), Tuganova (1988, 1991), Naldrett (1992), Fedorenko (1994), Czamanske et al. (1995), Naldrett et al. (1995), Arndt (2005, 2011), Lightfoot and Keays (2005), Li et al. (2009a), Diakov et al. (2002), Yakubchuk and Nikishin (2004), Gorbachev (2010), Keays and Lightfoot (2010); Starostin and Sorokhtin (2011), Malitch et al. (2013; 2014b), Krivolutskaya (2016)

In recent studies (e.g., Malitch et al., 2010a,b, 2012, 2013, 2014b; Malitch and Latypov, 2011), zircon, baddeleyite and sulfide samples were collected from nine mafic-ultramafic intrusions of the Noril'sk area. These intrusions were subdivided (Petrov et al., 2009) into three types in terms of sulfide mineralization style and economic significance (Table 7.3).

Type 1 comprises the economic ore-bearing intrusions that host commercial reserves, including the Noril'sk-1, Talnakh, and Kharaelakh intrusions, which contain well-defined horizons of plagioclase-bearing dunite, wehrlite (with elevated contents of Cr), and taxitic-textured rocks, with later hosting disseminated, veined, and massive ores. In the upper part of these intrusions a PGE-rich low-sulfide horizon is hosted by leucogabbro with lenses of ultramafic rocks. Rocks of these intrusions are characterized by "radiogenic" Sr isotopes ($^{87}Sr/^{86}Sr_i = 0.7055-0.7075$) against rather constant εNd values of $\sim +1$.

Type 2 comprises subeconomic intrusions (namely Chernogorsk, Zub-Marksheider, Vologochan, and Imangda) hosting noncommercial ore reserves. Interestingly, the rocks of the subeconomic intrusions have similar lithological, mineralogical, geochemical, and Nd-Sr isotopic features with that of the economic intrusions. The subeconomic intrusions mainly contain disseminated or locally rare vein-disseminated sulfide ores. They may contain small- to medium-sized Ni-Cu sulfide deposits, and medium-sized to large PGE deposits.

Type 3 comprises weakly mineralized mafic-ultramafic intrusions, the so-called Lower Talnakh type, as represented by the Nizhny Talnakh and Zelyonaya Griva intrusions. They contain low-grade disseminated Cu-Ni ores with ~ 0.2 wt.% of Cu and Ni, and low Cr and PGE (~ 5 ppb, rarely up to 20 ppb). Rocks from the Nizhny Talnakh and Zelyonaya Griva intrusions have more radiogenic initial Sr values ($^{87}Sr/^{86}Sr_i = 0.7076-0.7086$) together with negative εNd values of ~ -5.

Table 7.3 Summary of Investigated Samples From Ultramafic-Mafic Intrusions of the Noril'sk and Taimyr Provinces

Intrusion	Drill Hole	Type of Intrusion
Noril'sk-1	MN-2	Economic (comprising massive, vein-disseminated, disseminated, and low-sulfide ores forming world-class Ni-Cu-PGE deposits)
Talnakh (Main Talnakh)	OUG-2	
Kharaelakh (Western Talnakh)	KZ-844	
	KZ-963	
Chernogorsk (Mount Chyornaya)	MP-2bis	Subeconomic (containing disseminated ores that form noncommercial reserves equivalent to medium and small size Ni-Cu deposits along with medium- to large-size PGE deposits)
Zub-Marksheider	MP-27	
Vologochan (Vologochansky)	OV-29	
Imangda	KP-4	
Nizhny Talnakh (Lower Talnakh)	TG-31	Noneconomic (with disseminated ores with off grade BM contents)
Zelyonaya Griva	F-233	
Dyumtaley	TP-43	Prospective (containing disseminated ores with unidentified potential)
Binyuda	C-1	

Several names of intrusions in this study differ from those in earlier studies (shown in parentheses).

The most prospective targets for the massive Ni-Cu-PGE sulfide ores within the Taimyr Province are restricted to the Dyumtaley and Binyuda ultramafic-mafic intrusions, which show some compositional differences. The Dyumtaley intrusion consists mostly of ferrogabbro (i.e., gabbro abnormally high in Fe) with titanomagnetite ores in the upper part and troctolitic ferrogabbro with Ni-Cu-PGE sulfide mineralization in the lower part (Komarova et al., 1999; Malitch et al., 2002). The Binyuda intrusion is composed of dunite, plagiodunite, and plagiowehrlite in the upper and middle parts and of melanotroctolite in the lower part of intrusion (Dyuzhikov et al., 1995; Romanov et al., 2011; Malitch et al., 2016). The rocks from both intrusions have distinct Nd-Sr signatures (εNd = 4.2 ± 0.7 and ^{87}Sr/^{86}Sr = 0.70474 ± 0.00020 at Dyumtaley and εNd = − 3.8 ± 0.4 and ^{87}Sr/^{86}Sr = 0.70588 ± 0.00013 at Binyuda, Malitch et al., 2016). The rocks investigated in this study comprise sulfide-rich varieties of ferrogabbro and melanocratic troctolite occurring in the bottom parts of the Dyumtaley and Binyuda intrusions, respectively.

The geological setting of ultramafic-mafic intrusions within the Noril'sk and Taimyr provinces is shown in Figs. 7.1−7.3, with the details of the investigated samples given in Malitch et al. (2010a,b, 2013, 2014b, 2016), Malitch and Latypov (2011), and Appendix A. The little-known

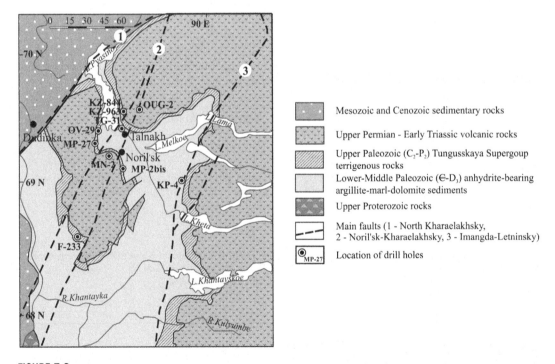

FIGURE 7.2

Schematic geological map of the Noril'sk Province showing locations of drill cores in the studied mafic-ultramafic intrusions.

Modified after data collected by the "Noril'skgeology" Prospecting Venture.

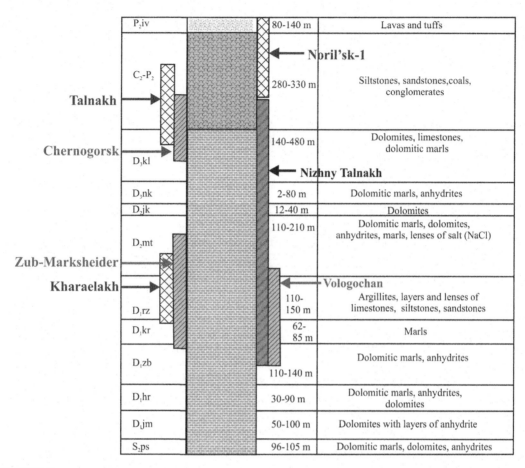

P₂iv		80-140 m	Lavas and tuffs
C₂-P₂		280-330 m	Siltstones, sandstones,coals, conglomerates
D₃kl		140-480 m	Dolomites, limestones, dolomitic marls
D₃nk		2-80 m	Dolomitic marls, anhydrites
D₂jk		12-40 m	Dolomites
D₂mt		110-210 m	Dolomitic marls, dolomites, anhydrites, marls, lenses of salt (NaCl)
D₁rz		110-150 m	Argillites, layers and lenses of limestones, siltstones, sandstones
D₁kr		62-85 m	Marls
D₁zb		110-140 m	Dolomitic marls, anhydrites
D₁hr		30-90 m	Dolomitic marls, anhydrites, dolomites
D₁jm		50-100 m	Dolomites with layers of anhydrite
S₂ps		96-105 m	Dolomitic marls, dolomites, anhydrites

Labels on figure: Talnakh, Chernogorsk, Zub-Marksheider, Kharaelakh, Noril'sk-1, Nizhny Talnakh, Vologochan

FIGURE 7.3

Stratigraphic section showing the positions of economic, subeconomic, and noneconomic ultramafic-mafic intrusions.

After Malitch, K.N., Belousova, E.A., Griffin, W.L., Badanina, I.Yu., 2013. Hafnium-neodymium constraints on source heterogeneity of the economic ultramafic-mafic Noril'sk-1 intrusion (Russia). Lithos 164−167, 36−46; modified after Czamanske, G.K., Zen'ko, T.E., Fedorenko, V.A., Calk, L.C., Budahn, J.R., Bullock, J.H., Jr., et al., 1995. Petrography and geochemical characterization of ore-bearing intrusions of the Noril'sk type, Siberia; with discussion of their origin: resource Geology Special 18, 1−48; Dyuzhikov, O.A., Distler, V.V., Strunin, B.M., Mkrtychyan, A.K., Sherman, M.L., Sluzhenikin, S.F., 1992. Geology and metallogeny of sulfide deposits, Noril'sk region, USSR. Soc. Econ. Geol. Special Publication, 241 p (Dyuzhikov et al., 1992).

subeconomic Zub-Marksheider intrusion in the Noril'sk Province is hosted in the Manturovsky and Razvedochninsky Formations, which are composed of anhydrite-rich rocks (Fig. 7.4), and extends along strike for about 5 km (Godlevsky, 1959). Unlike the economic intrusions, the Zub-Marksheider intrusion contains a thick zone (40−100 m) of hybrid rocks composed of

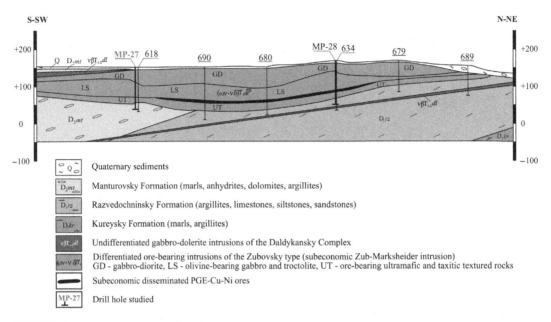

FIGURE 7.4

Geological cross section of the Zub-Marksheider intrusion showing location of the studied MP-27 drill hole.

After Malitch, K.N., Latypov, R.M., Badanina, I.Yu., Sluzhenikin, S.F., 2014b. Insights into ore genesis of Ni-Cu-PGE sulfide deposits of the Noril'sk Province (Russia): evidence from copper and sulfur isotopes. Lithos 204, 172–187.

gabbro-diorite, diorite, syenite, and alkaline metasomatitic rocks (Figs. 7.4 and 7.5) and minor ultramafic rocks. As revealed from the drill hole MP-27 (Fig. 7.5) and unpublished data, the ~40-m-thick upper horizon (samples 27-1–27-5) is an alkaline metasomatite composed of albite, K-feldspar, and quartz, and diorites or gabbroic diorite, which have been formed as a result of interaction of mafic magmas with the Devonian sedimentary rocks (Godlevsky, 1959). The lower part of the section (Fig. 7.5, samples 27-9–27-15) reveals a notable marginal reversal with whole-rock MgO progressively increasing upward from 10 to 22 wt.%. The reversal corresponds to an upward transition from the olivine-bearing gabbro, through gabbro and troctolite, to melatroctolite and plagioclase-bearing wehrlite. This is followed by an essentially gabbroic sequence composed of olivine-bearing gabbro, gabbro-diorite, diorite and Fe-Ti oxide-bearing gabbro with decreasing olivine contents and whole-rock MgO (from 22 to 4 wt.%). Primary minerals of the Zub-Marksheider intrusion include plagioclase, augite, and olivine. The sulfide-rich zone is situated in the olivine-rich part at the bottom of the intrusion; it is mainly composed of disseminated ore and ranges in thickness from 2 to 40 m. Exploration results show that the sulfide ores have an average grade of 0.44 wt.% Cu, 0.23 wt.% Ni, 0.015 wt.% Co and 2.23 ppm PGE. Sulfide-rich mineralization is represented by an assemblage of troilite, pentlandite, and chalcopyrite in plagiowehrlite, and by troilite, pyrrhotite, pentlandite, and cubanite in troctolite and gabbro (Fig. 7.5). Disseminated ores were sampled from plagioclase-bearing wehrlite, troctolite, and olivine gabbro in the bottom of the intrusion (Figs. 7.4 and 7.5).

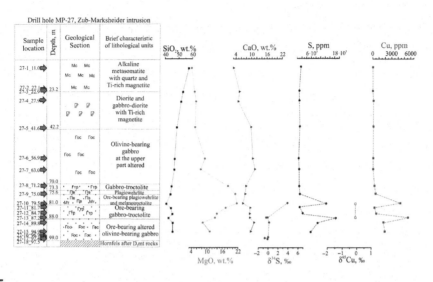

FIGURE 7.5

Variations in the contents of SiO_2, CaO, and MgO (in wt.%), S and Cu (in ppm) and sulfur and copper-isotope compositions through a section of the Zub-Marksheider subeconomic intrusion, drill hole MP-27.

After Malitch, K.N., Latypov, R.M., Badanina, I.Yu., Sluzhenikin, S.F., 2014b. Insights into ore genesis of Ni-Cu-PGE sulfide deposits of the Noril'sk Province (Russia): evidence from copper and sulfur isotopes. Lithos 204, 172–187.

7.3 U-Pb AND Re-Os CONSTRAINTS ON THE TEMPORAL EVOLUTION OF THE NORIL'SK-TYPE INTRUSIONS AND ASSOCIATED Ni-Cu-PGE SULFIDE DEPOSITS

Noril'sk-type intrusions are frequently interpreted as being genetically linked to the ∼250 Ma Siberian flood-basalt volcanism (e.g., Naldrett et al., 1995), which has been widely considered to have formed over a period of 1 Ma or less, despite the available evidence for older and younger ages (Dalrymple et al., 1995; Reichow et al., 2002; Malitch et al., 2010a,b, 2012, 2014a; Malitch and Petrov, 2010). Re-Os isochron ages of 245.7 ± 0.6 Ma for sulfide ores at Talnakh and Noril'sk-1, and 247.0 ± 3.8 Ma for sulfide ores at Kharaelakh (Walker et al., 1994) closely match those reported in the subsequent study (Malitch and Latypov, 2011). The Re-Os data for both massive and disseminated ore samples at Kharaelakh and Talnakh give an isochron age of around 247 Ma, indicating that Re-Os isotope systematics were not disturbed after sulfide crystallization. Widely-cited U-Pb ages of zircon and baddeleyite from pegmatitic leucogabbro of the Noril'sk-1 intrusion (i.e., 248.0 ± 3.7 Ma (Campbell et al., 1992) and 251.1 ± 3.6 Ma (Kamo et al., 1996)) were taken as evidence of close relationship between magmatic and ore-forming processes. However, recent detailed in situ U-Pb, REE, and Hf-isotope studies of zircon from the economic Kharaelakh, Talnakh, and Noril'sk-1 and other intrusions identified different zircon populations with ages ranging from 347 to 227 Ma (Table 7.4). Zircons are characterized by a fuzzy (smoky)

Table 7.4 Summary of Age Characteristics of Zircon From Ultramafic-Mafic Intrusions of the Noril'sk and Taimyr Provinces

Intrusion	Number of Analyses	Age, Myr				Reference
		From	To	Probability density distribution	Concordant, most common	
Noril'sk-1	129	213.7	280.5	252.3; 232.6	261.3; 245.7; 236.5; 226.7	Malitch et al. (2012)
Talnakh	148	213.3	294.4	260.9; 234.9	262.7; 256.2; 234.6	Malitch and Petrov (2010)
Kharaelakh	24	220.5	355.0	346.5; 252.2	347.0; 265.0; 253.9; 235.9	Malitch et al. (2010b)
Chernogorsk	12	241.0	293.0	292.5; 247.4	292.5; 248.4	Malitch and Petrov (2010)
Vologochan	20	222.1	331.6	331.6; 271.6; 246.0	331; 271.4; 246.0; 226.7	Malitch and Petrov (2010)
Imangda	18	219.0	262.4	261.4; 233.0	261.4; 238.8	Malitch and Petrov (2010)
Nizhny Talnakh	113	197.4	304.7	259.5; 232.3	266.2; 229.0	Malitch and Petrov (2010)
Zelyonaya Griva	9	237.5	279.0	266.2; 241.0	266.0; 241.0	Malitch and Petrov (2010)
Dyumtaley	3	255.0	258.0		256.2	Malitch et al. (2016)
Binyuda	2	238.0	255.0		245.7	Malitch et al. (2016)

cathodoluminescence, with a virtually total absence of zoning. Petrographic inspection, however, revealed at least two generations of zircon that show distinct solid-inclusion assemblages (dominated by minerals and glass, respectively). In the binary Th-U diagram (Fig. 7.6), the investigated zircons (Th/U = 0.8−11.5) are clearly different from zircons derived from various geological settings, but they partly overlap the field of zircon from glimmerite (Rudnick et al., 1998) and closely approach the composition of zircons from some MARID rocks (Kinny and Dawson, 1992), pointing to specific conditions for their generation. In REE discrimination diagrams on $(Sm/La)_N$−La and Ce/Ce^*−$(Sm/La)_N$ spaces, zircons mainly plot in the field of "magmatic" zircon (after Hoskin, 2005). At Kharaelakh, the U-Pb ages of four zircon populations cover a significant time span (from 347 ± 16 to 235.7 ± 6.1 Ma; Malitch et al., 2010b) suggesting multiple magmatic events clustering around 350 and 250 Ma, which are consistent with two recognized stages of active tectonism in the development of the Siberian Craton (Malitch, 1975). The oldest zircon population, however, represents a previously unknown stage of magmatic activity in the Noril'sk area. On the basis of these results it was proposed that mafic-ultramafic magmas were emplaced in the lithospheric mantle or deep crust c. 90 Myr before the flood basalts, and that the magmas generated in Early Mesozoic (around 235−265 Ma) perhaps inherited zircons from these earlier intrusions. However,

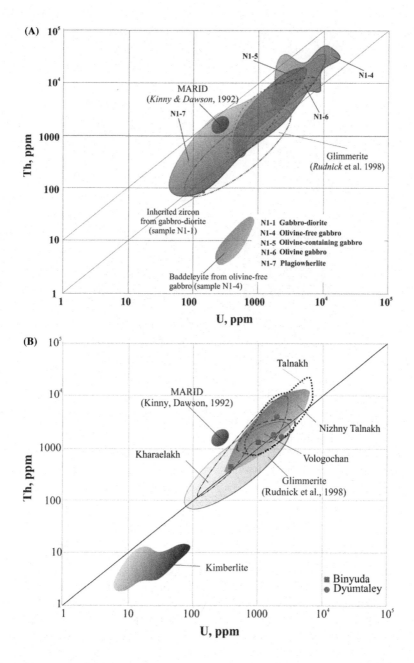

FIGURE 7.6

Th-U composition of (A) zircon and baddeleyite from the Noril'sk-1 intrusion, (B) zircon from variously mineralized intrusions of the Noril'sk and Taimyr Provinces, compared with fields of zircon from other mantle-derived rocks.

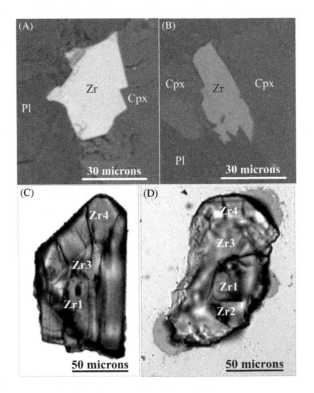

FIGURE 7.7

Examples of zircon in situ in polished sections (A, B; in back-scattered electrons) and mineral separates (C, D; in transmitted light) from rocks of the Noril'sk-1 intrusion; A and B, assemblages of zircon and rock-forming minerals from olivine gabbro (sample N1-6, 60) and taxitic pyroxene leucogabbro (sample N1-9, 74). *Zr*, zircon; *Cpx*, clinopyroxene; *Pl*, plagioclase; C and D, inner structure of polyphase grains, composed of "cores" (Zr1 and Zr2) and "rims" (Zr3 and Zr4).

After Malitch K.N., Badanina I.Yu., Belousova E.A., Tuganova E.V., 2012. Results of U-Pb dating of zircon and baddeleyite from the Noril'sk-1 ultramafic-mafic intrusion (Russia). Russian Geol. Geophys. 53 (2), 123–130.

evidence of *core—rim* overgrowths in complex grains (see Fig. 7.4a,c,e in Malitch et al., 2010b) suggests that the older zircons are not just accidentally "picked up" grains, but that they probably are derived from cogenetic, possibly ultramafic magmas, and that these magmas had contributed to the source of the later mafic magmas. In this scenario, the Kharaelakh intrusion might be viewed as a mixture of distinct magmatic pulses represented by earlier ultramafic rocks and later gabbroic lithologies, respectively.

A more recent U-Pb study of baddeleyites and zircons from the main lithological units of the Noril'sk-1 intrusion (Malitch et al., 2012) has confirmed the polyphase nature of the zircons (Fig. 7.7). Baddeleyite and four zircon populations define five age clusters that span a period from 290 ± 2.8 to 226.7 ± 0.9 Ma (i.e., 290 ± 2.8, 261.3 ± 1.6, 245.7 ± 1.1, 236.5 ± 1.8, and

226.7 ± 0.9 Ma). Multiple crystallization ages of baddeleyite and zircon were suggested to reveal several stages in the protracted evolution of the ore-forming magma and/or to characterize interaction between distinct magmas during formation of the Noril'sk-1 intrusion. The only exception, with the oldest concordant $^{206}Pb/^{238}U$ age of 1914 ± 92 Ma (Fig. 5 in Malitch et al., 2012), has been identified for a single zircon grain from the gabbro-diorite. This age suggests that this zircon has been inherited from 1.9 Ga old basement rocks, which may indicate the location of a deep-seated chamber for the magmatic protolith. The U-Pb data presented by Campbell et al. (1992) gave a range of $^{206}Pb/^{238}U$ ages from 243.8 ± 4.9 to 251.6 ± 5.0 Ma, with a mean of 248 ± 3.7 Ma. U-Pb results presented by Kamo et al. (1996) gave an even broader range of $^{206}Pb/^{238}U$ ages. Kamo et al. (1996) showed (see Fig. 2, p. 3508) that all the analyses ($n = 10$) gave a $^{206}Pb/^{238}U$ age of 256.5 ± 2.6 Ma, whereas nine tightly clustered analyses of zircon and baddeleyite produced an average $^{206}Pb/^{238}U$ age of 251.2 ± 0.3 Ma. The variability of the U-Pb ages obtained by both techniques on the same lithology is striking and may be due to the polyphase nature of the zircons studied. In fact, measured ages that are older or younger than *ca* 250 Ma in these rocks tend to be deemphasized in the literature because of the assumption that the Siberian volcanism was a Permo-Triassic boundary event (c. 250 Ma; Renne and Basu, 1991; Campbell et al., 1992), coincident with the largest extinction event in Earth's history (Kamo et al., 2003).

Broadly similar results (Malitch et al., 2010b), with the majority of concordant U-Pb ages lying in the range of 230−270 Ma, have been documented for zircons from the rocks of the third economic intrusion of the Noril'sk region (e.g., Talnakh). However, the dominant set of zircon grains from the main lithological units at Talnakh yielded relatively "old" concordant U-Pb ages (262.7 ± 0.4 Ma, MSWD = 0.033, $n = 39$), corresponding to that of population *ZR 1* at Noril'sk-1 (e.g., 261.3 ± 1.6 Ma, MSWD = 0.7, $n = 14$).

If the formation of the basalts was indeed restricted to the Permian-Triassic boundary as advocated by Reichow et al. (2009; 248 ± 0.6−250.3 ± 1.1 Ma), the U-Pb zircon data provide only partial supporting evidence for a genetic link between the chalcophile element-depleted basalts and the sulfide-rich Noril'sk-type intrusions, and imply that their relationship may be coincidental. Similar conclusions, arising from different lines of reasoning, have been reached previously by Godlevsky (1959), Tuganova (1991), Czamanske et al. (1995) and Latypov (2002, 2007), all of which are in conflict with the conduit model proposed by Naldrett et al. (1995).

Multiple magmatic events that have affected the Talnakh, Kharaelakh, and Noril'sk-1 intrusions preceded an extensive episode of sulfide-ore formation, as documented by the Re-Os data, and imply an extended time span for concentration of the ore components. This might be a key factor in the formation of economic deposits. This indirect evidence may suggest that the ultramafic-mafic magmas and the associated ores are unlikely to have been completely coeval with the relatively short formation time of the basalts.

The massive and disseminated sulfides represent the last liquid portion to crystallize and solidify in the chamber due to their low liquidus temperatures, so their Re-Os ages must be the best indicator of the emplacement age of these intrusions (about 247 Ma). In contrast, zircon and baddeleyite may have been entrained from a deep-staging chamber in which magma, parental to the intrusions, may have been stored and crystallized for a long time, producing several populations of zircon of different ages (effectively magmatic xenocrysts). The delivery of these early-crystallized populations of zircons (and baddeleyites at Noril'sk-1) with magma from the staging chamber to the present position of intrusions may well explain the observed "old" ages at Kharaelakh (e.g., 347 ± 16; 265 ± 11, and 253.9 ± 1.7 Ma; Malitch et al., 2010b), Talnakh (262.4 ± 0.9 and 256.4 ± 1.3 Ma; Malitch et al., 2010a) and Noril'sk-1 (290 ± 2.8 and 261.3 ± 1.6 Ma, Malitch et al., 2012).

However, this scenario does not eliminate the problem of the age discrepancy entirely, since it is necessary to explain the ages of zircons that are younger (e.g., 235.9 ± 6.1 Ma at Kharaelakh, Malitch et al., 2010b; 229.3 ± 3.4 Ma at Talnakh, Malitch et al., 2010b; 236.5 ± 1.8 and 226.7 ± 0.9 Ma at Noril'sk-1, Malitch et al., 2012) than the c. 247 Ma (Re-Os ages of the sulfides). We suggest that these young age peaks might be attributed to the thermal recrystallization and zircon overgrowth, consistent with observations on the internal morphologies of zircon grains. Indeed, the structural complexity of the grains shows evidence of recrystallization, and distinct inclusion populations in the cores and outer parts of zircons reflect several superimposed tectonothermal episodes, the last of which coincides with magmatic activity in the Noril'sk area (e.g., the Bolgohtokh granite intrusion at 228 ± 2 Ma; Kamo et al., 2003).

7.4 Hf-Nd-Sr ISOTOPE CONSTRAINTS FOR THE ORIGIN OF ULTRAMAFIC-MAFIC INTRUSIONS

Radiogenic Hf-Nd-Sr isotopic studies of the main lithological units of the economic Noril'sk-type intrusions and associated Ni-Cu-PGE sulfide ores (Wooden et al., 1992; Czamanske et al., 1994; Hawkesworth et al., 1995; Tuganova and Shergina, 1997, 2003; Arndt et al., 2003; Petrov et al., 2009; Malitch and Petrov, 2010; Malitch et al., 2010b, 2013, 2014a) have provided useful constraints on possible sources and substantially contributed to the understanding of the origin of Noril'sk-type intrusive host and associated ores.

According to their Sr-Nd isotopic systematics, the lithological units of economic and subeconomic intrusions manifest a clear increase in the radiogenic component expressed in $^{87}Sr/^{86}Sr_i$, which varies from 0.70552 to 0.7079 at relatively constant εNd values (c. $+1 \pm 0.5$, Fig. 7.8). In contrast, silicate materials from the Binyuda intrusion show a distinct Nd-Sr isotopic signature ($\varepsilon Nd = -3.8 \pm 0.4$ and $^{87}Sr/^{86}Sr_i = 0.70588 \pm 0.00013$) that deviates from the compositional trend "Depleted Mantle—Earth crust" (Fig. 7.8). This trend is indicated by the Dyumtaley intrusion from the Taimyr Province ($\varepsilon Nd = 4.2 \pm 0.7$ and $^{87}Sr/^{86}Sr_i = 0.70474 \pm 0.00020$) and noneconomic intrusions of the Noril'sk Province ($\varepsilon Nd = \sim -5$ and $^{87}Sr/^{86}Sr_i = 0.70765-0.70867$), respectively (Fig. 7.8). An important role for the juvenile component is clearly pronounced in the Dyumtaley intrusion, whereas a major contribution from a SCLM or essentially crustal source is inferred for the Binyuda intrusion.

Zircons from the economic intrusions with U-Pb ages between c. 230 and 340 Ma yield $\varepsilon Hf_{(t)}$ values in a range from 2.3 to 16.3 ($n = 24$) at Kharaelakh, from 0.1 to 16.7 ($n = 40$) at Talnakh and from -4.7 to $+15.5$ ($n = 54$) at Noril'sk-1, consistent with a model that involves an interaction between distinct magma sources (Fig. 7.9; Malitch et al., 2010b, 2013, 2014a). Zircons from subeconomic and prospective intrusions ($n = 38$) with U-Pb ages between c. 230 and 290 Ma show similar ranges in $\varepsilon Hf_{(t)}$ (mean $\varepsilon Hf_{(t)}$ +7.2 at Chernogorsk ($n = 11$), +8.5 at Vologochan ($n = 10$) and +9.5 at Dyumtaley ($n = 4$)). In contrast, zircons from the noneconomic Nizhny Talnakh and Zelyonaya Griva intrusions with U-Pb ages between c. 215 and 305 Ma yield $\varepsilon Hf_{(t)}$ values of

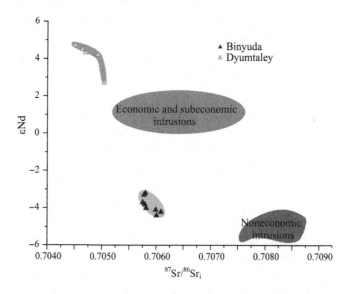

FIGURE 7.8

Variations of initial isotopic composition of neodymium and strontium in rocks from the Binyuda and Dyumtaley intrusions in coordinates εNd−^{87}Sr/^{86}Sr$_i$. Nd-Sr isotopic characteristics of variously mineralized intrusions of the Noril'sk region (Arndt et al., 2003; Petrov et al., 2008; Romanov et al., 2011; Malitch et al., 2013) are shown for comparison.

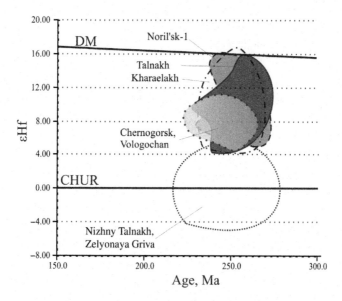

FIGURE 7.9

Plot of εHf (t) (deviation in parts in 10^4 of initial Hf-isotope values between the zircon sample and the chondrite uniform reservoir) versus U-Pb ages for variously mineralized intrusions from the Noril'sk Province.

+0.8 ($n = 76$) and +1.7 ($n = 11$), respectively (Fig. 7.9), while zircons from the prospective Binyuda intrusion show less radiogenic mean $\varepsilon Hf_{(t)}$ of −3.8 ($n = 3$). The Hf-isotope data suggest that the zircons from economic and subeconomic intrusions display the signature of juvenile mantle-derived magmas. The less radiogenic Hf-isotope values of zircons from noneconomic intrusions indicate a mixing between mantle and crustal magma sources (Malitch et al., 2008).

Much more nuanced information was revealed from a combined Hf-Nd isotope study. For the Noril'sk-1 intrusion, in situ Hf-isotope analyses of zircon and baddeleyite, combined with whole-rock Nd-isotope results (Malitch et al., 2013), identify three distinct clusters of Hf-Nd isotope values (Fig. 7.10A) restricted to different lithological units (e.g., gabbro-diorite, unmineralized layered gabbros, and mineralized portions, represented by ultramafic and taxitic-textured rocks). The Hf-isotope signatures of zircon and baddeleyite from unmineralized mafic rocks (ε_{Hf} from 7.3 ± 1.1 to 11.4 ± 0.3) reflect the dominant role of mantle-derived magmas and suggest that a juvenile mantle was one of the main sources for the ultramafic-mafic Noril'sk-1 intrusion. The less radiogenic Hf-isotope values for zircons from mineralized rocks (ε_{Hf} from 4.9 ± 1.4 to 6.4 ± 1.2) and gabbro-diorite (ε_{Hf} −1.2 ± 1.9) indicate the involvement of distinct source components. Thus, the Hf-isotope data for zircons from the ore-bearing rocks indicate that the Noril'sk-type magmas represent a mixing between magmas from a juvenile source equivalent to the Depleted Mantle, and another, possibly an enriched lithospheric mantle source as suggested by Yang et al. (2006), or the SCLM source proposed by Griffin et al. (2000); the latter is especially prominent in the mineralized portions of the intrusion.

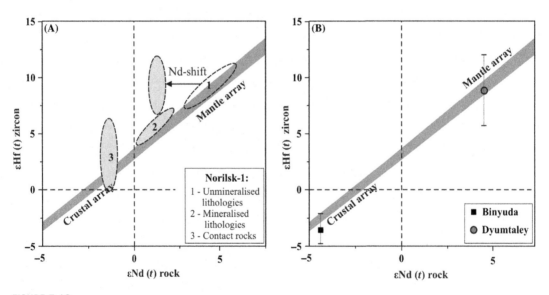

FIGURE 7.10

Plot of zircon εHf (t) versus whole-rock εNd (t) for (A) the Noril'sk-1 intrusion (modified after Malitch, K.N., Belousova, E.A., Griffin, W.L., Badanina, I.Yu., 2013. Hafnium-neodymium constraints on source heterogeneity of the economic ultramafic-mafic Noril'sk-1 intrusion (Russia). Lithos 164−167, 36−46) and (B) the Binyuda and Dyumtaley intrusions. Mantle and crustal arrays according to Vervoort et al. (1999).

A derivation from a mixed source, which includes both the OIB-type source that fed most of the flood-basalt system and a major component from the SCLM underlying the Siberian Craton, has been advocated by Horan et al. (1995) for high-MgO meimechites from the Guli area, characterized by γ_{Os} of -1.2 to -2.6 (calculated at 250 Ma) and ε_{Nd} of $+3.7$ to $+4.9$. Malitch and Latypov (2011) also proposed the involvement of the SCLM, which was documented by a long-term depletion in rhenium (γOs values of -5.7 and -2.5) for a subset of massive Ni-Cu-PGE sulfide ores from the Kharaelakh intrusion. Lithospheric interaction has been also suggested for the rocks of the Kharaelakh intrusion by Malitch et al. (2010b). A significant range in $^{176}Hf/^{177}Hf$ documented in four populations of zircon has been interpreted to represent mixing between a juvenile source equivalent to the depleted mantle, and a crustal source with ε_{Hf} less than 0, or a SCLM source (Griffin et al., 2000) with a Hf-isotope composition substantially less "radiogenic" than observed in the zircon populations studied there (Fig. 8 in Malitch et al., 2010b). A small population of zircons from the gabbro-diorite at Noril'sk-1 shows the least radiogenic Hf-isotope values (Fig. 7.10A), indicating the input of a distinctly older lithospheric, possibly crustal component, which is consistent with the hybrid nature of this lithology.

The depleted-mantle component had not been previously recognized by Nd- and Sr-isotope studies (Arndt et al., 2003; Petrov et al., 2007). All studied rocks at Noril'sk-1, from which zircons were separated, are characterized by ε_{Nd} values from -1.8 to $+2.0$ (Malitch et al., 2013), with two distinct clusters (Fig. 7.10A), characteristic of gabbro-diorite and contact gabbro ($\varepsilon_{Nd(250\ Ma)}$ about -1.4) and the majority of lithological units ($\varepsilon_{Nd(250\ Ma)}$ about $+1.1$). However, contributions from a highly-depleted mantle source (ε_{Nd} in the range of $+3.7$ and $+5.1$) have been confirmed (Kogarko et al., 1988; Horan et al., 1995) in picritic lavas, both at Noril'sk (Gudchihinsky suite) and in the Maimecha-Kotui (Maimechian suite) area. Therefore, the major set of ε_{Nd} values around $+1.0$ is typical of the ore-bearing "Noril'sk-type" ultramafic-mafic intrusions (Arndt et al., 2003; Petrov et al., 2007; Malitch et al., 2010b), and has been attributed to certain levels of crustal contamination, presumably in deep magma chambers, whereas the smaller set of negative ε_{Nd} values (about -1.3) in contact rocks has been interpreted as due to a crustal component, reflecting an interaction with the wall-rocks during the emplacement.

Hf-Nd isotope values (epsilon notations) for the majority of the Noril'sk-1 samples, shown in Fig. 7.10A, plot in the upper right quadrant (suprachondritic $^{176}Hf/^{177}Hf$ and $^{143}Nd/^{144}Nd$). The data from mineralized rocks (with disseminated and low sulfide ores) closely match the "mantle array." In contrast, the ε_{Hf} values of unmineralized layered-sequence rocks do not distribute along the "mantle-crust array" in Hf-Nd space, but trend vertically toward positive ε_{Hf} values (Fig. 7.10A). The difference between the Hf- and Nd-isotope compositions in unmineralized lithologies is expressed in a positive deviation of the zircon ε_{Hf} values from the whole-rock ε_{Nd} values relative to the "mantle-crustal array" (Fig. 7.10A). The systematics of the Lu-Hf decay system closely parallel those of the Sm-Nd system (Vervoort et al., 1999), but the latter is more prone to disturbance by metamorphism or alteration (Vervoort and Blichert-Toft, 1999). It has been suggested (Malitch et al., 2013) that Hf and Nd have been decoupled, perhaps via fluids flowing through the rocks of the layered sequence. The most likely reason is that Hf was locked up in zircon grains while the rock matrix was free to interact with crustally-derived fluids, which homogenized the Nd-isotope composition to a uniformly low ($\sim +1\ \varepsilon_{Nd}$) signature. In this case, whole-rock Nd-isotope values of unmineralized layered-sequence rocks before decoupling should have been much more radiogenic, as illustrated by "Nd-shift" in Fig. 7.10A. It is noteworthy that the latter Hf-Nd isotope signature is similar to that of the prospective Dyumtaley intrusion (Fig. 7.10B), which is consistent with an important role of the juvenile component (εHf = 9.5 ± 2.5, εNd = 4.2 ± 0.7). Thus, we conclude that much less radiogenic but still "mantle-like" εNd values for lithologies of the economic and subeconomic intrusions of the Noril'sk Province, in comparison with the resistant Hf-isotope values of zircon, indicate open-system behavior of the Nd-isotope system in the rocks, probably during a transition from mantle to crust.

A wide variation of Hf-isotope values in zircons in rocks with homogeneous whole-rock Nd-isotope data has been documented previously (e.g., Griffin et al., 2002; Yang et al., 2006; Belousova et al., 2006; Murgulov et al., 2008), where the Sm-Nd whole rock data reflect the homogenizing effect of whole-rock analysis, while the zircon data allow better constraints on the end-members of the mixing components. These examples illustrate the usefulness of the in situ analytical techniques; the Hf-isotope composition of the zircon records the evolution of the magma chamber, with input of magmas from different sources, or/and progressive contamination by wall-rocks. Whole-rock Nd-isotope data represent the homogenized end product of these processes; they provide cumulative information and as such have limited potential to unravel source contributions during the temporal evolution of ultramafic-mafic magmas.

7.5 Os-ISOTOPE CONSTRAINTS ON THE ORIGIN OF Ni-Cu-PGE SULFIDE ORES

There are at least four distinct osmium isotope reservoirs (Walker et al., 1994). These include: (1) long-term rhenium-depleted SCLM, characterized by negative γOs values (from 0 to −10); (2) depleted upper mantle (DM) with characteristics similar to chondrites (γOs values averaging to 0), including some mid-ocean ridge basalt (MORB) sources (with γOs values up to +4); (3) long-term rhenium-enriched ocean island basalt (OIB)-type plume-mantle (γOs values from −3 to +60), and (4) continental and oceanic crust, which is much more "radiogenic" (γOs values > 400 for average crust; Esser and Turekian, 1993) than any of the other reservoirs. Therefore, application of the Re-Os isotope system is particularly useful for distinguishing between mantle and crustal sources. Because crustal material is much more enriched in Re than the mantle (Hauri and Hart, 1997), the more crustal rocks are assimilated, the higher the radiogenic ^{187}Os concentrations and the corresponding initial ^{187}Os/^{188}Os value (Shirey and Walker, 1998).

Investigations of the main PGE deposits using osmium isotopes suggested that the ore components have been derived from: (1) mantle sources as in the ultramafic-mafic intrusions at Noril'sk (Walker et al., 1994; Malitch and Latypov, 2011) and ultramafic massifs at Kondyor and Guli (Malitch, 1999), and (2) mixed crustal and mantle sources, with various amounts of the crustal component incorporated in the mantle source, as recorded at Bushveld (Hart and Kinloch, 1989), Stillwater (Lambert et al., 1994), and Sudbury (Walker et al., 1991; Dickin et al., 1992).

In a Re-Os study of sulfide mineralization from the Noril'sk area, Walker et al. (1994) have documented differences in initial osmium isotope compositions for the three economic intrusions (e.g., Noril'sk-1, Talnakh and Kharaelakh). These results are consistent with the Pb-Pb study of Wooden et al. (1992), who showed differences between the ores from the Noril'sk-1 intrusion (^{206}Pb/^{204}Pb in the range 18.043−18.082) and those from the Talnakh and Kharaelakh intrusions

(^{206}Pb/^{204}Pb in the range 18.149–18.282). The authors have suggested that the host magmas do not provide convincing evidence of interactions between mantle melts and lithosphere. Walker et al. (1994) concluded that magmas preserve a feature of the OIB source plume material and, hence, the slightly "radiogenic" osmium isotopic compositions (initial γOs values that vary from +6 to +9) are not necessarily indicators of contributions from either crustal rocks or SCLM.

This interpretation can be reevaluated on the basis of Re-Os data presented by Malitch and Latypov (2011). The mean initial ^{187}Os/^{188}Os value for massive ore at Kharaelakh (0.1283 ± 0.0054, γOs$_{(247\ Ma)}$ = 1.1; Malitch and Latypov, 2011) suggests a near-chondritic source for the osmium in the sulfide mineralization, being only slightly higher than the average CHUR ^{187}Os/^{188}Os value of 0.12689 at 247 Ma. Furthermore, among eight samples of massive ore at Kharaelakh, two samples (e.g., 963–18 and 844–19) clearly indicate long-term rhenium depletion (γOs values of −5.7 and −2.5), suggesting possible involvement of the SCLM. This is consistent with the findings of Zhang et al. (2008), who argued that in "fertile" flood-basalt provinces, an ancient cratonic lithospheric mantle with preexisting Ni- and PGE-rich sulfides has contributed significantly to the Ni- and PGE budget during interaction with plume-related magmas.

In contrast to the massive ores at Kharaelakh, indicative of a more primitive nature of the source, initial Os-isotope values of other investigated ores (massive ore at Talnakh, disseminated ores at Kharaelakh and Talnakh) show similar variations in Re and Os contents, Re/Os ratios, and initial ^{187}Os/^{188}Os values lying between 0.1331 (γOs = +4.9, calculated at 247 Ma) and 0.1366 (γOs = +7.6, calculated at 247 Ma). The slightly radiogenic ^{187}Os/^{188}Os values were explained by Walker et al. (1994) as a typical feature of long-term rhenium-enriched mantle reservoir such as the source of hot spots or plumes. Alternatively, it may reflect a minor contribution from a crustal component, which might represent ancient recycled crust enriched in radiogenic osmium as proposed by Hauri and Hart (1993).

7.6 Cu-ISOTOPE CONSTRAINTS ON THE ORIGIN OF Ni-Cu-PGE SULFIDE ORES

Cu-isotope analysis has been applied to a wide variety of different geological environments, including porphyry deposits (Larson et al., 2003; Graham et al., 2004; Mathur et al., 2005, 2009, 2012; Asael et al., 2007), skarn deposits (Larson et al., 2003; Graham et al., 2004; Maher and Larson, 2007), volcanogenic massive sulfide deposits and modern black smokers (Zhu et al., 2000; Rouxel et al., 2004; Mason et al., 2005), other hydrothermal deposits (Michigan native copper deposits, Larson et al., 2003; Schwarzwald mining district, Markl et al., 2006), sedimentary Cu deposits (Asael et al., 2007), and rocks/sediments (Maréchal et al., 1999; Archer and Vance, 2004; Rouxel et al., 2004; Li et al., 2009b).

There are several common features of Cu-isotope compositions in different ore-forming systems, including: (1) δ^{65}Cu values of Cu-rich minerals are close to zero; (2) the range of δ^{65}Cu values in most geological environments is more than 1‰; (3) minerals that experienced low-temperature redox processes have more variable δ^{65}Cu values than minerals crystallized at high temperatures (Larson et al., 2003; Mathur et al., 2009, 2012, and references therein).

Few Cu-isotope data are available for world-class PGE-rich deposits such as Bushveld, Stillwater and Sudbury (Zhu et al., 2000; Larson et al., 2003; Malitch et al., 2014b). Zhu et al. (2000) reported a restricted range of $\delta^{65}Cu$ values for igneous rocks, where chalcopyrite has a total variation of about 1‰, with a major cluster from −0.23‰ to +0.13‰. Larson et al. (2003) reported a variation of 1.5‰ for the $\delta^{65}Cu$ values of chalcopyrite in mafic-ultramafic intrusions, with most values clustering between −0.20‰ and −0.10‰. Similar or even tighter ranges in $\delta^{65}Cu$ were reported in mantle peridotite (0.0‰−0.18‰; Ben Othman et al., 2006), basalt (−0.10‰ to −0.03‰; Luck et al., 2003) and granite (−0.46‰−1.51‰, with the main cluster within −0.14‰ to +0.25‰ and a mean $\delta^{65}Cu$ of 0.01‰ ± 0.30‰ ($n = 30$, two outlier samples excluded; Li et al., 2009b). These results indicate that fractionation of Cu isotopes is insignificant during terrestrial differentiation processes. However, carbonaceous chondrites have $\delta^{65}Cu$ values ranging from −1.5‰ to 0.0‰ and ordinary chondrites along with iron meteorites range from −0.5‰ to 0.5‰ (Luck et al., 2003, 2005), indicating the fractionation of Cu-isotope compositions. This variation is attributed to mixing of two isotopically distinct primordial reservoirs (Luck et al., 2003, 2005). It is worth noting that seawater has $\delta^{65}Cu$ of 0.75‰−1.35‰, much heavier than expected (Bermin et al., 2006).

Unexpectedly high $\delta^{65}Cu$ variations of over 3‰ (−2.3‰ to +1.0‰) were observed in the economic intrusions of the Noril'sk and Taimyr Provinces (Fig. 7.11), exceeding the typical range of $\delta^{65}Cu$ values in high-temperature sulfides formed from magmas (−1.0‰−1.0‰). There is a gradual progression from lighter isotope compositions for the sulfides of the Kharaelakh (−2.3‰ to −0.9‰) and Talnakh (−1.1‰−0.0‰) intrusions to heavier compositions for those of the Noril'sk-1 intrusion (−0.1‰ to +0.6‰). The low-sulfide ore in the Noril'sk-1 intrusion has the heaviest $\delta^{65}Cu$ value of 1.0‰ ± 0.15‰.

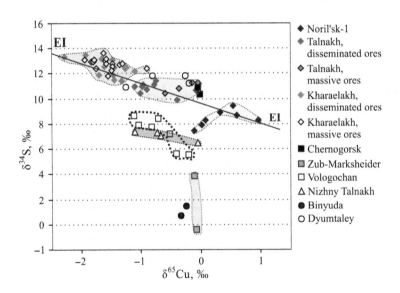

FIGURE 7.11

$\delta^{34}S-\delta^{65}Cu$ systematics of Ni-Cu-PGE sulfide ores and accumulations from economic, subeconomic, noneconomic, and prospective intrusions of the Noril'sk and Taimyr Provinces.

The samples of massive and disseminated sulfide ores define three major clusters of $\delta^{65}Cu$ values (Fig. 7.11) that span the distinct reservoirs defined by iron meteorites and carbonaceous chondrites (Luck et al., 2003, 2005). Disseminated ores of the Noril'sk-1, Zub-Marksheider, Chernogorsk intrusions, massive ores of the Talnakh intrusion and sulfide accumulations of the Binyuda intrusion form the first cluster around 0.0‰, close to that of Bulk Silicate Earth and iron meteorites. The disseminated ores of the Talnakh, Vologochan, Nizhny Talnakh and Dyumtaley intrusions form the second cluster around −0.7‰ (Fig. 7.11), matching the carbonaceous chondrites. The massive or disseminated ores of the Kharaelakh intrusion make up the third cluster with mean $\delta^{65}Cu$ values of $-1.52‰ \pm 0.24‰$ and $-1.60‰ \pm 0.31‰$, respectively, close to the light-isotope end of the range in carbonaceous chondrites. These results show that the different types of ore textures have indistinguishable Cu isotopic compositions.

Various processes could potentially produce the observed shift in the Cu-isotope composition of sulfide assemblages, including the emplacement of magmas/sulfide liquids with different isotopic composition, or magmatic and nonmagmatic fractionation of Cu isotopes. It has been shown that redox reactions play an important role in the fractionation of Cu isotopes at low temperatures (Zhu et al., 2002; Graham et al., 2004; Rouxel et al., 2004; Asael et al., 2006; Markl et al., 2006). In general, the variations of Cu-isotope ratios may be due to fluid-mineral fractionation during precipitation, physicochemical fluid conditions (e.g., redox changes), overprinting by hydrothermal processes (Graham et al., 2004; Rouxel et al., 2004), and fractionation between different complex species in solution (Maréchal and Albarède, 2002). It has also been shown that redox reactions produce Cu-rich minerals in different valence states, i.e., Cu(I) and Cu(II), and may additionally vary as a function of solution salinity (Mason et al., 2005). In porphyry copper deposits, the $\delta^{65}Cu$ values span a significant range from −16.96‰ to +9.98‰ (Mathur et al., 2009 and references therein); chalcopyrite from primary high-temperature mineralization has a relatively restricted cluster of $\delta^{65}Cu$ values from +1‰ to −1‰, whereas secondary minerals formed by low-temperature processes have $\delta^{65}Cu$ values from −16.96‰ to +9.98‰ (Mathur et al., 2009). Secondary chalcocite is relatively heavy, with $\delta^{65}Cu$ varying from −0.3‰ to 6.5‰, whereas the leach cap minerals dominated by Fe-oxides (jarosite, hematite, and goethite) are relatively light, ranging from −9.9‰ to 0.14‰ (Mathur et al., 2009), indicating the secondary modification of copper-isotope compositions. Oxidation−reduction processes are considered to be responsible for the isotopically-light copper of chalcopyrite in the stratiform sediment-hosted copper deposits where Cu sulfides formed through interaction of small, disconnected Cu solution reservoirs with H_2S, which was formed by bacterial reduction of sulfate-containing pore waters (Asael et al., 2007).

From the existing evidence, significant variation in Cu isotopic signatures appears to be a result of secondary processes, none of which are relevant to the high-temperature Ni-Cu-PGE sulfide ores of the Noril'sk and Taimyr Provinces. We suggest that the shift of $\delta^{65}Cu$ values for the ores of the Kharaelakh intrusion (from −2.3‰ to −0.9‰) may be attributed to the magmatic fractionation of Cu isotopes. A few samples of the intrusion have isotopically-light copper, similar to that of the native copper from the Arylakh intrusion (−1.9‰ ± 0.15‰), which can be attributed to addition from an external source (Malitch et al., 2014b). However, the indistinguishable $\delta^{65}Cu$ values of disseminated and massive ores of the Kharaelakh intrusion may indicate that the parental magma of the intrusion had a low-$\delta^{65}Cu$ signature, differing from that of the Talnakh and Noril'sk-1 intrusions. Thus we suggest that the $\delta^{65}Cu$ values of the ores from the economic intrusions reflect their primary signature rather than a mixing of sources, although the magmatic fractionation of Cu isotopes and/or assimilation of external material, particularly in case of Kharaelakh ores, cannot be ruled out.

7.7 S-ISOTOPE CONSTRAINTS ON THE ORIGIN OF Ni-Cu-PGE SULFIDE ORES

The key question of why the Ni-Cu-PGE sulfide ores of the Noril'sk-type economic deposits contain isotopically-heavy sulfur has been debated since the 1960s (Godlevsky and Grinenko, 1963; Vinogradov and Grinenko, 1966; Grinenko, 1966, 1984, 1985, 1990; Kuz'min and Tuganova, 1977; Godlevsky and Likhachev, 1986; Distler et al., 1998; Pushkarev, 1997, 1999; Tuganova, 2000; Li et al., 2003; Ripley et al., 2003, 2010; Ripley and Li, 2003, 2013; Malitch and Latypov, 2011; Krivolutskaya, 2014a,b, 2016; Malitch et al., 2014b).

The elevated $\delta^{34}S$ values of sulfides (8‰−14‰) in the Noril'sk-type deposits have been used as an argument for crustal contamination (Godlevsky and Grinenko, 1963; Vinogradov and Grinenko, 1966; Grinenko, 1966, 1990; Li et al., 2003; Ripley and Li, 2003, 2013; Ripley et al., 2003, 2010). However, assimilation is not universally accepted as a key process in the formation of the Noril'sk deposits (Kuz'min and Tuganova, 1977; Grinenko et al., 1984, 1985; Godlevsky and Likhachev, 1986; Wooden et al., 1992; Distler et al., 1998; Pushkarev, 1997, 1999; Malitch and Latypov, 2011; Krivolutskaya, 2014a,b, 2016; Malitch et al., 2014b).

Massive and disseminated ores of the Khraelakh and Talnakh intrusions have $\delta^{34}S$ values of 12.6‰ ± 0.5‰ and 10.9‰ ± 0.6‰, respectively, indicating that crustal contamination is essential in their formation. In this scenario, the high $\delta^{34}S$ values of the sulfides (8‰−14‰) in the ore-bearing intrusions are due to contamination of the parental magma with ^{34}S-enriched crustally-derived sulfur (e.g., Grinenko, 1966). The crustal source of sulfur is likely to be the anhydrite-bearing sediments that are spatially associated with the Kharaelakh intrusion, which have high $\delta^{34}S$ values (18‰−22‰ reported by Ripley et al., 2010; and 17.8‰ reported by Malitch and Latypov, 2011). The contamination could have occurred either during the emplacement of upwelling magma (Arndt et al., 2003) or in situ by fluids (Li et al., 2003). In situ contamination is considered to be unlikely, however, since some of the ore-bearing intrusions, such as Noril'sk-1 and Talnakh, were emplaced well above the evaporite strata (Malitch and Latypov, 2011; Krivolutskaya et al., 2012; Ripley and Li, 2013). An alternative explanation is that there was a mantle source with unusually heavy sulfur beneath the Noril'sk-type intrusions (Godlevsky and Likhachev, 1986; Wooden et al., 1992). Grinenko (1984, 1985) proposed that the mafic magmas were sulfurized by gases such as H_2S, presumably with isotopically-heavy sulfur ($\delta^{34}S = +10‰$).

Subduction-related metasomatism has been used to explain the heavy sulfur isotopes (up to + 14‰) in compositionally-similar sulfide inclusions (pyrrhotite, chalcopyrite, and pentlandite) in diamonds (Chaussidon et al., 1987, 1989; Eldridge et al., 1991). The heavy sulfur in the sulfide assemblage at Noril'sk would be consistent with models of Dodin et al. (2001), Ivanov (2007), and Starostin and Sorokhtin (2011) who speculated that the origin of the ore-bearing magmas parental to the Noril'sk-type intrusions may have been due to the subduction processes.

Distler et al. (1988) argued that the sediments could serve as a source of sulfur for the Noril'sk-Talnakh deposits, but not of metals such as Ni and PGE. It is unclear then why spatially and temporally associated mafic-ultramafic intrusions in the Noril'sk region, emplaced in the same sedimentary strata, contain highly variable proportions of crustal S (Grinenko, 1966, 1985, 1990; Kuz'min and Tuganova, 1977; Ripley et al., 2003; Malitch et al., 2009b, 2014b; Krivolutskaya, 2014a). The mean ^{34}S value of the disseminated ores of the subeconomic intrusions spans a range over 10.5‰ (Fig. 7.12). Furthermore, the ores of the Zub-Marksheider intrusion have δ^{34}S values of −0.4‰ to +3.9‰ with a mean of 0.39‰ and standard deviation of 1.55‰, which is inconsistent with a significant crustal sulfur component (Malitch et al., 2014b; Fig. 7.12), even though the intrusion was emplaced in the thickest series of hybrid rocks (Figs. 7.4 and 7.5). Likewise, the disseminated ores of the Gorozubovsky intrusion, which experienced the assimilation of large amounts of country rock (Godlevsky, 1959), also have low δ^{34}S values with a mean of 2.7‰ ± 2.3‰ (Kuz'min and Tuganova, 1977). The findings at Gorozubovsky (Kuz'min and Tuganova, 1977) and Zub-Marksheider (Malitch et al., 2014b) challenge a model demanding assimilation of crustal sulfur as a prerequisite to the formation a magmatic deposit. These results suggest that the "mantle-like" S isotopic composition reflects the primary sulfide assemblage, which has not been changed by interaction in a deep-staging chamber, during passage to the surface or by in situ assimilation, as is frequently suggested (Arndt et al., 2003; Li et al., 2003, among others). Furthermore, rocks of the noneconomic Nizhny Talnakh intrusion have a mean ^{34}S value of 6.38 ± 1.89‰, but they still have the most radiogenic ^{187}Os/^{188}Os (Walker et al., 1994; Malitch et al., 2009a; Petrov et al., 2009) and ^{87}Sr/^{86}Sr$_i$ (Arndt et al., 2003; Petrov et al., 2007, 2009) and the least radiogenic ^{187}Hf/^{188}Hf$_{(t)}$ values of zircons (Malitch et al., 2008), which is consistent with high degrees of crustal contamination. As an alternative hypothesis, the low δ^{34}S values of these rocks may be due to the interaction of mantle sulfides with crustal material that had a δ^{34}S close to 0‰ (Keays and Lightfoot, 2010).

The segregation of large amounts of sulfides may have occurred in the deep-seated staging chamber due to crustal contamination (Hawkesworth et al., 1995; Arndt et al., 2003; Lightfoot and Keays, 2005; Ripley and Li, 2013). In the model by Li et al. (2009a), it is suggested that the sulfides were redissolved by new magmas, transported towards the present position of the intrusions and segregated again due to the interaction of magmas with anhydrite-bearing evaporates. However, this scenario is difficult to reconcile with the remarkable homogeneity of δ^{34}S and δ^{65}Cu values of the sulfide ores in different intrusions in the Noril'sk Province (see Figs. 7 and 8 in Malitch et al., 2014b), which appear to require a long-lived staging chamber to achieve compositional homogeneity. Pushkarev (1997, 1999) invoked mantle-crust interaction that took place in the mantle to reach high levels of isotopic compositional homogeneity.

The issue of whether or not crustal sulfur must be involved in the formation of Ni-Cu-PGE sulfide deposits has been critically evaluated by Keays and Lightfoot (2010) and Ripley and Li (2013). The fact that many large sulfide Ni-Cu-PGE deposits show evidence for external sulfur addition implies that efficient collection of immiscible sulfide derived solely from mantle sulfur is rare. However, if the magmatic system is large enough, economic deposits could have been formed even without the addition of external sulfur (Ripley and Li, 2013). Thus, the data from Zub-Marksheider and Kharaelakh illustrate two important points. They suggest that the immediate country rocks may have very little influence on the mineralization in igneous rocks, and that an interaction of magma with host rocks can take place long before the final emplacement.

Finally, the Hf-Nd-Sr-Cu-S isotope study of the Binyuda and Dyumtaley intrusions allows characterization of an isotopic paradox, which is difficult to explain within the framework of existing ideas on the formation of sulfide ores in the ultramafic-mafic intrusions (Naldrett, 1992; Ripley

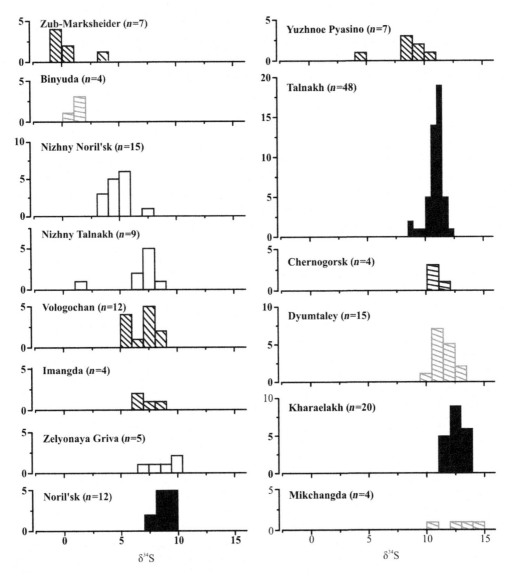

FIGURE 7.12

Variations in the sulfur-isotope composition of disseminated Ni-Cu-PGE sulfide ores from economic (in black), subeconomic (dashed black), prospective (dashed gray), and noneconomic (white) intrusions of the Northern Siberia.

Modified after Malitch, K.N., Petrov, O.V., Tuganova, E.V., Shevchenko, S.S., Bocharov, S.N., Kapitonov, I.N., 2009b. Isotope-geochemical criteria in exploration for Ni-Cu-PGE sulfide ores of the Noril'sk-type intrusions (Russia): insights from S and Cu isotope data. Mafic-ultramafic complexes of folded regions and related deposits. Abstracts of the 3rd International Conference 2. Ekaterinburg: Institute of geology and geochemistry of Russian Academy of Sciences, 38–41 (in Russian).

et al., 2003; Li et al., 2009a; Arndt, 2011). It is defined by the negative correlation between the Hf-Nd-Sr isotope signatures of different lithologies and the S-isotope composition of the sulfide they host. At Binyuda, sulfides of mantle origin (δ^{34}S in the range of 0.7‰—2.0‰ with a mean of 1.5‰ ± 0.4‰) are typical for the rocks that carry the so-called crustal Hf-Nd-Sr isotope signatures. At Dyumtaley, sulfide with heavy S-isotope composition (δ^{34}S in the range of 9.9‰—12.9‰ with a mean of 11.4‰ ± 0.6‰) is typical for rocks with Hf-Nd-Sr isotope signatures that show an important role for the juvenile component.

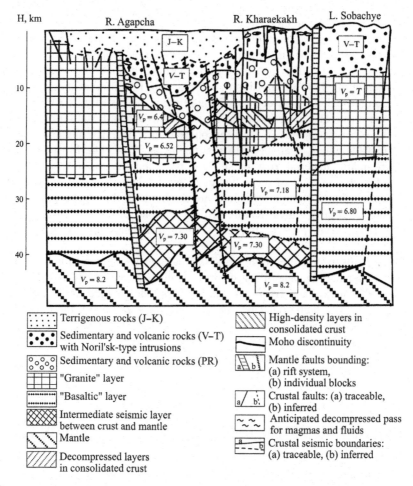

FIGURE 7.13

Geoseismic section along the profile Dikson—Khilok

Modified after Egorkin, A.V., Zyuganov, S.K., Chernyshev, N.M., 1984. Upper mantle of Siberia. 27-th Intern. Geol. Congress, Moscow. Geophysics 8, 27—42; Mitrofanov F.P., Bayanova T.B., Korchagin A.U., Groshev N.Yu., Malitch K.N., Zhirov D.V., Mitrofanonov A.F., 2013. East Scandinavian and Noril'sk plume mafic large igneous provinces of Pt-Pd ores: geological and metallogenic comparison. Geology of Ore Deposits 55 (5), 305—319.

7.8 GEOPHYSICAL FINGERPRINTS IN THE SEARCH FOR ECONOMIC Ni-Cu-PGE SULFIDE ORES

There is a general consensus that massive ores represent concentrations of immiscible magmatic sulfide liquid, which occupy depressions along the floor of the economic intrusions (Kharaelakh, Talnakh, and Noril'sk-1) or lie below the intrusion floor, separated from it by a screen of metasomatized sedimentary rock. It is further assumed that these intrusions resulted from contemporaneous emplacement of ultramafic-mafic magma laden with immiscible sulfide liquid and large volumes of almost pure sulfide liquid. The reasons for enormous concentration of ore metals and the mechanism for lifting large volumes of very dense sulfide liquid remain poorly defined.

Clues to the location of one of the richest ore camps based on geophysical data have been summarized by Tuganova (1991, 2000) and are briefly outlined here. Using seismic data (Egorkin et al., 1984; Malitch et al., 1987), a transition zone has been identified at the base of the crust in the Noril'sk region, with higher velocities of longitudinal waves (Vp; 7.3 km/s) compared to those typical of the crust. This zone has been considered as a manifestation of ultramafic material injected in the subcrustal levels of the lithosphere (Fig. 7.13). The thickness of this zone is 5−10 km, with a spatial extent of about 500 km at depths of 32−43 km. Most likely, a large volume of ultramafic-mafic material rich in sulfides preceded the tholeiite-basaltic volcanism, which contributed to the removal of these bodies to form relatively small hybrid layered intrusions. It is noteworthy that economic intrusions are restricted to the region above this transition zone, which is considered as a typical feature of the basal part of the crust in oceanic and continental paleorifts (e.g., Krylov, 1976; Belousov, 1982; Ramberg and Morgan, 1984; Malitch et al., 1987).

The analysis of the deep structure of the Noril'sk area suggests a connection between the formation of Ni-Cu-PGE deposits in paleorift structures (Malitch et al., 1988; Tuganova and Popov, 1989) and particular geological and geophysical parameters. These include: (1) high-gradient downfolds in the basement, (2) a high abundance of horst-graben structures in the crust, and (3) large volumes of intruded mantle-derived material between the crust and mantle. In general, the deep structure of the Noril'sk area shows noncratonic features with a differentiated crust of transitional type (Malitch et al., 1988). Therefore, identification of deep structures as exemplified at Noril'sk is an important criterion for regional tectonic forecasting.

7.9 ISOTOPIC-GEOCHEMICAL FINGERPRINTS IN THE SEARCH FOR ECONOMIC Ni-Cu-PGE SULFIDE ORES

Table 7.5 provides a summary of isotope-geochemical data for the economic, subeconomic, noneconomic, and prospective intrusions and associated Ni-Cu-PGE sulfide ores.

7.9.1 OS-ISOTOPE SIGNATURES

One of the most effective indicators of the economic scale of a deposit is the initial isotopic composition of osmium. This is primarily because the massive Ni-Cu-PGE sulfide deposits are

Table 7.5 Summary of Isotopic-Geochemical Data for Economic, Subeconomic, Prospective and Noneconomic Ultramafic-Mafic Intrusions of the Noril'sk and Taimyr Provinces

Intrusion	εHf	εNd	$^{87}Sr/^{86}Sr_i$	$\delta^{65}Cu$	$\delta^{34}S$	γOs
Noril'sk-1	6.0 ± 3.7	0.7 ± 1.1	0.70591 ± 0.00022	0.23 ± 0,28	9.2 ± 1.8	
	− 4.7−15.5	− 1.8−2.0	0.70552−0.70625	− 0.10−0.61	7.5−13.8	
Talnakh	9.1 ± 3.2	0.3 ± 1.3	0.70638 ± 0.00039	− 0.55 ± 0.41	10.9 ± 0.6	6.9 ± 0.9
	0.1−16.7	− 3.3−1.2	0.70584−0.70709	− 1.1−0.0	7.8−12.1	5.3−8.2
Kharaelakh	9.4 ± 3.8	1.0 ± 0.2	0.70666 ± 0.00076	− 1.56 ± 0.27	12.7 ± 0.5	2.5 ± 3.4
	2.3−16.3	0.8−1.4	0.70555−0.70798	− 2.3− − 0.9	11.5−13.6	− 5.7−5.8
Chernogorsk	7.2 ± 2.1	1.4 ± 0.6	0.70711 ± 0.00088	− 0.03 ± 0.06	10.9 ± 0.4	9.6 ± 0.1
	4.7−10.9	0.9−2.6	0.70640−0.70898	− 0.1−0.0	10.4−11.2	9.5−9.6
Zub-Marksheider		1.2 ± 0.5	0.70685 ± 0.00127	− 0.1 ± 0.0	0.4 ± 1.6	20.7 ± 15.5
		0.4−2.1	0.70570−0.70908	− 0.1− − 0.1	− 0.7−3.9	9.7−31.6
Vologochan	8.5 ± 1.7	1.2 ± 0.8	0.70634 ± 0.00068	− 0.72 ± 0.29	6.9 ± 1.2	6.2 ± 3.2
	6.5−11.1	− 1.7−2.4	0.70560−0.70808	− 1.1− − 0.4	5.1−8.5	4.3−9.9
Binyuda	− 3.8 ± 1.3	− 3.8 ± 0.4	0.70588 ± 0.00013	− 0.4 ± 0.1	1.5 ± 0.4	8.8 ± 1.0
	− 4.9− − 1.9	− 4.4− − 3.2	0.70566−0.70609	− 0.5− − 0.3	0.7−2.0	8.1−9.5
Dyumtaley	9.5 ± 2.5	4.2 ± 0.7	0.70474 ± 0.00020	− 0.66 ± 0.42	11.4 ± 0.6	
	6.0−12.4	2.7−4.7	0.70451−0.70504	− 1.24− − 0.25	9.9−12.9	
Nizhny Talnakh	0.8 ± 2.8	− 4.6 ± 1.1	0.70824 ± 0.00030	-0.60 ± 0.42	6.4 ± 1.9	65.0 ± 33.0
	− 5.2−5.6	− 5.5− − 1.2	0.70765−0.70863	− 1.0−0.0	1.8−7.6	35.6−117.8
Zelyonaya Griva	1.7 ± 2.6	− 5.2 ± 0.4	0.70836 ± 0.00024	0.28 ± 0.47	8.4 ± 1.3	54.5
	− 5.9−3.3	− 5.9− − 4.6	0.70800−0.70867	− 0.26−1.21	6.8−9.7	

Isotopic-geochemical data are after Malitch, K.N., Latypov, R.M., 2011. Re-Os and S isotope constraints on timing and source heterogeneity of PGE-Cu-Ni sulfide ores: A case study at the Talnakh ore junction, Noril'sk Province, Russia. Can. Mineral. 49, 1653−1677; Malitch, K.N., Krymsky, R.S., Petrov, O.V., Pushkarev, Yu.D., Tuganova, E.V., 2009a. Re-Os isotope systematics of Ni-Cu-PGE sulfide ores of ultramafic-mafic intrusions, Noril'sk Province. Isotope systems and time of geological processes. Abstracts of the IV Russian conference on isotope geochronology 2, IP Katalkina, St. Petersburg, 20−22 (in Russian); Malitch, K.N., Belousova, E.A., Griffin, W.L., Badanina, I.Y., Pearson, N.J., Presnyakov, S.L., and et al., 2010b. Magmatic evolution of the ultramafic-mafic Kharaelakh intrusion (Siberian craton, Russia): insights from trace-element, U-Pb and Hf-isotope data on zircon. Contrb. Mineral. Petrol. 159, 753−768; Malitch, K.N., Belousova, E.A., Griffin, W.L., Badanina, I.Yu., 2013. Hafnium-neodymium constraints on source heterogeneity of the economic ultramafic-mafic Noril'sk-1 intrusion (Russia). Lithos 164−167, 36−46; Malitch, K.N., Latypov, R.M., Badanina, I.Yu., Sluzhenikin, S.F., 2014b. Insights into ore genesis of Ni-Cu-PGE sulfide deposits of the Noril'sk Province (Russia): evidence from copper and sulfur isotopes. Lithos 204, 172−187; Malitch, K.N., Badanina, I.Yu., Romanov, A.P., Sluzhenikin, S.F., 2016. U-Pb age and Hf-Nd-Sr-Cu-S isotope systematics of the Binyuda and Dyumtaley ore-bearing intrusions (Taimyr, Russia). Litosphera 1, 107−128 (in Russian); Petrov, O.V., Malitch, K.N., Tuganova E.V., Pushkarev, Yu.D., Badanina, I.Yu., et al., 2009. Test-methodical investigations on elaborating isotope-geochemical criteria for prospecting platinum-group metals, gold, copper, nickel and cobalt in the layered intrusions of the northern Siberia (Krasnoyarsk region). Proceedings of VSEGEI − Year 2008, VSEGEI Press, St. Petersburg 8(56), 248−262 (in Russian) and unpublished data.

characterized by Os-isotope values corresponding to a mixture of mantle and crustal material with a predominance of the mantle component. In the disseminated ores of the same deposits, the mantle signature is somewhat diluted by the crustal component and the proportion of the mantle component is minimal in the noneconomic Nizhny Talnakh and Nizhny Noril'sk intrusions (Walker et al., 1994; Malitch et al., 2009a). Correspondingly, noneconomic intrusions have much more variable and higher γOs values (e.g., 35−118, Arndt et al., 2003, Malitch et al., 2009a) than those of economic intrusions (γOs in the range from −5 to +8).

7.9.2 Sr AND Nd-ISOTOPE SIGNATURES

In the Sr-Nd isotope systematics, lithological units of economic intrusions manifest a clear increase in initial $^{86}Sr/^{87}Sr$ value, which varies from 0.7055 to 0.7080, at rather constant subchondritic Nd isotope values (εNd about +1). In contrast, the Nd-Sr isotope compositions of noneconomic intrusions show the least radiogenic εNd values (εNd = −5) and most radiogenic $(^{87}Sr/^{86}Sr)_i$, in the range of 0.7077−0.7087. The most radiogenic strontium is found in Cu-Ni ores, in which $(^{87}Sr/^{86}Sr)_i$ varies from 0.7085 to 0.7112 (Petrov et al., 2007, 2008). This feature can be employed as a prospecting tool for the presence of significant quantities of the ore contaminant, which has isotopic signatures close to that of Ni-Cu-PGE sulfide ores.

7.9.3 Hf-ISOTOPE SIGNATURES

We suggest that a radiogenic Hf-isotope composition of zircon along with a significant range in initial $^{176}Hf/^{177}Hf$ values is indicative of mixing between juvenile mantle and lithospheric sources, and can be employed as an effective fingerprint for identifying prospective intrusive host rocks (Malitch et al., 2010b, 2013). Consequently, it is a useful indicator in exploration for sulfide-rich ores associated with "Noril'sk-type" intrusions.

7.9.4 Cu AND S-ISOTOPE SIGNATURES

Different intrusions in the Noril'sk and Taimyr Provinces have a very limited range of $\delta^{65}Cu$. In contrast, the $\delta^{34}S$ values are highly variable in economic (from 7.5‰ to 13.3‰), subeconomic (−0.1‰−11.2‰) and noneconomic (6.5‰−7.6‰) intrusions (Fig. 7.11). The sulfide ores in the three economic intrusions show a negative correlation between $\delta^{34}S$ and $\delta^{65}Cu$ (trend EI−EI, Fig. 7.11). Samples from the other intrusions show a decoupling of S and Cu isotopes (Fig. 7.11), with $\delta^{65}Cu$ ranging from 0‰ to −1‰, but $\delta^{34}S$ between 0‰ and 12‰. The disseminated ores of the subeconomic Chernogorsk and prospective Dyumtaley intrusions have $\delta^{34}S$ and $\delta^{65}Cu$ values overlapping the trend for the economic intrusions; they may be the next prospective targets for massive ores. The other intrusions, in which stable isotope compositions do not correspond to those in economic deposits, show low potential for the occurrence of massive Ni-Cu-PGE sulfide ores.

Listed according to increasingly high $\delta^{34}S$, disseminated ores from ultramafic-mafic intrusions of Polar Siberia fall into the following sequence (Fig. 7.12): Zub-Marksheider (−0.7‰ to +3.9‰), Binyuda (0.7‰−2.0‰), Nizhny Noril'sk (3.8‰−7.7‰), Nizhny Talnakh (1.8‰−8.0‰), Vologochan (5.1‰−8.5‰), Imangda (6.4‰−8.7‰), Zelyonaya Griva (6.8‰−9.7‰), Noril'sk-1 (7.5‰−9.4‰), Yuzhnoe Pyasino (4.3‰−10.5‰), Talnakh (8.7‰−12.0‰), Chernogorsk (10.4‰−11.2‰), Dyumtaley (9.9‰−12.9‰), Kharaelakh (11.9‰−13.3‰), and Mikchangda (11.0‰−14.0‰). The crustal (sulfate) isotopic composition of S frequently has been postulated as evidence of high economic potential, but it is contradicted by mantle-like S isotopic compositions of disseminated Ni-Cu-PGE sulfide mineralization from the subeconomic Zub-Marksheider intrusion. The field evidence for latter intrusion also suggests that it has experienced significant assimilation of sulfate-rich Devonian sediments. Following Kuz'min and Tuganova (1977), we question the validity of isotopically-heavy sulfur as a feature indicative of economic Ni-Cu-PGE sulfide ore. It has been noted (Malitch et al., 2009b) that an increase in the range of S-isotope values correlates negatively with ore-deposit reserves. Thus, we propose that a restricted range of S-isotope values, and the negative trend of coupled S-Cu isotope compositions can be employed as useful guides to assess the economic potential of a Ni-Cu-PGE sulfide deposit.

CONCLUDING REMARKS

Multiple magmatic events have been recorded by U-Pb data acquired on zircons and baddeleyites (347−235 Ma, Malitch et al., 2010a,b, 2012, 2014a; Malitch and Petrov, 2010). They indicate a prolonged evolution of magmas parental to the economic intrusions (Noril'sk-1, Talnakh, and Kharaelakh) in deep-seated chambers, and a history of later recrystallization of the intrusive rocks in the crust. A case study of the Noril'sk-1 intrusion has revealed a decoupling of the Hf-isotope composition of zircon and the Nd-isotope compositions of unmineralized lithologies. These data also suggest that an extended period for concentration of the ore components in staging chambers during this interaction is a key factor for the formation of economic deposits, and invite a reevaluation of the genetic characteristics of Ni-Cu-PGE sulfide mineralization. Sulfide ores, thus, had a prehistory responsible for the concentration of sulfides from the large volume of ultramafic-mafic magmas parental to Noril'sk-type intrusions. This was followed by partial assimilation of crustal material and isolation of significant amounts of sulfide liquid, resulting in the possibility of its capture by a later magma that served as the vehicle that facilitated their passage to the surface. We propose that the models for the origin of the Norilsk-type intrusions and their relationship to the Siberian flood basalts (reviewed by Arndt, 2005, 2011; Ivanov, 2007; Dobretsov et al., 2008; Zhang et al., 2008) should be revised in the light of new evidence, indicative of a complex geological history. Following previous studies (Petrov et al., 2007, 2008, 2009; Malitch et al., 2008, 2009b, 2013, 2014b; Malitch and Latypov, 2011), we suggest a set of geophysical and isotopic-geochemical fingerprints to guide an exploration for Ni-Cu-PGE sulfides. The useful prospective indicators for Ni-Cu-PGE sulfide deposits are considered to be: (1) the deep structure of the Noril'sk area with noncratonic features and a differentiated crust of transitional type, (2) the wide range of εHf and $^{87}Sr/^{86}Sr_i$ values combined with a restricted range of εNd values, and (3) the negative correlation of S- and Cu-isotope compositions and a restricted range of $\delta^{34}S$ and $\delta^{65}Cu$ values for an individual intrusion.

ACKNOWLEDGEMENTS

Financial support from the Academy of Finland (grant 131619 to KNM) and Russian Foundation of Basic Research (grant 13-05-00671-a to KNM) is gratefully acknowledged. The analytical work was supported by Agency of Natural Resources of the Taymyr Autonomous Area and Agency of Natural Resources of the Krasnoyarsk Territory (Russian Federation) through contract 7F-TAO/2005 and by Belousova's ARC FT1100100685 grant. Valuable discussions with Tamara B. Bayanova, Vadim V. Distler, Andrey E. Izokh,

Nadezhda A. Krivolutskaya, Wolfgang Maier, Felix P. Mitrofanov, Anthony J. Naldrett, Norman J. Pearson, Yury D. Pushkarev, Oleg V. Petrov, Andrey P. Romanov, Sergey S. Shevchenko, Oleg N. Simonov, and Evgeniya V. Tuganova are greatly appreciated. We thank Edward Ripley and the anonymous reviewer for constructive comments and useful suggestions. Editorial input by Sisir Mondal and Tasha Frank are much appreciated. This is contribution 958 from the ARC Centre of Excellence for Core to Crust Fluid Systems (http://www.ccfs.mq.edu.au) and 1148 from the GEMOC Key Centre (http://www.gemoc.mq.edu.au).

REFERENCES

Archer, C., Vance, D., 2004. Mass discrimination correction in multiple-collector plasma source mass spectrometry: an example using Cu and Zn isotopes. J. Anal. Atomic Spectrom. 19, 656−665.

Arndt, N.T., 2005. The conduits of magmatic ore deposits: Ottawa. Mineral. Assoc. Canada Short Course 35, 161−182.

Arndt, N.T., 2011. Insights into the geologic setting and origin of Ni−Cu−PGE sulfide deposits of the Norilsk-Talnakh Region, Siberia. Rev. Econ. Geol. 17, 199−215.

Arndt, N.T., Czamanske, G.K., Walker, R.J., Chauvel, C., Fedorenko, V.A., 2003. Geochemistry and origin of the intrusive hosts of the Noril'sk-Talnakh Cu-Ni-PGE sulfide deposits. Econ. Geol. 98, 495−515.

Arndt, N.T., Lesher, C.M., Czamanske, G.K., 2005. Magmas and magmatic Ni-Cu-(PGE) deposits. Econ. Geol. 100th Anniversary Volume 5−23.

Asael, D., Matthews, A., Butler, I., Rickard, A.D., Bar-Matthews, M., Halicz, L., 2006. $^{65}Cu/^{63}Cu$ fractionation during copper sulphide formation from iron sulphides in aqueous solution. Geochim. Cosmochim. Acta. 70 (18 Suppl. 1), A23.

Asael, D., Matthews, A., Bar-Matthews, M., Halicz, L., 2007. Copper isotope fractionation in sedimentary copper mineralization (Timna Valley, Israel). Chem. Geol. 243, 238−254.

Badanina, I.Yu, Malitch, K.N., Romanov, A.P., 2014. Isotopic-geochemical characteristics of the ore-bearing ultramafic-mafic intrusions of Western Taimyr, Russia. Doklady Earth Sci. 458 (1), 1165−1167.

Belousov, V.V., 1982. Transition Zones Between Continents and Oceans. Nedra Press, Moscow (in Russian).

Belousova, E.A., Griffin, W.L., O'Reilly, S.Y., 2006. Zircon crystal morphology, trace-element signatures and Hf-isotope composition as a tool for petrogenetic modelling: examples from eastern Australian granitoids. J. Petrol. 47, 329−353.

Ben Othman, D., Luck, J.M., Bodinier, J.L., Arndt, N.T., Albarède, F., 2006. Cu-Zn isotopic variations in the Earth's mantle. Geochim. Cosmochim. Acta. 70 (18), A46.

Bermin, J., Vance, D., Archer, C., Statham, P.J., 2006. The determination of the isotopic composition of Cu and Zn in seawater. Chem. Geol. 226 (3-4), 280−297.

Burgess, S.D., Bowring, S.A., 2015. High-precision geochronology confirms voluminous magmatism before, during, and after Earth's most severe extinction. Sci. Adv. 1, e1500470.

Cabri, L.J., Sylvester, P.J., Tubrett, M.N., Peregoedova, A., Laflamme, J.H.G., 2003. Comparison of LAM-ICP-MS and micri-PIXE results for palladium and rhodium in selected samples of Noril'sk and Talnakh sulfides. Can. Mineral. 41 (2), 321−329.

Campbell, I.H., Czamanske, G.K., Fedorenko, V.A., Hill, R.I., Stepanov, V., 1992. Synchronism of the Siberian traps and the Permian-Triassic boundary. Science 255, 1760−1763.

Chaussidon, M., Albarede, F.L., Sheppard, S.M.F., 1987. Sulphur isotope heterogeneity in the mantle from ion microprobe measurements of sulphide inclusions in diamond. Nature 330, 242−244.

Chaussidon, M., Albarede, F.L., Sheppard, S.M.F., 1989. Sulphur isotope variations in the mantle from ion microprobe analyses of micr-sulphide inclusions. Earth Palnet. Sci. 92, 144−156.

Czamanske, G.K., Wooden, J.L., Zientek, M.L., Fedorenko, V.A., Zen'ko, T.E., Kent, J., et al., 1994. Geochemical and isotopic constraints on the petrogenesis of the Noril'sk-Talnakh ore-forming system. In: Lightfoot, P.C., Naldrett, A.J. (Eds.), Proceedings of the Sudbury − Noril'sk Symposium: Spec. Publ, 5. Geological Survey, Ontario, pp. 313−342.

Czamanske, G.K., Zen'ko, T.E., Fedorenko, V.A., Calk, L.C., Budahn, J.R., Bullock Jr., J.H., et al., 1995. Petrography and geochemical characterization of ore-bearing intrusions of the Noril'sk type, Siberia; with discussion of their origin. Resour. Geol. Special 18, 1−48.

Dalrymple, G.B., Czamanske, G.K., Fedorenko, V.A., Simonov, O.N., Lanphere, M.A., Likhachev, A.P., 1995. A reconnaissance ^{40}Ar/^{39}Ar study of ore-bearing and related rocks, Siberian Russia. Geochim. Cosmochim. Acta. 59, 2071−2083.

Diakov, S., West, R., Schissel, D., 2002. Recent advances in the Noril'sk model and its application for exploration of Ni-Cu-PGE sulfide deposits. Soc. Econ. Geol. Special Publ. 9, 203−226.

Dickin, A.P., Richardson, J.M., Crocket, J.H., McNutt, R.H., Peredery, W.V., 1992. Osmium isotope evidence for a crustal origin platinum-group elements in the Sudbury nickel ore. Geochim. Cosmochim. Acta. 56, 3531−3537.

Distler, V.V., 1994. Platinum mineralization of the Noril'sk deposits. In: Lightfoot, P.C., Naldrett, A.J. (Eds.), Proceedings of the Sudbury-Noril'sk Symposium, 5. Ontario Geological Survey Special, Sudbury, pp. 243−260.

Distler, V.V., Smirnov, A.V., Grokhovskaya, T.L., Philimonova, A.A., Muravizkaya, G.N., 1979. The stratification, cryptic variation and the formation of the sulfide mineralization of the differentiated trap intrusions. In: Smirnov, V.I. (Ed.), The Formation of the Magmatic Ore Deposits. Nauka, Moscow, pp. 211−269. (in Russian).

Distler, V.V., Grokhovskaya, T.L., Evstigneeva, T.L., Sluzhenikin, S.F., Filimonova, A.A., Dyuzhikov, O.A., et al., 1988. Petrology of Magmatic Sulfide Ore Formation. Nauka, Moscow, 232 p. (in Russian).

Distler, V.V., Sluzhenikin, S.F., Cabri, L.J., Krivolutskaya, N.A., Turovtsev, D.M., Golovanova, T.A., et al., 1999. Platinum ores of the Noril'sk layered intrusions: magmatic and fluid concentration ratio of noble metals. Geol. Ore Deposits 41, 214−237.

Dobretsov, N.L., Kirdyashkin, A.A., Kirdyashkin, A.G., Vernikovsky, V.A., Gladkov, I.N., 2008. Modelling of thermochemical plumes and implications for the origin of the Siberian traps. Lithos 100, 66−92.

Dodin, D.A., Batuev, B.N., 1971. Geology and petrology of the Talhakh differentiated intrusions and their metamorphic aureole. Petrology and Ore Resource Potential of the Talnakh and Noril'sk Differentiated Intrusions. Nedra, Leningrad, pp. 31−100 (in Russian).

Dodin, D.A., Chernyshev, N.M., Cherednikova, O.I., 2001. Metallogeny of Platinum-Group Elements of Large Regions of Russia. Geoinformark, Moscow, 302 p. (in Russian).

Dyuzhikov, O.A., Distler, V.V., Strunin, B.M., Mkrtychyan, A.K., Sherman, M.L., Sluzhenikin, S.F., 1992. Geology and metallogeny of sulfide deposits, Noril'sk region, USSR. Soc. Econ. Geol. Special Publ. 241 p.

Dyuzhikov, O.A., Kurbatov, I.I., Laputina, I.P., Mkrtych'jan, A.K., Romanov, A.P., Sluzhenikin, S.F., 1995. Platinum-bearing plagioolivinite − the new ore-bearing magmatic formation of Taimyr. Doklady Akademii Nauk 340 (2), 212−217 (in Russian).

Egorkin, A.V., Zyuganov, S.K., Chernyshev, N.M., 1984. Upper mantle of Siberia. 27-th Intern. Geol. Congress, Moscow. Geophysics 8, 27−42.

Eldridge, C.S., Compston, W., Williams, I.S., Harris, J.W., Bristow, J.W., 1991. Isotopic evidence for the involvement of recycled sediments in diamond formation. Nature 353, 649−653.

Esser, B.K., Turekian, K.K., 1993. The osmium isotopic composition of the continental crust. Geochim. Cosmochim. Acta. 57, 3093−3104.

Fedorenko, V.A., 1991. Tectonic control of magmatism and localization of Ni-bearing areas in the northwestern Siberian Platform. Geol.Geofizika 32 (1), 48–56 (in Russian).

Fedorenko, V.A., 1994. Evolution of magmatism as reflected in volcanic sequence of the Noril'sk region. In: Lightfoot, P.C., Naldrett, A.J. (Eds.), Proceedings of the Sudbury-Noril'sk Symposium. Ontario Geological Survey, Sudbury, pp. 171–183.

Genkin, A.D., Evstigneeva, T.L., 1986. Associations of platinum-group minerals of the Noril'sk copper-nickel sulfide ores. Econ. Geol. 81, 1203–1212.

Genkin, A.D., Distler, V.V., Gladyshev, G.D., Filimonova, A.A., Evstigneeva, T.L., Kovalenker, V.A., et al., 1981. Sulfide copper-nickel ores of the Noril'sk deposit. Nauka Press, Moscow, 234 p. (in Russian).

Godlevsky, M.N., 1959. Traps and ore-bearing intrusions of the Noril'sk region. Moscow, Gostekhmetizdat 68 p. (in Russian).

Godlevsky, M.N., Grinenko, L.N., 1963. Some data on the isotopic composition of sulphur in the sulphides of the Noril'sk deposit. Geokhimiya 1, 35–39 (in Russian).

Godlevsky, M.N., Likhachev, A.P., 1986. Types and distinctive features of ore-bearing formations of copper-nickel deposits. In: Friedrich, G.H., Genkin, A.D., Naldrett, A.J., Ridge, J.D., Sillitoe, R.H., Vokes, F.M. (Eds.), Geology and metallogeny of copper deposits. Springer-Verlag, Berlin, pp. 111–123.

Gorbachev, N.S., 2010. Experimental study of interaction between fluid-bearing basaltic melts and peridotite: a mantle-crustal source of trap magmas in the Norilsk area. Petrology 18 (4), 416–431.

Graham, S., Pearson, N., Jackson, S., Griffin, W., O'Reilly, S.Y., 2004. Tracing Cu and Fe from source to porphyry: in situ determination of Cu and Fe isotope ratios in sulfides from the Grasberg Cu-Au deposit. Chem. Geol. 207, 147–169.

Griffin, W.L., Pearson, N.J., Belousova, E.A., Jackson, S.E., van Achtenbergh, E., O'Reilly, S.Y., et al., 2000. The Hf isotope composition of cratonic mantle: LAM-MC-ICPMS analysis of zircon megacrysts in kimberlites. Geochim. Cosmochim. Acta. 64, 133–147.

Grinenko, L.N., 1966. Sulphur isotope composition of sulphides of the Talnakh copper-nickel deposit (Noril'sk region) in light of its origin. Geokhimiya 4, 15–30 (in Russian).

Grinenko, L.N., 1984. Hydrogen sulfide-containing gas deposits as a source of sulphur for sulphurization of magma in ore-bearing intrusions of the Noril'sk area. Dokl. Akad. Nauk. SSSR. 278, 730–732 (in Russian).

Grinenko, L.N., 1985. Sources of sulfur of the nickeliferous and barren gabbro-dolerite intrusions of the northwest Siberian platform. Inter. Geol. Rev. 28, 695–708.

Grinenko, L.N., 1990. Substanse sources and conditions of formation of sulphide copper-nickel ores deduced from isotope-geochemical data. Geology of copper-nickel deposits of the U.S.S.R.. Nauka Press, Leningrad, pp. 57–66 (in Russian).

Hart, S.R., Kinloch, E.D., 1989. Osmium isotope systematics in Witwatersrand and Bushveld ore deposits. Econ. Geol. 84, 1651–1655.

Hauri, E.H., Hart, S.R., 1997. Rhenium abundances and systematics in oceanic basalts. Chem. Geol. 139, 185–205.

Hawkesworth, C.J., Lightfoot, P.C., Fedorenko, V.A., Blake, S., Naldrett, A.J., Doherty, W., et al., 1995. Magma differentiation and mineralisation in the Siberian flood basalts. Lithos 61–88.

Horan, M.F., Walker, R.J., Fedorenko, V.A., Czamanske, G.K., 1995. Osmium and neodymium isotopic constraints on the temporal and spatial evolution of Siberian flood basalt sources. Geochim. Cosmochim. Acta. 59, 5159–5168.

Hoskin, P.W.O., 2005. Trace-element composition of hydrothermal zircon and the alteration of Hadean zircon from the Jack Hills, Australia. Geochim. Cosmochim. Acta. 69, 637–648.

Ivanov, A.V., 2007. Evaluation of different models for the origin of the Siberian traps. In: Fougler, G.R., Jurdy, D.M. (Eds.), The Origin of Melting Anomalies: Plates, Plumes and Planetary Processes, Geological Society of America Special Paper 430, 669–691.

Kamo, S.L., Czamanske, G.K., Amelin, Y., Fedorenko, V.A., Davis, D.W., Trofimov, V.R., 2003. Rapid eruption of Siberian flood-volcanic rocks and evidence for coincidence with the Permian-Triassic boundary and mass extinction at 251 Ma. Earth Planetary Sci. Lett 214, 75−91.

Kamo, S.L., Czamanske, G.K., Krogh, T.E., 1996. A minimum U-Pb age for Siberian flood-basalt volcanism. Geochim. Cosmochim. Acta. 60, 3505−3511.

Keays, R.R., Lightfoot, P.C., 2010. Crustal sulfur is required to form magmatic Ni-Cu sukfide deposits: evidence from chalcophile element signatures of Siberian and Deccan Trap basalts. Mineral. Deposita 45, 241−257.

Kinny, P.D., Dawson, J.B., 1992. A mantle metasomatic injection event linked to late Cretaceous kimberlite magmatism. Nature. 360, 726−728.

Kogarko, L.N., Karpenko, S.F., Lyalikov, A.V., Teptelev, M.P., 1988. Isotope criteria of the meimechite genesis. Doklady Academii Nauk SSSR 301, 939−942 (in Russian).

Komarova, M.Z., Kozyrev, S.M., Kokorin, N.I., Knauf, V.V., 1999. Layered intrusion of the Dyumtaley River: petrology, ore potential. In: Simonov, O.N. (Ed.), Natural Resources of Taimyr 3. VSEGEI Press, Noril'sk, 42−67 (in Russian).

Komarova, M.Z., Kozyrev, S.M., Lyul'ko, V.A., Vilinsky, S.A., 2000. Precious-metal mineralization of the disseminated ores of the Noril'sk ore-bearing district. In: Simonov, O.N. (Ed.), Natural Resources of Taimyr 4. VSEGEI Press, Noril'sk, pp. 122−136 (in Russian).

Komarova, M.Z., Kozyrev, S.M., Simonov, O.N., Lyul'ko, V.A., 2002. The PGE mineralization of disseminated sulfide ores of the Noril'sk-Taimyr Region. In: Cabri, L.J. (Ed.), The Geology, Geochemistry, Nineralogy and Mineral Beneficiation of Platinum-Group Elements. Canadian Institute of Mining, Metallurgy and Petroleum, 54. pp. 547−567 Special Volume.

Kotulsky, V.K., 1946. About the origin of magmatic copper-nickel deposits. Doklady Acad. Sci. USSR 51, 381−384 (in Russian).

Krivolutskaya, N.A., 2014a. Evolution of Trap Magmatism and Processes Producing PGE-Cu-Ni Mineralization in the Noril'sk Area. KMK Scientific Press, Moscow, 305 p. (in Russian).

Krivolitskaya, N.A., 2014b. Mantle origin of heavy isotopes of sulfur in ores of the Noril'sk deposits. Doklady Earth Sci. 454 (1), 76−78.

Krivolutskaya, N.A., 2016. Siberian Traps and Pt-Cu-Ni Deposits in the Noril'sk Area. Springer International Publishing, Switzerland, 364 p.

Krivolutskaya, N.A., Gongalskiy, B.I., Yushin, A.A., Schlychkova, T.B., Kononkova, N.N., Tushentsova, I.N., 2011. Mineralogical and geochemical characteristics of PGE-Cu-Ni ores of the Maslovsky deposit in the Noril'sk area, Russia. Can. Mineral. 49, 1649−1674.

Krivolutskaya, N.A., Sobolev, A.V., Snisar, S.G., Gongalskiy, B.I., Hauff, B., Kuzmin, D.V., et al., 2012. Mineralogy, geochemistry and stratigraphy of the Maslovsky Pt-Cu-Ni sulfide deposit, Noril'sk Region, Russia: implications for relationship of ore-bearing intrusions and lavas, Mineral. Deposita, 47. pp. 69−88.

Krylov, S.V., 1976. About the reasons of anomalous properties of the mantle in rift zones. Geol. Geophyzika 17, 3−18 (in Russian).

Kuz'min, V.K., Tuganova, E.V., 1977. New data on the isotope composition of sulfur in the copper-nickel sulfide ores of the northwestern part of the Siberian platform. Soviet Geol. Geophys. 18, 98−100.

Lambert, D.D., Walker, R.J., Morgan, J.W., Shirey, S.B., Carlson, R.W., Zientek, M.L., et al., 1994. Re-Os and Sm-Nd isotope geochemistry of the Stillwater Complex, Montana: implications for the petrogenesis of the J-M Reef. J. Petrol. 35, 1717−1753.

Larson, P.B., Maher, K., Ramos, F.C., Chang, Z.S., Gaspar, M., Meinert, L.D., 2003. Copper isotope ratios in magmatic and hydrothermal ore-forming environments. Chem. Geol. 201 (3-4), 337−350.

Latypov, R.M., 2002. Phase equilibria constraints on relations of ore-bearing intrusions with flood basalts in the Noril'sk region, Russia. Contrb. Mineral. Petrol. 143, 438−449.

Latypov, R.M., 2007. Noril'sk- and Lower Talnakh-type intrusions are not conduits for overlying flood basalts: insights from residual gabbroic sequence of intrusions. Trans. Inst. Mining Metal. 116, B215−B225.

Li, C., Ripley, E.M., Naldrett, A.J., 2003. Compositional variations of olivine and sulfur isotopes in the Noril'sk and Talnakh intrusions, Siberia: implications for ore-forming processes in dynamic magma conduits. Econ. Geol. 98, 69−86.

Li, C., Ripley, E.M., Naldrett, A.J., 2009a. A new genetic model for the giant Ni-Cu-PGE sulfide deposits associated with the Siberian flood basalts. Econ. Geol. 104, 291−301.

Li, W.-Q., Jackson, S.E., Pearson, N.J., Alard, O., Chappell, B.W., 2009b. The Cu isotopic signature of granites from the Lachlan Fold Belt, SE Australia. Chem. Geol. 258, 38−49.

Lightfoot, P.C., Hawkesworth, C.J., Hergt, J., Naldrett, A.J., Gorbachev, N.S., Fedorenko, V.A., et al., 1993. Remobilization of the continental lithosphere by mantle plumes: major-, trace-element and Sr-, Nd-, and Pb-isotope evidence from picritic and tholeiitic lavas of the Noril'sk district Siberian Trap, Russia. Contrib. Mineral. Petrol. 114, 171−188.

Lightfoot, P.C., Keays, R.R., 2005. Siderophile and chalcophile metal variations in flood basalts from the Siberian Trap, Noril'sk Region: implications for the origin of the Ni-Cu-PGE sulfide ores. Econ. Geol. 100, 439−462.

Likhachev, A.P., 1978. Conditions of formation of ore-bearing and barren mafic/ultramafic magmas. Doklady Acad. Sci. USSR 238 (2), 447−450 (in Russian).

Likhachev, A.P., 1994. Ore-bearing intrusions of the Noril'sk region. In: Naldrett, A.J., Lightfoot, P.C. (Eds.), Proceedings of the Sudbury-Noril'sk Symposium. Ontario Geological Survey, Sudbury, pp. 185−201.

Luck, J.-M., Ben Othman, D., Barrat, J.A., Albarède, F., 2003. Coupled ^{63}Cu and ^{16}O excesses in chondrites. Geochim. Cosmochim. Acta. 67, 143−151.

Luck, J.-M., Ben Othman, D., Albarède, F., 2005. Zn and Cu isotopic variations in chondrites and iron meteorites: early solar nebula reservoirs and parent-body processes. Geochim. Cosmochim. Acta. 69, 5351−5363.

Lyul'ko, V.A., Fedorenko, V.A., Distler, V.V., Sluzhenikin, S.F., Kunilov, V.E., Stekhin, A.I., et al., 1994. Geology and ore deposits of the Norilsk region. Guidebook of the VII International Platinum Symposium (Distler, V.V., Kunilov, V.E., eds.). Moskovsky contact Press, Moscow-Norilsk. 43 p.

Lyul'ko, V.A., Amosov, Y.N., Kozyrev, S.M., Komarova, M.Z., Ryabikin, V.A., Rad'ko, V.A., et al., 2002. The state of the ore base of non-ferrous and noble metals in the Norilsk region, with guidelines of top-priority geological and exploration works. Rudy i Metally 5, 66−82 (in Russian).

Maher, K.C., Larson, P.B., 2007. Variation in copper isotope ratios and controls on fractionation in hypogene skarn mineralization at Coroccohuayco and Tintaya, Peru. Econ. Geol. 102, 225−237.

Malich K., Badanina I., Belousova E., Pearson N., Romanov A. and Sluzhenikin S., 2014. Radiogenic and stable isotope study of the Dyumtaley and Binyuda ultramafic-mafic intrusions and associated Ni-Cu-PGE sulfide ores (Russian Arctic). Geophysical Research Absracts 16, EGU2014-13102-1. EGU General Assembly 2014 (CD-ROM).

Malitch, K.N., 1999. Platinum-Group Elements in Clinopyroxenite-Dunite Massifs of the Eastern Siberia (Geochemistry, Mineralogy, and Genesis). Saint Petersburg Cartographic Factory VSEGEI Press, St. Petersburg, 296 p. (in Russian).

Malitch, K.N. and Petrov, O.V., 2010. Geochronology and Hf-Nd-Sr-Os-S isotope systematics of the Noril'sk-type intrusions: new insights for prolonged evolution and source heterogeneity. Giant Ore Deposits Down-Under. 13th Quadrennial IAGOD Symposium Proceedings, Government of South Australia, pp. 234−236.

Malitch, K.N., Latypov, R.M., 2011. Re-Os and S-isotope constraints on age and source heterogeneity of Ni-Cu-PGE sulfide ores: a case study at the Talnakh ore junction (Russia). Can. Mineral. 49 (6), 1653−1677.

Malitch, K.N., Belousova, E.A., Griffin, W.L., Badanina, I.Yu, Petrov, O.V., Pearson, N.J., 2008. Contrasting magma sources in ultramafic-mafic intrusions of the Noril'sk area (Russia): Hf isotope evidence from zircon. Geochim. Cosmochim. Acta. 72 (12S), A589.

Malitch, K.N., Krymsky, R.S., Petrov, O.V., Pushkarev, Yu.D. and Tuganova, E.V., 2009a. Re-Os isotope systematics of Ni-Cu-PGE sulfide ores of ultramafic-mafic intrusions, Noril'sk Province. Isotope systems and time of geological processes. Abstracts of the IV Russian conference on isotope geochronology 2, IP Katalkina, St. Petersburg, pp. 20−22 (in Russian).

Malitch, K.N., Petrov, O.V., Tuganova, E.V., Shevchenko, S.S., Bocharov, S.N. and Kapitonov, I.N., 2009b. Isotope-geochemical criteria in exploration for Ni-Cu-PGE sulfide ores of the Noril'sk-type intrusions (Russia): insights from S and Cu isotope data. Mafic-ultramafic complexes of folded regions and related deposits. Abstracts of the 3rd International Conference 2. Ekaterinburg: Institute of geology and geochemistry of Russian Academy of Sciences, pp. 38−41 (in Russian).

Malitch, K.N., Badanina, I., Yu., Tuganova, E.V., 2010a. Magmatic evolution of the ultramafic-mafic intrusions of the Noril'sk Province (Russia): insights from compositional and geochronological data. Lithosphera 5, 37−63 (in Russian).

Malitch, K.N., Belousova, E.A., Griffin, W.L., Badanina, I.Y., Pearson, N.J., Presnyakov, S.L., et al., 2010b. Magmatic evolution of the ultramafic-mafic Kharaelakh intrusion (Siberian craton, Russia): insights from trace-element, U-Pb and Hf-isotope data on zircon. Contrb. Mineral. Petrol. 159, 753−768.

Malitch, K.N., Badanina, I.Yu, Belousova, E.A., Tuganova, E.V., 2012. Results of U-Pb dating of zircon and baddeleyite from the Noril'sk-1 ultramafic-mafic intrusion (Russia). Russian Geol. Geophys. 53 (2), 123−130.

Malitch, K.N., Belousova, E.A., Griffin, W.L., Badanina, I.Yu, 2013. Hafnium-neodymium constraints on source heterogeneity of the economic ultramafic-mafic Noril'sk-1 intrusion (Russia). Lithos 164-167, 36−46.

Malitch, K.N., Badanina, I.Yu, Khiller, V.V., Belousova, E.A., Bocharov, S.N., et al., 2014a. Age and Hf-Nd isotope systematics of ultramafic-mafic intrusions from the Noril'sk Province based on study of monazite, baddeleyite and zircon in ore-bearing and barren rocks. Trans. Inst. Geol. Geochem. Ural Branch Russian Acad. Sci. 161, 191−197 (in Russian).

Malitch, K.N., Latypov, R.M., Badanina, I., Yu., Sluzhenikin, S.F., 2014b. Insights into ore genesis of Ni-Cu-PGE sulfide deposits of the Noril'sk Province (Russia): evidence from copper and sulfur isotopes. Lithos 204, 172−187.

Malitch, K.N., Badanina, I., Yu., Romanov, A.P., Sluzhenikin, S.F., 2016. U-Pb age and Hf-Nd-Sr-Cu-S isotope systematics of the Binyuda and Dyumtaley ore-bearing intrusions (Taimyr, Russia). Litosphera 1, 107−128 (in Russian).

Malitch, N.S., 1975. Tectonic Evolution of the Cover of the Siberian Craton. Nedra Press, Moscow, 215 p. (in Russian).

Malitch, N.S., Grinson, A.S., Tuganova, E.V., Chernyshev, N.M., 1988. Rifting of the Siberian platform. 28th Session of Intern. Geol. Congress. Tectonic Processes. Nauka Press, Moscow, pp. 184−193 (in Russian).

Malitch, N.S., Mironyuk, E.P., Tuganova, E.V. (Eds.), 2002. Geology and Mineral Resources of Russia. Volume 3. Eatern Siberia. VSEGEI Press, St. Petersburg, 396 p. (in Russian).

Malitch, N.S., Mironyuk, E.P., Tuganova, E.V., Grinson, A.S., Masaitis, V.L., Surkov, V.S., et al., 1987. Geological structure of the USSR and regularities of distribution of mineral deposits. In: Malitch, N.S., Masaitis, V.L., Surkov, V.S. (Eds.), The Siberian Platform 4. Nedra Press, Leningrad (in Russian).

Maréchal, C., Albarède, F., 2002. Ion-exchange fractionation of copper and zinc isotopes. Geochim. Cosmochim. Acta. 66, 1499−1509.

Maréchal, C.N., Telouk, P., Albarède, F., 1999. Precise analysis of copper and zinc isotopic compositions by plasma-source mass spectrometry. Chem. Geol. 156, 251−273.

Markl, G., Lahaye, Y., Schwinn, G., 2006. Copper isotopes asmonitors of redox processes in hydrothermal mineralization. Geochim. Cosmochim. Acta. 70, 4215−4228.

Mason, T.F.D., Weiss, D.J., Chapman, J.B., Wilkinson, A.J., Tessalina, V.G., Spiro, A., et al., 2005. Zn and Cu isotopic variability in the Alexandrinka volcanic-hosted massive sulphide (VHMS) ore deposit, Urals, Russia. Chem. Geol. 221, 170−187.

Mathur, R., Ruiz, J., Casselman, M.J., Megaw, P., van Egmond, R., 2012. Use of Cu isotopes to distinguish primary and secondary Cu mineralization in the Canariaco Norte porphyry copper deposit, Northern Peru. Mineral. Deposita 47, 755−762.

Mathur, R., Ruiz, J., Titley, S., Liermann, L., Buss, H., Brantley, S., 2005. Cu isotopic fractionation in the supergene environment with and without bacteria. Geochim. Cosmochim. Acta. 69 (22), 5233−5246.

Mathur, R., Titley, S., Barra, F., Brantley, S., Wilson, M., Phillips, A., et al., 2009. Exploration potential of Cu isotope fractionation in porphyry copper deposits. J. Geochem. Expl. 102, 1−6.

Mitrofanov, F.P., Bayanova, T.B., Korchagin, A.U., Groshev, N., Yu, Malitch, K.N., et al., 2013. East Scandinavian and Noril'sk plume mafic large igneous provinces of Pt-Pd ores: geological and metallogenic comparison. Geol. Ore Deposits 55 (5), 305−319.

Murgulov, V., O'Reilly, S.Y., Griffin, W.L., Blevin, P.L., 2008. Magma sources and gold mineralisation in the Mount Leyshon and Tuckers igneous Complexes, Queensland, Australia: U-Pb and Hf isotope evidence. Lithos 101, 281−307.

Naldrett, A.J., 1992. A model for the Ni-Cu-PGE ores of the Noril'sk region and its application to other areas of flood basalt. Econ. Geol. 87, 1945−1962.

Naldrett, A.J., 2004. Magmatic Sulfide Deposits: Geology, Geochemistry and Exploration. Springer Verlag, Heidelberg, Berlin, 728 p.

Naldrett, A.J., Lightfoot, P.C., Fedorenko, V.A., Doherty, W., Gorbachev, N.S., 1992. Geology and geochemistry of intrusions and flood basalts of the Noril'sk region, USSR, with implications for the origin of the Ni-Cu ores. Econ. Geol. 87, 975−1004.

Naldrett, A.J., Fedorenko, V.A., Lightfoot, P.C., Kunilov, V.I., Gorbachev, N.S., Doherty, W., et al., 1995. Ni-Cu-PGE deposits of Noril'sk region, Siberia: their formation in conduits for flood basalt volcanism. Trans. Inst. Mining Metal. 104, B18−B36.

Naldrett, A.J., Fedorenko, V.A., Asif, M., Shushen, L., Kunilov, V.E., Stekhin, A.I., et al., 1996. Controls on the composition of Ni-Cu sulfide deposits as illustrated by those at Noril'sk, Siberia. Econ. Geol. 91, 751−773.

Petrov, O.V., Malitch, K.N., Pushkarev, Yu.D., Bogomolov, E.S., 2007. Isotope-geochemical criterion in search for the Noril'sk-type massive Ni-Cu-PGE sulfide ores: constraints from Pb, Nd and Sr isotope data. Geochim. Cosmochim. Acta. 71 (15S), A782.

Petrov, O.V., Malitch, K.N., Lokhov, K.I., Bogomolov, E.S. and Khalenev, V.O., 2008. Isotope-geochemical criteria in exploration for the Noril'sk-type PGE-Cu-Ni sulfide ores: insights from Nd, Sr, S and Cu isotope data. Abstracts of the 33rd International Geological Congress, 6−14 August 2008, Oslo, Norway (CD-ROM).

Petrov, O.V., Malitch, K.N., Tuganova E.V., Pushkarev, Yu.D., Badanina, I.Yu., et al., 2009. Test-methodical investigations on elaborating isotope-geochemical criteria for prospecting platinum-group metals, gold, copper, nickel and cobalt in the layered intrusions of the northern Siberia (Krasnoyarsk region). Proceedings of VSEGEI − Year 2008, VSEGEI Press, St. Petersburg 8 (56), 248−262 (in Russian).

Pokrovskii, B.G., Sluzhenikin, S.F., Krivolutskaya, N.A., 2005. Interaction conditions of Noril'sk trap intrusions with their host rocks: isotopic (O, H, and C) evidence. Petrology 13 (1), 49−72.

Proskurnin, V.F., Vernikovsky, V.A., Metelkin, D.V., Petrushkov, B.S., Vernikovskaya, A.E., Gavrish, A.V., et al., 2014. Rhyolite-granite association in the Central Taimyr zone: evidence of accretionary-collisional events in the Neoproterozoic. Russian Geol. Geophys. 55, 18−32.

Pushkarev, Yu.D., 1997. Two types of interaction of crustal and mantle materials and new approach to problems of deep seated ore formation. Doklady RAN 355 (4), 524−526 (in Russian).

Pushkarev, Yu.D., 1999. The nature of super large ore deposits: a conception of crust-mantle interaction within the mantle. Mineral Deposits: Processes to Processing. Balkema, Rotterdam, pp. 1291−1294.

Rad'ko, V.A., 1991. Model of dynamic differentiation of intrusive traps in the northwestern Siberian platform. Soviet Geol. Geophys. 32, 15−20.

Ramberg, I., Morgan, P., 1984. Physical characteristics and evolutionary trends of continental rifts. 27th International Geological Congress. Tectonics. Nauka Press, Moscow, pp. 78−109.

Reichow, M.K., Saunders, A.D., White, R.V., Pringle, M.S., Al'Mukhamedov, A.I., Medvedev, A., et al., 2002. New ^{40}Ar-^{39}Ar data on basalts from the West Siberian basin: extent of the Siberian flood basalt province doubled. Science 296, 1846−1849.

Reichow, M.K., Pringle, M.S., Al'mukhamedov, A.I., Allen, M.B., Andreichev, V.L., Buslov, M.M., et al., 2009. The timing and extent of the eruption of the Siberian Traps large igneous province. Implications for the end-Permian environmental crisis. Earth. Planet. Sci. Lett. 277, 9–20.

Renne, P.R., Basu, A.R., 1991. Rapid eruption of the Siberian Traps flood basalts at the Permo-Triassic boundary. Science 253, 176–179.

Ripley, E.M., Li, C., 2003. Sulfur isotope exchange and metal enrichment in the formation of magmatic Cu-Ni-(PGE) deposits. Econ. Geol. 98, 635–641.

Ripley, E.M., Li, C., 2013. Sulfide saturation in mafic magmas: is external sulfur required for magmatic Ni-Cu-(PGE) ore genesis? Econ. Geol. 108, 45–58.

Ripley, E.M., Lightfoot, P.C., Li, C., Elswick, E.R., 2003. Sulfur isotopic studies of continental flood basalts in the Noril'sk region: implications for the association between lavas and ore-bearing intrusions. Geochim. Cosmochim. Acta. 67, 2805–2817.

Ripley, E.M., Li, C., Moore, C.H., Schmitt, A.K., 2010. Micro-scale S isotope studies of the Kharaelakh intrusion, Noril'sk region, Siberia: constraints on the genesis of coexisting anhydrite and sulfide minerals. Geochim. Cosmochim. Acta. 74, 634–644.

Ripley, E.M., Dong, S., Li, C., Wasylenki, L.E., 2015. Cu isotope variations between conduit and sheet-style Ni-Cu-PGE sulfide mineralization in the Midcontinent Rift System, North America. Chem. Geol. 414, 59–68.

Romanov, A.P., Kurbatov, I.I., Malitch, K.N., Snisar, S.G., Borodina, E.V., Erykalov, S.P., et al., 2011. In: Dodin, D.A. (Ed.), The Resource Potential of the Platinum Metals in Western Taimyr, Volume VII. Tipografija "Znak", Krasnojarsk, pp. 135–160 (in Russian).

Rouxel, O., Fouquet, Y., Ludden, J.N., 2004. Copper isotope systematics of the Lucky Strike, Rainbow, and Logatchev seafloor hydrothermal fields on the Mid-Atlantic Ridge. Econ. Geol. 99, 585–600.

Rudnick, R.L., Ireland, T.R., Gehrrels, G., Irving, A.J., Chesley, J.T. and Hanchar, J.M. 1998. Dating mantle metasomatism: U-Pb geochronology of zircons in cratonic mantle xenoliths from Montana and Tanzania. Extended Abstracts. 7th International Kimberlite Conference. Cape Town, pp. 754–756.

Ryabov, V.V., Shevko, A.Y., Gora, M.P., 2000. Magmatic Formations of the Noril'sk Region. 1. Petrology of Traps. Nonparel' Press, Novosibirsk, 408 p. (in Russian).

Sharma, M., Basu, A.R., Nesterenko, G.V., 1992. Temporal Sr-, Nd- and Pb-isotopic variations in the Siberian flood basalts: implications for the plume-source characteristics. Earth. Planet. Sci. Lett. 113, 365–381.

Shirey, S.B., Walker, R.J., 1998. Re-Os isotopes in cosmochemistry and high-temperature geochemistry. Ann. Rev. Earth Planet. Sci. 26, 423–500.

Simonov, O.N., Lyul'ko, V.A., Amosov, Yu.N. and Salov, V.M., 1994. Geological structure of the Noril'sk region. In: Lightfoot, P.C., Naldrett, A.J. (Eds.), Proceedings of the Sudbury-Noril'sk Symposium: Sudbury, Ontario Geological Survey Special Volume 5, 161–170.

Sluzhenikin, S.F., Mokhov, A.V., 2015. Gold and silver in PGE-Cu-Ni and PGE ores of the Noril'sk deposits, Russia. Mineral. Deposita 50, 465–492.

Sluzhenikin, S.F., Distler, V.V., Dyuzhikov, O.A., Kravtsov, V.E., Kunilov, V.E., Laputina, I.P., et al., 1994. Low sulfide platinum mineralization of the Noril'sk differentiated intrusions. Geol. Rudnykh Mestorozhdenii 36, 195–217 (in Russian).

Sluzhenikin, S.F., Krivolutskaya, N.A., 2015. Pyasino-Vologochan intrusion: geological structure and platinum-copper-nickel ores (Noril'sk Region). Geol. Ore Deposits 57 (5), 381–401.

Sluzhenikin, S.F., Krivolutskaya, N.A., Rad'ko, V.A., Malitch, K.N., Distler, V.V. and Fedorenko, V.A., 2014, Ultramafic-mafic intrusions, volcanic rocks and PGE-Cu-Ni sulfi de deposits of the Noril'sk Province, Polar Siberia. Field trip guidebook. 12th International Platinum Simposium (Simonov, O.N., ed.). Yekaterinburg, IGG UB RAS, 80 p.

Starostin, V.I., Sorokhtin, O.G., 2011. A new assessment of the Noril'sk-type deposits origin. Moscow Univ. Geol. Bull. 66 (2), 73–83.

Torgashin, A.S., 1994. Geology of the massive and copper ores of the western part of the Oktyabr'sky deposit. In: Lightfoot, P.C., Naldrett, A.J. (Eds.), Proceedings of the Sudbury-Noril'sk Symposium: Sudbury, Ontario Geological Survey Special Volume 5, 231−241.

Tuganova, E.V., 1988. Genetic model for the Noril'sk type Ni-Cu sulfide deposits. The Formation of Ores And Genetic Models of the Endogenic Ore Deposits. Nauka, Novosibirsk, pp. 197−204.

Tuganova, E.V., 1991. Petrological geodynamical model of formation of sulfide copper-nickel deposits. Soviet Geol. Geophys. 32, 1−7.

Tuganova, E.V., 2000. Petrographic Types, Genesis and Occurrence of Ni-Cu-PGE Sulfide Deposits. VSEGEI Press, St. Petersburg, 102 p. (in Russian).

Tuganova, Ye.V., Malich, K.N., 1990. Platinum potential of Noril'sk-type intrusions. Trans. (Doklady) USSR Acad. Sci./Earth Sci. Sect. 313 (4), 267−271 (translated from Doklady Akademii Nauk SSSR. 313 (1), 1990, 178-183).

Tuganova, E.V., Popov, V.E., 1989. Connecting link between rifting and Ni-bearing magmatism. Magmatism of Rifts. Nauka Press, Moscow, pp. 133−139 (in Russian).

Tuganova, E.V., Shergina, Yu.P., 1997. Isotope-geochemical peculiarities of rocks from Noril'sk-type intrusions. In: Simonov, O.N., Malitch, N.S. (Eds.), Nedra Taimyra (Resources of Taimyr) 2. VSEGEI Press, Noril'sk, pp. 114−122. (in Russian)

Tuganova, E.V., Shergina, Yu.P., 2003. Isotope-geochemical lithological heterogeneity in ore-bearing Noril'sk-type intrusions: genetic considerations. Regional Geol. Metall. 17, 140−146 (in Russian)

Turovtsev, D.M., 2002. Contact metamorphism of the Noril'sk intrusions. Moscow, Nauchny Mir, p. 318 (in Russian).

Urvantsev, N.N., 1967. Petrology of the Traps of the Siberian Platform. Nedra Press, Leningrad, p. 187 (in Russian).

Vernikovsky, V.A., Vernikovskaya, A.E., 2001. Central Taimyr accretionary belt (Arctic Asia): Meso-Neoproterozoic tectonic evolution and Rodinia breakup. Precamb. Res. 110, 127−141.

Vervoort, J.D., Blichert-Toft, J., 1999. Evolution of the depleted mantle: Hf isotope evidence from juvenile rocks through time. Geochim. Cosmochim. Acta. 63, 533−556.

Vervoort, J.D., Patchett, P.J., Blichert-Toft, J., Albarede, F., 1999. Relationships between Lu-Hf and Sm-Nd isotopic systems in the global sedimentary system. Earth Planet. Sci. Lett. 168, 79−99.

Vinogradov, A.P., Grinenko, L.N., 1966. Sulphur isotope composition of sulphides of copper-nickel deposits and occurrences from the Noril'sk region and problem of their genesis. Geokhimiya 1, 3−14, (in Russian).

Walker, R.J., Morgan, J.W., Horan, M.F., Czamanske, G.F., Krogstad, E.J., Fedorenko, V.A., et al., 1994. Re-Os isotopic evidence for an enriched-mantle source for the Noril'sk-type ore-bearing intrusions, Siberia. Geochim. Cosmochim. Acta. 58, 4179−4197.

Walker, R.J., Morgan, J.W., Naldrett, A.J., Li, C., Fassett, J.D., 1991. Re-Os isotope systematics of Ni-Cu sulfide ores, Sudbury igneous complex: evidence for a major crustal component. Earth Planet. Sci. Lett. 105, 416−429.

Wooden, J.L., Czamanske, G.K., Bouse, R.M., Likhachev, A.P., Kunilov, V.E., Lyul'ko, V., 1992. Pb isotope data indicate a complex mantle origin for the Norilsk-Talnakh ores, Siberia. Econ. Geol. 87, 1153−1165.

Wooden, J.L., Czamanske, G.K., Fedorenko, V.A., Arndt, N.T., Chauvel, C., Bouse, R.M., et al., 1993. Isotopic and trace-element constraints on mantle and crustal contributions to Siberian continental flood basalts, Noril'sk area, Siberia. Geochim. Cosmochim. Acta. 57, 3677−3704.

Yakubchuk, A., Nikishin, A., 2004. Norilsk-Talnakh Cu-Ni-PGE deposits: a revised tectonic model. Mineral. Deposita 39, 125−142.

Yang, J.-H., Wu, F.-Y., Chung, S.-L., Wilde, S.-A., Chu, M.F., 2006. A hybrid origin for the Qianshan A-type granite, northeast China: geochemical and Sr-Nd-Hf isotopic evidence. Lithos 89, 89−106.

Zen'ko, T.E., 1983. Mechanism of formation of the Noril'sk ore-bearing intrusions. Izvestiya Akademii nauk SSSR 11, 21−39.

Zen'ko, T.E., Czamanske, G.K., 1994a. Spatial and petrological aspects of the intrusions of the Noril'sk and Talnakh ore junctions. In: Lightfoot, P.C., Naldrett, A.J. (Eds.), Proceedings of the Sudbury-Noril'sk Symposium: Sudbury, Ontario Geological Survey. pp. 263–281.

Zen'ko, T.E., Czamanske, G.K., 1994b. Tectonic controls on ore-bearing intrusions of the Talnakh ore junctions: position, morphology and ore distribution. Inter. Geol. Rev. 36, 1033–1067.

Zhang, M., O'Reilly, S.Y., Wang, K.-L., Hronsky, J., Griffin, W.L., 2008. Flood basalts and metallogeny: the lithospheric connection. Earth Sci. Rev. 86, 145–174.

Zhu, X.K., Guo, Y., Williams, R.J.P., O'Nions, R.K., Matthews, A., Burgess, B.K., et al., 2002. Mass fractionation processes of transition metal isotopes. Earth. Planet. Sci. Lett. 200, 47–62.

Zhu, X.K., O'Nions, R.K., Guo, Y., Belshaw, N.S., Rickard, D., 2000. Determination of natural Cu-isotope variation by plasma-source mass spectrometry: implications for use as geochemical tracers. Chem. Geol. 163, 139–149.

Zolotukhin, V.V., Ryabov, Y.A., Vasiliev, Y.R., Shatkov, V.A., 1975. Petrology of the Talnakh Ore-Bearing Differentiated Trap Intrusion. Nauka, Novosibirsk, 432 p. (in Russian).

Zotov, I.A., 1979. Genesis of Trap Intrusions and Metamorphic Formations of Talnakh. Nauka, Moscow, 155 p. (in Russian).

APPENDIX A **GEOLOGICAL BACKGROUND OF THE STUDIED ORE SAMPLES**

A large proportion of the samples that include massive and disseminated ores were collected from a series of drill holes KZ-844 and KZ-963 from the Oktyabr'sky ore deposit, which lies at the base of the western limb of the Kharaelakh intrusion, and a drill hole OUG-2 through the Talnakh intrusion, both located within the Talnakh ore junction (Fig. 2; see for details Fig. 1, page 1656 in Malitch and Latypov, 2011). In some previous studies (e.g., Naldrett et al., 1992, 1996) the Kharaelakh and Talnakh intrusions were recognized as the Northwest and Main bodies that formed the Talnakh intrusion.

At Kharaelakh, the massive sulfide orebody covers an area of about 2.5 km^2, and is up to 40 m thick (Torgashin, 1994). Distinctive features of the Kharaelakh intrusion include the presence of two horizons of ultramafic rocks and the widespread presence of the disseminated and massive sulfide ores throughout the entire section (drill holes KZ-844, Fig. 4 in Malitch and Latypov (2011), and KZ-963, Fig. 5 in Malitch et al. (2014b)). In both drill cores the massive ore is separated from the intrusion by a screen of metamorphosed sedimentary rocks.

Studied samples are representative of pyrrhotite-rich massive ores located close to the bottom part of the intrusion in drill holes KZ-844 and KZ-963. Samples of chalcopyrite-rich massive ore are dominant within the middle and upper parts of the intrusion (drill hole KZ-963, Fig. 5 in Malitch and Latypov, 2011). Pyrrhotite, chalcopyrite, and pentlandite are by far the predominant minerals among other sulfides and as such represent a widespread assemblage in the ores (Fig. 6A–D,F in Malitch and Latypov, 2011). PGM, being minor constituents of the ores, frequently comprise Pd-Pt sulfides (braggite (Pt,Pd)S, cooperate (PtS), sperrylite (PtAs$_2$), and a solid solution series of atokite (Pd$_3$Sn) and rustenburgite (Pt$_3$Sn)).

At Talnakh, disseminated Ni-Cu-PGE sulfide ores occur within the lower part of intrusion, composed of plagiowehrlite (e.g., samples T-13, T-14, and T-15) and taxitic-textured mafic rocks with "patches" of ultramafic rocks (T-16, T-17, and T-18), whereas massive Ni-Cu-PGE sulfide ores form *ca* 30 meter thick orebody, similarly to Kharaelakh separated from the intrusion by a

2-meter thick alkaline-rich metasomatite (Fig. 6 in Malitch et al., 2014b). The inset of Fig. 3 (Malitch and Latypov, 2011) details the position of disseminated ore samples within the lower part of the intrusion (samples D26 −...− D69) and massive ore samples from the underlying ore-body (samples M71 −...− M101).

Disseminated ores at Talnakh are characterized by various sulfide mineral assemblages. From top to bottom (Fig. 6 in Malitch et al., 2014b), an assemblage of hexagonal pyrrhotite (Po_h), Fe-rich pentlandite (Pn^{Fe}), chalcopyrite (Cp), and cubanite (Cub) is replaced by $Po_h + Pn + Cp$ and by monoclinic pyrrhotite (Po_m), Ni-rich pentlandite (Pn^{Ni}), Cp, and Cub. Massive ores are pyrrhotite-rich; they are dominated by hexagonal pyrrhotite, with chalcopyrite and pentlandite in descending order. Predominantly pyrrhotitic ore is the most abundant of the massive Ni-Cu-PGE sulfide ore types.

Disseminated and low-sulfide ores from the Noril'sk-1 intrusion were sampled from the Medvezhy Creek open pit and drill hole MN-2 (Fig. 3 in Malitch et al., 2013). Location of samples KN 97-1, KN 97-2 and KN 97-3 derived from the western part of the open pit of the Medvezhy Creek Mine, 180 m level of the Noril'sk-1 deposit are shown in Distler et al. (1999, Fig. 5) and in Cabri et al. (2002, Fig. 1). Samples from the drill hole MN-2 (see for details Malitch et al., 2012, Fig. 1), comprise the PGE-rich sulfide assemblage from a LS horizon (i.e., N1-3) in the upper part of the intrusion, whereas disseminated ores came from olivine gabbro (N1-6), plagiowehrlite (N1-7), underlying taxitic-textured rocks (N1-8 and N1-9), and contact gabbro (N1-10). In the drill hole MN-2 through the Noril'sk-1 intrusion the top and bottom parts cut the Syverminsky (T_1sv) and subalkaline Ivakinsky suite (P_2iv) basalts respectively. Several sulfide assemblages characterize the studied samples at Noril'sk-1, comprising pentlandite (Pn), millerite (Mil), chalcopyrite (Cp), and pyrite (Py) from the low sulfide horizon (sample N1-3), followed by assemblages $Po + Cp + Pn$ in olivine gabbro (samples KN97-1 and N1-6), $Po + Pn + Cp$ in plagiowehrlite (samples KN97-2 and N1-7) and $Po + Pn + Cp + Py$ in taxitic textured rocks (KN97-3; N1-8 and N1-9) and contact gabbro (N1-10) at the bottom part of intrusion.

Other samples of disseminated Ni-Cu-PGE sulfide ore, composed of $Po + Pn + Cp$, were sampled from the subeconomic Chernogorsk (drill hole MP-2bis, Fig. 2), Imangda (drill hole KP-4, Fig. 2), Vologochan (drill hole OV-29, Fig. 2; see for details Fig. 3, page 40 in Malitch et al. (2010a) and Zub-Marksheider (drill hole MP-27, Fig. 2; see for details Fig. 4, page 177 in Malitch et al. (2014b) intrusions. They were sampled from plagiowehrlite, melanotroctolite, gabbro-troctolite, and olivine gabbro in the bottom parts of the intrusions. The remaining samples of sulfide mineralization came from the noneconomic Zelyonaya Griva and Nizhny Talnakh intrusions (drill holes F-233 and TG-31, Fig. 2; see for details Fig. 4, page 41 in Malitch et al., 2010a). At Nizhny Talnakh sulfide minerals were sampled from (a) olivine gabbro (sample 31-1, assemblage $Po + Pn + Cp$) in the upper part of the intrusion, while the main suite of samples were derived from melanotroctolite (samples 31-3, 31-11, and 31-16) and plagiowehrlite (31-9, 31-10, and 31-13) from both the middle and bottom parts of the intrusion. Sulfide assemblages are dominated either by $Tr + Pn + Cp$ (samples 31-3 and 31-16) or $Tr + Pn + Cp + Cub$ (31-9, 31-10, 31-11, and 31-13). Further details on the lithologies with sulfide mineralization are given in Malitch et al. (2010a).

MAGMATIC SULFIDE AND FE-TI OXIDE DEPOSITS ASSOCIATED WITH MAFIC-ULTRAMAFIC INTRUSIONS IN CHINA

Kwan-Nang Pang[1] and J. Gregory Shellnutt[2]

[1]Academia Sinica, Taipei, Taiwan [2]National Taiwan Normal University, Taipei, Taiwan

CHAPTER OUTLINE

8.1 INTRODUCTION

Magmatic ore deposits are important sources of many FRTEs and in some cases PGEs. The link between these deposits and mantle-derived magmas stems from the fact that the abundances of these elements in mantle rocks are in general much higher than in rocks present in the continental crust. A mass transfer process in the form of magmatism is the first step leading to the potential formation of magmatic ore deposits. Another important perspective is that ore minerals such as sulfide and oxide phases do not generally form in amounts large enough for mining when mantle-derived magmas solidify and freeze in volcanic environments. Magma chamber processes associated with intrusive facies of

Processes and Ore Deposits of Ultramafic-Mafic Magmas through Space and Time. DOI: http://dx.doi.org/10.1016/B978-0-12-811159-8.00009-3

magmatism play an important role in the concentration of ore minerals or immiscible liquids from which these minerals solidified. However, the picture is not so simple. Not all mantle-derived magmas form ore deposits. In many cases, it is very common that a given phase of magmatism generates both ore-bearing and ore-barren intrusions. Undoubtedly, magmatic ore deposits must owe their existence to a combination of factors favorable to ore genesis. The important question is: what are these factors?

In order to identify the salient metallogenic factors, it is logical to perform research on existing magmatic ore deposits. Over the past two decades, numerous investigations were initiated on ore deposits associated with mafic-ultramafic rocks in China, in parallel with rapid economic growth. This chapter provides an overview of the current knowledge with respect to Ni-Cu-(PGE) sulfide and Fe-Ti-(V)-(P) oxide deposits in China, focusing on those that originated from mafic-ultramafic magmas. Its scope is thus different from earlier reviews focused on regional tectonics and metallogeny in a broader context (e.g., Pirajno et al., 2009; Pirajno, 2013; Pirajno and Santosh, 2014), for a specific type of magmatic ore deposit (Zhou et al., 2002d), for a localized region (Munteanu et al., 2013), or a combination of the latter two (Pang et al., 2010). This review begins with background information on the geology of China and brief accounts of the major phases of magmatism with which the magmatic ore deposits are associated. What follows are brief descriptions of individual ore deposits and related research work.

8.2 GEOLOGICAL BACKGROUND

China contains three major Precambrian blocks separated and sutured by Phanerozoic orogenic belts (Fig. 8.1). The NCC (or the synonymous equivalents North China Block, Sino-Korean Craton) and the Tarim craton are bounded to the north by the CAOB and to the south by the orogenic belts of Kunlun, Qilianshan, Qinling-Dabie, and Su-Lu. The SCC, consisting of the Yangtze Block to the north and the Cathaysia Block to the south, is bounded to the north by the Qinling-Dabie and Su-Lu orogenic belts, and to the west by the Longmenshan and Ailaoshan-Song Ma fold belts. The Tibetan plateau is a topographic high bounded to the north by the Kunlun orogenic belt and to the south by the Himalayan fold belt. The plateau has been assembled by collision of a series of Gondwana-derived terranes during the opening and closure of the Tethys Ocean during the Phanerozoic (Zhu et al., 2013).

8.3 BASALTIC MAGMATISM AND METALLOGENY
8.3.1 MAGMATISM ASSOCIATED WITH THE XIONG'ER LIP

The Late Paleoproterozoic Xiong'er LIP occurs in the southern margin of the NCC. It consists of the Xiong'er volcanic province and the North China giant mafic dyke swarm, which are coeval at ~ 1.78 Ga (Peng et al., 2007, 2008). The volcanic sequence ranges in thickness from 3 to 7 km in places, cropping out in a region of $>6000\, km^2$. It consists of basaltic andesite, andesite, trachyandesite, dacite, rhyodacite, and rhyolite, with continental sedimentary intercalations and pyroclastic units. The North China giant mafic dyke swarm is widespread in the central NCC, with probable extensions in the eastern and western parts of the craton. The dyke swarm has an areal extent of at least $0.1 \times 10^5\, km^2$. The dykes vary in composition from basaltic through andesitic to dacitic, although basaltic compositions dominate. The dyke swarm occurs on the largest scale in the

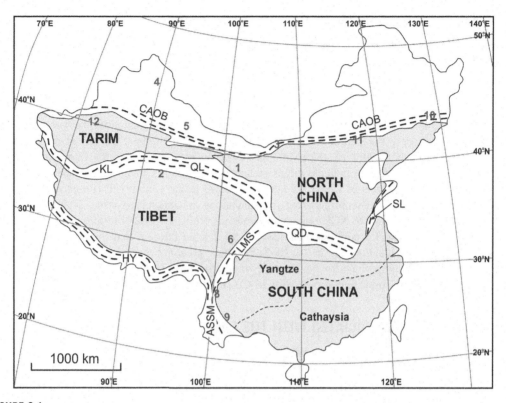

FIGURE 8.1

Simplified map of China illustrating the major tectonic domains and major fold and thrust belts between them (modified after Zhao, G., Wilde, S.A., Cawood, P.A., Sun, M., 2001. Archean blocks and their boundaries in the North China Craton: lithological, geochemical, structural and P-T path constraints and tectonic evolution. Precamb. Res. 107, 45−73 (Zhao et al., 2001)). Geographic locations of the major magmatic ore deposits are labeled as follows: (1) Jinchuan, (2) Xiarihamu, (3) Erbutu, (4) Kalatongke, (5) Huangshandong, Huangshanxi, and Poyi, (6) Yangliuping, (7) Panzhihua, Hongge, Baima, Taihe, and Limahe, (8) Jinbaoshan, (9) Baimazhai, (10) Hongqiling, (11) Damiao, (12) Wajilitag and Puchang. *KL*, Kunlun; *QL*, Qilianshan; *QD*, Qinling-Dabie; *SL*, Sulu; *LMS*, Longmenshan; *ASSM*, Ailaoshan-Song Ma; *HY*, Himalayan.

Taihang mountains, extending mainly NNW for over 1000 km. The dykes have individual widths up to 100 m and some of them can be traced continuously along strike for about 60 km (Peng et al., 2007). They are characterized by enrichments of Fe, Ti, and P. Petrologic and geochemical studies established a cogenetic relationship between the Xiong'er volcanic province and the North China giant mafic dyke swarm, which together define a LIP of $>0.1 \times 10^5$ km^3 in magmatic volume. The compositional differences between them have been ascribed to large-scale immiscibility between iron-rich and silica-rich liquids (Peng et al., 2015). The ~ 1.7 Ga Damiao anorthosite complex and associated Fe-Ti-(P) oxide ores in the eastern NCC might have been related to this phase of intracontinental magmatism (e.g., Zhou et al., 2002d).

8.3.2 NEOPROTEROZOIC MAGMATISM IN SOUTH CHINA

Neoproterozoic magmatic activity was widespread in South China. Metamorphic core complexes, granitic plutons and mafic-ultramafic complexes along the western and northwestern margin of the Yangtze block have been dated by the SHRIMP zircon U-Pb technique at 865 to 760 Ma (Zhou et al., 2002c; Wang et al., 2016). The age results were taken to indicate a period of arc magmatism (i.e., the Hannan arc) generated by subduction beneath the western and northern margins of the Yangtze block. On the other hand, widespread bimodal magmatism was active between 830 and 795 Ma within not only the peripheral but also the central part of the SCC (Li et al., 2003a,b, 2005a,b). These, together with the ~ 827 Ma mafic dyke swarm identified in the Guangxi Province (Li et al., 1999), were taken to indicate the existence of a ~ 825 Ma mantle plume beneath South China, the magmatic products of which are later referred to as the Guibei LIP (Pirajno, 2013). The dual tectonic setting of the Neoproterozoic magmatism is an unresolved issue. Nonetheless, the Jinchuan deposit in the western NCC is contemporaneous with the ~ 825 Ma magmatism in South China, and might represent a common magmatic event despite the fact that they occupy different present-day geographic locations (Li et al., 2005a,b). Other potential magmatic ore systems include the ~ 780 Ma Bijigou intrusion with Fe-Ti oxide mineralization (Zhou et al., 2002a; Wang et al., 2016) and the Tongde-Gaojiacun complex with Ni-Cu-(PGE) mineralization.

8.3.3 MAGMATISM ASSOCIATED WITH THE CAOB

The CAOB is a collage of terranes bounded by the Siberian craton in the north, the Tarim craton and NCC in the south, the Urals in the west, and the Sea of Japan in the east (Jahn, 2004; Windley et al., 2007; Xiao and Santosh, 2014; Kröner et al., 2014). It is considered to be one of the most significant sites of crustal generation during the Phanerozoic but there is considerable debate on the relative proportions of newly created juvenile crust and reworked crust (Jahn, 2004; Kröner et al., 2014). Accretion of island arcs, ophiolites and microcontinents probably began during the Mesoproterozoic at about 1000 Ma and ended during the Late Paleozoic to Early Mesozoic at about 250 Ma. The CAOB was widely affected by postorogenic magmatism following terrane accretion (Kröner et al., 2014; Dostal et al., 2015). In the CAOB, granitic plutons with strong mantle isotopic fingerprints predominate over subordinate mafic magmatic rocks (Jahn, 2004). Magmatic Ni-Cu-(PGE) sulfide deposits are known to occur in the Chinese part of the CAOB, in association with either subduction-related or postorogenic magmatism. For example, those at Erbutu and Xiarihamu probably are related to subduction-related, boninitic magmatism (Peng et al., 2013; Li et al., 2015a,b) whereas those at Kalatongke, Huangshan and many others might be related to extension-related, basaltic magmatism (Zhou et al., 2002d, 2004; Song and Li, 2009). Recently, some researchers suggested that a ~ 275 Ma mantle plume might have affected not only the Tarim basin but also the Beishan-Tianshan region in the CAOB (Zhang et al., 2010a,c), and might be responsible for the genesis of some sulfide-bearing intrusions therein (Qin et al., 2011; Su et al., 2011). However, others maintained that a nonplume model is more applicable to explain the protracted basaltic magmatism in the region over about 20 Myr (Xue et al., 2016a,b).

8.3.4 MAGMATISM ASSOCIATED WITH THE TARIM LIP

The Permian Tarim LIP in the western Xinjiang province, NW China, consists of diverse lithologies, including kimberlites, lamprophyres, flood basalts, mafic-ultramafic intrusions, bimodal dyke swarms, granitoids, rhyolites, and pyroclastic rocks, spanning a period of about 20 to 25 Myr. The estimated areal extent of these magmatic products exceeds 2.5×10^6 km^2. Perovskite and baddeleyite from kimberlitic intrusions in the Wajilitag area yield U-Pb ages of ~ 300 Ma, signifying the onset of the Tarim LIP at the Carboniferous-Permian boundary (Zhang et al., 2013). Flood basalts with OIB-like signatures were reported in the Keping and Bachu area and in drillcores elsewhere in the central-western part of the Tarim basin (Zhou et al., 2009; Zhang et al., 2010d; Yu et al., 2011). Radiometric dating by ^{40}Ar/^{39}Ar method of the basaltic rocks yields a range of eruptive ages from about 285 Ma to 262 Ma (Zhang et al., 2010d). The basalts also contain zircon crystals dated by SHRIMP at ~ 290 Ma, an age generally taken as indicative of the eruption of the flood basalts (Yu et al., 2011). Recently, some researchers suggested that the Tarim LIP might have had another pulse of magmatic activity at ~ 280 Ma, affecting not only the Bachu region but also the eastern Tianshan terrane in the CAOB (Zhang et al., 2010a,b). A plume incubation model was proposed to account for the spatial-temporal distribution of magmatic rocks and the magmatic longevity of the Tarim LIP (Xu et al., 2014). Magmatic Fe-Ti-(V) oxide ores were reported in mafic-ultramafic intrusions in the Bachu region (Li et al., 2012; Cao et al., 2014; Zhang et al., 2014, 2016a,b).

8.3.5 MAGMATISM ASSOCIATED WITH THE EMEISHAN LIP

The Late Permian Emeishan LIP is situated in the provinces of Sichuan, Guizhou and Yunnan, SW China, with displaced units in northern Vietnam. Tectonically, it occupies a region between the western margin of the Yangtze Block and the eastern margin of the Tibetan plateau (Ali et al., 2005; Shellnutt, 2014). Flood basalts are the dominant rock type of the Emeishan LIP with erosional remnants covering an area of about 3×10^5 km^2. Small volumes of picritic and silicic lavas are reported in some of the volcanic sections (Chung and Jahn, 1995; Xu et al., 2001, 2010; Zhang et al., 2006; Anh et al., 2011). The basalts and picrites are divided into low-Ti and high-Ti affinities, which were considered to represent derivation from compositionally distinct mantle sources (Xu et al., 2001; Xiao et al., 2004; Kamenetsky et al., 2012). However, later studies revealed that the two groups of basalts are not separated by a clear compositional gap and rocks with intermediate Ti contents also exist (Shellnutt and Jahn, 2011; Hou et al., 2011a,b). The Emeishan LIP is divided into three concentric zones: the inner zone, the middle zone, and the outer zone, which probably correspond to decreasing crustal thickness. The inner zone is considered to be where the fossilized plume head was emplaced (Xu et al., 2004), and where the intrusive facies of the LIP, in the form of mafic-ultramafic intrusions and granitoids, is exposed. There is a diverse spectrum of peraluminous, metaluminous, and peralkaline A-type granitoids with different proposed origins (Shellnutt and Zhou, 2007; Shellnutt et al., 2011). The mafic-ultramafic intrusions host several magmatic Ni-Cu-(PGE) sulfide and Fe-Ti-(V) oxide deposits (Zhou et al., 2002b,d, 2008; Pang et al., 2010; Munteanu et al., 2013).

8.4 GEOLOGY AND RECENT WORK ON INDIVIDUAL DEPOSITS

8.4.1 NI-CU-(PGE) SULFIDE DEPOSITS

8.4.1.1 Jinchuan

The Jinchuan Ni-Cu-(PGE) sulfide deposit in northern central China is the third largest magmatic sulfide deposit in the world (Fig. 8.2). The deposit is hosted in a 6 km-long, 1 km-deep, and

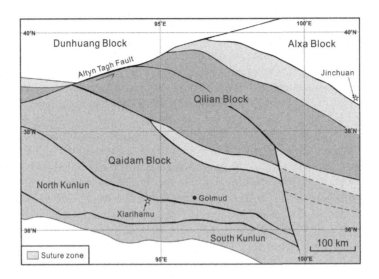

FIGURE 8.2

Schematic map of the Kunlun-Qilianshan orogenic belt illustrating the geographic locations of the Jinchuan and Xiarihamu magmatic sulfide deposits.

After Zhang, Z., Tang, Q., Li, C., Wang, Y., Ripley, E.M., 2016b. Sr-Nd-Os-S isotope and PGE geochemistry of the Xiarihamu magmatic sulfide deposit in the Qinghai-Tibet plateau, China. Mineral. Deposita, doi:10.1007/s00126-016-0645-0; Zhang, D., Zhang, Z., Mao, J., Huang, H., Cheng, Z., 2016a. Zircon U-Pb ages and Hf-O isotopic signatures of the Wajilitag and Puchang Fe-Ti oxide-bearing intrusive complexes: Constraints on their source characteristics and temporal-spatial evolution of the Tarim large igneous province. Gond. Res. 37, 71–85.

400 m-wide ultramafic intrusion emplaced in Proterozoic granitic migmatite, gneiss, and marble. The intrusion consists, in decreasing order of abundance, of lherzolite, dunite, wehrlite, harzburgite, and troctolite. More than a hundred sulfide orebodies have been identified and the largest one exists in the central part of the intrusion at about 300 m below ground surface. Sulfide ores are either disseminated or net-textured; massive ores are very rare.

The Jinchuan deposit has been extensively studied from various perspectives, including radiometric dating (Li et al., 2005a,b; Yang et al., 2008; Zhang et al., 2010d), geochemistry and ore genesis (Chai and Naldrett, 1992a,b; Song et al., 2006, 2011, 2012; Lehmann et al., 2007; Su et al., 2008; Chen et al., 2013, 2015a,b), chrome spinel chemistry (Barnes and Tang, 1999), platinum-group minerals (Prichard et al., 2013), S-O-H isotopes (Ripley et al., 2005), multiple S isotopes (Duan et al., 2016), and noble gas isotopes (Zhang et al., 2013). SHRIMP zircon and baddeleyite U-Pb ages of ~825 Ma obtained for ultramafic rocks of the Jinchuan intrusion and dolerite dykes cutting the intrusion by Li et al. (2005a,b) were later refined to 831.8 ± 0.6 Ma by Zhang et al. (2010d) using ID-TIMS. Yang et al. (2008) obtained highly variable Re-Os ages for the Jinchuan sulfides and attributed the inconsistency to postsegregation diffusion of Os. Most geochemical investigations indicate that the Jinchuan intrusion represents a magma conduit system with multiple events of sulfide segregation, presumably related to one or more deep-seated staging chambers.

A strong role of crustal contamination is indicated by trace element systematics and Sr-Nd-Hf isotopes (Li et al., 2005a,b; Song et al., 2006; Lehmann et al., 2007; Duan et al., 2016). Other studies point to additional complexities of the magma plumbing system, including the probable involvement of both PGE-undepleted and PGE-depleted magmas (Chen et al., 2013), and the recognition that different subchambers of the intrusion might not have been originally connected (Song et al., 2011). It was also proposed that fractionation of sulfide liquid occurred prior to segregation from silicate magmas, as indicated by PGE data obtained by bulk and in situ geochemical analyses of sulfide ores (Su et al., 2008; Chen et al., 2015a,b). The Jinchuan deposit underwent strong hydrothermal alteration, as shown by stable isotopic and noble gas isotopic data (Ripley et al., 2005; Zhang et al., 2013). Sulfur isotopic data by Ripley et al. (2005) did not indicate extensive assimilation of S from the country rocks, but a recent study of multiple S isotopes identified anomalous $\Delta^{33}S$ that was ascribed to the involvement of external S from sedimentary rocks.

8.4.1.2 Xiarihamu

The Xiarihamu Ni-Co sulfide deposit is a magmatic sulfide deposit discovered in the East Kunlun orogenic belt in recent years and is the second largest Ni deposit in China (Fig. 8.2). The deposit is hosted in an oval-shaped mafic-ultramafic intrusion 2 $\times$ 1.4 km in plan. It intruded Proterozoic granitic gneiss, schist and marble. The intrusion is composed of a weakly mineralized gabbroic portion and an ultramafic portion where sulfide ores are present. No chilled margins or grain-size variation are present.

Li et al. (2015a,b) presented the first age and geochemical study of the Xiarihamu deposit, showing that ultramafic rocks therein might have been emplaced about 20 Myr earlier than gabbroic rocks. However, Song et al. (2016) provided age data for one gabbronorite and two websterites from the deposit and they yield similar Early Devonian ages of $\sim$406 Ma. A genetic relationship of the deposit to subduction-related, boninitic magmatism was established based on olivine- and bulk-rock geochemical data (Li et al., 2015a,b). Low PGE tenors of rocks from the Xiarihamu intrusion were attributed to sulfide retention in the mantle source with or without low-pressure sulfide segregation (Song et al., 2016; Zhang et al., 2016a,b). Crustal contamination of the parental magma was suggested to be minor (Li et al., 2015a,b), but subsequently revised to 5−30% on the basis of Sr-Nd-Os isotopic data (Zhang et al., 2016a,b). Sulfur-isotope data ($\delta^{34}S$ = 2.2 to 6.8‰) point to involvement of external S in the genesis of the sulfide ores (Li et al., 2015a,b; Zhang et al., 2016a,b).

8.4.1.3 Erbutu

The Erbutu Ni-Cu sulfide deposit occurs in the eastern CAOB, close to the border between China and Mongolia (Fig. 8.3). The deposit is hosted in a small ultramafic intrusion with surface exposure of less than 200 m across and depth of about 200 m. The intrusion was emplaced in metavolcanic and metasedimentary rocks. A chilled margin of dolerite and a hornfels zone were identified in the eastern part of the intrusion. Major rock types are olivine orthopyroxenite and orthopyroxenite within which disseminated sulfide mineralization is identified. Currently, data on tonnage and grade of ore reserves have not been available to the public.

Peng et al. (2013) undertook a geochronological and geochemical investigation of the Erbutu deposit. Zircon grains separated from an orthopyroxenite sample yield a crystallization age of 294 $\pm$ 2.7 Ma. Geochemical and S isotopic data indicate that crustal contamination and external S

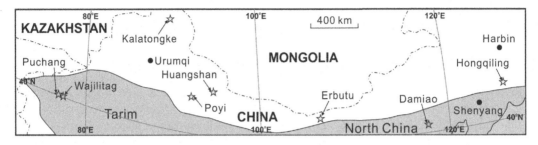

FIGURE 8.3

Schematic map of the northern part of North China craton and the CAOB illustrating the geographic locations of the magmatic ore deposits.

were involved during magma evolution and ore genesis. Collectively, the authors suggested that the deposit represents a magmatic sulfide deposit related to boninitic magmatism in an arc setting.

8.4.1.4 Kalatongke

The Kalatongke (also known as Karatungk) Ni-Cu sulfide deposit is located about 380 km north of Urumqi in the central CAOB (Fig. 8.3). The deposit is hosted in a mafic intrusion emplaced in Lower Carboniferous tuffaceous rocks. It is divided into three segments with only one of them exposed, and many small outcrops in the region might be connected to these segments at depth. Major rock types are norite, troctolite, gabbro, and diorite. Massive and disseminated sulfide ores occur in norites and gabbros, and in some cases in the country rocks.

The geochemistry of the Kalatongke deposit was investigated by Zhang et al. (2009a,b), Song and Li (2009) and Gao et al. (2012a). Depleted Sr-Nd isotopes and elevated $\delta^{18}O$, coupled with trace element indicators, were taken by Zhang et al. (2009a,b) to indicate significant crustal contamination of an asthenosphere-derived magma in the formation of the intrusion. It was also proposed that flow differentiation concentrated dense sulfide melt and olivine crystals in the center of the intrusion. Song and Li (2009) emphasized that the parental magma of the Kalatongke intrusion was derived from metasomatized mantle composed of garnet lherzolite, and the PGE-depleted nature of the parental magma indicated earlier S saturation and sulfide segregation. Gao et al. (2012a) proposed, based on additional geochemical data for different lithologies, a revised model involving two pulses of PGE-poor basaltic magmas that underwent different degrees of crustal contamination and fractional crystallization. The Os isotopes of the intrusion were investigated by Gao et al. (2012b) and Qu et al. (2013). The former ascribed the isotopic variability observed between massive and disseminated sulfide ores to different degrees of crustal contamination in magma pulses, and the latter to postsegregation diffusion of Os, in a manner similar to that proposed for the Jinchuan deposit.

8.4.1.5 Huangshan

Magmatic Ni-Cu-(PGE) sulfide ores occur in various mafic-ultramafic intrusions in the Tianshan-Beishan region the Huangshan district (Zhou et al., 2002d, 2004; Qin et al., 2011; Su et al., 2013)

(Fig. 8.3). The sulfide ores are hosted mainly in two small mafic-ultramafic bodies, known as Huangshandong and Huangshanxi, about 10 km apart along the Huangshan Fault, but many intrusions with subeconomic sulfide mineralization exist in the region. The Huangshandong intrusion is a lensoidal intrusive body with about 3.5 km $\times$ 1.2 km (2.8 km^2) in size, intruding Mesoproterozoic schist and amphibolite. The intrusion can be divided into a massive gabbronorite unit, a layered sequence and a marginal diorite zone surrounding the layered sequence in places. The layered sequence is composed of two or three upward-concave layers of ultramafic rocks overlain by olivine gabbro, gabbro, and gabbrodiorite. The ultramafic rocks are lherzolite, wehrlite, and olivine websterlite. Disseminated sulfides and net-textured sulfide ores are associated with ultramafic rocks of the layered sequence and the massive gabbronorite unit, but rarely with the gabbroic rocks of the layered sequence. The Huangshanxi intrusion is a funnel-shaped body about 3.8 km $\times$ 800 m. It is a multiphase intrusion consisting of at least three separate intrusive units formed by distinct pulses of magma. The oldest is an ultramafic unit composed of a sheet-like body of plagioclase lherzolite and a dyke-like body of gabbronorite. A layered ultramafic-mafic unit dominated by gabbro was emplaced after the ultramafic unit. The youngest unit is a dyke-like, layered ultramafic body in the western part of the intrusion. Economic disseminated sulfide mineralization occurs at the base of this ultramafic body.

Ages and geochemical and Sr-Nd isotopic data of the Huangshandong and Huangshanxi intrusion were presented by Zhou et al. (2004). It is suggested that the parental magma of the intrusion was derived from a mantle source previously contaminated by subduction and variably contaminated by continental upper-crustal rocks. The intrusion presumably formed by plume-related activity in a continental setting, and thus probably has a postorogenic origin. The PGE and Sr-Nd-Hf-Pb isotopic data documented by Sun et al. (2013) for the Huangshandong rocks are consistent with mixing between a depleted mantle-derived magma and a granitic melt formed by partial melting of juvenile arc crust. It is also argued that the parental magma was PGE-depleted and contained less than 7 wt% MgO, due to sulfide segregation in a staging chamber. Gao et al. (2013) identified oxide-rich and oxide-poor sulfide mineralization in the Huangshandong intrusion based on PGE concentrations and ascribed the two types of mineralization to the amount of sulfides removed in the first S-saturation event. Mao et al. (2014) documented mineral compositions and whole-rock geochemical and PGE data for rocks of the Huangshanxi intrusion. Based on these data, these authors proposed a two-stage model for sulfide ore formation: the first stage during which immiscible sulfide droplets and olivine crystals were brought up by an ascending magma and a second stage of in situ differentiation where the sulfide droplets and olivine crystals settled to form the sulfide-olivine cumulate zones at the base of the magmatic system.

8.4.1.6 Poyi

The Poyi Ni-Cu sulfide deposit occurs in the Beishan terrane of the CAOB (Fig. 8.3). It is hosted in an ultramafic body in the Pobei mafic-ultramafic complex that intruded in Proterozoic schists, gneiss, and gneissic granite. The Poyi intrusion is a pipe-like body with surface exposure of about 3 km^2, extending vertically for more than 2.4 km. Its wall is highly irregular and discordant with the gently dipping modal layering in the gabbroic host rocks of the Pobei complex. Chilled margins were not observed along the contact zone. Major rock types of the intrusion include dunite, lherzolite, wehrlites, troctolite, olivine websterite, and olivine clinopyroxenite. Sulfide mineralization is dominated by disseminated sulfides occurring as concordant layers or zones within the steeply-dipping lherzolite layers. The sulfide assemblages are composed of pyrrhotite, pentlandite, and chalcopyrite.

The geochemistry of the Poyi deposit was studied by Xia et al. (2013) and Xue et al. (2016a,b). Rocks of the Poyi intrusion display an arc-like signature and depleted mantle-like Sr-Nd isotopic compositions, which were collectively taken by Xia et al. (2013) to indicate involvement of a subduction-modified lithospheric mantle or a granitic melt derived from a juvenile arc crust in magma genesis. A role played by external S in sulfide segregation is hinted at by differences in Os isotopes between sulfide-poor samples and sulfide separates, but this is inconsistent with mantle-like S isotopes. Xue et al. (2016a,b) documented additional PGE data to argue that the parental magma for the deposit was PGE-depleted as a result of sulfide segregation at depth. It was also suggested that the deposit represents a dynamic conduit used by multiple pulses of olivine- and sulfide-charged magma.

8.4.1.7 Yangliuping

The Yangliuping Ni-Cu-(PGE) sulfide deposit is about 200 km west of Chengdu (Fig. 8.4). It is hosted in a series of mafic-ultramafic sills that are considered to be part of the Emeishan LIP. The sills have variable sizes but are typically about 2 km long and 300 m thick, and intruded Devonian marble and graphitic schist. The sills underwent substantial postemplacement alteration and meta-morphism and now consist of serpentinite, talc schist, tremolite schist, and metagabbro. Disseminated sulfides occur in the serpentinite in the lower parts of the sills. Massive sulfide ores are located at the base of the sills and in the footwall along fractures beneath the mineralized ser-pentinite. Flood basalts of the Emeishan LIP also crop out widely in the region.

The elemental and PGE systematics of the Yangliuping deposit and associated basalts were documented by Song et al. (2003). Geochemical data demonstrate a genetic link between the mineralized sills and the basalts that are related to a common parental magma. It is suggested that sulfide immiscibility was triggered by fractional crystallization and introduction of S and CO_2 by wallrock assimilation and/or contamination. It was also proposed that the sills acted as conduits for the flood basalts with the sulfide liquid, along with early-crystallizing olivine and pyroxene, segregating from the magma as it passed through the conduits prior to eruption. Song et al. (2004) studied the relations between base-metal sulfides and PGMs of the Yangliuping deposit. The major finding is that the PGMs and cobaltite-gersdorffite solid solution have mag-matic origin, and were precipitated from residual sulfide melt after crystallization of monosulfide solid solution. Song et al. (2006) carried out a detailed geochemical study of a flood basalt sequence in the region and found that the basalts are variably depleted in PGEs. It was proposed that S-undersaturated primary magmas were processed in lower or upper crustal staging chambers and that the Middle Unit of the sequence that shows strong PGE depletion is complementary to the mineralized sills of the Yangliuping deposit.

8.4.1.8 Limahe

The Limahe Ni-Cu sulfide deposit is hosted in a small, funnel-shaped mafic-ultramafic intrusion, part of the intrusive facies of the Emeishan LIP (Fig. 8.4). The intrusion crops out in an area of about 900 m × 180 m, extending about 200 m below ground surface. The country rocks are Proterozoic pyrite-bearing quartzites, graphitic slates, and siliceous limestones. The intrusion consists of a mafic unit, which accounts for two third of its volume, and an ultramafic unit. The contacts between these

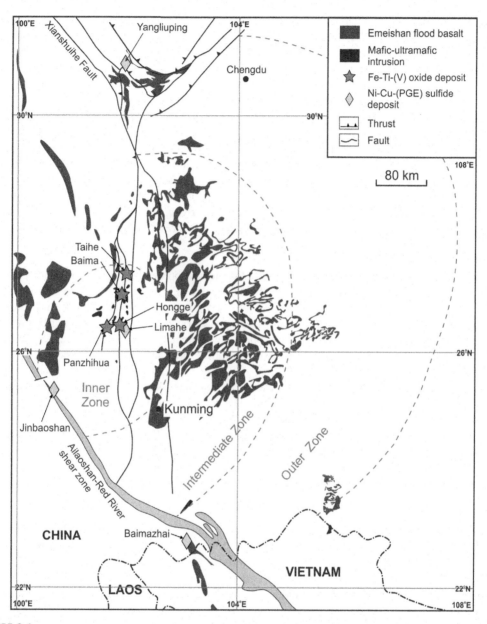

FIGURE 8.4

Schematic map of the Emeishan LIP illustrating the geographic locations of the magmatic ore deposits.

two units are sharp but no obvious chilled margins between them were documented. The mafic unit comprises a diorite zone at the top in gradational contact against a gabbro zone at the bottom. The ultramafic unit comprises wehrlite and olivine websterite with gradational contact between them. In many places the ultramafic unit has fault contacts with the country rocks. Sulfide mineralization, ads disseminated, net-textured and massive sulfides, is restricted to the ultramafic unit. Several massive sulfide lenses occur along the contacts with sedimentary country rocks.

Geochemical, PGE, and Sr-Nd-S isotopic data for the Limahe intrusion were presented by Tao et al. (2008). These data indicate that (i) the intrusion was derived from picritic magma by olivine fractionation and contamination in a staging chamber at mid-crustal levels, (ii) at least two separate pulses of magma were involved in the development of the intrusion, and (iii) sulfide mineralization is related to second-stage sulfide segregation after the fractionated magmas acquired external S from pyrite-bearing country rocks during magma ascent. It was proposed that the Limahe intrusion was once a wider part of a dynamic conduit that fed magma to the overlying subvolcanic dykes, sills, or lavas. These findings are consistent with Os-isotope data subsequently presented by Tao et al. (2010).

8.4.1.9 Baimazhai

The Baimazhai Ni-Cu-(PGE) sulfide deposit is close to the border between China and Vietnam, southwest of the Ailaoshan-Red River strike-slip fault zone (Fig. 8.4). The deposit is hosted in a small mafic−ultramafic intrusion about 530×190 m and $24-64$ m thick that intruded Ordovician sandstone and slate. The intrusion is a concentric body with an orthopyroxenite core surrounded successively by websterite and gabbro. Disseminated sulfides are present in websterites and gabbros whereas massive and net-textured sulfide ores are restricted to the orthopyroxenite core.

The geochemistry of the Baimazhai intrusion and associated sulfide ores were investigated by Wang et al. (2006) and Wang and Zhou (2006). The results indicate that all sulfides in the rocks formed in a single S-saturation event, and that these rocks formed from a magma contaminated by crustal rocks in a staging chamber. It was also suggested that flow differentiation preferentially concentrated dense sulfide melt and orthopyroxene in the central part of the intrusion. Low PGE concentrations and general absence of PGMs in the massive ores were ascribed to a relatively large mass fraction of the sulfide melts (Wang and Zhou, 2006).

8.4.1.10 Jinbaoshan

The Jinbaoshan Pt-Pd deposit is hosted in an ultramafic intrusion about 50 km southeast from Dali (Fig. 8.4). The intrusion is about 5×1 km and $20-150$ m thick, and is part of the intrusive facies of the Emeishan LIP. It is a sheet-like body that intruded Devonian dolomite which unconformably overlies Proterozoic metamorphic rocks. Marble hornfels locally occurs close to the contact with dolomites. The Jinbaoshan intrusion is composed essentially of wehrlite with minor hornblende pyroxenites at its margins. Gabbroic rocks are found in the upper portions of the intrusion but their relationship to the ultramafic rocks remains unclear. Discontinuous layers of disseminated sulfides, with thickness ranging from several centimeters to several meters, were found to be rich in Pt and Pd. These PGE-rich horizons are transitional to their host rocks.

The mineralogy and geochemistry of the Jinbaoshan intrusion were investigated by Tao et al. (2007). No conclusive evidence was found for crustal contamination based on S, Nd, and Os isotopes; this finding led to the proposal that sulfide saturation occurred at depth and immiscible sulfide droplets were transported by ascending magma to accumulate in the Jinbaoshan conduit, where they reacted with new magma passing through the conduit to achieve high PGE concentrations. It is also suggested that the wehrlites represent residual assemblages formed by dissolution of plagioclase by passing magmas in a dynamic conduit. Wang et al. (2008) studied the mineralogy of the deposit in detail and concluded that whereas the intrusion underwent extensive hydrothermal alteration, enrichment in PGE is a primary magmatic feature. The PGE data presented by Wang et al. (2010), in addition to those by Tao et al. (2007), do not alter the existing model that the Jinbaoshan intrusion represents an open magma conduit system where PGE-rich magma formed by multistage dissolution-upgrading processes.

8.4.1.11 Hongqiling

Numerous mafic-ultramafic intrusions with magmatic sulfide ores or mineralization occur in the eastern Xing'an-Mongolian orogenic belt (Fig. 8.3). In the Hongqiling region, about 100 km southeast of Changchun, 33 intrusions have been identified; seven of them contain Ni-Cu sulfide mineralization and two of them, Daling and Fujia (or the Nos. 1 and 7 intrusions, respectively), host the most important sulfide ores. They represent the second largest Ni producer in China following the Jinchuan deposit described above. The deposits are hosted by a series of mafic-ultramafic complexes intruding Paleozoic gneiss, schist, and marble. The Daling intrusion is about 980 × 150−280 m in size and 560 m thick. It is composed mainly of lherzolite and minor olivine websterite, pyroxenite, gabbro, hornblendite, and leucogabbro. The contact relationships between different rock types are unclear. The Fujia intrusion is about 700 × 35 m and 600 m thick. It is composed mainly of orthopyroxenite and minor norite, gabbro, and lherzolite.

Wu et al. (2004) undertook a geochronological and geochemical study of the Hongqiling complex. Its magmatic age of ∼216 Ma postdates Early Triassic syn-orogenic granitic magmatism and regional metamorphism, implying that the complex is postorogenic in origin. Geochemical data indicate that most rocks of the complex are cumulates from a juvenile basaltic parental magma derived from the lithospheric mantle that underwent minor crustal contamination. These findings are in line with Os-isotope data reported by Lü et al. (2011) for sulfide ores from the complex, but a higher degree of crustal contamination is required by the data. Based on new PGE data obtained from the Fujia intrusion (No. 7), Wei et al. (2013) proposed that sulfide ores therein formed from a magma that became depleted in PGE due to crustal contamination in a staging chamber. It is proposed that the PGE-depleted magma became S-undersaturated again and a second sulfide segregation took place when it was emplaced at the Hongqiling chamber.

8.4.2 FE-TI-(V)-(P) OXIDE DEPOSITS

8.4.2.1 Damiao

The Damiao Fe-Ti-P oxide deposit occurs in the northern margin of the North China craton, about 150 km northeast of Beijing (Fig. 8.3). The deposit is hosted in an anorthosite complex with an exposed area of about 80−120 km^2 and is the only known massif-type anorthosite in China. The complex was emplaced into Late Archean high-grade metamorphic rocks and is unconformably overlain by Jurassic sedimentary rocks. Associated spatially and temporally with alkali granitoids and rapakivi granites, the complex is considered part of a typical anorthosite-mangerite-charnockite-granite suite. The Damiao anorthosite complex crops out in three bodies and major rock types include anorthosite, oxide-apatite gabbronorite, melanorite, leuconorite, mangerite, and troctolite. Different rock units of the complex are separated by intrusive contacts with anorthosite as the earliest unit. Massive Fe-Ti-(P) oxide ores are observed. At the margin of one body (i.e., West Body), anorthositic host rocks are cut by the gabbroic, ferrodioritic, and felsic dykes.

Models put forward to explain the petrogenesis of the Damiao anorthosite complex are diverse. Zhao et al. (2009) suggested that the parental magma of the Damiao complex formed from high-degree melting of ancient lower crust. It is also suggested that the magma evolved by polybaric crystallization and gave rise to the complex at mid-crustal depths. Chen et al. (2015a, b) favored a model involving melting of depleted mantle and significant crustal contamination in a deep-seated magma chamber. Teng and Santosh (2015) proposed that the parental magma was sourced from spinel-bearing subcontinental lithospheric mantle and assimilated Archean crustal components before ascent. Opinions regarding ore genesis are equally diverse. Li et al. (2015a, b) suggested that massive Fe-Ti-(P) oxide ores segregated from Fe-Ti-P-rich pyroxenites, which show rhythmic layering with "late-stage" anorthosites. He et al. (2016) proposed that oxide ores and mangerites are the crystallized products of immiscible Fe-rich and Si-rich melts, respectively. Their interpretations rely heavily on structures such as irregular dykes or veins of oxide ores cutting anorthosites and gabbronorites. These structures, however, were taken by Li et al. (2014) as reflecting hydrothermal remobilization. A recent experimental study by Wang et al. (2017) indicates that nelsonites in the Damiao complex do not melt at 1200 °C, implying that the Fe-Ti-(P) oxide ores do not represent frozen immiscible melt. Moreover, melt inclusions hosted in apatite, although spanning a wide range of major element compositions, were used as indicative of a hydrated late-stage immiscible Fe-rich melt formed by silicate liquid immiscibility.

8.4.2.2 Wajilitag

The Wajilitag Fe-Ti-(V) oxide deposit is hosted in a mafic-ultramafic intrusion near the city of Bachu in the northwestern part of the Tarim craton (Fig. 8.3). The intrusion is about 5 $\times$ 1.5$-$3 km and 110$-$270 m thick, with an exposed area of $\sim$15 km^2. It was emplaced in Devonian sedimentary rocks. Major rock types include olivine clinopyroxenite, clinopyroxenite, and gabbro. Oxide ores occur as lenses or pods associated with clinopyroxenitic rocks. Disseminated ores are dominant over massive ores, which were identified in boreholes. Most contacts between the disseminated ores and the adjacent silicate rocks are transitional, whereas most, if not all, massive ores always have sharp magmatic boundaries with the barren silicate rocks and disseminated ores. The mafic-ultramafic complex is associated with diorites and syenites, which probably represent separate intrusions based on contact relations.

The Wajilitag intrusion was dated by the zircon U-Pb method at $\sim$283 Ma (Zhang et al., 2016a,b), consistent with alkali mafic dykes cutting the clinopyroxenite unit that were dated at 281.4 $\pm$ 1.7 Ma, but older than the diorite unit dated at 275.2 $\pm$ 1.2 Ma (Zou et al., 2015). The mineralogy and geochemistry of the Wajilitag intrusion were investigated by Li et al. (2012), Cao et al. (2014), Zhang et al. (2014), Cao et al. (2016), and Zhang et al. (2016a,b). A general consensus is that an Fe-Ti-rich parental magma was involved in the petrogenesis of the intrusion. Moreover, magnetite included in olivine and clinopyroxene apparently has higher Cr contents than that not completely surrounded by these silicate phases, indicating crystallization of magnetite at two different stages (Li et al., 2012; Cao et al., 2014). The PGE-depleted nature of the Wajilitag rocks was ascribed to low-degree mantle melting (Cao et al., 2016) or earlier sulfide segregation (Zhang et al., 2014). Zhang et al. (2016a,b) proposed that the Wajilitag intrusion formed by partial melting of subcontinental lithospheric mantle refertilized by recycled subduction-related materials.

8.4.2.3 Piqiang

The Piqiang (also known as Puchang) deposit is about 100 km northwest of the Wajilitag deposit (Fig. 8.3). The Piqiang intrusion has an outcrop area of about 25 km^2 and intruded Lower Carboniferous calcareous slate and marble. Partially-assimilated marble enclaves are present in the contact zone. The complex consists mainly of gabbro and minor clinopyroxenite and anorthosite. The contacts between these rock types are gradational and no visible igneous layering or foliation is present therein. The Fe-Ti oxide ores are hosted exclusively in the gabbroic rocks. Orebodies mainly are irregular lenses, layers or pods, of which the layered bodies are the most common. The Fe-Ti oxide ores generally have sharp contacts with or within the barren gabbro. The massive and disseminated ores also have sharp contacts with each other.

The Piqiang intrusion was dated by zircon the U-Pb method at ~275 Ma (Zhang et al., 2016a,b). The 8 Myr magmatic gap between the formation of the Wajilitag and Piqiang deposits was taken to indicate that the two deposits represent discrete magmatic events at the waning stage of the Tarim LIP magmatism. Compared with the Wajilitag intrusion, the Piqiang intrusion exhibits lower zircon εHf values and higher bulk-rock δ^{18}O. Zhang et al. (2016a,b) considered these features as indicative of higher proportions of recycled subduction-related materials in the mantle source.

8.4.2.4 Panzhihua

The Panzhihua Fe-Ti oxide deposit is hosted in a 19 km-long and 2 km-thick layered gabbroic intrusion with exposed area of about 30 km^2 (Fig. 8.4). The intrusion was cut by faults into fault-bounded blocks, but remains largely intact, unmetamorphosed, and undeformed. It was emplaced into Neoproterozoic dolomitic limestone and resulted in the formation of a ~300-m-thick metamorphic aureole consisting of marble and skarns along its lower contact (Ganino et al., 2008). The intrusion is divided from the base upwards into the Marginal zone, Lower zone, Middle zone, and Upper zone. The Marginal zone is a hornblende-rich unit composed mainly of microgabbros, and is suggested to be the chilled base of the intrusion. The Lower zone and the lower part of the Middle zone consist of gabbroic, oxide gabbroic, and Fe-Ti oxide cumulates with varying amounts of olivine, clinopyroxene, plagioclase, and titanomagnetite (exsolved into magnetite and ilmenite). The upper part of the Middle zone consists of leucogabbroic cumulates with apatite and ilmenite in addition to the above cumulus assemblage and olivine is present throughout this zone. The Upper zone is dominated by gabbroic cumulates. Oxide ores occur as conformable lenses and layers, up to ~60-m-thick, in the Marginal zone, Lower zone, and the lower part of the Middle zone, either in sharp or gradational contacts against their host rocks.

The Panzhihua intrusion was dated by the SHRIMP zircon U-Pb method on a leucogabbro that yielded an age of 263 ± 3 Ma (Zhou et al., 2005). The mineralogy and geochemistry of the intrusion were studied by Zhou et al. (2005), Pang et al. (2008a,b), Pang et al. (2009), Zhang et al. (2009a,b), Pang et al. (2013), Song et al. (2013), and Howarth and Prevec (2013), yielding a diversity of views on the ore genesis. Zhou et al. (2005) proposed that the oxide ores formed by very late-stage crystallization of titanomagnetite from an immiscible oxide liquid. Pang et al. (2008a,b) and Pang et al. (2009) investigated the oxide- and silicate-mineral chemistry and suggested that the primary oxide, which oxyexsolved into ilmenite and hercynitic spinel, was at high temperature magnetite with 40 mol% ulvospinel in solid solution. This phase, found in places as euhedral crystals in cumulus clinopyroxene, was one of the cumulus phases in the ferrobasaltic magma at a relatively early stage and its gravitational accumulation resulted in formation of the ore layers in the lower parts of the intrusion. Zhang et al. (2009a,b) proposed that the Panzhihua intrusion formed by interaction between plume-derived magma and lithospheric mantle rich in Fe and Ti, consistent with the noble-gas isotopic data reported by Hou et al. (2012a,b) and a geochemical study of picritic dykes by Hou et al. (2013). Song et al. (2013) claimed that the unusually thick stratiform Fe-Ti oxide layers resulted from coupling of gravity settling and sorting of the Fe-Ti oxide crystals from Fe-Ti-enriched magmas and frequent magma replenishment along the floor of the magma chamber. Howarth et al. (2013) and Howarth and Prevec (2013) argued that Sr-Nd isotopic and PGE data show the intrusion represents a feeder or plumbing system to the volcanic rocks of the Emeishan LIP. Cheng et al. (2014) studied the compositions of plagioclase megacrysts in the lower part of the intrusion and suggested that the combination of large volumes of Fe-rich magma flux and efficient metal precipitation led to ore formation in a short period. Xing et al. (2014) examined the trace element geochemistry of apatite from the MZb of the intrusion and attributed the change from negative to positive Eu anomalies to interaction with upward-migrating Si-rich melt as a result of silicate liquid immiscibility. Ganino et al. (2008) and Ganino et al. (2013a,b) studied the contact aureole of the intrusion and suggested that CO_2 released during decarbonation of carbonate country rocks might have triggered early and abundant crystallization of Fe-Ti oxides in the magma. However, this hypothesis is not supported by the Sr-Nd-O isotopic data documented by Yu et al. (2015).

8.4.2.5 Hongge

The Hongge Fe-Ti-(V) oxide deposit is hosted in a layered lopolith with an exposed area of about 60 km^2 and thickness of about 1.7 km (Fig. 8.4). It was emplaced into Neoproterozoic dolomitic limestone and Precambrian basement rocks, and is divided from the base upwards into the lower olivine-clinopyroxenite zone, middle clinopyroxenite zone, and an upper gabbro zone (Zhong et al., 2002). The lower olivine-clinopyroxenite zone is composed of peridotitic cumulates with varying amounts of olivine, clinopyroxene, and titanomagnetite. The middle clinopyroxenite zone consists mostly of clinopyroxenitic cumulates with the above cumulus assemblage; plagioclase appears in the upper part of this zone. The upper gabbro zone is composed of clinopyroxenitic and gabbroic cumulates with olivine, clinopyroxene, plagioclase, and titanomagnetite as major minerals and apatite appears in the upper part of this zone. Six Fe-Ti oxide ore horizons, ranging from about 14 to 84 m in thickness, have been identified in the middle clinopyroxenite zone and upper gabbro zone.

The Hongge intrusion was dated by ID-TIMS zircon U-Pb at 259.3 ± 1.3 Ma (Zhong and Zhu, 2006). The geochemistry of the intrusion was investigated by Zhong et al. (2002), Zhong et al. (2003), Bai et al. (2012a,b), Luan et al. (2014) and Liao et al. (2015). The results of these investigations indicate that the intrusion developed from multiple pulses of primitive ferrobasaltic magma that were saturated in titanomagnetite, which settled and formed the oxide ore layers.

However, Wang and Zhou (2013) considered that liquid immiscibility played a role in the formation of oxide ores based on magnetite textures and compositions. Zhong et al. (2005) and Bai et al. (2012a,b) suggested that the intrusion might have acted as an open-system conduit through which the high-Ti basalts of the Emeishan LIP erupted. Bai et al. (2014) highlighted the compositional differences between the parental magmas of the Hongge and the Panzhihua intrusions, which were ascribed to selective assimilation of newly subducted, stagnant oceanic crust by ascending mantle plume-derived magmas.

8.4.2.6 Baima

The Baima Fe-Ti oxide deposit is hosted by a layered gabbroic intrusion covering an area of over 50 km^2 and with a thickness of over 1 km (Fig. 8.4). The intrusion was divided from the base upwards into the Lower cumulate zone, Oxide ore zone, Olivine gabbro zone, and Upper gabbro zone (Fig. 8.3). Igneous layering is generally west-dipping at shallow to medium angles. Oxide orebodies occur as conformable masses in the Oxide ore zone (up to ~70-m-thick) and, to a lesser extent, in the Olivine gabbro zone. The Lower cumulate zone, Olivine gabbro zone, and Upper gabbro zone consist of gabbroic and oxide gabbroic cumulates with varying amounts of olivine, clinopyroxene, plagioclase, and titanomagnetite. Cumulus apatite is restricted to the upper part of the Olivine gabbro zone. The Oxide ore zone is composed mainly of Fe-Ti oxide cumulates with titanomagnetite as the main cumulus mineral.

The Baima intrusion was dated by SHRIMP zircon U-Pb at 261 ± 2 Ma (Shellnutt et al., 2009). The mineralogy and geochemistry of the intrusion were studied by Shellnutt et al. (2009), Shellnutt and Pang (2012), Zhang et al. (2012), Zhang et al. (2013), and Liu et al. (2014a). Shellnutt et al. (2009) demonstrated that the Baima intrusion and associated syenites are related by fractionation to a common parental magma. Zhang et al. (2012) asserted that frequent magma replenishment and gravitational sorting and settling were crucial to form thick Fe-Ti oxide ore layers at the base of the intrusion. Liu et al. (2014a,b) favored a model involving silicate liquid immiscibility in an open and isotopically heterogeneous, periodically recharged magma chamber. Iron-isotope investigations of the intrusion led to diverse opinions on its genesis. Liu et al. (2014c) identified disequilibrium fractionation of Fe isotopes between olivine, clinopyroxene, and titanomagnetite, which were attributed to crystallization of silicate and oxide minerals from immiscible Si-rich and Fe-rich liquids. Chen et al. (2014) held a different view and attributed the isotopic variations to subsolidus reequilibration.

8.4.2.7 Taihe

The Taihe Fe-Ti oxide deposit is hosted in a small intrusion with exposed area of about 13 km^2 (Fig. 8.4). The intrusion is about 3×2 km, extends more than 3 km southeastward underground and has a thickness of about 1.2 km. Although the Taihe intrusion is expected to have been originally emplaced in the Neoproterozoic strata, like the other oxide ore-bearing layered intrusions in the Emeishan LIP, it is completely surrounded by a coeval syenitic pluton (Xu et al., 2008; Shellnutt et al., 2011). The intrusion was divided into a lower zone, a middle zone, and an upper zone, based on mineral assemblages and petrography of samples from the drill core (She et al., 2014). The lower zone is made up of olivine clinopyroxenite, olivine gabbro, and gabbro. The middle zone consists of magnetite clinopyroxenites, apatite-magnetite clinopyroxenites, and apatite gabbro and gabbro. The dominant rock type of the upper zone is apatite gabbro and a thin layer of apatite clinopyroxenite occurs in its lower part. Massive Fe-Ti oxide ores occur in the upper part of the lower zone.

The Taihe intrusion was dated by zircon U-Pb at 264 ± 3 Ma (Shellnutt et al., 2011). The geochemistry of the intrusion was studied by Shellnutt et al. (2011) and Hou et al. (2012a,b). The former drew attention to the similarity of the Taihe intrusion to the Panzhihua intrusion and the mafic unit of the BIC, and should thus share a common petrogenesis. Negative HFSE anomalies in rocks from the intrusion were argued by the latter as a signature of subducted material. These authors proposed a model involving interaction between a mantle plume and subduction-modified lithospheric mantle for the petrogenesis and ore genesis. Other studies focus on the mineralogy and mineral chemistry of the intrusion. She et al. (2014) presented similar data for a set of drill-core samples indicating that the intrusion developed from at least two pulses of magmas, from which Fe-Ti oxides, clinopyroxene, and apatite crystallized and accumulated to form oxide ore layers. The model was then substantiated by in situ trace element analysis of magnetite and ilmenite (She et al., 2015). Bai et al. (2016) combined cumulus-mineral chemistry and MELTS modeling to argue that oxide ores in the intrusion formed by gravitational accumulation of Fe-Ti oxides crystallized from a basaltic magma at high fO_2, not an Fe-Ti-P-rich immiscible liquid segregated from such magma.

8.5 DISCUSSION AND SUMMARY WORDS

We have glimpsed at the Chinese magmatic sulfide and Fe-Ti oxide deposits with general conclusions and open questions. Clear differences exist between them. Magmatic Ni-Cu-(PGE) sulfide deposits are associated with a wide range of settings including within-plate, arcs, or postorogenic. In many cases, the deposits are hosted in small and zoned intrusions representing parts of magma plumbing systems at relatively shallow paleodepths. Sulfide ores or mineralization therein are associated with lithologies made up of olivine and/or pyroxene. In contrast, magmatic Fe-Ti-(V) oxide deposits are identified almost exclusively with intraplate magmatism related to the formation of LIPs. The deposits are hosted either in large Proterozoic massif-type anorthosites or layered intrusions emplaced in middle to upper crustal paleodepths. The dominant rocks associated with the oxide ores consist mainly of pyroxene and/or plagioclase. The differences between the two types of orthomagmatic deposits imply that factors relevant to the genesis of the sulfide deposits might not be equally applied to the oxide deposits.

Following abundant studies on world-class magmatic sulfide deposits, the mechanisms and key factors responsible for ore genesis are generally well established. These involve: (1) a mantle plume to generate a sulfide-undersaturated, metal-rich magma; (2) a rifted continental or arc system to provide access through crustal rocks; (3) an external, crustal source of S; and (4) a dynamic plumbing system to promote thermomechanical erosion or devolatilization of wall rocks and interaction of the sulfides with large masses of magma. It is apparent from the above review that most investigations of the Chinese magmatic sulfide deposits are based on this existing framework. Differences in sulfide abundance and composition are generally attributed to stratigraphic architecture and dynamics of the system (e.g., Song et al., 2009; Tao et al., 2010). However, some Chinese magmatic sulfide deposits, such as those along the CAOB, are not associated with periods of mantle plume activity and the formation of LIPs. An external source of S, which should be indicated by S isotopes, is not established for all of the deposits. In addition, magmatic sulfide deposits related to the same magmatic event may have different sulfide content and PGE tenor (e.g., the Emeishan LIP), which need to be explored in more detail.

The origin of magmatic Fe-Ti-(V) oxide deposits has been and continues to be enigmatic. One of the long-lasting and fundamental debates is in what form the Fe-Ti oxides accumulated in ferrobasaltic magma systems. Opinions are twofold: accumulation and sorting of crystalline Fe-Ti spinel from mafic silicates in ferrobasaltic magmas (Pang et al., 2008a,b, 2010; Song et al., 2013; Bai et al., 2016), or formation of immiscible, oxide-rich melt (Zhou et al., 2005, 2013; Liu et al., 2014; Wang et al., 2017). While the debate is difficult to resolve in this review, it is important with respect to the differentiation of ferrobasaltic magmas. If early saturation of Fe-Ti spinel can be demonstrated for ferrobasaltic compositions pertinent to the target intrusion, then the oxide ores should have an origin involving settling and sorting of crystalline Fe-Ti spinel, an intrusion-wide processes that should affect the liquid line of descent of the evolving magma. On the contrary, if the ores formed by downward percolation and solidification of immiscible Fe-Ti oxide melts, then the silicate minerals within and between the ore layers will have little relevance to the point where the oxide melt separated from the differentiating ferrobasaltic magma. Without resolving this fundamental debate, new geochemical and isotopic data will lead to diverse interpretations (e.g., Liu et al., 2014; Chen et al., 2014) (Table 8.1).

8.6 NOMENCLATURE

CAOB Central Asian orogenic belt
FRTE first-row transition element
ID isotope dilution
LIP large igneous province
NCC North China craton
OIB ocean island basalt
PGE platinum-group element
PGM platinum-group mineral
SCC South China craton
SHRIMP sensitive high-resolution ion microprobe
TIMS thermal ionization mass spectrometry

ACKNOWLEDGEMENTS

Comments and suggestions from an anonymous reviewer are helpful but we take blame for any biases. Efficient editorial handling and comments by Sisir Mondal and Bill Griffin is appreciated. This project was supported by funding from the Ministry of Science and Technology, Taiwan (102-2628-M-003-001-MY4 to JGS and 104-2116-M-001-001-MY2 to KNP).

Table 8.1 Summary of Key Information of Major Chinese Orthomagmatic Ore Deposits

Name	Province	Lat N/Long E	Age	Commodity	Grade and Tonnage
Magmatic Ni-Cu-(PGE) Sulfide Deposits					
Jinchuan	Gansu	38°28′/102°10′	~825 Ma	Ni, Cu, PGEs	~664 Mt sulfide ore at ~1.25% Ni, ~0.81% Cu, ~0.343% Co, ~0.028–0.21 g/t Pt, ~0.014–0.12 g/t Pd
Xiarihamu	Qinghai	36°28′/93°19′	~406 Ma	Ni, Co	~157 Mt of sulfide ore at ~0.65% Ni, ~0.14% Cu, ~0.013% Co
Erbutu	Inner Mongolia	41°27′/105°40′	~294 Ma	Ni, Cu	Not given
Kalatongke	Xinjiang	46°45′/89°41′	~287 Ma	Ni, Cu	~33 Mt sulfide ore at ~0.8% Ni, ~1.3% Cu
Huangshandong	Xinjiang	42°14′/95°01′	~274 Ma	Ni, Cu, Co	~384 Kt Ni, ~188 Kt Cu, ~17 Kt Co at ~0.52% Ni, ~0.27% Cu, ~0.024% Co
Huangshanxi	Xinjiang	42°16′/94°37′	~270 Ma	Ni, Cu, PGEs	~334 Kt Ni, ~208 Kt Cu, ~20 Kt Co at ~0.47% Ni, ~0.31% Cu, ~0.026% Co
Poyi	Xinjiang	40°36′/91°35′	~271 Ma	Ni, Cu	~1.3 Mt Ni, ~220 Kt Cu, 60 Kt Co
Yangliuping	Sichuan	30°42′/101°48′	~260 Ma	Ni, Cu, Co, PGEs	~275 Kt Ni, ~102 Kt Cu, ~10 Kt Co, ~35 t PGE at ~0.45% Ni, ~0.16% Cu, ~0.016% Co, ~0.55 ppm PGE
Limahe	Sichuan	26°28′/102°03′	~260 Ma	Ni, Cu, PGEs	~28 Kt Ni, ~14 Kt Cu at ~1% Ni, ~0.51% Cu
Baimazhai	Sichuan	22°53′/102°59′	~260 Ma	Ni, Cu, PGEs	~43 Kt Ni, ~27 Kt Cu, ~2 Kt Co
Jinbaoshan	Yunman	25°06′/100°51′	~260 Ma	Ni, Cu, Pt, Pd	~58 Kt Ni, ~49 Kt Cu, ~45 t PGE
Hongqiling	Jilin	42°55′/126°19′	~220 Ma	Ni, Cu	~204 Kt Ni, ~39 Kt Cu, ~3 Kt Co at ~2.3% Ni, ~0.63% Cu, ~0.05% Co
Magmatic Fe-Ti-(V)-(P) Oxide Deposits					
Damiao	Hebei	41°10′/117°55′	~1.7 Ga	Fe, Ti, P	~33 Mt oxide ore at ~36% Fe_2O_3, ~7% TiO_2, ~0.3% V_2O_5, ~2% P_2O_5
Wajilitag	Xinjiang	39°32′/78°56′	~283 Ma	Fe, Ti, V	~146 Mt oxide ore at ~17% total FeO, ~7% TiO_2, ~0.2% V_2O_5
Puchang	Xinjiang	40°26′/77°38′	~275 Ma	Fe, Ti, V	~120 Mt oxide ore at ~20% total FeO, ~11% TiO_2, ~0.8% V_2O_5
Panzhihua	Sichuan	26°35′/101°40′	~260 Ma	Fe, Ti, V	~1.33 Bt oxide ore at ~33% total Fe, ~12% TiO_2 and ~0.3% V_2O_5
Hongge	Sichuan	26°39′/101°59′	~260 Ma	Fe, Ti, V	~4.57 Bt oxide ore at ~27% total Fe, ~11% TiO_2 and ~0.24% V_2O_5
Baima	Sichuan	27°05′/102°06′	~260 Ma	Fe, Ti, V	~1.50 Bt oxide ore at ~26% total Fe, ~7% TiO_2 and ~0.21% V_2O_5
Taihe	Sichuan	27°54′/102°07′	~260 Ma	Fe, Ti, V	~99 Mt TiO_2, ~2 Mt V_2O_5

(1) http://www.gsj.jp/Map/EN/docs/overseas_doc/DataSheet.xls, (2) Ma et al. (2003).

REFERENCES

Ali, J.R., Thompson, G.M., Zhou, M.-F., Song, X.Y., 2005. Emeishan large igneous province, SW China. Lithos 79, 475–489.

Anh, T.V., Pang, K.-N., Chung, S.-L., Lin, H.-M., Tran, T.H., Tran, T.A., et al., 2011. The Song Da magmatic suite revisited: A petrologic, geochemical and Sr-Nd isotopic study on picrites, flood basalts and silicic volcanic rocks. J. Asian Earth Sci. 42, 1341–1355.

Bai, Z.-J., Zhong, H., Li, C., Zhu, W.-G., Hu, W.-J., 2016. Association of cumulus apatite with compositionally unusual olivine and plagioclase in the Taihe Fe-Ti oxide ore-bearing layered mafic-ultramafic intrusion: petrogenetic significance and implications for ore genesis. Am. Mineral. 101, 2168–2175.

Bai, Z.-J., Zhong, H., Naldrett, A.J., Zhu, W.-G., Xu, G.-W., 2012a. Whole-rock and mineral composition constraints on the genesis of the giant Hongge Fe-Ti-V oxide deposit in the Emeishan large igneous province, southwest China. Econ. Geol. 107, 507–524.

Bai, Z.-J., Zhong, H., Li, C., Zhu, W.-G., Xu, G.-W., 2012b. Platinum-group elements in the oxide layers of the Hongge mafic–ultramafic intrusion, Emeishan Large Igneous Province, SW China. Ore Geol. Rev. 46, 149–161.

Bai, Z.-J., Zhong, H., Li, C., Zhu, W.-G., He, D.-F., Qi, L., 2014. Contrasting parental magma compositions for the Hongge and Panzhihua magmatic Fe-Ti-V oxide deposits, Emeishan large igneous province, SW China. Econ. Geol. 109, 1763–1785.

Barnes, S.J., Tang, Z.-L., 1999. Chrome Spinel from the Jinchuan Ni-Cu sulfide deposit, Gansu Province, People's Republic of China. Econ. Geol. 94, 343–356.

Cao, J., Wang, C.Y., Xing, C.-M., Xu, Y.-G., 2014. Origin of the early Permian Wajilitag igneous complex and associated Fe-Ti oxide mineralization in the Tarim large igneous province, NW China. J. Asian Earth Sci. 84, 51–68.

Cao, J., Wang, C.-Y., Xu, Y.-G., Xing, C.-M., Ren, M.-H., 2016. Triggers on sulfide saturation in Fe-Ti oxide-bearing, mafic-ultramafic layered intrusions in the Tarim large igneous province, NW China. Mineral. Deposita . Available from: http://dx.doi.org/10.1007/s00126-016-0670-z.

Chai, G., Naldrett, A.J., 1992a. Characteristics of Ni-Cu-PGE Mineralization and genesis of the Jinchuan deposit, Northwest China. Econ. Geol. 87, 1475–1495.

Chai, G., Naldrett, A.J., 1992b. The Jinchuan ultramafic intrusion: Cumulate of a high-Mg basaltic magma. J. Petrol. 33, 277–303.

Cheng, L., Zeng, L., Ren, Z., Wang, Y., Luo, Z., 2014. Timescale of emplacement of the Panzhihua gabbroic layered intrusion recorded in giant plagioclase at Sichuan Province, SW China. Lithos 204, 203–219.

Chen, W.T., Zhou, M.-F., Gao, J.-F., Zhao, T.-P., 2015b. Oscillatory Sr isotopic signature in plagioclase megacrysts from the Damiao anorthosite complex, North China: Implication for petrogenesis of massif-type anorthosite. Chem. Geol. 393-394, 1–15.

Chen, L.-M., Song, X.-Y., Keays, R.R., Tian, Y.-L., Wang, Y.-S., Deng, Y.-F., et al., 2013. segregation and fractionation of magmatic Ni-Cu-PGE sulfides in the Western Jinchuan intrusion, Northwestern China: Insights from platinum group element geochemistry. Econ. Geol. 108, 1793–1811.

Chen, L.-M., Song, X.-Y., Zhu, X.-K., Zhang, X.-Q., Yu, S.-Y., Yi, J.-N., 2014. Iron isotope fractionation during crystallization and sub-solidus re-equilibration: constraints from the Baima mafic layered intrusion, SW China. Chem. Geol. 380, 97–109.

Chen, L.-M., Song, X.-Y., Danyushevsky, L.V., Wang, Y.-S., Tian, Y.-L., Xiao, J.-F., 2015a. A laser ablation ICP-MS study of platinum-group and chalcophile elements in base metal sulfide minerals of the Jinchuan Ni-Cu sulfide deposit, NW China. Ore Geol. Rev. 65, 955–967.

Chung, S.-L., Jahn, B.-M., 1995. Plume-lithosphere interaction in generation of the Emeishan flood basalts at the Permian-Triassic boundary. Geology 23, 889–892.

Dostal, J., Owen, J.V., Shellnutt, J.G., Keppie, J.D., Gerel, O., Corney, R., 2015. Petrogenesis of the Triassic Bayan-Ulan alkaline granitic pluton in the North Gobi rift of central Mongolia: Implications for the evolution of Early Mesozoic granitoid magmatism in the Central Asian Orogenic Belt. J. Asian Earth Sci. 109, 50–62.

Duan, J., Li, C., Qian, Z., Jiao, J., Ripley, E.M., 2016. Multiple S isotopes, zircon Hf isotopes, whole-rock Sr-Nd isotopes, and spatial variations of PGE tenors in the Jinchuan Ni-Cu-PGE deposit, NW China. Mineral. Deposita 51, 557–574.

Ganino, C., Arndt, N.T., Zhou, M.-F., Gaillard, F., Chauvel, C., 2008. Interaction of magma with sedimentary wall rock and magnetite ore genesis in the Panzhihua mafic intrusion, SW China. Mineral. Deposita 43, 677–694.

Ganino, C., Arndt, N.T., Chauvel, C., Alexandre, J., Athurion, C., 2013a. Melting of carbonate wall rocks and formation of the heterogeneous aureole of the Panzhihua intrusion, China. Geosci. Front. 4, 535–546.

Ganino, C., Harris, C., Arndt, N.T., Prevec, S.A., Howarth, G.H., 2013b. Assimilation of carbonate country rock by the parent magma of the Panzhihua Fe-Ti-V deposit (SW China): Evidence from stable isotopes. Geosci. Front. 4, 547–554.

Gao, J.-F., Zhou, M.-F., Lightfoot, P.C., Wang, C.Y., Qi, L., 2012a. Origin of PGE-poor and Cu-rich magmatic sulfides from the Kalatongke deposit, Xinjiang, Northwest China. Econ. Geol. 107, 481–506.

Gao, J.-F., Zhou, M.-F., Lightfoot, P.C., Qu, W., 2012b. Heterogeneous Os isotope compositions in the Kalatongke sulfide deposit, NW China: the role of crustal contamination. Mineral. Deposita 47, 731–738.

Gao, J.-F., Zhou, M.-F., Lightfoot, P.C., Wang, C.Y., Qi, L., Sun, M., 2013. Sulfide saturation and magma emplacement in the formation of the Permian Huangshandong Ni-Cu sulfide deposit, Xinjiang, Northwestern China. Econ. Geol. 108, 1833–1848.

He, H.-L., Yu, S.-Y., Song, X.-Y., Du, Z.-S., Dai, Z.-H., Zhou, T., et al., 2016. Origin of nelsonite and Fe-Ti oxides ore of the Damiao anorthosite complex, NE China: Evidence from trace element geochemistry of apatite, plagioclase, magnetite and ilmenite. Ore Geol. Rev. 79, 367–381.

Hou, T., Zhang, Z., Kusky, T., Du, Y., Liu, J., Zhao, Z., 2011a. A reappraisal of the high-Ti and low-Ti classification of basalts and petrogenetic linkage between basalts and mafic-ultramafic intrusions in the Emeishan Large Igneous Province, SW China. Ore Geol. Rev. 41, 133–143.

Hou, T., Zhang, Z., Ye, X., Encarnacion, J., Reichow, M.K., 2011b. Noble gas isotopic systematics of Fe-Ti-V oxide ore-related mafic-ultramafic layered intrusions in the Pan-Xi area, China: The role of recycled oceanic crust in their petrogenesis. Geochim. Cosmochim. Acta 75, 6727–6741.

Hou, T., Zhang, Z., Encarnacion, J., Santosh, M., 2012a. Petrogenesis and metallogenesis of the Taihe gabbroic intrusion associated with Fe-Ti-oxide ores in the Panxi district, Emeishan Large Igneous Province, southwest China. Ore Geol. Rev. 49, 109–127.

Hou, T., Zhang, Z., Pirajno, F., 2012b. A new metallogenic model of the Panzhihua giant V-Ti-iron oxide deposit (Emeishan Large Igneous Province) based on high-Mg olivine-bearing wehrlite and new field evidence. Inter. Geol. Rev. 54, 1721–1745.

Hou, T., Zhang, Z., Encarnacion, J., Santosh, M., Sun, Y., 2013. The role of recycled oceanic crust in magmatism and metallogeny: Os-Sr-Nd isotopes, U-Pb geochronology and geochemistry of picritic dykes in the Panzhihua giant Fe-Ti oxide deposit, central Emeishan large igneous province, SW China. Contrib. Mineral. Petrol. 165, 805–822.

Howarth, G.H., Prevec, S.A., 2013. Hydration vs. oxidation: Modelling implications for Fe-Ti oxide crystallisation in mafic intrusions, with specific reference to the Panzhihua intrusion, SW China. Geosci. Front. 4, 555–569.

Howarth, G.H., Prevec, S.A., Zhou, M.-F., 2013. Timing of Ti-magnetite crystallisation and silicate disequilibrium in the Panzhihua mafic layered intrusion: Implications for ore-forming processes. Lithos 170-171, 73–89.

Jahn, B.-M., 2004. The Central Asian Orogenic Belt and growth of the continental crust in the Phanerozoic. In: Malpas, J., Fletcher, C.J.N., Ali, J.R., Aitchison, J.C. (Eds.), Aspects of the Tectonic Evolution of China, 226. Geological Society London Special Publications, pp. 73−100.

Kamenetsky, V.S., Chung, S.-L., Kamenetsky, M.B., Kuzmin, D.V., 2012. Picrites from the Emeishan large igneous province, SW China: a compositional continuum in primitive magmas and their respective mantle sources. J. Petrol. 53, 2095−2113.

Kröner, A., Kovach, V., Belousova, E., Hegner, E., Armstrong, R., Dolgopolova, A., et al., 2014. Reassement of continental growth during the accretionary history of the Central Asian Orogenic Belt. Gond. Res. 25, 103−125.

Lehmann, J., Arndt, N., Windley, B., Zhou, M.-F., Wang, C.Y., Harris, C., 2007. Field relationships and geochemical constraints on the emplacement of the Jinchuan intrusion and its Ni-Cu-PGE sulfide deposit, Gansu, China. Econ. Geol. 102, 75−94.

Li, C., Zhang, Z., Li, W., Wang, Y., Sun, T., Ripley, E.M., 2015a. Geochronology, petrology and Hf-S isotope geochemistry of the newly-discovered Xiarihamu magmatic Ni-Cu sulfide deposit in the Qinghai-Tibet plateau, western China. Lithos 216-217, 224−240.

Li, H., Li, L., Zhang, Z., Santosh, M., Liu, M., Cui, Y., et al., 2014. Alteration of the Damiao anorthosite complex in the northern North China Craton: Implications for high-grade iron mineralization. Ore Geol. Rev. 57, 574−588.

Li, L.-X., Li, H.-M., Li, Y.-Z., Yao, T., Yang, X.-Q., Chen, J., 2015b. Origin of rhythmic anorthositic-pyroxenitic layering in the Damiao anorthosite complex, China: Implications for late-stage fractional crystallization and genesis of Fe-Ti oxide ores, J. Asian Earth Sci. 113. pp. 1035−1055.

Li, W.-X., Li, X.-H., Li, Z.-X., 2005a. Neoproterozoic bimodal magmatism in the Cathaysia Block of South China and its tectonic significance. Precamb. Res. 136, 51−66.

Li, X.-H., Su, L., Chung, S.-L., Li, Z.X., Liu, Y., Song, B., et al., 2005b. Formation of the Jinchuan ultramafic intrusion and the world's third largest Ni-Cu sulfide deposit: Associated with the ∼825 Ma south China mantle plume? Geochem. Geophys. Geosys. 6, Q11004.

Li, X.-H., Li, Z.-X., Ge, W., Zhou, H., Li, W., Liu, Y., et al., 2003a. Neoproterozoic granitoids in South China: crustal melting above a mantle plume at ca. 825 Ma? Precamb. Res. 122, 45−83.

Li, Y.-Q., Li, Z.-L., Chen, H.-L., Yang, S.-F., Yu, X., 2012. Mineral characteristics and metallogenesis of the Wajilitag layered mafic−ultramafic intrusion and associated Fe-Ti-V oxide deposit in the Tarim large igneous province, northwest China. J. Asian Earth Sci. 49, 161−174.

Li, Z.X., Li, X.H., Kinny, P.D., Wang, J., 1999. The breakup of Rodinia: did it start with a mantle plume beneath South China? Earth Planet. Sci. Lett. 173, 171−181.

Li, Z.X., Li, X.-H., Kinny, P.D., Wang, J., Zhang, S., Zhou, H., 2003b. Geochronology of Neoproterozoic syn-rift magmatism in the Yangtze Craton, South China and correlations with other continents: evidence for a mantle superplume that broke up Rodinia. Precamb. Res. 122, 85−109.

Liao, M., Tao, Y., Song, X., Li, Y., Xiong, F., 2015. Multiple magma evolution and ore-forming processes of the Hongge layered intrusion, SW China: insights from Sr-Nd isotopes, trace elements and platinum-group elements. J. Asian Earth Sci. 113, 1082−1099.

Lü, L., Mao, J., Li, H., Pirajno, F., Zhang, Z., Zhou, Z., 2011. Pyrrhotite Re-Os and SHRIMP zircon U-Pb dating of the Hongqiling Ni-Cu sulfide deposits in Northeast China. Ore Geol. Rev. 43, 106−119.

Luan, Y., Song, X.-Y., Chen, L.-M., Zheng, W.-Q., Zhang, X.-Q., Yu, S.-Y., et al., 2014. Key factors controlling the accumulation of the Fe-Ti oxides in the Hongge layered intrusion in the Emeishan Large Igneous Province, SW China. Ore Geol. Rev. 57, 518−538.

Liu, P.-P., Zhou, M.-F., Wang, C.Y., Xing, C.-M., Gao, J.-F., 2014a. Open magma chamber processes in the formation of the Permian Baima mafic-ultramafic layered intrusion, SW China. Lithos 184-187, 194−208.

Liu, P.-P., Zhou, M.-F., Chen, W.T., Boone, M., Cnudde, V., 2014b. Using multiphase solid inclusions to constrain the origin of the Baima Fe-Ti-(V) oxide deposit, SW China. J. Petrol 55, 951–976.

Liu, P.-P., Zhou, M.-F., Luais, B., Cividini, D., Rollion-Bard, C., 2014c. Disequilibrium iron isotopic fractionation during the high-temperature magmatic differentiation of the Baima Fe-Ti oxide-bearing mafic intrusion, SW China. Earth Planet. Sci. Lett. 399, 21–29.

Ma, Y., Ji, X.T., Li, J.C., Huang, M., Kan, Z.Z., 2003. Mineral Resources of the Panzhihua Region. Chengdu: Sichuan Sci. Technol. Press 275 pp. (in Chinese).

Mao, Y.-J., Qin, K.-Z., Li, C., Xue, S.-C., Ripley, E.M., 2014. Petrogenesis and ore genesis of the Permian Huangshanxi sulfide ore-bearing mafic-ultramafic intrusion in the Central Asian Orogenic Belt, western China. Lithos 200-201, 111–125.

Munteanu, M., Yao, Y., Wilson, A.H., Chunnett, G., Luo, Y., He, H., et al., 2013. Panxi region (South-West China): Tectonics, magmatism and metallogenesis. A review. Tectonophysics 608, 51–71.

Pang, K.-N., Li, C., Zhou, M.-F., Ripley, E.M., 2008a. Abundant Fe-Ti oxide inclusions in olivine from the Panzhihua and Hongge layered intrusions, SW China: Evidence for early saturation of Fe-Ti oxides in ferrobasaltic magma. Contrib. Mineral. Petrol. 156, 307–321.

Pang, K.-N., Zhou, M.-F., Lindsley, D.H., Zhao, D., Malpas, J., 2008b. Origin of Fe-Ti oxide ores in mafic intrusions: evidence from the Panzhihua intrusion. J. Petrol. 49, 295–313.

Pang, K.-N., Li, C., Zhou, M.-F., Ripley, E.M., 2009. Mineral compositional constraints on petrogenesis and ore genesis of the Panzhihua layered gabbroic intrusion, SW China. Lithos 110, 199–214.

Pang, K.-N., Zhou, M.-F., Qi, L., Shellnutt, G., Wang, C.Y., Zhao, D., 2010. Flood basalt-related Fe-Ti oxide deposits in the Emeishan large igneous province, SW China. Lithos 119, 123–136.

Pang, K.-N., Zhou, M.-F., Qi, L., Chung, S.-L., Chu, C.-H., Lee, H.-Y., 2013. Petrology and geochemistry at the LZ-MZa transition of the Panzhihua intrusion, SW China: implications for differentiation and oxide ore genesis. Geosci. Front. 4, 517–533.

Peng, P., Zhai, M.-G., Guo, J.-H., Kusky, T., Zhao, T.-P., 2007. Nature of mantle source contributions and crystal differentiation in the petrogenesis of the 1.78 Ga mafic dykes in the central North China craton. Gond. Res. 12, 29–46.

Peng, P., Zhai, M., Ernst, R.E., Guo, J., Liu, F., Hu, B., 2008. A 1.78 Ga large igneous province in the North China craton: The Xiong'er Volcanic Province and the North China dyke swarm. Lithos 101, 260–280.

Peng, R., Zhai, Y., Li, C., Ripley, E.M., 2013. The Erbutu Ni-Cu deposit in the Central Asian orogenic belt: A Permian magmatic sulfide deposit related to boninitic magmatism in an arc setting. Econ. Geol. 108, 1879–1888.

Peng, P., Wang, X., Lai, Y., Wang, C., Windley, B., 2015. Large-scale liquid immiscibility and fractional crystallization in the 1780 Ma Taihang dyke swarm: Implications for genesis of the bimodal Xiong'er volcanic province. Lithos 236-237, 106–122.

Pirajno, F., 2013. The Geology and Tectonic Settings of China's Mineral Deposits. Springer, Berlin, 679 pp.

Pirajno, F., Ernst, R.E., Borisenko, A.S., Fedoseev, G., Naumov, E.A., 2009. Intraplate magmatism in Central Asia and China and associated metallogeny. Ore Geol. Rev. 35, 114–136.

Pirajno, F., Santosh, M., 2014. Rifting, intraplate magmatism, mineral systems and mantle dynamics in central-east Eurasia: An overview. Ore Geol. Rev. 63, 265–295.

Prichard, H.M., Knight, R.D., Fisher, P.C., McDonald, I., Zhou, M.-F., Wang, C.Y., 2013. Distribution of platinum-group elements in magmatic and altered ores in the Jinchuan intrusion, China: an example of selenium remobilization by postmagmatic fluids. Mineral. Deposita 48, 767–786.

Qin, K.-Z., Su, B.-X., Asamoah Sakyi, P., Tang, D.-M., Li, X.-H., Sun, H., et al., 2011. SIMS zircon U-Pb geochronology and Sr-Nd isotopes of Ni-Cu-bearing mafic-ultramafic intrusions in eastern Tianshan and Beishan in correlation with flood basalts in Tarim Basin (NW China): constraints on a ca. 280 Ma mantle plume. Am. J. Sci. 311, 237–260.

Qu, W.J., Chen, J.F., Wang, L.B., Li, C., Du, A.D., 2013. Re-Os pseudo-isochron of disseminated ore from the Kalatongke Cu-Ni sulfide deposit, Xinjiang, Northwest China: Implications for Re−Os dating of magmatic Cu-Ni sulfide deposits. Ore Geol. Rev. 53, 39−49.

Ripley, E.M., Sarkar, A., Li, C., 2005. Mineralogic and stable isotope studies of hydrothermal alteration at the Jinchuan Ni-Cu deposit, China. Econ. Geol. 100, 1349−1361.

She, Y.-W., Yu, S.-Y., Song, X.-Y., Chen, L.-M., Zheng, W.-Q., Luan, Y., 2014. The formation of P-rich Fe-Ti oxide ore layers in the Taihe layered intrusion, SW China: Implications for magma-plumbing system process. Ore Geol. Rev. 57, 539−559.

She, Y.-W., Song, X.-Y., Yu, S.-Y., He, H.-L., 2015. Variations of trace element concentration of magnetite and ilmenite from the Taihe layered intrusion, Emeishan large igneous province, SW China: implications for magmatic fractionation and origin of Fe-Ti-V oxide ore deposits. J. Asian Earth Sci. 113, 1117−1131.

Shellnutt, J.G., 2014. The Emeishan large igneous province: a synthesis. Geosci. Front. 5, 369−694.

Shellnutt, J.G., Jahn, B.-M., 2011. Origin of Late Permian Emeishan basaltic rocks from the Panxi region (SW China): Implications for the Ti-classification and spatial−compositional distribution of the Emeishan flood basalts. J. Volcanol. Geother. Res. 199, 85−95.

Shellnutt, J.G., Pang, K.-N., 2012. Petrogenetic implications of mineral chemical data for the Permian Baima igneous complex, SW China. Mineral. Petrol. 106, 75−88.

Shellnutt, J.G., Zhou, M.-F., 2007. Permian peralkaline, peraluminous and metaluminous A-type granites in the Pan-Xi district, SW China: Their relationship to the Emeishan mantle plume. Chem. Geol. 243, 286−316.

Shellnutt, J.G., Zhou, M.-F., Zellmer, G., 2009. The role of Fe-Ti oxide crystallization in the formation of A-type granitoids with implications for the Daly gap: An example from the Permian Baima igneous complex, SW China. Chem. Geol. 259, 204−217.

Shellnutt, J.G., Wang, K.-L., Zellmer, G.F., Iizuka, Y., Jahn, B.-M., Pang, K.-N., et al., 2011. Three Fe-Ti oxide ore-bearing gabbro-granitoid complexes in the Panxi region of the Permian Emeishan large igneous province, SW China. Am. J. Sci. 311, 773−812.

Song, X.-Y., Li, X.-R., 2009. Geochemistry of the Kalatongke Ni-Cu-(PGE) sulfide deposit, NW China: implications for the formation of magmatic sulfide mineralization in a postcollisional environment, Mineral. Deposita, 44. pp. 303−327.

Song, X.-Y., Zhou, M.-F., Cao, Z.-M., Sun, M., Wang, Y.-L., 2003. Ni-Cu-(PGE) magmatic sulfide deposits in the Yangliuping area, Permian Emeishan igneous province, SW China. Mineral. Deposita 38, 831−843.

Song, X.-Y., Zhou, M.-F., Cao, Z.-M., 2004. Genetic relationships between base-metal sulfides and platinum-group minerals in the Yangliuping Ni-Cu-(PGE) sulfide deposit, southwestern China. Can. Min 42, 469−483.

Song, X.-Y., Zhou, M.-F., Wang, C.Y., Qi, L., Zhang, C.-J., 2006. Role of crustal contamination in formation of the Jinchuan Intrusion and its world-class Ni-Cu-(PGE) sulfide deposit, Northwest China. Inter. Geol. Rev. 48, 1113−1132.

Song, X.-Y., Keays, R.R., Zhou, M.-F., Qi, L., Ihlenfeld, C., Xiao, J.-F., 2009. Siderophile and chalcophile elemental constraints on the origin of the Jinchuan Ni-Cu-(PGE) sulfide deposit, NW China. Geochim. Cosmochim. Acta 73, 404−424.

Song, X., Wang, Y., Chen, L., 2011. Magmatic Ni-Cu-(PGE) deposits in magma plumbing systems: features, formation and exploration. Geosci. Front 2, 375−384.

Song, X.-Y., Danyushevsky, L.V., Keays, R.R., Chen, L.-M., Wang, Y.-S., Tian, Y.-L., et al., 2012. Structural, lithological, and geochemical constraints on the dynamic magma plumbing system of the Jinchuan Ni−Cu sulfide deposit, NW China. Mineral. Deposita 47, 277−297.

Song, X.-Y., Qi, H.-W., Hu, R.-Z., Chen, L.-M., Yu, S.-Y., Zhang, J.-F., 2013. Formation of thick stratiform Fe-Ti oxide layers in layered intrusion and frequent replenishment of fractionated mafic magma: Evidence from the Panzhihua intrusion, SW China. Geochem. Geophys. Geosys. 14, 712−732.

Song, X.-Y., Yi, J.-N., Chen, L.-M., She, Y.-W., Liu, C.-Z., Dang, X.-Y., et al., 2016. The Giant Xiarihamu Ni-Co Sulfide Deposit in the East Kunlun Orogenic Belt, Northern Tibet Plateau, China. Econ. Geol. 111, 29−55.

Su, B.-X., Qin, K.-Z., Asamoah Sakyi, P., Li, X.-H., Yang, Y.-H., Sun, H., et al., 2011. U-Pb ages and Hf-O isotopes of zircons from Late Paleozoic mafic−ultramafic units in the southern Central Asian Orogenic Belt: Tectonic implications and evidence for an Early-Permian mantle plume. Gond. Res. 20, 516−531.

Su, B.-X., Qin, K.-Z., Tang, D.-M., Asamoah Sakyi, P., Liu, P.-P., Sun, H., et al., 2013. Late Paleozoic mafic-ultramafic intrusions in southern Central Asian Orogenic Belt (NW China): Insight into magmatic Ni-Cu sulfide mineralization in orogenic setting. Ore Geol. Rev. 51, 57−73.

Su, S., Li, C., Zhou, M.-F., Ripley, E.M., Qi, L., 2008. Controls on variations of platinum-group element concentrations in the sulfide ores of the Jinchuan Ni-Cu deposit, western China. Mineral. Deposita 43, 609−622.

Sun, T., Qian, Z.-Z., Deng, Y.-F., Li, C., Song, X.-Y., Tang, Q., 2013. PGE and isotope (Hf-Sr-Nd-Pb) Constraints on the origin of the Huangshandong magmatic Ni-Cu sulfide deposit in the Central Asian orogenic belt, northwestern China. Econ. Geol. 108, 1849−1864.

Tao, Y., Li, C., Hu, R., Ripley, E.M., Du, A., Zhong, H., 2007. Petrogenesis of the Pt-Pd mineralized Jinbaoshan ultramafic intrusion in the Permian Emeishan large igneous province, SW China. Contrib. Mineral. Petrol. 153, 321−337.

Tao, Y., Li, C., Song, X.-Y., Ripley, E.M., 2008. Mineralogical, petrological, and geochemical studies of the Limahe mafic-ultramafic intrusion and associated Ni-Cu sulfide ores, SW China. Mineral. Deposita 43, 849−872.

Tao, Y., Li, C., Hu, R., Qi, L., Qu, W., Du, A., 2010. Re-Os isotopic constraints on the genesis of the Limahe Ni−Cu deposit in the Emeishan large igneous province, SW China. Lithos 119, 137−146.

Teng, X., Santosh, M., 2015. A long-lived magma chamber in the Paleoproterozoic North China Craton: Evidence from the Damiao gabbro-anorthosite suite. Precamb. Res. 256, 79−101.

Wang, C.Y., Zhou, M.-F., 2006. Genesis of the Permian Baimazhai magmatic Ni-Cu-(PGE) sulfide deposit, Yunnan, SW China. Mineral. Deposita 41, 771−783.

Wang, C.Y., Zhou, M.-F., Keays, R.R., 2006. Geochemical constraints on the origin of the Permian Baimazhai mafic−ultramafic intrusion, SW China. Contrib. Mineral. Petrol. 152, 309−321.

Wang, C.Y., Prichard, H.M., Zhou, M.-F., Fisher, P.C., 2008. Platinum-group minerals from the Jinbaoshan Pd-Pt deposit, SW China: evidence for magmatic origin and hydrothermal alteration. Mineral. Deposita 43, 791−803.

Wang, C.Y., Zhou, M.-F., Qi, L., 2010. Origin of extremely PGE-rich mafic magma system: An example from the Jinbaoshan ultramafic sill, Emeishan large igneous province, SW China. Lithos 119, 147−161.

Wang, C.Y., Zhou, M.-F., 2013. New textural and mineralogical constraints on the origin of the Hongge Fe-Ti-V oxide deposit, SW China. Mineral. Deposita 48, 787−798.

Wang, M., Nebel, O., Wang, C.Y., 2016. The flaw in the crustal 'zircon archive': mixed Hf isotope signatures record progressive contamination of late-stage liquid in mafic−ultramafic layered intrusions. J. Petrol. 57, 27−52.

Wang, M., Veksler, I., Zhang, Z., Hou, T., Keiding, J.K., 2017. The origin of nelsonite constrained by melting experiment and melt inclusions in apatite: The Damiao anorthosite complex, North China Craton. Gond. Res. 42, 163−176.

Wei, B., Wang, C.Y., Li, C., Sun, Y., 2013. Origin of PGE-depleted Ni-Cu sulfide mineralization in the Triassic Hongqiling No. 7 orthopyroxenite Intrusion, Central Asian orogenic belt, northeastern China. Econ. Geol. 108, 1813−1831.

Windley, B.F., Alexeiv, D., Xiao, W., Kroner, A., Badarch, G., 2007. Tectonic models for accretion of the Central Asian Orogenic Belt. J. Geol. Soc. London 164, 31−47.

Wu, F.-Y., Wilde, S.A., Zhang, G.-L., Sun, D.-Y., 2004. Geochronology and petrogenesis of the post-orogenic Cu-Ni sulfide-bearing mafic-ultramafic complexes in Jilin Province, NE China. J. Asian Earth Sci. 23, 781–797.

Xia, M.-Z., Jiang, C.-Y., Li, C., Xia, Z.-D., 2013. Characteristics of a newly discovered Ni-Cu sulfide deposit hosted in the Poyi ultramafic intrusion, Tarim craton, NW China. Econ. Geol. 108, 1865–1878.

Xiao, L., Xu, Y.G., Mei, H.J., Zheng, Y.F., He, B., Pirajno, F., 2004. Distinct mantle sources of low-Ti and high-Ti basalts from the western Emeishan large igneous province, SW China: implications for plume–lithosphere interaction. Earth Planet. Sci. Lett. 228, 525–546.

Xiao, W., Santosh, M., 2014. The western Central Asian Orogenic Belt: a window to accretionary orogenesis and continental growth. Gond. Res. 25, 1429–1444.

Xing, C.-M., Wang, C.Y., Li, C., 2014. Trace element compositions of apatite from the middle zone of the Panzhihua layered intrusion, SW China: insights into the differentiation of a P- and Si-rich melt. Lithos 204, 188–202.

Xu, Y.G., Chung, S.L., Jahn, B.M., Wu, G.Y., 2001. Petrologic and geochemical constraints on the petrogenesis of Permian-Triassic Emeishan flood basalts in southwestern China. Lithos 58, 145–168.

Xu, Y.-G., He, B., Chung, S.-L., Menzies, M.A., Frey, F.A., 2004. Geologic, geochemical, and geophysical consequences of plume involvement in the Emeishan flood-basalt province. Geology 32, 917–920.

Xu, Y.-G., Luo, Z.-Y., Huang, X.-L., He, B., Xiao, L., Xie, L.-W., et al., 2008. Zircon U-Pb and Hf isotope constraints on crustal melting associated with the Emeishan mantle plume. Geochim. Cosmochim. Acta 72, 3084–3104.

Xu, Y.-G., Chung, S.-L., Shao, H., He, B., 2010. Silicic magmas from the Emeishan large igneous province, Southwest China: Petrogenesis and their link with the end-Guadalupian biological crisis. Lithos 119, 47–60.

Xu, Y.-G., Wei, X., Luo, Z.-Y., Liu, H.-Q., Cao, J., 2014. The Early Permian Tarim Large Igneous Province: Main characteristics and a plume incubation model. Lithos 204, 20–35.

Xue, S.-C., Li, C., Qin, K.-Z., Tang, D.-M., 2016a. A non-plume model for the Permian protracted (266–286 Ma) basaltic magmatism in the Beishan-Tianshan region, Xinjiang, Western China. Lithos 256-257, 243–249.

Xue, S., Qin, K., Li, C., Tang, D., Mao, Y., Qi, L., et al., 2016b. Geochronological, Petrological, and geochemical constraints on Ni-Cu sulfide mineralization in the Poyi ultramafic-troctolitic intrusion in the northeast rim of the Tarim craton, western China. Econ. Geol. 111, 1465–1484.

Yang, S., Qu, W., Tian, Y., Chen, J., Yang, G., Du, A., 2008. Origin of the inconsistent apparent Re-Os ages of the Jinchuan Ni-Cu sulfide ore deposit, China: Post-segregation diffusion of Os. Chem. Geol. 247, 401–418.

Yu, X., Yang, S.-F., Chen, H.-L., Chen, Z.-Q., Li, Z.-L., Batt, G.E., et al., 2011. Permian flood basalts from the Tarim Basin, Northwest China: SHRIMP zircon U-Pb dating and geochemical characteristics, Gond. Res. 20. pp. 485–497.

Yu, S.-Y., Song, X.-Y., Ripley, E.M., Li, C., Chen, L.-M., She, Y.-W., et al., 2015. Integrated O-Sr-Nd isotope constraints on the evolution of four important Fe-Ti oxide ore-bearing mafic-ultramafic intrusions in the Emeishan large igneous province, SW China. Chem. Geol. 401, 28–42.

Zhang, C.-L., Li, Z.-X., Li, X.-H., Xu, Y.-G., Zhou, G., Ye, H.-M., 2010a. A Permian large igneous province in Tarim and Central Asian orogenic belt, NW China: Results of a ca. 275 Ma mantle plume? Geol. Soc. Am. Bull. 122, 2020–2040.

Zhang, C.-L., Xu, Y.-G., Li, Z.-X., Wang, H.-Y., Ye, H.-M., 2010b. Diverse Permian magmatism in the Tarim Block, NW China: Genetically linked to the Permian Tarim mantle plume? Lithos 119, 537–552.

Zhang, D., Zhang, Z., Santosh, M., Cheng, Z., Huang, H., Kang, J., 2013. Perovskite and baddeleyite from kimberlitic intrusions in the Tarim large igneous province signal the onset of an end-Carboniferous mantle plume. Earth Planet. Sci. Lett. 361, 238–248.

Zhang, D., Zhang, Z., Huang, H., Encarnación, J., Zhou, N., Ding, X., 2014. Platinum-group elemental and Re-Os isotopic geochemistry of the Wajilitag and Puchang Fe-Ti-V oxide deposits, northwestern Tarim Large Igneous Province. Ore Geol. Rev. 57, 589−601.

Zhang, D., Zhang, Z., Mao, J., Huang, H., Cheng, Z., 2016a. Zircon U-Pb ages and Hf-O isotopic signatures of the Wajilitag and Puchang Fe-Ti oxide-bearing intrusive complexes: Constraints on their source characteristics and temporal-spatial evolution of the Tarim large igneous province. Gond. Res. 37, 71−85.

Zhang, M., Kamo, S.L., Li, C., Hu, P., Ripley, E.M., 2010c. Precise U-Pb zircon-baddeleyite age of the Jinchuan sulfide ore-bearing ultramafic intrusion, western China. Mineral. Deposita 45, 3−9.

Zhang, M., Tang, Q., Hu, P., Ye, X., Cong, Y., 2013. Noble gas isotopic constraints on the origin and evolution of the Jinchuan Ni-Cu-(PGE) sulfide ore-bearing ultramafic intrusion, Western China. Chem. Geol. 339, 301−312.

Zhang, X.-Q., Song, X.-Y., Chen, L.-M., Xie, W., Yu, S.-Y., Zheng, W.-Q., et al., 2012. Fractional crystallization and the formation of the thick Fe-Ti-V oxide layers in the Baima layered intrusion, SW China. Ore Geol. Rev. 49, 96−108.

Zhang, Y., Liu, J., Guo, Z., 2010d. Permian basaltic rocks in the Tarim basin, NW China: Implications for plume-lithosphere interaction. Gond. Res. 18, 596−610.

Zhang, Z., Mahoney, J.J., Mao, J., Wang, F., 2006. Geochemistry of picritic and associated basalt flows of the western Emeishan flood basalt province, China. J. Petrol. 47, 1997−2019.

Zhang, Z., Mao, J., Chai, F., Yan, S., Chen, B., Pirajno, F., 2009a. Geochemistry of the Permian Kalatongke mafic intrusions, northern Xinjiang, northwest China: Implications for the genesis of magmatic Ni-Cu sulfide deposits. Econ. Geol. 104, 185−203.

Zhang, Z., Mao, J., Saunders, A.D., Ai, Y., Li, Y., Zhao, L., 2009b. Petrogenetic modeling of three mafic-ultramafic layered intrusions in the Emeishan large igneous province, SW China, based on isotopic and bulk chemical constraints. Lithos 113, 369-292.

Zhang, Z., Tang, Q., Li, C., Wang, Y., Ripley, E.M., 2016b. Sr-Nd-Os-S isotope and PGE geochemistry of the Xiarihamu magmatic sulfide deposit in the Qinghai-Tibet plateau, China. Mineral. Deposita. Available from: http://dx.doi.org/10.1007/s00126-016-0645-0.

Zhao, G., Wilde, S.A., Cawood, P.A., Sun, M., 2001. Archean blocks and their boundaries in the North China Craton: lithological, geochemical, structural and P-T path constraints and tectonic evolution. Precamb. Res. 107, 45−73.

Zhao, T.-P., Chen, W., Zhou, M.-F., 2009. Geochemical and Nd-Hf isotopic constraints on the origin of the ~1.74-Ga Damiao anorthosite complex, North China Craton. Lithos 113, 673−690.

Zhong, H., Zhu, W.-G., 2006. Geochronology of layered mafc intrusions from the Pan-Xi area in the Emeishan large igneous province, SW China. Mineral. Deposita 41, 599−606.

Zhong, H., Zhou, X.H., Zhou, M.F., Sun, M., Liu, B.G., 2002. Platinum-group element geochemistry of the Hongge Fe-V-Ti deposit in the Pan-Xi area, southwestern China. Mineral. Deposita 37, 226−239.

Zhong, H., Yao, Y., Hu, S.-F., Zhou, X.-H., Liu, B.G., Sun, M., et al., 2003. Trace-element and Sr-Nd isotopic geochemistry of the PGE-Bearing Hongge layered intrusion, Southwestern China. Int. Geol. Rev. 45, 371−382.

Zhong, H., Hu, R.-Z., Wilson, A.H., Zhu, W.-G., 2005. Review of the link between the Hongge layered intrusion and the Emeishan flood basalts, southwest China. Int. Geol. Rev. 47, 971−985.

Zhou, M.-F., Kennedy, A.K., Sun, M., Malpas, J., Lesher, C.M., 2002a. Neoproterozoic arc-related mafic intrusions along the northern margin of South China: implications for the accretion of Rodinia. J. Geol. 110, 611−618.

Zhou, M.-F., Malpas, J., Song, X.-Y., Robinson, P.T., Sun, M., Kennedy, A.K., et al., 2002b. A temporal link between the Emeishan large igneous province (SW China) and the end-Guadalupian mass extinction. Earth Planet. Sci. Lett. 196, 113−122.

Zhou, M.-F., Yan, D.-P., Kennedy, A.K., Li, Y.-Q., Ding, J., 2002c. SHRIMP U-Pb zircon geochronological and geochemical evidence for Neoproterozoic arc-magmatism along the western margin of the Yangtze Block, South China. Earth Planet. Sci. Lett. 196, 51−67.

Zhou, M.-F., Yang, Z.-X., Song, X.-Y., Lesher, C.M., Keays, R.R., 2002d. Magmatic Ni-Cu-(PGE) sulfide deposits in China. In: Cabri, L.J. (Ed.), The geology, geochemistry, mineralogy and mineral beneficiation of platinum-group elements. Canadian Institute of Mining, Metallurgy and Petroleum, Special, Volume, 54. pp. 619−636.

Zhou, M.-F., Lesher, C.M., Yang, Z., Li, J., Sun, M., 2004. Geochemistry and petrogenesis of 270 Ma Ni-Cu-(PGE) sulfide-bearing mafic intrusions in the Huangshan district, Eastern Xinjiang, Northwest China: implications for the tectonic evolution of the Central Asian orogenic belt, Chem. Geol., 209. pp. 233−257.

Zhou, M.-F., Robinson, P.T., Lesher, C.M., Keays, R.R., Zhang, C.-J., Malpas, J., 2005. Geochemistry, petrogenesis and metallogenesis of the Panzhihua gabbroic layered intrusion and associated Fe-Ti-V oxide deposits, Sichuan Province, SW China. J. Petrol. 46, 2253−2280.

Zhou, M.-F., Arndt, N.T., Malpas, J., Wang, C.Y., Kennedy, A., 2008. Two magma series and associated ore deposit types in the Permian Emeishan large igneous province, SW China. Lithos 103, 352−368.

Zhou, M.-F., Zhao, J.-H., Jiang, C.-Y., Gao, J.-F., Wang, W., Yang, S.-H., 2009. OIB-like, heterogeneous mantle sources of Permian basaltic magmatism in the western Tarim Basin, NW China: Implications for a possible Permian large igneous province. Lithos 113, 583−594.

Zhou, M.-F., Chen, W.T., Wang, C.Y., Prevec, S.A., Liu, P.-P., Howarth, G.H., 2013. Two stages of immiscible liquid separation in the formation of Panzhihua-type Fe-Ti-V oxide deposits, SW China. Geosci. Front. 4, 481−502.

Zhu, D.-C., Zhao, Z.-D., Niu, Y., Dilek, Y., Hou, Z.-Q., Mo, X.-X., 2013. The origin and pre-Cenozoic evolution of the Tibetan Plateau. Gond. Res. 23, 1429−1454.

Zou, S.-Y., Li, Z.-L., Song, B., Ernst, R.E., Li, Y.-Q., Ren, Z.-Y., et al., 2015. Zircon U-Pb dating, geochemistry and Sr-Nd-Pb-Hf isotopes of the Wajilitag alkali mafic dikes, and associated diorite and syenitic rocks: Implications for magmatic evolution of the Tarim large igneous province. Lithos 212-215, 428−442.

ALASKAN-TYPE COMPLEXES AND THEIR ASSOCIATIONS WITH ECONOMIC MINERAL DEPOSITS

9

Joyashish Thakurta

Western Michigan University, Kalamazoo, MI, United States

CHAPTER OUTLINE

9.1 INTRODUCTION

Alaskan-type complexes constitute a special category of igneous bodies composed of zoned ultramafic to mafic lithological units. These complexes are commonly oriented in groups along the trends of convergent tectonic margins (Taylor, 1967; Irvine, 1974; Himmelberg and Loney, 1995). Lithologically the complexes show an outward gradation from silica-depleted rocks in the central region to silica-enriched rocks in the periphery. A representative sequence of rocks from the center outward is: dunite, wehrlite, olivine clinopyroxenite, hornblende-magnetite clinopyroxenite, hornblendite, and gabbro (Taylor, 1967; Himmelberg and Loney, 1995; Johan, 2002). Ideally, these

Processes and Ore Deposits of Ultramafic-Mafic Magmas through Space and Time. DOI: http://dx.doi.org/10.1016/B978-0-12-811159-8.00010-X

rock units form individual concentric zones. Typically, outcrops of Alaskan-type complexes are small relative to other types of mafic-ultramafic intrusive bodies, such as layered complexes. Most Alaskan-type complexes vary in surface area from 12 to 40 km^2 although a few may extend to 70−80 km^2 (Johan, 2002). Owing to their ultramafic compositions, Alaskan-type complexes are highly susceptible to chemical weathering and consequently their surface outcrops commonly are highly eroded. Extensive growth of secondary minerals such as serpentine, caused by low-temperature hydrothermal alteration (Thakurta et al., 2009) as well as surficial chemical weathering makes characterization of these complexes difficult. Moreover, the accessibility of several Alaskan-type complexes is limited because of their locations in mobile belts. Finally, the close association of these complexes with ophiolite complexes along convergent tectonic boundaries, shear zones, and thrust belts (Saleeby, 1992; Gehrels et al., 1987) has made their study more difficult. This chapter describes the salient characteristics of Alaskan-type complexes in terms of their petrological and geochemical features and their tectonic occurrence, and investigates the potential of these complexes as target locations for economic Cu-Ni-PGE mineralizations. Three Alaskan-type complexes: Duke Island, Salt Chuck, and Union Bay have been used as case studies.

9.2 SALIENT CHARACTERISTICS OF ALASKAN-TYPE COMPLEXES

Ultramafic bodies in southeastern Alaska (Fig. 9.1) have been studied by several early workers. These studies noted the concentric nature of the lithological units which comprise them (Buddington and Chapin, 1929; Kennedy and Walton, 1946; Walton, 1951; Ruckmick and Noble, 1959). Subsequent studies (Taylor and Noble, 1960; Taylor, 1967; Wyllie, 1967; Jackson, 1971) confirmed these early observations. Taylor (1967) highlighted the similarities of Alaskan-type zoned ultramafic complexes with those in the Ural Mountain Belt in Russia. The terms, "Ural(ian)-Alaskan type complex," and "Alaskan-type complex" have been used by many authors (Irvine, 1974; Garuti et al., 2003; Johan, 2002; Thakurta et al., 2008a,b, 2014; Guillou-Frottier et al., 2014) for this special group of intrusive bodies. In the present work the name, Alaskan-type complexes is used.

Alaskan-type complexes have been mined for platinum group elements (PGEs) for more than a century. Before the discovery of PGE mineralization in South Africa, most of the world's platinum came from placer deposits associated with Alaskan-type complexes around the world, particularly those in the Ural Mountains of Russia (Fominykh and Khvostova, 1970; Mochalov and Khoroshilova, 1998). Additional PGE-producing Alaskan-type complexes include the Goodnews Bay Complex in southwestern Alaska (Foley et al., 1997) and the Fifield Platinum complex in southeastern Australia (Suppel and Barron, 1986). In southeastern Alaska, PGEs have been mined along with Cu from the Salt Chuck Alaskan-type complex. In other well-known Alaskan-type complexes discussed in this chapter, such as the Duke Island Complex and the Union Bay Complex, the occurrences of PGE have been reported and studied, but none have emerged as economic or feasible for mining. At present, small PGE mining operations continue in the Urals (Nizhny Tagil Complex; Tessalina et al., 2016). However, present global PGE production is dominated by layered mafic intrusive bodies such as the Bushveld Complex (Naldrett et al., 1986) in South Africa and the Stillwater Complex (Czamanske and Zientek, 1985) in the United States, and the flood basalt-related mafic intrusive bodies such as Noril'sk in Russia (Naldrett et al., 1992).

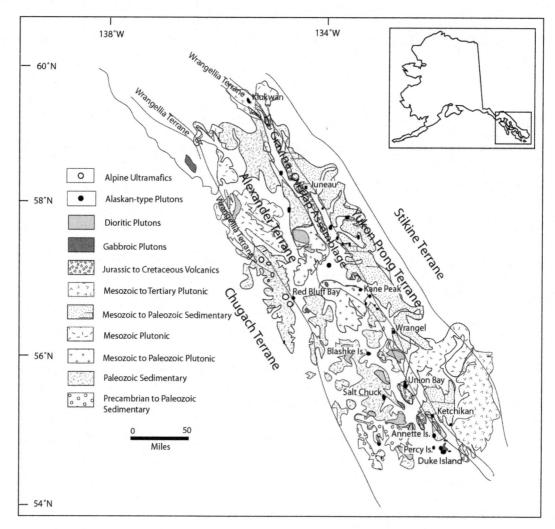

FIGURE 9.1

Geological map of southeastern Alaska, showing some of the major Alaskan-type complexes in the area including Duke Island Complex, Union Bay Complex, and Salt Chuck Complex.

After Thakurta, J., Ripley, E.M., Li, C., 2008a. Geochemical constraints on the origin of sulfide mineralization in the Duke Island Complex, southeastern Alaska. Geochem. Geophys. Geosyst. 9, 34 p., Q07003, doi:10.1029/2008GC001982.

Many Alaskan-type complexes have been described as structures with funnel-shaped cross-sections (Irvine, 1974; Himmelberg and Loney, 1995; Johan, 2002). Although the structural and petrogenetic relationships of lithological units in Alaskan-type complexes, and their associations with volcanic rocks, remain controversial, a number of theories have been developed to explain the observed relationships. Based on funnel-like cross-sections, and their associations with volcanic activity, Alaskan-type complexes have been proposed to be mid-crustal storage reservoirs for subduction-zone volcanoes (Murray, 1972). This hypothesis is consistent with the location of Alaskan-type complexes, which occur in linear clusters parallel to volcanic belts in convergent tectonic zones as seen in southeastern Alaska. The ages of Alaskan-type complexes in southeastern Alaska fall in two groups: 100−118 Ma and 400−440 Ma (Lanphere and Eberlain, 1966; Himmelberg and Loney, 1995). These dates correspond to the episodes of tectonic convergence and magmatic activity in the area (Rubin and Saleeby, 1992).

One problem in the identification of Alaskan-type complexes is their similarity to ophiolite sequences, which are located in similar tectonic provinces, often along the same geographic trend. This is particularly common in southeastern Alaska where several ophiolite complexes, also called Alpine-type peridotite bodies, are located close to Alaskan-type complexes. Although broadly similar in lithology and tectonic setting, ophiolite bodies have distinctly different modes of origin from the Alaskan-type complexes. The former represents slices of oceanic crust and lithospheric mantle obducted as thrust sheets, while the latter are magmatic intrusive bodies. Alaskan-type complexes show clear evidence of crystallization from pulses of mafic-ultramafic magma. Secondly, harzburgite, orthopyroxenite, and norite are common constituents of ophiolite complexes but such orthopyroxene-rich rocks are absent in Alaskan-type complexes. Thirdly, hydrous minerals such as hornblende are quite common in Alaskan-type complexes but are not usually found in ophiolite complexes. And finally, undeformed Alaskan-type complexes have near-vertical funnel-shaped or pipe-like cross-sections, with concentrically zoned rock units. Such structures are absent in ophiolite sequences.

Although most Alaskan-type complexes occur in convergent tectonic zones, or mobile belts, there are a few which do not. The Kondyor Complex (Burg et al., 2009) in the Aldan Shield of Russia is one example. Thus, in a broader sense, the Alaskan-type complexes could be deemed to represent volcanic flow-through systems and magma storage reservoirs regardless of their tectonic settings.

As stated before, lithological zoning has been regarded to be an important characteristic of Alaskan-type complexes. In rare instances, such as Blashke Islands in southeastern Alaska (Himmelberg et al., 1986) and Kondyor (Burg et al., 2009), the lithological zoning is symmetrical, well-defined and concentric. In these occurrences, the ultramafic and mafic lithological units have dunite at the center and wehrlite, olivine clinopyroxenite, clinopyroxenite, and gabbro in concentric rims around the central dunite massif. However, concentric lithological zoning is not a universal characteristic of Alaskan-type complexes. In some instances, such as the Union Bay Complex (Ruckmick and Noble, 1959) and the Duke Island Complex (Irvine, 1974) the zonal patterns have been extensively deformed, so that any original concentricity of the lithological units is no longer observed. There are a few Alaskan-type intrusions, such as the Salt Chuck Complex in southeastern Alaska (Loney and Himmelberg, 1992) in which only two lithological units occur as layered bands, with no observed concentric alignment.

In summary, there is a large variety of Alaskan-type complexes in terms of the number, type, and shape of constituent lithological zones, although in all instances there is a progressive outward gradation toward more siliceous compositions. The relationships between the ultramafic and the mafic lithological units in Alaskan-type complexes are controversial (Findlay, 1969; Irvine, 1974) and poorly understood. In some instances, such as the Duke Island Complex, the gabbroic rocks have been dated to be older than the ultramafic units, consistent with intrusive relationships (Saleeby, 1992). Based on geochemical considerations, Irvine (1974) suggested that the two groups of rocks represent two distinct petrogenetic events. In the Union Bay Complex and Tulameen Complex (British Columbia, Canada), however, the gabbroic units are intimately banded with the ultramafic rocks and follow the same structural style (Findlay, 1969), and gabbroic dikes intrude the ultramafic rocks (Van Treeck, 2009).

9.3 AGES OF ALASKAN-TYPE COMPLEXES

Ages of Alaskan-type complexes around the world span geological time. To the knowledge of the present author, the Neo-Archean Quetico Intrusion of Western Superior Province in Canada, dated at ~2690 Ma (Pettigrew and Hattori, 2006) is the oldest known Alaskan-type complex, while the Tertiary Alto Condoto Complex in Colombia, dated at ~20 Ma (Tistl, 1994) is the youngest. In southeastern Alaska, most Alaskan-type complexes are between 100 and 110 Ma old (Lanphere and Eberlain, 1966; Taylor, 1967). A few complexes, such as Salt Chuck, are dated between 400 and 440 Ma (Loney et al., 1987). The Goodnews Bay Complex in southwestern Alaska formed in Middle Jurassic (Plafker and Berg, 1994) while the Tulameen Complex in British Columbia is dated approximately to the Upper Triassic (Findlay, 1969). In the Ural Mountain Belt, the Alaskan-type complexes are dated between 340 and 355 Ma (Fominykh and Khvostova, 1970).

9.4 OCCURRENCES AND PROSPECTS OF ECONOMIC MINERALIZATIONS

Several Alaskan-type complexes such as Nizhny Tagil (Ural belt, Russia), Goodnews Bay (Alaska, USA) and Alto Condoto (Colombia) have been mined for PGE. However, none of these mineralizations are associated with sulfide segregations. Owing to the apparent scarcity of sulfide minerals, PGE-rich Alaskan-type complexes have been historically regarded as sulfide-absent systems and traditionally have not been considered as exploration targets for magmatic sulfide Cu-Ni-PGE mineralization. Johan (2002) identified three types of PGE occurrences in sulfide-absent Alaskan-type complexes:

1. PGE concentrations associated with chromitites in the dunite cores of complexes, such as Nizhny Tagil (Russia), Tulameen (Canada), Fifield (Australia), Yubdo (Ethiopia).
2. Pt-Fe alloys occurring as accessory minerals in the ultramafic rocks of complexes, such as Kondyor (Russia) and Alto Condoto (Columbia).
3. PGE-rich minerals found in the magnetite-rich clinopyroxenite units forming a rim around dunite core in complexes, such as Katchkanar (Russia) and Owendale (Australia).

However, in recent years the presence of sulfide minerals has been reported from several Alaskan-type complexes and some of them are under active exploration for Cu-Ni-PGE mineralizations (Thakurta et al., 2008b). Cu-Ni mineralization has been recently reported in the Duke Island Complex (Fig. 9.2) in southeastern Alaska (Thakurta et al., 2008a); Ni-sulfide mineralization has

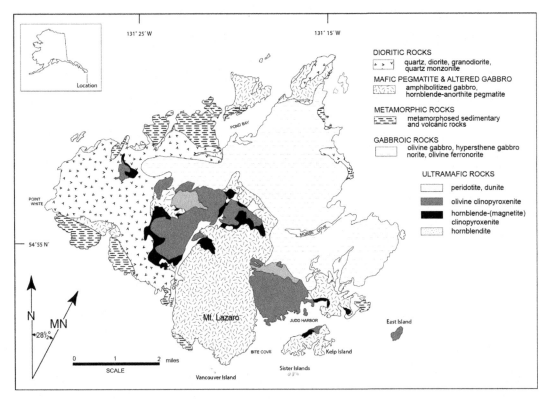

FIGURE 9.2

Geological map of the Duke Island Complex, showing the Hall Cove and Judd Harbor intrusions at the center and southeast respectively.

been identified in the Turnagain Complex in northwestern British Columbia, Canada (Nixon, 1998; Scheel et al., 2009); economic Pd-sulfide mineralization has been reported from the Salt Chuck Complex in southeastern Alaska (Loney and Himmelberg, 1992; Watkinson and Melling, 1992). Several other complexes such as Union Bay in southeastern Alaska are under exploration for sulfide deposits.

The primary magmas responsible for most Alaskan-type complexes are known to contain reasonably high concentrations of PGEs as exemplified by PGE deposits associated with such complexes around the world (Johan, 2002). In addition, if the Alaskan-type complexes represent feeder channels to extensive arc-type andesitic volcanism (e.g., the Gravina arc volcanic sequence) it seems probable that the intrusions were parts of conduit systems through which large volumes of magma may have passed. Thus, the immiscible sulfide liquid generated in the magmas by oversaturation of sulfide, may have interacted with quantities of metal-rich silicate magma sufficient to produce economic Cu-Ni-PGE-rich intersects in the sulfide zones of the complexes. However, the true potential of Alaskan-type complexes as hosts of economic PGE-sulfide deposits still remains to be evaluated in more locations around the world (Naldrett, 2010).

9.5 CASE STUDIES

This section provides case studies from three representative Alaskan-type complexes in southeastern Alaska: Duke Island Complex, Salt Chuck, and Union Bay Complex (Fig. 9.1). Southeastern Alaska has been divided into six terranes (Himmelberg and Loney, 1995) based on structure and tectonic history: (1) Chugach terrane, composed of Early Jurassic to Late Cretaceous flysch and mafic metavolcanic rocks; (2) Wrangellia terrane, comprising Permian and Triassic graywacke, limestone, and metavolcanic rocks; (3) Alexander terrane, composed of a continuous succession of Late Proterozoic to Late Triassic graywacke turbidites, limestone, and mafic to felsic volcanic rocks; (4) Gravina overlap sequence, composed of metamorphosed and deformed late Jurassic to Cretaceous marine argillite, graywacke and conglomerate, interbedded andesitic to basaltic rocks and volcaniclastic sediments and intrusive rocks which include quartz diorite, dunite, and peridotite (Gehrels and Berg, 1994); (5) Yukon Prong terrane, comprising Early to Paleozoic metapelite, metabasalt, marble, and quartzite; and (6) Stikine terrane, composed of Devonian metasediments and Mississippian and Permian volcaniclastic rocks, mafic to felsic volcanic rocks and limestones.

9.5.1 THE DUKE ISLAND COMPLEX

Irvine (1963, 1967, 1974) wrote detailed field and petrological descriptions of the Duke Island Complex, which is the southernmost in a long chain of mafic-ultramafic intrusions emplaced along the 550 km length of the Alaskan panhandle (Fig. 9.1). The Percy Islands Complex is about 6 km west of Duke Island and the Annette Island Complex is located about 16 km northwest of it.

Duke Island is located in the Alexander terrane, which is noted for Early Ordovician to Early Silurian arc-type volcanic and plutonic activity. The magmatism resulted in large dioritic to granitic rock bodies (Gehrels et al., 1996). The Duke Island Complex formed at the end of the Early Cretaceous extension, just before the onset of Mid-Cretaceous thrust faulting in the Alexander terrane (Saleeby, 1992).

Apart from ultramafic rocks, several other units are exposed at Duke Island (Fig. 9.2), ranging in age from Ordovician to Late Triassic (Buddington and Chapin, 1929; Berg et al., 1972; Irvine, 1974; Gehrels et al., 1987; Saleeby, 1992). The Ordovician to Early Silurian Descon Formation consists of volcanic and volcaniclastic rocks along with graywackes, mudstones, and turbidites and ranges in metamorphic grade from greenschist to amphibolite. Silurian dioritic rocks in the western part of the island and associated with these metamorphosed sediments yield a U-Pb zircon age of 429 ± 20 Ma (Gehrels et al., 1987). Similar ages have been reported for trondhjemitic to granitic rocks on the island. A Late Triassic gabbro, dated by U-Pb zircon method at 226 ± 3 Ma (Gehrels et al., 1987), is the most widespread unit in the island.

There are two main outcrops of ultramafic rock at Duke Island: the Hall Cove and the Judd Harbor intrusions (Fig. 9.2). The Hall Cove intrusion is located near the center of the island; the Judd harbor intrusion is in the southeastern part of the island. Based on petrological similarities, Irvine (1974) argued that the two intrusions are connected at depth and form two different lobes of a single magmatic system. The ultramafic rocks exposed at East and Kelp Islands were also interpreted as parts of the same conduit system. Based on structural interpretations and aeromagnetic datasets, Butler et al. (2001), confirmed the three-pronged structure of the magmatic conduit

(Fig. 9.2). It was proposed that the three lobes are combined results of multiple magmatic outlet channels, as well as structural deformation such as post-intrusion faulting.

The concentric lithological structure is poorly developed at the Duke Island Complex. In the Hall Cove intrusion, some degree of concentric lithological zoning can be recognized, but in the Judd Harbor intrusion, concentric zoning is absent. Several ultramafic rock units in the Duke Island complex, as described below, are surrounded by gabbroic rocks in the east, north, and south, and dioritic rocks in the west (Fig. 9.2). Olivine clinopyroxenite is the predominant rock type in both the Hall Cove and Judd Harbor intrusions. In the interior region of the Hall Cove intrusion, wehrlite is exposed, with small isolated occurrences of dunite. Wehrlite is rimmed by a moderately thick zone of olivine clinopyroxenite. Along the outer fringes of the intrusion, a thin zone of hornblende-magnetite clinopyroxenite is exposed. This does not show prominent outcrops in the field, but has been mapped by Irvine (1974) based on drill core data and magnetic anomalies. The contact between olivine clinopyroxenite and hornblende-magnetite clinopyroxenite is interpreted from drill core studies to be gradational. In the Judd Harbor intrusion, dunite, with minor wehrlite, is found in the northern area, while olivine clinopyroxenite is exposed in the rest of the intrusion. The wehrlite-dunite unit is intrusive into the olivine clinopyroxenite unit in both Hall Cove and Judd Harbor locations (Irvine, 1974; Thakurta et al., 2008a).

In the Hall Cove intrusion, gradational layering with respect to grain size is very well developed in the wehrlite and olivine clinopyroxenite units. These layered rocks show conspicuous repetitive upward-fining sequences, and each layer ranges from a few centimeters to one meter thick (Fig. 9.3B). At the bottom of each layer, there are several coarse crystals and xenocrysts of clinopyroxene or xenoliths of clinopyroxenite with sharp contacts with the underlying layer. Irvine (1974) interpreted these graded layers as products of crystal mush that rolled down the inward-sloping walls of the funnel-shaped magmatic conduit system.

Unequivocal occurrences of xenoliths of clinopyroxenite are seen in the central wehrlite unit of the intrusion (Fig. 9.4). Spectacular repetitive layering and the occurrence of large varieties of

FIGURE 9.3

(A and B) Upward fining repetitive layered structures defined by xenoliths and xenocrysts of olivine clinopyroxenite, in wehrlite unit at the Hall Cove intrusion at Duke Island.

FIGURE 9.4

Xenoliths of olivine clinopyroxenite of different sizes and shapes included in the wehrlite unit indicate that the magma crystalizing dunite-wehrlite was younger than the surrounding olivine clinopyroxenite unit and was intrusive into it.

xenoliths in the central wehrlite unit of the Hall Cove intrusion suggest that the Duke Island Complex is a remnant of highly dynamic magma chamber, characterized by mass movements of wall-rock and crystal slurries. It follows that the magma plumbing system at the Duke Island Complex was an active dynamic flow-through system along which large volumes of magma have moved up to the surface.

With the wehrlite intrusive into the olivine clinopyroxenite and the gradational contact between olivine clinopyroxenite and hornblende-magnetite clinopyroxenite, it can be argued that the hornblende-magnetite clinopyroxenite and olivine clinopyroxenite crystallized from magmatic pulses derived from a continuously evolving magma chamber. However, the magma responsible for the wehrlite unit must have been more primitive and relatively fresh from the mantle source. The well-accepted notion of compositionally more primitive rocks being younger and located at the center of the sequence holds true for the Duke Island Complex (Irvine, 1974; Himmelberg and Loney, 1995; Johan, 2002).

As with all Alaskan-type complexes, in which glass or volcanic rock compositions representative of magma are absent, the composition of the parental magma of the Duke Island Complex has remained enigmatic. Numerical modeling with major element compositions, using MELTS (Ghiorso and Sack, 1995) by Thakurta et al. (2008a) suggests that the Bridget Cove ankaramite is a suitable parental magma composition for the Duke Island Complex, as was proposed by Irvine (1973).

Thakurta et al. (2008a) described and modeled the occurrence of sulfide mineral deposits in the Duke Island Complex. Most of the sulfide minerals were found in olivine clinopyroxenite, although small disseminations and veins of sulfide minerals have also been found in the neighboring wehrlite of the Hall Cove intrusion. The sulfide mineral occurrences can be classified into three textural categories: massive, net-textured, and disseminated. Pyrrhotite, the principal sulfide mineral, comprises approximately 95% by volume of the massive sulfide domains. Chalcopyrite, often intergrown with cubanite, comprises ~4% by volume and pentlandite occurs in traces (~1%). Occasionally, sulfide rims up to 5 mm thick surround grains of olivine and clinopyroxene. Rounded inclusions of sulfide, representing droplets of sulfide liquid, are included in olivine and clinopyroxene crystals. Sulfide inclusions comprise up to 20 vol.% chalcopyrite and up to 5 vol.% pentlandite. The average composition of sulfides from the Duke Island Complex, recalculated to 100% sulfide, is 1.4% Cu, 0.48% Ni and 1 ppm combined Pt and Pd.

Although Alaskan-type complexes are typically sulfide-depleted and platinum-group mineral (PGM)-enriched, the Duke Island Complex is a stark exception in this regard. Thakurta et al. (2014) noted that that the sulfide-rich zones at Duke Island are characterized by strong depletions in Ir, Os, and Ru (IPGE) and relative enrichments in Pt, Pd, and Cu. Crystallization modeling using MELTS (Thakurta et al., 2008a) suggests that sulfide-hosting olivine clinopyroxenite formed from an evolved magma which had fractionally crystallized olivine and chromite. PGE-rich domains are well-known in Alaskan-type complexes, but in most cases in the Ural Mountain Belt and in other occurrences around the world, the PGE are concentrated in Fe- and Cr-alloys associated with chromian spinels in the dunitic cores of the intrusions (Johan, 2002). However, in the Duke Island Complex, the concentrations of PGE in the chromian spinels in dunite are low, and Fe- and Cr-alloys are rare. Instead, all PGEs are concentrated in sulfide minerals, primarily in olivine clinopyroxenite. Thakurta et al. (2014) reported concentrations of 400 ppb Pt and 700 ppb Pd in massive sulfide domains. In net-textured sulfide domains Pt and Pd concentrations reach 300 and 400 ppb, respectively. In disseminated sulfide domains, Pt and Pd concentrations are up to 300 and 250 ppb, respectively. When normalized to 100% sulfide, disseminated sulfides show higher PGE concentrations than massive and net-textured domains, with Pt concentrations up to 1200 ppb and Pd concentrations up to 1800 ppb.

However, the distribution of sulfide minerals in the intrusion is not uniform (Thakurta et al., 2008a). It is important to note that the observations and interpretations made by Thakurta et al. (2008a,b, 2014) were based on sulfides from drilling operations in the central and northwestern parts of the Hall Cove intrusion in 2001 and 2005. The drill locations were in target areas chosen after ground-based and airborne geophysical surveys. There are zones in the complex, particularly in the olivine clinopyroxenite and wehrlite units at the Hall Cove area, where sulfide horizons were intercepted at depths of less than a hundred meters. There is no consistent relationship between depth and the concentration of sulfide minerals in all of the drill cores in the Hall Cove intrusion.

Geochemical modeling by Thakurta et al. (2008a) suggests the possible existence of a Ni-rich sulfide mineral deposit at depth in the complex. However, this hypothetical zone was not intercepted by the drill cores which extended to a maximum depth of 245 m. The occurrence of the predicted chalcophile element-enriched sulfide zone at depth was modeled based on the identification of two olivine populations in the olivine clinopyroxenite. One population contained sulfide minerals, while the other was devoid of sulfides. Sulfide-free olivine pyroxenite contains relatively Ni-rich olivine (570−1189 ppm), whereas olivine from the sulfide-bearing olivine pyroxenite has remarkably low levels of Ni (63−246 ppm). Low Ni contents in the sulfide-free population cannot be explained by the measured Ni content of the sulfide minerals in the host rocks. A substantial portion of the Ni extracted from the magma must have been concentrated in a sulfide liquid, which has not been found in the associated rocks. Thus, Thakurta et al. (2008a) proposed multiple magmatic events even within individual rock units of the complex. In olivine clinopyroxenite, the sulfide-present and sulfide-absent domains represent upheavals of sulfide-saturated and sulfide-undersaturated magmatic pulses. The very low levels of Ni in olivine relative to Fo-content are indicative of much larger degrees of Ni extraction than what has been observed in the associated sulfide deposits in the complex (Thakurta et al, 2008a). Further, it can be hypothesized that this deeper sulfide zone, if present, will contain significant concentrations of other chalcophile elements such as Cu and PGE.

9.5.2 THE SALT CHUCK COMPLEX

The Salt Chuck Complex is a mafic-ultramafic Alaskan-type complex located on Prince of Wales Island in southeastern Alaska. The two major rock types in the complex are gabbro and clinopyroxenite (Watkinson and Melling, 1992; Loney and Himmelberg, 1992). In addition, there are local occurrences of diorite and a pegmatitic gabbro. The Salt Chuck Complex is located in the Alexander terrane and is surrounded by metasedimentary country rocks of Descon Formation which have been metamorphosed to greenschist facies (Watkinson and Melling, 1992). The complex hosts a Cu-Au-Pd-Ag deposit which has been mined intermittently between 1905 and 1941 and has produced 0.3 million metric tons of ore at an average grade of 0.95% Cu, 1 ppm Au, 2 ppm Pd, and 6 ppm Ag (Watkinson and Melling, 1992; Holt et al., 1948).

The gabbro consists of primary clinopyroxene, plagioclase, minor biotite, intergrown magnetite and ilmenite, and occasional apatite (Fig. 9.6). Secondary metamorphic/hydrothermal minerals include chlorite, epidote, actinolite, sericite, calcite, and sphene. On outcrop scale, the gabbro is mostly massive, although localized centimeter- to decimeter-scale banding, defined by medium- to coarse-grained augite and plagioclase crystals, has been reported (Watkinson and Melling, 1992). Rhythmic layering, with layers up to 20 cm thick, is occasionally present. Small-scale "cross-bedding" and "scour marks" have also been identified locally. In some locations, primary cumulus clinopyroxene and plagioclase are rare; most of the primary igneous textures have been modified by subsequent metamorphic recrystallization. Clinopyroxene comprises approximately 35% of the rock by volume, and mostly occurs as subhedral grains ranging in size between 1 and 3 mm. Magnetite inclusions rimmed by sericite and chlorite are seen in clinopyroxene.

Clinopyroxenite is medium to coarse-grained and consists primarily of cumulus clinopyroxene with postcumulus plagioclase, amphibole, mica, magnetite, and apatite (Watkinson and Melling, 1992). Phlogopite oikocrysts up to 1 cm in diameter are also seen in this rock. With the exception of a few centimeter-thick massive magnetite bands, layering has not been reported. Clinopyroxenite locally displays pegmatitic texture, defined by the occurrence of coarse feldspar crystals surrounded by coarse magnetite. These pegmatitic aggregates are 5−10 cm wide and more abundant near the contact with gabbro. Clinopyroxene crystals are 2−4 mm across and have an average composition of diopside. A coarse mosaic of clinopyroxene crystals showing triple junctions, with small anhedral intercumulus grains of sericitized and saussuritized plagioclase (anorthite), amphibole (magnesiohastingsite), and mica (phlogopite-eastonite), define the adcumulate texture of the clinopyroxenite. There are rare actinolite overgrowths on clinopyroxene.

The Salt Chuck Complex does not show any concentric zoning. The petrological relationship between gabbro and clinopyroxenite, both of which occur as bands, has been debated. Watkinson and Melling (1992) reported an intrusive contact between the gabbro and the clinopyroxenite, whereas Loney and Himmelberg (1992) interpreted a gradational contact between them. A gradational contact would indicate a gradually evolving melt composition crystallizing a more primitive member (clinopyroxenite) and then fractionating to crystallize a more evolved member (gabbro). An intrusive contact would indicate two separate pulses of magma from an evolving but periodically active magma chamber.

In the absence of concentric zoning, the identification of the Salt Chuck Complex as an Alaskan-type complex remains questionable. However, Salt Chuck was included as one among ~30 Alaskan-type mafic-ultramafic complexes in southeastern Alaska by Loney et al. (1987), who reported a K-Ar date of 429 Ma for biotite in the Salt Chuck intrusion. Similar dates were obtained for comparable ultramafic bodies in Dall Island (400 Ma) and Sukkwan Island (440 Ma). Loney and Himmelberg (1992) interpreted these dates as representative of Paleozoic magmatic arc activity similar to the Cretaceous (100−110 Ma) subduction-zone magmatism that was responsible for most of the other Alaskan-type complexes in the area. Thus, two distinct generations of Alaskan-type complexes, one Paleozoic and another Cretaceous, have been recognized.

Sulfide mineralization is observed in the eastern part of the Salt Chuck intrusion, around the Salt Chuck mine area (Fig. 9.5). Sulfide deposits occur as irregular disseminations and fine veins in clinopyroxenite. Rare spheroidal aggregates of sulfides are less than 1 cm in diameter. Small sulfide blebs interstitial to clinopyroxene suggest that some sulfides formed as result of liquid immiscibility. However, the existence of sulfide veins and secondary hydrous phases such as epidote,

FIGURE 9.5

The gabbroic rock exposed near the old Salt Chuck Mine. It is composed of coarse clinopyroxene and plagioclase with minor phlogopite, magnetite, and ilmenite.

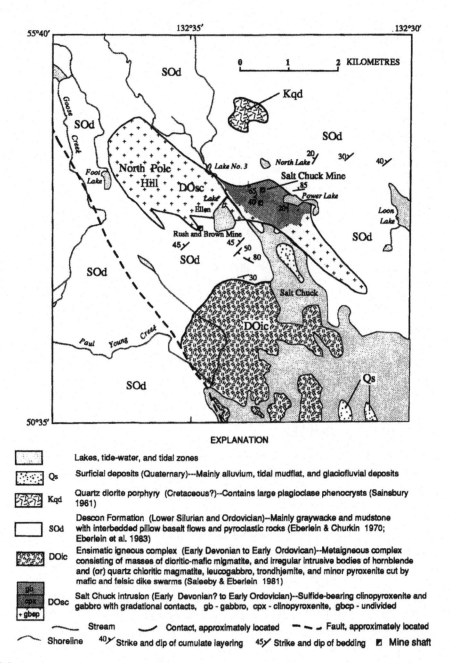

FIGURE 9.6

Geological map of the Salt Chuck Complex.

Reproduced with permission from Loney, R.A., Himmelberg, G.R., 1992. Petrogenesis of the Pd-rich intrusion of the Salt Chuck, Prince of Wales Island: an early Paleozoic Alaskan-type ultramafic body. Can. Mineral. 30, 1005–1022.

actinolite, and chlorite in close association with sulfide veins strongly suggest remobilization and reprecipitation of initial magmatic sulfide by a secondary hydrothermal fluid phase. Watkinson and Melling (1992) pointed out that magmatic sulfide mineral deposits are characterized by pyrrhotite-rich assemblages with high Ni/Cu ratios. In contrast, sulfide minerals at Salt Chuck are Cu-rich and Ni-Fe poor and consist of phases such as chalcopyrite, bornite, and digenite. This raises the possibility of a secondary origin for the Cu-Pd mineralization.

Mineralization occurs primarily within clinopyroxenite, mostly along the contact region with the gabbro (Watkinson and Melling, 1992). The principal ore minerals for Pd are kotulskite (PdTe) and speryllite ($PdAs_2$), while Cu is concentrated in bornite, chalcopyrite, chalcocite, and covellite. Native Au is also present. As reported by Watkinson and Melling (1992) and Loney and Himmelberg (1992), the mode of occurrence of the ore deposit suggests hydrothermal remobilization and precipitation of material derived from a preexisting sulfide mineral deposit.

Apart from the known mineralizations of Cu and Pd, the recent discovery of enriched zones with concentrations up to 128 ppm of Au and 58 ppm of Ag has helped to identify a new front of hydrothermal mineralization away from the original mine site (Pure Nickel, 2014). New Pd-enriched zones with concentrations up to 234 ppb seem to represent primary concentrations of magmatic sulfides at deeper levels of the intrusion.

While the Salt Chuck Complex is not concentrically zoned, it shows similar characteristics to other Alaskan-type intrusions in the area with respect to lithology, location, and enrichment in PGE. Additionally, the Salt Chuck Complex contains sulfide deposits which are known to host mineable grades of Cu, Pd, Au, and Ag. Although the process that led to metallic enrichment in the sulfide-rich zones of the intrusions is not well constrained, it may reflect hydrothermal remobilization of a preexisting magmatic sulfide mineral deposit at depth.

9.5.3 THE UNION BAY COMPLEX

The Union Bay Complex is the largest Alaskan-type complex in southeastern Alaska (Figs. 9.1 and 9.7). It is ~11.5 km long and ~8 km wide and shows a distorted concentrically zoned structure. It is located ~56 km north of Ketchikan at the northeast end of Cleveland Peninsula. The intrusion is situated within the narrow northwest-southeast trending Late Jurassic to Cretaceous Gravina tectonic belt. The Gravina Belt was tectonically deformed in Early Cretaceous and was regionally metamorphosed from greenschist to amphibolite facies during subduction-related intrusive activity which resulted in the formation of the Union Bay Complex.

Kennedy and Walton (1946) published the first comprehensive geological map of Union Bay and identified the mafic-ultramafic units of the complex. Ruckmick and Noble (1959) mapped the intrusion in greater detail and interpreted the rock units as parts of an overturned fold (Fig. 9.1). Ruckmick and Noble (1959) divided the complex into a western subhorizontal lopolithic domain, predominantly composed of clinopyroxenite, and an eastern vertical cylinder composed predominantly of dunite. Subsequent work by Clark and Greenwood (1972), Maas et al. (1995), and Himmelberg and Loney (1995) led to a better understanding of the petrological, structural, and geochemical characteristics of the intrusion. Van Treeck (2009) produced an updated geological map of the intrusion (Fig. 9.7) and highlighted the occurrences of platinum- and palladium-enriched zones in the intrusion. Petrological descriptions used in this chapter are based on this work.

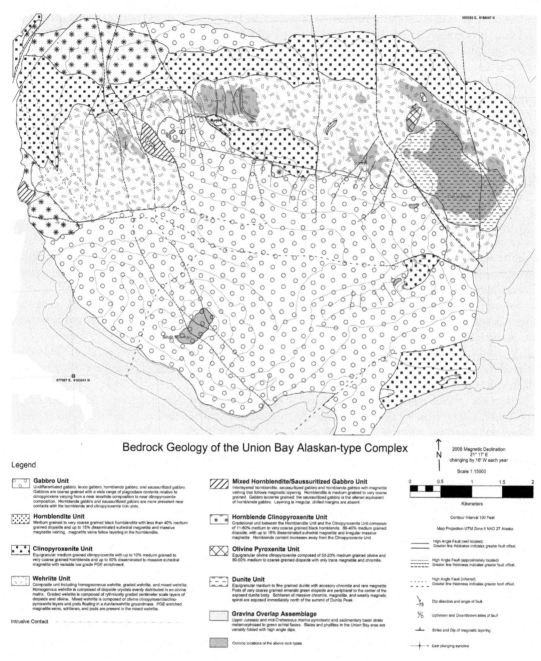

Bedrock Geology of the Union Bay Alaskan-type Complex

FIGURE 9.7

Geological map of the Union Bay Complex.

Reproduced with permission from Van Treeck, C.J., 2009. Platinum Group Element Enriched Hydrothermal Magnetite of the Union Bay Alaskan-type Ultramafic Intrusion, Southeast Alaska (Unpublished M.S. thesis). University of Alaska, Fairbanks, AK, 188 p.

Dunite (Fig. 9.7) occurs as a barren red dome called Dunite Peak in the northeastern part of the intrusion. It is composed of >90% olivine, about 9%−10% of euderal to subhedral diopside with minor chromite and magnetite. As in the Duke Island Complex, the dunite contains abundant xenoliths and xenocrysts of wehrlite and clinopyroxenite.

The wehrlite unit is described by Van Treeck (2009) as a collection of several lithological types termed homogeneous, graded, and mixed. The homogeneous wehrlite consists of >10 vol.% diopside, with large olivine crystals evenly distributed in a matrix of finer grained olivine. The graded wehrlite member has a higher content of coarse-grained diopside with respect to the finer grained crystals. Graded layering, similar to the Duke Island Complex, is seen in this member. The mixed wehrlite member is the most abundant lithology and it consists of interlayered clinopyroxene-rich and olivine-rich bands.

Olivine pyroxenite is homogenous and composed of 20%−50% medium- to coarse-grained olivine and 50%−80% medium- to coarse-grained diopside with minor chromite and magnetite.

Clinopyroxenite is composed of equigranular diopside with <10% hornblende and up to 50% disseminated, massive, and euhedral magnetite, often forming layers. In some outcrops west of the wehrlite unit, clinopyroxenite becomes richer in olivine and the rock grades into hornblende clinopyroxenite. Clinopyroxenite is often cross-cut by veins rich in magnetite and phlogopite, which also are enriched in PGE (Van Treeck, 2009).

Hornblende clinopyroxenite consists of 10%−60% euhedral hornblende, 40%−90% euhedral diopside, and up to 15% euhedral magnetite. Near its contact with gabbro, hornblende clinopyroxenite is intruded by gabbroic dikes.

A small outcrop of the hornblendite (Fig. 9.7) is composed of >60% euhedral to subhedral hornblende, <40% euhedral to subhedral diopside and <15% disseminated euhedral magnetite. Magnetite veins follow layering in the hornblendite.

There are several varieties of gabbroic rocks which are exposed over large areas of the intrusion. Van Treeck (2009) reported the occurrences of gabbro, orthopyroxene gabbro, leucogabbro, hornblende gabbro, and saussuritized gabbro. The gabbroic rocks, in general, show a large range in relative abundances of plagioclase and clinopyroxene and vary between anorthosite and clinopyroxenite. Orthopyroxene gabbro contains about 15% of orthopyroxene (Ruckmick and Noble, 1959). Hornblende gabbro is interlayered with bands of pegmatitic hornblendite. The contact between gabbroic rocks and ultramafic rocks is not seen in outcrop, and their petrogenetic relationships have remained poorly understood. However, ubiquitous intrusions of gabbroic dikes into the ultramafic units without any observed chilled margins suggests that ultramafic and gabbroic rocks developed contemporaneously.

No significant occurrences of sulfide minerals have been reported from the Union Bay Complex. However, PGE-rich magnetite veins in the ultramafic units have attracted considerable attention in the exploration community. Magnetite veins in wehrlite contain up to 15.9 ppm Pt and 2.9 ppm Pd (Avalon, 2001). Exploratory sampling and drilling in the intrusion has revealed a higher abundance of PGE, primarily hosted in the irregular magnetite veins in dunite, clinopyroxenite, and hornblende clinopyroxenite (Van Treeck, 2009). Pt/Pd ratios greater than 1 were measured in magnetite veins in olivine-rich lithological units, whereas Pt/Pd is ≤1 is calculated from magnetite veins in pyroxene- and hornblende-rich units.

Magnetite veins, bands, and schlierens in wehrlite contain the highest PGE concentrations. In the North Zone prospect, 10.6 ppm of Pt was measured over a half-meter interval in drill core (Van Treeck, 2009). Despite high concentrations of PGE in the magnetite veins, the nonsystematic and sporadic nature of mineralization challenges the economic viability of the complex.

Using Fe-Ti thermometry, Van Treeck (2009) determined that the PGE-enriched magnetite veins formed at a temperature range of $675-475°C$ with fO_2 conditions around the quartz-fayalite-magnetite (QFM) buffer. Therefore, the veins are indicative of circulation of hydrothermal fluids at submagmatic temperatures; these fluids precipitated magnetite and the associated ilmenite and spinel, along with the PGMs. Secondary hornblende and phlogopite indicate saline fluids that were enriched in Cl. The occurrence of Cl in hydrothermal fluids is positively correlated with the solubility and transport of PGE (Wood, 2002).

In general, Alaskan-type complexes are known to have higher concentrations of Pt relative to Pd and are extremely enriched in Pt with respect to Ru and Ir in whole-rock analyses (Johan, 2002). Garuti et al. (1997) pointed out that Alaskan-type complexes have strong enrichments in Pt and Au and depletions in Ni, Os, and Ir relative to primitive mantle while contents of Pd, Rh, and Cu vary from enriched to strongly depleted. The Union Bay Complex has an average Pt/Pd ratio of 5 and Os/Ir ratio of 2.25. The Pt/Pd ratio of the Union Bay Complex is lower than most sulfide-absent Alaskan-type complexes in the Urals but its Os/Ir ratio is much higher (Van Treeck, 2009). In the Ural belt and in most other Alaskan-type complexes, PGEs are present as alloys in chromite from dunite cores. However, at Union Bay the PGE content is heterogeneous and is distributed over large areas inside the complex. Such differences in the relative abundance of PGEs—as well as the observation of PGM in secondary magnetite veins cross-cutting ultramafic rock units of Union Bay—lead to the conclusion that the occurrence and enrichment of PGEs are controlled by processes that are distinct from most other known PGE occurrences in Alaskan-type complexes. The hydrothermal process of dissolution and reprecipitation of older ore minerals seems to be the most probable mechanism.

The question naturally arises as to the source of the PGEs and the mineralizing fluids. It can be hypothesized that the hydrothermal PGE mineralization in magnetite veins is a product of remobilization of preexisting PGMs at depth. The hypothetical mineralization at depth could be of primary magmatic origin, and could even include sulfide segregation as has been proposed for the Duke Island Complex (Thakurta et al., 2008a). Thus, the concentrations of PGE reflect their solubilities in the hydrothermal fluid rather than their concentrations in the primary magmatic assemblage. However, the depth of this prospective deposit is uncertain.

Hydrothermal alteration of the ultramafic rocks in the Alaskan-type complexes is well known. At Duke Island, extensive serpentinization of olivine-rich rocks has occurred (Thakurta et al., 2009). Oxygen and hydrogen isotopic ratios of the serpentine are consistent with the circulation of meteoric water. Thus, it can be hypothesized that the mineralizing fluid for the Union Bay complex had a comparable source. The large-scale and pervasive deformational structures in the Gravina Tectonic Belt, such as faults and shear zones developed during and after formation of the complex (Ruckmick and Noble, 1959), might have generated interconnected pathways for wide circulation of hydrothermal fluids.

The economic potential and feasibility of the Union Bay Complex remain uncertain, but the PGE-enriched secondary magnetite veins found during preliminary exploration have opened up possibilities for future exploration projects. These veins are distinct from the common modes of PGE mineralization in Alaskan-type complexes, which include PGE-alloys in the dunitic cores, placer deposits in the surrounding areas, and sulfide-enriched horizons in selected rock units. The identification of the new mode of PGE mineralization at the Union Bay Complex may enhance the development of new exploration strategies in Alaskan-type complexes around the world.

9.6 ORIGIN OF ALASKAN-TYPE COMPLEXES AND THE ASSOCIATED MINERAL DEPOSITS

The origin of Alaskan-type complexes has remained controversial (Himmelberg and Loney, 1995; Johan, 2002). In order to create a comprehensive model for the origin of this group of magmatic complexes, the following issues need to be addressed:

1. The origin of lithological zoning, which is sometimes concentric, with the most MgO-rich, SiO_2-poor lithologies at the center and the most MgO-poor, SiO_2-rich lithologies in the periphery
2. The nature of the parental magma(s)
3. Their modes of emplacement
4. The occurrence of PGE enrichment in these complexes and its origin
5. The occurrence of sulfide minerals in some of these complexes and their absence in the others

These issues are systematically discussed in the following sections.

9.6.1 ORIGIN OF LITHOLOGICAL ZONING

The existence of lithological domains in igneous intrusive bodies is well-known. Lithological zones in large layered igneous complexes such as Bushveld (Cawthorn et al, 1991) in South Africa, Stillwater (McCallum, 1996) in the United States, Muskox (Irvine, 1975) in Canada, and Skaergard (McBirney and Noyes, 1979) in Greenland have been extensively studied. In most cases, layering is classified as: (1) modal layering: caused by variation in relative proportions of constituent minerals; (2) phase layering: caused by appearance or disappearance of minerals in the crystallization sequence; and (3) cryptic layering: systematic variation in the chemical compositions of minerals across a layered sequence, which is not determinable by visual inspection. The layered patterns have also been classified into rhythmic and intermittent (Maaløe, 1978), based on the regularity of layering. In very large layered intrusions, such as the Bushveld and Stillwater, individual horizons of a particular rock type can be continuous over tens of kilometers. Some horizons, such as the Merensky Reef in the Bushveld Complex and the J-M Reef in the Stillwater complex are well known for economic concentrations of PGEs in sulfide minerals.

However, in Alaskan-type complexes, the nature of lithological zoning and the relationships between the zones differ from large layered intrusions. In special cases, such as Blashke Islands, the lithological zoning is concentric as seen in outcrop patterns and each zone is inferred to have a funnel-shaped three-dimensional structure, dipping inward toward the center. At Duke Island, there is geophysical evidence for the concentric funnel-shaped structure of the layered horizons (Butler et al., 2001). In cases such as Union Bay, where initial structures have been largely deformed by an overturned fold, the primary igneous structure is less discernable. Owing to the small sizes of Alaskan-type intrusive complexes and the structural distortion caused by subsequent deformation in convergent tectonic settings, the original distribution of lithological domains has been rarely preserved in outcrops.

Zoned lithological units include dunite, wehrlite, and olivine clinopyroxenite. Very frequently xenoliths of the outermost unit are seen preserved in the inner units, suggesting that the lithological zones in most Alaskan-type complexes were formed from the margin inwards (Fig. 9.4). In fact, the existence of xenoliths and xenocrysts in the neighboring unit can often be used as evidence for the establishment of relative times of formation between the units. The Duke Island Complex is a good example in this regard. In such cases the intrusive relationship of one unit with respect to the other is obvious.

However, the boundaries of lithological units are always not sharp and well-defined. In many cases they are gradational. For instance, at the Duke Island Complex, olivine clinopyroxenite grades outward into magnetite-hornblende clinopyroxenite. The transition from one unit to another is manifested by gradual changes in modal abundances of key minerals such as olivine and hornblende. The more primitive olivine clinopyroxenite crops out in the interior of the complex. Such gradational changes in lithology are indicative of fractional crystallization and progressive chemical evolution of the magma as temperature drops. Assimilation of Si-rich and Na-rich crustal rocks might also have led to the saturation of the melt in hornblende.

The intrusive relationships between units and differentiation caused by fractional crystallization are keys to developing a better understanding of the origin of Alaskan-type complexes. The most peculiar characteristic, as revealed by primary igneous structural relationships and modal and chemical compositions of lithological units, is that the more Mg-rich, silica-poor, and thus more primitive, members of the crystallization sequence tend to lie at the center while the more evolved Mg-poor, silica-rich members lie in the periphery. Thus, the primary question is: Why are the most primitive (highest-temperature) mineral assemblages found in the interiors of Alaskan-type intrusions, which are inferred to crystallize relatively late? In other words, how can the apparent discrepancy in the sequence of crystallization of Alaskan-type complexes indicated by field observations (youngest at the core) be reconciled with the widely accepted concept of cooling-induced crystallization (oldest at the core)? Did the magmatic system start cooling from the center outward? If the zoning was caused by crystallization differentiation of a single large batch of magma, it is intuitive that the cooling would begin at the margins and progress toward the center. That would make the marginal rocks most primitive and central rocks most evolved. This is contrary to what has been observed.

Numerous models have been proposed to explain the lithological zoning, as briefly described below.

Early workers such as Wyssotsky (1913) and Zavaritsky (1928) proposed that the olivine clinopyroxenite and hornblende clinopyroxenite zones were formed as metasomatic rims caused by the intrusion of peridotite into the gabbroic basement rocks. This model is unacceptable because the two clinopyroxenite variants are clearly products of magmatic crystallization as revealed by petrological and geochemical studies.

Walton (1951) studied the Blashke Islands Complex and theorized large-scale diffusion of Si, Ca, and H_2O outward from the center and Mg toward the center through a supercritical gas phase permeating through the interstitial pore spaces of a peridotitic magma undergoing crystallization. There are several problems with this model. Firstly, the compositional gradients necessary for such large-scale transport of chemical species over long distances are difficult to achieve through a crystal mush with magma in the interstitial pore spaces (Taylor, 1967). Secondly, the flow of magma and the turbulence caused by it will destroy the development of pervasive compositional gradients from the center to the periphery of the magma chamber. Thirdly, the theory does not account for complexes such as Haines and Klukwan in southeastern Alaska where dunite or peridotitic rocks are absent (Himmelberg and Loney, 1995).

Murray (1972) proposed that lithological zoning in Alaskan-type complexes was caused by flow differentiation in crystallizing magmas during passage through conduit systems under volcanoes. Flow differentiation is a hydraulic process that occurs during passage of slurries such as melted chocolate and tar through pipelines, and has been studied by engineers. Baragar (1960) used the model of flow differentiation to explain magmatic processes in the Labrador Trough of Canada. Bhattacharji and Smith (1964) demonstrated by experiments that monomineralic segregation and cryptic zonation, as seen in the picritic dikes of Skye (Scotland) and in the feeder dike of the Muskox intrusion in Canada, could be produced by concentration of early-formed crystals along the central regions of the conduit systems during magmatic flow. During flow of a liquid through a conduit, suspended particles concentrate along the central axis of the conduit. Thus, in a vertical magmatic conduit, early-formed Mg-rich olivine crystals will tend to concentrate along the central axis, followed by clinopyroxene and so on. The progressive and sequential development of crystals upon fractional crystallization in a dynamic magmatic conduit system, as supported by the process of flow differentiation, has been used to explain the concentric lithological zones in Alaskan-type complexes by Murray (1972) and by Himmelberg and Loney (1995) but has been questioned by Irvine (1974).

Although simple in principle, there are multiple inconsistencies in this model. Firstly, even though the process of flow differentiation can be demonstrated in laboratory experiments, in large magmatic conduit systems such processes are unlikely. Actual magmatic conduits are not smooth pipelines, but complex interlinked flow channels where a central linear axis may not develop. Moreover, the turbulent flows of wall rock fragments and convectional movements of magma charged with crystals will tend to obliterate any systematic concentration of early-formed crystals along the central regions of the conduits. Although concentration caused by flow differentiation in magmatic conduits could occur on local scales, this process is unlikely in large magmatic flow systems such as the Alaskan-type complexes. Moreover for flow differentiation to be responsible for the entire zonal structures as seen in Blashke Islands and Duke Island, the large diameter of the magma conduit and the enormous quantity of magma flow through them as required by the mechanism seem unlikely.

Ruckmick and Noble (1959), Taylor (1967), Taylor and Noble (1969), and Irvine (1959, 1963, 1967) proposed the mechanism of multiple intrusion. According to Taylor and Noble (1969), "Essentially all the observable field, chemical and mineralogical relationships in these ultramafic bodies can be explained by successive intrusions of magma in the order: magnetite pyroxenite—olivine pyroxenite—peridotite and/or dunite." They considered the possibility that these magmas could be derived by differentiation of a deeper magma reservoir close to the source region in the upper mantle. The gabbroic envelope of rocks is considered to be the first product of crystallization by an early magmatic pulse. This was followed in sequence by the magmas that crystallized the hornblende and olivine clinopyroxenites, wehrlite and dunite. It is reasonable to believe that if the ultramafic magmas responsible for these intrusions were derived by melting in the upper mantle or by fractional crystallization in deeper level reservoirs in the conduit system, the lowest-melting temperature fraction would be formed first. This would crystallize the gabbro. Successive surges of magma with higher temperatures and more primitive compositions would crystallize higher temperature minerals. The later melt fractions would intrude the preexisting rocks and partially incorporate them as xenoliths, as seen in the wehrlite zone of Duke Island Complex. The intrusive contacts at Duke Island between the ultramafic rocks and the gabbroic rocks, as well as between the wehrlite and dunite, are well documented.

Taylor and Noble (1969) discussed the applicability of fractional crystallization as a sole mechanism responsible for lithological zonation in Alaskan-type complexes. From a theoretical standpoint, fractional crystallization of an ultramafic magma could potentially generate a sequence of rock types as seen in Alaskan-type complexes. However as mentioned before, the expected sequence of rocks from the center outward is opposite to what is consistent with the model. Irvine (1974) combined the model of fractional crystallization with the model of diapiric updoming of semiconsolidated crystal mush. According to this theory, layers of fractionally crystallized cumulate minerals are domed upward in a compressive tectonic regime such that the lowest and earliest crystallized fraction, i.e., the dunite, is moved to the center of the dome, followed by wehrlite, and so on. The formation of this recumbent anticline structure in response to lateral compression of crystal mush in a compressional tectonic setting seems to be a reasonable explanation for the origin of zoning as well. The nappe structure, as seen at Union Bay, is a clear evidence of structural updoming and overturn formed in response to laterally compressive stress.

The main item of dissidence between the models of diapiric updoming and multiple intrusions lies in the nature of contacts between the clinopyroxenite, wehrlite, and dunite units in complexes such as Union Bay and Duke Island. If these contacts represent intrusive infiltration of the magma crystallizing the dunite, then the operative mechanism must be multiple intrusions. However, if the contacts represent rupture and solid-state displacements of previously-crystallized minerals, the mechanism is more akin to diapiric updoming. To the present author, the observed widespread occurrences of xenoliths in the wehrlite unit of Duke Island as well as repetitive grain-size layering are indicative of movements of liquids. This liquid cannot just be the interstitial and residual liquid between crystals; the observed nature of distortions and layering are indicative of large volumes of liquid in which the crystals and xenoliths of the wall rocks were entrained. Thus, the multiple-intrusive model seems more acceptable than the model of diapiric updoming.

9.6.2 NATURE OF PARENTAL MAGMAS

The ultramafic complexes clearly represent cumulate rocks formed by crystals precipitated by fractional crystallization of magmas. According to the classifications by Irvine (1982), "A cumulate is defined as an igneous rock characterized by a cumulus framework of touching mineral crystals or grains that were evidently formed and concentrated primarily through fractional crystallization." There are multiple lines of evidence for the cumulate nature of these rocks. Firstly, in most cases the rocks represent clinopyroxene and olivine grains with close interlocking contacts, sometimes with mosaic texture. There is little or no evidence of interstitial liquid. Secondly, the bulk compositions as determined from common rock types such as dunite and wehrlite (Table 9.1) cannot represent known types of magma. Thirdly, the concentrations of incompatible elements such as Zr, Y, and REE are extremely low relative to MORB. Thus, it is most probable that the high-Mg and low-SiO_2 rocks are early accumulations of crystals separated from a magma early in the crystallization sequence.

So, in the absence of rocks representative of actual magmatic compositions and of chilled margins along the contact areas of the intrusions with host rocks, the determination of the parental-magma composition is problematic. Intuitively, if the Alaskan-type complexes represent storage reservoirs for subduction-zone volcanoes, the volcanic rocks at the surface might represent the parental-magma composition. However, since the lavas erupted from the volcanic vents have undergone substantial degrees of crystallization differentiation in the conduit channels, they can no longer be regarded as representative of primitive magmatic compositions.

As stated Section 9.6.1, one of the flaws of the multiple intrusions model is the assumption that each rock unit is a crystallization product of a specific magmatic type. However, that might not necessarily be true. The ultramafic rock zones could be products of fractional crystallization of the

Table 9.1 Major-Element Oxide (in wt%) for ultramafic rocks From Duke Island (Thakurta et al., 2008a), Blashke Island (Himmelberg and Loney, 1995), and Union Bay (Himmelberg and Loney, 1995) Complexes

Complex	Duke					Blashke			UB		
Rock	Dunite	Wehrlite	Ol-Cpxnite	Ol-Cpxnite	Hbl-Mt-Cpxnite	Dunite	Wehrlite	Ol-Cpxnite	Dunite	Wehrlite	Ol-Cpxnite
SiO_2	34.34	42.27	47.04	49.16	35.95	35.60	37.30	48.60	38.20	39.00	51.60
TiO_2	0.06	0.22	0.32	0.33	1.99	0.00	0.00	0.31	0.00	0.00	0.12
Al_2O_3	2.18	0.97	2.08	3.79	14.43	0.31	0.41	3.66	0.00	0.16	1.38
FeO^a	15.68	15.30	11.83	8.51	21.21	9.40	10.23	7.82	10.53	11.01	4.42
MnO	0.23	0.23	0.14	0.12	0.17	0.17	0.19	0.14	0.19	0.20	0.10
MgO	33.78	32.61	19.65	18.93	11.62	43.20	38.50	18.00	45.80	41.70	20.20
CaO	0.04	7.39	17.79	19.14	11.14	0.23	2.81	20.10	0.25	2.34	21.70
Na_2O	1.74	0.05	0.17	0.18	1.84	0.00	0.00	0.22	0.21	0.18	0.21
K_2O	0.00	0.00	0.01	0.01	0.70	0.00	0.00	0.00	0.00	0.00	0.00
P_2O_5	0.00	0.01	0.00	0.00	0.00	0.00	0.00	0.00	0.00	0.00	0.00
LOI/H_2O	10.71	2.39	0.99	0.90	0.82	9.26	9.54	0.85	4.93	5.37	0.35
Total	98.76	101.43	100.01	101.08	99.88	98.17	98.98	99.70	100.11	99.96	100.08

UB, Union Bay.
[a]Total Fe as FeO.

same kind of parental magma at different temperatures and stages of differentiation. Moreover, according to Irvine (1974), "similar ultramafic rocks can form as early differentiates of dissimilar liquids that may eventually yield strongly contrasting end products." So, similarities in rocks types such as wehrlite and olivine clinopyroxenite in a large number of Alaskan-type complexes do not necessarily mean identical magmatic compositions but indicate a restricted compositional range which is consistent with the crystallization of these rock types.

A group of workers which include Murray (1972), Loucks (1990), and Himmelberg and Loney (1995), support the theory that the Alaskan-type complexes represent early differentiates from a basaltic parental magma. According to the interpretations of Himmelberg and Loney (1995), this magma is hypersthene-normative and similar to island arc basalts. Although most lithologies found in Alaskan-type complexes are devoid of orthopyroxene, Himmelberg and Loney (1995) report minor orthopyroxene in wehrlite and olivine clinopyroxenite at Blashke Island and Kane Peak. According to Loucks (1990), the activity of essenite is diminished in hydrous basaltic magmas, resulting in increased activity of aluminous pyroxene components. The presence of water, therefore, leads to preferential crystallization of clinopyroxene instead of orthopyroxene. Considering the close genetic similarities with island arc basalts in terms of tectonic setting, such an interpretation seems plausible, but a silica-saturated parental magma might not necessarily explain all the observed characteristics of Alaskan-type complexes.

Irvine (1974) proposed that the parental magmas of Alaskan-type complexes are liquids that are critically undersaturated in silica. Such magmas with high MgO and low SiO_2 contents are consistent with the observed compositions and contents of mineral phases like olivine and clinopyroxene. Irvine (1973) reported the occurrence of primitive (11.9 wt% MgO, 47.9 wt% SiO_2, 2.4 wt% Na_2O, and 1.3 wt% K_2O) augite-phyric (ankaramitic) lava flows near Juneau, Alaska. The ankaramitic magmas were proposed to be parental to Alaskan-type complexes in southeastern Alaska. Thakurta et al. (2008a) used the program MELTS (Ghiorso and Sack, 1995) with a starting composition of the ankaramitic rocks near Juneau, under fO_2 condition of QFM + 2, and at 7.5% crystallization; upon clinopyroxene saturation, the fractionated magma with 49.5 wt% SiO_2, 11.5 wt% Al_2O_3, 9.6 wt% MgO could potentially have crystallized the wehrlite unit. Thakurta et al. (2008a) also discussed the possibility of a more primitive mantle melt of picritic composition as the parental magma. In fact, an ankaramitic magma could be demonstrated to be a differentiated member of a picritic parental liquid at 27% crystallization at an fO_2 of QFM + 2.

However, it seems unreasonable to assume a single magmatic composition to be representative of parental magma compositions of all Alaskan-type complexes. It is quite possible that each complex crystallized from several batches of magma. From the overall subduction zone setting it seems obvious that most Alaskan-type complexes were crystallized from magmas enriched in H_2O and thus under oxidizing conditions were possibly at or around the QFM + 2 buffer. The ultramafic cumulate sequences in most Alaskan-type complexes can be interpreted to have crystallized at pressures greater than 3 kbar (Himmelberg and Loney, 1995). Experimental studies on hydrous basalts at the NNO is Ni-NiO buffer show that at $P \geq 3$ kbar hornblende crystallizes at a much higher temperature than plagioclase (Holloway and Burnham, 1972). Thus, the absence of plagioclase in the ultramafic portions of the complexes (Himmelberg and Loney, 1995) is attributed to the relatively high oxidation state ($fO_2 \geq$ QFM + 2) of magmas. The relative abundance of oxide minerals such as chromite and magnetite is also attributed to the high oxidation state of the magmas (Buddington and Lindsley, 1964). However, as discussed in Section 9.6.5, conditions of relatively high oxygen fugacity are not conducive to crystallization of sulfide minerals (Jugo et al., 2005). This might be the reason why Alaskan-type complexes have been traditionally regarded as sulfide-depleted systems. Thus, a dramatic change in the oxidation state of magma is necessary to stabilize sulfide minerals in Alaskan-type complexes. Such a change may be triggered by selective assimilation of reducing crustal rocks (Nixon, 1998; Thakurta et al., 2008a).

9.6.3 MECHANISM OF EMPLACEMENT OF ALASKAN-TYPE COMPLEXES

The principal debate on the emplacement mechanism of Alaskan-type complexes has hinged on whether the complexes represent products of magmatic crystallization at the levels where they are found or solid-state diapiric upheavals of preexisting rocks from the upper mantle. Burg et al. (2009) studied the Kondyor Alaskan-type complex in eastern Russia and proposed a model of trans-lithospheric mantle diapirism based on numerical calculations of the rheological properties of the lithological units and the surrounding wall rocks. As seen in Fig. 9.8, the Kondyor Complex is a near-perfect circular body of ultramafic rock with a chromite-bearing dunite core surrounded by successive rims of wehrlite, olivine clinopyroxenite, and magnetite-amphibole with minor plagioclase. There is a metasomatic domain composed of phlogopite-amphibole-apatite-carbonate and Fe-Ti oxide minerals which occur as veins and stockworks in the central dunite massif. The complex has been dated between Late Jurassic and Early Cretaceous; radiometric ages are between 115 and 132 Ma, as determined by phlogopite K-Ar in dunite and pyroxenite, whole rock K-Ar in alkaline gabbros, and whole rock Rb-Sr in pyroxenite (Burg et al., 2009 and references cited therein). The complex is surrounded by metasedimentary hornfels of Lake Proterozoic age and by Archean gneiss, quartzite, and marble.

The contact region between the ultramafic rocks of the complex and the country rocks is near-vertical. Layering in the surrounding hornfelsed metasedimentary rocks dips ~ 60 degrees away from the intrusion and quickly changes to horizontal farther from the complex. These structural characteristics along with petrological and geochemical properties of the complex led Burg et al. (2009) to propose that the dunite massif represents mantle rocks which reacted with Jurassic to Cretaceous lamproiitic magmas and formed pyroxenite rims by peritectic melt-consuming reactions. Thus, according to Burg et al. (2009) the Kondyor Complex represents vertically remobilized rocks from the asthenospheric mantle which moved under fluid pressure, channelized along the dunite core.

The circular outcrop, the conical cross-section, and the deformational pattern of the country rocks in the immediate vicinity of the Kondyor Complex suggest that it formed as a vertically uplifted plug of solidified crystals. However, it is difficult to model vertical movement of a solid crystal mush without the involvement of interstitial liquid. In the opinion of the present author, the interstitial residual liquid in the crystal mush cannot be explained by minor involvement of the extraneous lamproiitic melts, although these melts could have added to the overall melt content in the system and could also have formed dike-like intrusive structures within the massif. Thus, the olivine-rich central plug and the surrounding pyroxenite and gabbroic rims can be explained as uplifted and updomed masses of semiconsolidated crystal layers emplaced and deformed by an upward-directed magmatic pressure from depth along a near-vertical conduit.

From this standpoint, the zonal structures and the near-concentric outcrop patterns could be explained, at least in part, by the degree of crystallinity of the rising magmatic pulse. A large crystal mush on top would deform the crustal rocks and effectively block the pathway for upward movement of liquid. The ascending magma would then thermally erode the country rock and seep

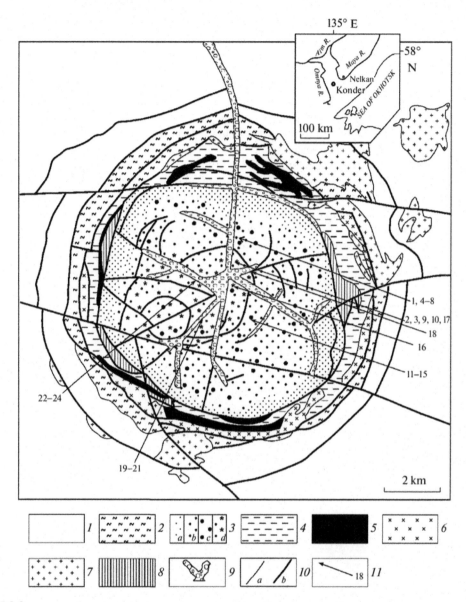

FIGURE 9.8

Geological map of the Kondyor Complex. (1) Middle Riphean sedimentary rocks: siltstone, sandstone; (2) Archean granites and metamorphic rocks (gneisses, schists, marbles, etc.); (3) dunite: a, fine-grained; b, fine-, medium-, and coarse-grained; c, coarse-grained and pegmatitic, d, metasomatically altered under the influence of intrusions of kosvite, mafic, alkaline, and granitic rocks; (4) pyroxenite; (5) kosvite; (6) gabbro; (7) subalkaline diorite, monzodiorite; (8) olivine—diopside metasomatic rocks; (9) placers of PGE; (10) boundaries and faults: a, geological boundaries; b, faults; (11) sampling sites and their number.

Reproduced with permission from Mochalov, A.G., Yakubovichb, O.V., Bortnikovb, N.S., 2016. 190Pt–4He age of PGE ores in the Alkaline–Ultramafic Kondyor Massif (Khabarovsk District, Russia). Doklady Earth Sci. 469 (Part 2), 846–850 (Mochalov et al., 2016).

out from the space between the crystal plug and the country rocks. In Alaskan-type complexes such as Duke Island there is ample evidence of direct crystallization from the liquid along with rearrangement of clasts during convective flow of liquid. Thus, the participation of a large volume of liquid in the crystallization process of the Duke Island Complex is obvious. Therefore, systems like Duke Island could be regarded as magma chambers along the liquid flow channels of volcanoes on the surface. In contrast, the Kondyor Complex could not have acted as a flow-through channel for a surficial volcanic system. It is understandable that the supply of magma for the Kondyor complex in the stable Aldan Shield could have been much lower than in a subduction zone setting where most Alaskan-type complexes are located.

The Annette Island Complex (Taylor and Noble, 1969; Li et al., 2011), about 20 km northwest of Duke Island, in southeastern Alaska, is composed almost entirely of a large dunite plug (Fig. 9.9), with minor occurrence of hornblende-magnetite clinopyroxenite in a narrow zone at the western periphery. This intrusion, even though in the subduction environment, could be comparable to the Kondyor Complex in terms of the large dunite massif and the probable mechanism of formation by the emplacement of a mush of accumulated olivine crystals.

FIGURE 9.9

Outcrop of the dunite massif at the Annette Island Complex in southeastern Alaska, seen from a helicopter.

Guillou-Frottier et al. (2014) studied 46 Alaskan-type complexes around the world and classified them into three categories: (1) single circular or elliptical bodies; (2) twin bodies with circular or elliptical shapes; and (3) dismembered dunite bodies, out of which the second category has been noted for PGE-enrichments. The second category of complexes has been explained as horizontal elliptical cross-sections of dunite pipes in a near-vertical "Y"-shaped three-dimensional structure. Based on high-resolution thermo-numerical modeling experiments the authors determined the mechanical properties of the lithosphere to allow the emplacement of Alaskan-type complexes by forceful diapiric ascent. The models indicate that in order to keep a vertical dike-like conduit, with the development of zones of weakness in the upper crust, the lower crust needs to possess strong rheological properties with local strain-softening processes such as tensile failure in brittle and ductile zones. It was also concluded that since the deformation of the upper crust is controlled solely by rheological properties, there is no specific requirement for the Alaskan-type complexes to be restricted to subduction zone tectonic settings. Thus, rising magmatic pulses and/or crystal mushes mobilized by upward fluid pressure could potentially form Alaskan-type complexes regardless of tectonic setting. However, the authors noted that subcontinental lithospheric mantle next to a subduction zone is a very likely location for the origin and storage of hydrous ultramafic magmas.

The study of Guillou-Frottier et al. (2014) identified an interesting relationship between the calculated strain-rate values in the complexes and the degree of PGE mineralization. Y-shaped Alaskan-type complexes, such as Goodnews Bay, Kachkanar, which contain highest PGE concentrations (up to 7902 ppb Pt and 55 ppb Pd; Augé et al., 2005) also show highest strain-rate values. In contrast, single or cylindrical complexes such as Kondyor and Tulameen show lowest strain-rate values.

A critical factor to explain the diapiric uplift and emplacement of solid crystal mushes is the relationship between the densities of the rising material and the surrounding rocks. Higher densities of minerals such as olivine and clinopyroxene in relation to crustal rocks would make the rise of crystal mushes difficult. Thus, an upward-directed force acting on the crystal mush and a thin crust, immediately above the mantle wedge, are necessary for the proposed diapiric emplacement of Alaskan-type intrusions. Thin crust is consistent with the location of these complexes along subduction zones, sometimes in accreted island arcs, as seen in southeastern Alaska. The proposed occurrences of volcanoes above these rock masses could explain the periodic upward movements of magma. Thrust faults and related fracture systems, which developed due to tectonic compression, could also have helped the developments of these volcanic plumbing systems.

9.6.4 PGE-ENRICHMENT AND POSSIBLE ORIGIN

The PGE-rich placer deposits associated with some Alaskan-type complexes are thought to originate from chromite-rich dunite units within central portions (Augé et al., 2005; Ivanov et al., 2008; Garuti, 2011).

Razin (1976) proposed a metasomatic process in the formation of PGE-rich chromitite in the Urals Pt-belt. According to this model, host dunites belong to a "special magmatic type of forsterite dunite" and the concentration of the PGEs occurred during the cooling of the dunite body when the PGEs underwent magmatic, pneumatolytic, and hydrothermal stages of mobilization. Zoloev et al. (2001), Volchenko and Koroteev (2002, 2003) explained that the PGEs may have been metasomatically leached from the host dunites and concentrated in the chromite-rich zones, leaving behind a PGE-depleted dunite. Augé et al. (2005) argued that the mechanism of PGE-enrichment of the chromitite within the dunite bodies is magmatic and proposed a dynamic model of accumulation of PGEs in the thin chromite lenses within the semiconsolidated dunite body by repetitive passage of primitive magma derived from the mantle. Since the volumetric abundance of chromitite is extremely small in comparison to the volume of magma passing by and crystallizing the platinum-group metals, the concentration of PGE collected by chromite increased with time. The primary PGMs thus precipitated could have been subsequently modified by low-temperature hydrothermal alteration and metasomatic addition of metals like Cu, Fe, and Ni.

With the assumption that PGEs in the mantle are selectively concentrated in the sulfide minerals, it is a combination of the degree of partial melting and the oxidation state of the mantle rocks that determine the release of PGEs into the magma (Mungall et al., 2006). Owing to high-fO_2 conditions in the fluid-enriched mantle wedge underneath a subduction zone, sulfide phases become unstable (Fig. 9.10) and the sulfide minerals readily disintegrate and release the PGEs into the melt.

Since PGE partition coefficients between silicate and sulfide liquids range between 10^3 and 10^6 (Peach et al., 1990), sulfide saturation will concentrate PGEs in the immiscible sulfide liquid.

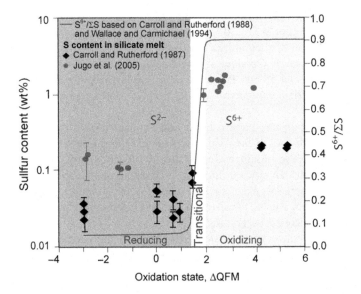

FIGURE 9.10

Changes in the solubility of sulfur in magma with respect to oxidation state. Immiscible sulfide liquids are stable in the system at oxidation states below QFM + 1.5, while sulfate is stable at oxidation states above QFM + 2.

After Jugo, P.J., Luth, R.W., Richards, J.P., 2005. Experimental data on the speciation of sulfur as a function of oxygen fugacity in basaltic melts. Geochim. Cosmochim. Acta 69, 497–503; Carroll, M.R., Rutherford, M.J., 1987. The stability of igneous anhydrite; experimental results and implications for sulfur behaviour in the 1982 El Chichon trachyandesite and other evolved magmas. J. Petrol. 28, 781–801; Carroll, M.R., Rutherford, M.J., 1988. Sulfur speciation in hydrous experimental glasses of varying oxidation-state—results from measured wavelength shifts of sulfur X-rays. Am. Mineral. 73, 845–849 (Carroll and Rutherford, 1988); Wallace, P., Carmichael, I.S.E., 1994. S speciation in submarine basaltic glasses as determined by measurement of SKa X-ray wavelength shifts. Am. Mineral. 79, 161–167 (Wallace and Carmichael, 1994).

Following the separation of an immiscible sulfide liquid, the residual melt will be depleted in the PGEs. For high PGE-contents to be stable in the magma until the late stage, and for the oxide minerals such as chromite to concentrate the PGEs, separation of a sulfide liquid must not have happened.

In sulfide-absent Alaskan-type complexes, Pt seems to be a predominant PGE. Garuti et al. (1997) regarded this as a distinctive characteristic of Alaskan-type complexes, indicative of the nature of the parental magmas and possibly the anomalously Pt-enriched mantle source rocks. Chromitites, in the dunite cores, contain the highest abundances of Pt and Ir and have high Pt/Pd ratios which suggests selective accumulation of discrete platinum group metals, particularly Pt-Fe alloys (Cabri, 1981). The Pt/Pd ratio decreases in the wehrlite and clinopyroxenite units, indicating progressive depletion of Pt by magmatic differentiation.

9.6.5 SULFIDE MINERALS IN ALASKAN-TYPE COMPLEXES

Elevated fO_2 is believed to be responsible for the absence of sulfides in most Alaskan-type complexes (Thakurta et al., 2008b). Owing to the possible metasomatic transfer of large quantities of sulfur from the subducted oceanic crust, the content of sulfur in magmas generated in the overlying mantle wedge may be substantial. Large volumes of sulfur-rich gases in active volcanoes in subduction zones (de Hoog et al., 2001) provide evidence for the availability of sulfur in such magmatic systems.

The influx of hydrous fluids from the descending slab causes high degrees of partial melting in the mantle source rocks and raises the ambient mantle fO_2 to $\sim QFM + 2$. This is consistent with high MgO compositions of the generated melts for Alaskan-type complexes. Experimental studies by Carroll and Rutherford (1987) and Jugo et al. (2005) have shown that at oxidizing conditions about 1.5 to 2 log units above the QFM buffer, sulfate replaces sulfide as the stable sulfur-bearing species in magma (Fig. 9.10) and the solubility of sulfur is also increased. Thus, for sulfide species to be stable, the oxidation state of magma needs to be lowered during magmatic evolution. This can be achieved by assimilation of large quantities of crustal rocks rich in minerals such as graphite. In the Duke Island and Turnagain complexes the assimilation of graphite is indicated by rafts of graphite-rich metasedimentary basement rocks hosted in the ultramafic units (Fig. 9.11). The presence of graphite at pressures between 3 and 4 kb requires reducing condition of $\log fO_2 \sim QFM - 2$ (Mathez et al., 1989).

In conclusion, the observed occurrences of sulfide mineralized zones at the Salt Chuck, Duke Island, and Turnagain complexes suggest that Alaskan-type complexes may be regarded as potential targets in future explorations for economic Cu-Ni-PGE deposits.

ACKNOWLEDGEMENTS

This chapter is a product of an ongoing investigation on the origin of Alaskan-type complexes and several researchers such Edward Ripley, Chusi Li, and industry professionals such as Curt Freeman and Chris van Treeck have contributed to this work in various capacities. Dejan Milidragovic and Bill Griffin provided thoughtful reviews to improve the quality of this chapter. Sisir Mondal was responsible for useful editorial suggestions.

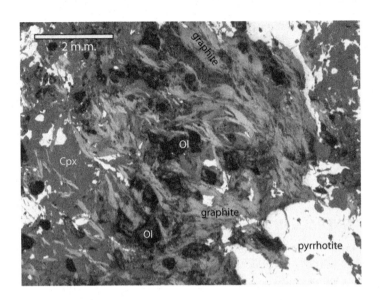

FIGURE 9.11

Graphite flakes locally seen in the olivine clinopyroxene unit at the Duke Island Complex.

From Thakurta, J., Ripley, E.M., Li, C., 2008a. Geochemical constraints on the origin of sulfide mineralization in the Duke Island Complex, southeastern Alaska. Geochem. Geophys. Geosyst. 9, 34 p., Q07003, doi:10.1029/2008GC001982.

REFERENCES

Avalon, 2001. Summary report for the Union Bay PGE prospect, Ketchikan Mining District, Alaska: Geol. Rept. UB01-1, unpublished final report for 2001 prepared by Avalon Development Corp. for Quaterra Resources Inc., 27 p.

Augé, T., Genna, A., Legendre, O., Ivanov, K.S., Volchenko, Y.A., 2005. Primary Platinum mineralization in the Nizhny Tagil and Kachkanar ultramafic complexes, Urals, Russia: a genetic model for PGE concentration in chromite-rich zones. Econ. Geol. 100, 707−732.

Baragar, W.R.A., 1960. Petrology of basaltic rocks in part of the Labrador Trough. Geol. Soc. Am. Bull. 71, 1589−1643.

Berg, H.C., Jones, D.L. and Richter, D.H., 1972, Gravina-Nutzotin Belt − Tectonic Significance of an Upper Mesozoic Sedimentary and Volcanic Sequence in Southern and Southeastern Alaska, U.S. Geological Survey Professional Paper 800-D, D1-D24.

Bhattacharji, S., Smith, C., 1964. Flowage differentiation. Science 145, 150−153.

Buddington, A.F., Chapin, T., 1929. Geology and Mineral Deposits of Southeastern Alaska. U.S. Geological Survey Bulletin, 800, 398 p.

Buddington, A.F., Lindsley, D.H., 1964. Iron-titanium oxide minerals and synthetic equivalents. J. Petrol. 5, 310−357.

Burg, J.-P., Bodinier, J.-L., Greya, T., Bedini, R.M., Boudier, F., Dautria, J.M., et al., 2009. Translithospheric mantle diapirism: geological evidence and numerical modeling of the Kondyor Zoned Ultramafic Complex (Russian Sar-East). J. Petrol. 50, 289−321.

Butler, R.F., Gehrels, G.E., Saleeby, J.B., 2001. Paleomagnetism of the Duke Island, Alaska, ultramafic complex revisited. J. Geophys. Res. 106, 259−269, no. B9, 19.

Cabri, L.J., 1981. Nature and distribution of platinum-group element deposits. Episodes 2, 31−35.

Carroll, M.R., Rutherford, M.J., 1987. The stability of igneous anhydrite; experimental results and implications for sulfur behaviour in the 1982 El Chichon trachyandesite and other evolved magmas. J. Petrol. 28, 781−801.

Carroll, M.R., Rutherford, M.J., 1988. Sulfur speciation in hydrous experimental glasses of varying oxidation-state—results from measured wavelength shifts of sulfur X-rays. Am. Mineral. 73, 845−849.

Clark, A.L., Greenwood, W.R., 1972. Geochemistry and Distribution of Platinum-Group Metals in the Mafic to Ultramafic Complexes of Southern and Southeastern Alaska. U.S. Geological Survey Professional Paper 800-C, pp. C157−C160.

Cawthorn, R.G., Meyer, P.S., Kruger, F.J., 1991. Major addition of magma at the Pyroxenite marker in the Western Bushveld Complex, South Africa. J. Petrol. 32, 739−763.

Czamanske, G.K., Zientek, M.L., 1985. The Stillwater Complex Montana: Geology and Guide. Montana Bureau of Mines and Geology Special Publication 92, 396 p.

de Hoog, J.C.M., Mason, P.R.D., van Bergen, M.J., 2001. Sulfur and chalcophile elements in subduction zones: constraints from a laser ablation ICP-MS study of melt inclusions from Galunggung Volcano, Indonesia. Geochim. Cosmochim. Acta 65, 3147−3164.

Findlay, D.C., 1969. Origin of the Tulameen ultramafic-gabbro complex, southern British Columbia. Can. J. Earth Sci. 6, 399−425.

Foley, J.Y., Light, T.D., Nelson, S.W., Harris, R.A., 1997. Mineral occurrences associated with mafic-ultramafic and related alkaline complexes in Alaska. In: Goldfarb, R.A., Miller, L.D. (Eds.), Mineral Deposits of Alaska. Economic Geology Monographs 9, pp. 396−449.

Fominykh, V.G., Khvostova, V.P., 1970. Platinum content of Ural dunite. Doklady Akad. Nauk SSSR 191, 443−445 (in Russian).

Garuti, G., 2011. Global tectonics and chromite—platinum mineralization monitoring genesis and evolution of Ural−Alaskan type complexes. Rev. Macla 15, 15−16.

Garuti, G., Fershtater, G., Bea, F., Montero, P., Pushkarev, E.V., Zaccarini, F., 1997. Platinum-group elements as petrological indicators in mafic-ultramafic complexes of the central and southern Urals, preliminary results. Tectonophysics 276, 181−194.

Garuti G., Pushkarev, E.V., Zaccarini, F., Cabella, R. & Anikina, E., 2003, Chromite composition and platinum-group mineral assemblage in the Uktus Uralian-Alaskan-type complex (Central Urals, Russia). Mineral. Deposita 38, 312−326.

Gehrels, G.E., Saleeby, J.B., Berg, H.C., 1987. Geology of Annette, Gravina, and Duke islands, southeastern Alaska. Can. J. Earth Sci. 24, 866−881.

Gehrels, G.E., Butler, R.F., Bazard, D.R., 1996. Detrital Zircon Geochronology of the Alexander Terrane, Southeastern Alaska. Geological Society of America Bulletin vol. 108, 722−734.

Gehrels, G.E., and Berg, H.C., 1994, Geology of southeastern Alaska, in Plafker, George, and Berg, H.C., eds., The Geology of Alaska: Geological Society of America, p. 451−467.

Ghiorso, M.S., Sack, R.O., 1995. Chemical mass transfer in magmatic processes; IV, A revised and internally consistent thermodynamic model for the interpolation and extrapolation of liquid-solid equilibria in magmatic systems at elevated temperatures and pressures. Contrib. Mineral. Petrol. 119, 197−212.

Guillou-Frottier, L., Burov, E., Augé, T., Gloaguen, E., 2014. Rheological conditions for emplacement of Ural−Alaskan-type ultramafic complexes. Tectonophysics 631, 130−145.

Himmelberg, G.R., Loney, R.A., and Craig, J.T., 1986, Petrogenesis of the Ultramafic Complex at the Blashke Islands, Southeastern Alaska: U.S. Geological Survey Bulletin 1662, 14 p.

Himmelberg, G.R., Loney, R.A., 1995. Characteristics and Petrogenesis of Alaskan-Type Ultramafic-Mafic Intrusions, Southeastern Alaska. United States Geological Survey Professional Paper, No.1564, 47 p.

Holloway, J.R., Burnham, C.W., 1972. Melting relations of basalt with equilibrium water pressure less than total pressure. J. Petrol. 13, 1−29.

Holt, S.P., Shepard, J.G., Thorne, R.L., Tolonen, A.W., Fosse, E.L., 1948. Investigation of the Salt Chuck Copper Mine, Prince of Wales Island, Southeastern Alaska. U.S. Bureau of Mines, Report of Investigations 4358, 16 p.

Irvine, T.N., 1959. The Ultramafic and Related Rocks of Duke Island, Southeastern Alaska (Ph.D. thesis). California Institute of Technology, Pasadena, 320 p.

Irvine, T.N., 1963. Origin of the Ultramafic Complex at Duke Island, Southeastern Alaska. Mineralogical Society of America Special Paper 1, pp. 36–45.

Irvine, T.N., 1973. Bridget Cove volcanics, Juneau area, Alaska: possible parental magma of Alaskan-type ultramafic complexes. Carnegie Inst. Yearb. 72 (1972–1973), 478–491.

Irvine, T.N., 1974. Petrology of the Duke Island Ultramafic Complex, Southeastern Alaska. Geological Society of America Memoir, 138, 240 p.

Irvine, T.N., 1975. Crystallization sequences in the Muskox intrusion and other layered intrusions—II. Origin of chromitite layers and similar deposits of other magmatic ores. Geochim. Cosmochim. Acta 39, 1009–1020.

Irvine, T.N., 1982. Terminology of layered intrusions. J. Petrol. 23, 127–162.

Ivanov, A.V., Palesskii, S.V., Demonterova, S.V., Nikolaeva, E.I., Ashchepkov, I.V., Ashchepkov, I.V., et al., 2008. Platinum-group elements and rhenium in mantle xenoliths from the East Sayan volcanic field (Siberia, Russia): evaluation of melt extraction and refertilization processes in lithospheric mantle of the Tuva-Mongolian massif. Terra Nova 20, 504–511.

Jackson, E.D., 1971. The origin of ultramafic rocks by cumulus processes. Fortschr. Mineral. 48, 128–174.

Johan, Z., 2002. Alaskan-type complexes and their platinum-group element mineralization. In: Cabri, L.J. (Ed.), The Geology, Geochemistry, Mineralogy and Mineral Beneficiation of Platinum-Group Elements, Special Volume, 54. Canadian Institute of Mining and Metallurgy, pp. 299–319.

Jugo, P.J., Luth, R.W., Richards, J.P., 2005. Experimental data on the speciation of sulfur as a function of oxygen fugacity in basaltic melts. Geochim. Cosmochim. Acta 69, 497–503.

Kennedy, G.C., Walton Jr., M.S., 1946. Geology and Associated Mineral Deposits of Some Ultrabasic Rock Bodies in Southeastern Alaska. U.S. Geological Survey Bulletin 947-D, pp. 65–84.

Lanphere, M.A., Eberlein, G.D., 1966. Potassium-Argon Ages of Magnetite-Bearing Ultramafic Complexes in Southeastern Alaska, Special Paper—Geological Society of America, 87, 94 p.

Li, C., Thakurta, J., Ripley, E.M., 2011. Low-Ca content and kink-banded texture are not unique for mantle olivine: evidence from the Duke Island Complex, Alaska. Mineral. Petrol. 104, 147–153. Available from: http://dx.doi.org/10.1007/s00710-011-0188-0.

Loney, R.A., Himmelberg, G.R., 1992. Petrogenesis of the Pd-rich intrusion of the Salt Chuck, Prince of Wales Island: an early Paleozoic Alaskan-type ultramafic body. Can. Mineral. 30, 1005–1022.

Loney, R.A., Himmelberg, G.R., Shew, N., 1987. Salt Chuck palladium-bearing ultramafic body, Prince of Wales Island. In: Hamilton, T.D., Galloway, J.P. (Eds.), Geologic Studies in Alaska by the U.S. Geological Survey During 1986. U.S. Geological Survey Circular 998, pp. 126–127.

Loucks, R.R., 1990. Discrimination of ophiolitic from nonophiolitic ultramafic-mafic allochthons in orogenic belts by the Al/Ti ratio in clinopyroxene. Geology 18, 346–349.

Maas, K.M., Bittenbender, P.E., Still, J.C., 1995. Mineral Investigations of the Ketchikan Mining District, Southeastern Alaska. U.S. Bureau of Mines Open-File Report, pp. 11–95.

Maaløe, S., 1978. The origin of rhythmic layering. Mineral. Mag. 42, 337–345.

Mathez, E.A., Dietrich, V.J., Holloway, J.R., Boudreau, A.E., 1989. Carbon distribution in the Stillwater Complex and evolution of vapor during crystallization of Stillwater and Bushveld magmas. J. Petrol. 30, 153–173.

McBirney, A.R., Noyes, R.M., 1979. Crystallization and layering of the Skaergaard Intrusion. J. Petrol. 20, 487–554.

McCallum, I.S., 1996. Layered Intrusions: The Stillwater Complex. Dev. Petrol. 15, 441–483. Available from: http://dx.doi.org/10.1016/S0167-2894(96)80015-7.

Mochalov, A.G., Khoroshilova, T.S., 1998. The Konder alluvial placer of platinum metals. In: Laverov, N.P., Distler, V.V. (Eds.), International Platinum. Theophrastus Publications, St. Petersburg-Athens, pp. 206−220.

Mochalov, A.G., Yakubovichb, O.V., Bortnikovb, N.S., 2016. 190Pt−4He age of PGE ores in the Alkaline−Ultramafic Kondyor Massif (Khabarovsk District, Russia). Doklady Earth Sci. 469 (Part 2), 846−850.

Mungall, J.E., Hanley, J.J., Arndt, N.T., Debecdelievre, A., 2006. Evidence from meimechites and other low-degree mantle melts for redox controls on mantle-crust fractionation of platinum-group elements. Proc. Natl. Acad. Sci. U.S.A. 103, 12695−12700.

Murray, C.G., 1972. Zoned ultramafic complexes of the Alaskan type: feeder pipes of andesitic volcanoes. In: Shagam, R.E., et al., (Eds.), Studies in Earth and Space Sciences (Hess Volume). Geological Society of America Memoir, 132, pp. 313−335.

Naldrett, A.J., 2010. Secular variation of magmatic sulfide deposits and their source magmas. Econ. Geol. 105, 669−688.

Naldrett, A.J., Gasparrini, E.C., Barnes, S.J., von Gruenewaldt, G., Sharpe, M.R., 1986. The upper critical zone of the Bushveld Complex and a model for the origin of Merensky-type ores. Econ. Geol. 81, 1105−1118.

Naldrett, A.J., Lightfoot, P.C., Fedorenko, V.A., Gorbachev, N.S., Doherty, W., 1992. Geology and geochemistry of intrusions and flood basalts of the Noril'sk region USSR with implications for the origin of the Ni-Cu ores. Econ. Geol. 87, 975−1004.

Nixon, G.T., 1998. Ni-Cu Sulfide Mineralization in the Turnagain Alaskan-Type Complex: A Unique Magmatic Environment. British Columbia Ministry of Employment and Investment, Geological Fieldwork, Report: 1998-1, pp. 18.1−18.11.

Peach, C.L., Mathez, E.A., Keays, R.R., 1990. Sulphide melt−silicate melt distribution coefficients for the noble metals and other chalcophile metals as deduced from MORB; implications for partial melting. Geochim. Cosmochim. Acta 54, 3379−3389.

Pettigrew, N.T., Hattori, K.H., 2006. The Quetico intrusions of western Superior Province; Neo-Archean examples of Alaskan/Ural-type mafic-ultramafic intrusions. Precambrian Res. 149, 21−42.

Plafker, G., Berg, H.C. (Eds.), 1994. The Geology of Alaska. Geological Society of America, 1068 p.

Pure Nickel, 2014. <http://www.purenickel.com/s/NewsReleases.asp?ReportID = 673710>.

Razin, L.V., 1976. Geological and genetic features of forsterites dunites and their platinum-group mineralization. Econ. Geol. 71, 1371−1376.

Rubin, C.M., Saleeby, J.B., 1992. Tectonic history of the eastern edge of the Alexander terrane, southeast Alaska. Tectonics 11, . Available from: http://dx.doi.org/10.1029/91TC02182 ISSN: 0278-7407.

Ruckmick, J.C., Noble, J.A., 1959. Origin of the ultramafic complex at Union Bay, southeastern Alaska. Bull. Geol. Soc. Am. 70, 981−1018.

Scheel, J.E., Scoates, J.S., Nixon, G.T., 2009. Chromian spinel in the Turnagain ultramafic intrusion northern British Columbia, Canada. Can. Mineral. 47, 63−80.

Saleeby, J.B., 1992. Age and tectonic setting of the Duke Island ultramafic intrusion, southeast Alaska. Can. J. Earth Sci. 29, 506−522.

Suppel, D., Barron, L.M., 1986. Platinum in Basic to Ultrabasic Intrusive Complexes at Fifield: A Preliminary Report. Quarterly notes, Geological Survey of New South Wales, 65, pp. 1−8.

Taylor, H.P., 1967. The zoned ultramafic complexes of southeastern Alaska. In: Wyllie, P.J. (Ed.), Ultramafic and Related Rocks. John Wiley and Sons Incorporated, New York, NY, pp. 96−118.

Taylor, H.P., Noble, J.A., 1960, Origin of the ultramafic complexes in southeastern Alaska. In: International Geological Congress, 21st, Copenhagen, 1960, Report, Part 13, pp. 175−187.

Taylor, H.P., Noble, J.A., 1969. Origin of magnetite in the zoned ultramafic complexes of southeastern Alaska. In: Wilson, H.D.B. (Ed.), Economic Geology Monograph 4, pp. 209−230.

Tessalina, S.G., Malitch, K.N., Auge, T., Puchkov, V.N., Belousova, E., McInnes, B.I.A., 2016. Origin of the Nizhny Tagil clinopyroxenite − dunite massif, Uralian Platinum Belt, Russia: insights from PGE and Os isotope systematics. J. Petrol. 56, 2297−2318.

Thakurta, J., Ripley, E.M., Li, C., 2008a. Geochemical constraints on the origin of sulfide mineralization in the Duke Island Complex, southeastern Alaska. Geochem. Geophys. Geosyst. 9, 34 p., Q07003, doi:10.1029/2008GC001982.

Thakurta, J., Ripley, E.M., Li, C., 2008b. Pre-requisites for sulfide-poor PGE and sulfide-rich Cu−Ni−PGE mineralization in Alaskan-type complexes. J. Geol. Soc. India 72, 611−622.

Thakurta, J., Ripley, E.M., Li, C., 2009. Oxygen isotopic variability associated with multiple stages of serpentinization, Duke Island Complex, southeastern Alaska. Geochim. Cosmochim. Acta 73, 6298−6312.

Thakurta, J., Ripley, E.M., Li, C., 2014. Platinum group element geochemistry of sulfide-rich horizons in the Ural-Alaskan type ultramafic complex of Duke Island, southeastern Alaska. Econ. Geol. 109, 643−659.

Tistl, M., 1994. Geochemistry of platinum-group elements of the zoned ultramafic Alto Condoto complex, northwest Colombia. Econ. Geol. 89, 158−167.

Van Treeck, C.J., 2009. Platinum Group Element Enriched Hydrothermal Magnetite of the Union Bay Alaskan-type Ultramafic Intrusion, Southeast Alaska (Unpublished M.S. thesis). University of Alaska, Fairbanks, AK, 188 p.

Volchenko, Y.A., Koroteev, V.A., 2002. The Urals platinum polygon—new data [abs.]. In: International Platinum Symposium, 9th, 21-25 July, 2002, Billings, Montana, Duke University, Abstracts with Programs, pp. 465−468.

Volchenko, Y.A., Koroteev, V.A., 2003. Structural and Substantial Evolution of Chromite-Platinum Ore Formation in Deposits of the Urals Type (in Russian). Ekaterinburg, Ezhegodnik (Yearbook) 2002 of the Institute of Geology and Geochemistry of the Ural Branch of the Russian Academy of Sciences, pp. 261−270.

Wallace, P., Carmichael, I.S.E., 1994. S speciation in submarine basaltic glasses as determined by measurement of SKa X-ray wavelength shifts. Am. Mineral. 79, 161−167.

Walton Jr., M.S., 1951. The Blashke Island Ultrabasic Complex, with Notes on Related Areas in Southeastern Alaska. U.S. Geological Survey Open-File Report [1586], 266 p.

Watkinson, D.H., Melling, D.R., 1992. Hydrothermal origin of platinum-group mineralization in low-temperature, copper sulfide-rich assemblages, Salt Chuck intrusion, Alaska. Econ. Geol. 87, 175−184.

Wood, S.A., 2002. The aqueous geochemistry of the platinum-group elements with applications to ore deposits. In: Cabri, L.J. (Ed.), The Geology, Geochemistry, Mineralogy and Mineral Beneficiation of Platinum-Group Elements. Canadian Institute of Mining, Metallurgy and Petroleum, Special Volume 54, pp. 211−249.

Wyllie, P.J. (Ed.), 1967. Ultramafic and Related Rocks. John Wiley, New York, NY, 464 p.

Wyssotzky, N., 1913. Die Platinseifengebiete von Iss- und Nischny-Tagil im Ural. (The platinum placer districts of Iss- and Nizhnyi-Tagil in Urals). Mémoires du Comité Géologique, Nouvelle Série, Livraison 62, St. Petersburg.

Zavarisky, A., 1928. Primary platinum deposits of the Urals, Matériaux pour la géologie générale et appliqués, 108. Comité géologique, 53 p. (in Russian).

Zoloev, K.K., Volchenko, Y.A., Koroteev, V.A., Malakhov, I.A., Mardirosian, A.N., Hripov, V.N., 2001. Platinum-metal mineralization in the Urals geological complexes. Department of Mineral Resources of the Urals region, Ekaterinburg, 200 p. (in Russian).

FURTHER READING

Li, C., Ripley, E.M., 2009. Sulfur contents at sulfide-liquid or anhydrite saturation in silicate melts: empirical equations and example applications. Econ. Geol. 104, 405−412.

Malitch, K.N., Kadik, A.A., Badanina, I.Y., Zharkova, E.V., 2011. Redox conditions of formation of osmium-rich minerals from the Guli Massif, Russia. Geochem. Int. 49, 726−730.

EXPERIMENTAL ASPECTS OF PLATINUM-GROUP MINERALS

10

Anna Vymazalová[1] and Dmitriy A. Chareev[2]

[1]Czech Geological Survey, Prague, Czech Republic [2]Institute of Experimental Mineralogy RAS,
Chernogolovka, Russia

CHAPTER OUTLINE

Processes and Ore Deposits of Ultramafic-Mafic Magmas through Space and Time. DOI: http://dx.doi.org/10.1016/B978-0-12-811159-8.00011-1

10.1 INTRODUCTION

The Platinum-group elements (PGEs) are of significant technological importance. PGEs are used primarily in industrial applications (e.g., catalysts) and have become widely established in chemical, electrical, and electronic engineering. The PGEs and the compounds they form have attracted considerable attention in recent years due to their specific properties and their application in new technologies. They exhibit various interesting physical, chemical, and structural properties that place these compounds at the interface of chemistry, mineralogy, solid-state physics, and material science. Therefore, there is still a demand for exploration and mining and to advance the understanding of the natural process that lead to their formation.

This chapter consolidates the experimental methods that can be applied for the synthesis of PGE compounds. Dry synthesis and the major experimental methods used to try to obtain single crystals are summarized and some examples of the synthesis of PGE phases are discussed. A list of approved platinum-group minerals (PGMs) recognized up to the end of 2016, is presented with a focus on newly-named minerals and their experimental aspects. The up-to-date list of recognized PGMs is given in Tables 10.1−10.3. This chapter is written to encourage the application of experimental studies for a better understanding of PGM, their formation and occurrence under natural conditions. The available knowledge is often dispersed across the chemical, physical, mineralogical, and metallurgical literature. Additionally, there is often an information gap between investigations of synthetic systems and those of natural minerals. A number of PGE-bearing compounds have been determined and studied due to their specific (e.g., electrical or electrochemical) properties rather than their mineralogical significance.

There is a wide range of data on the unary and binary systems available in the literature (e.g., a compendium of binary diagrams by Massalski, 1990). The systems containing PGEs were consolidated by Berlincourt et al. (1981), who thoroughly summarized the unary, binary, and ternary systems with PGEs. Later, Makovicky (2002) reported on further ternary and quaternary systems containing PGE. Since then, a few other ternary and quaternary systems of mineralogical significance containing PGEs were covered, involving Os, Pd, Pt, and Rh (Os-Mo-S: Drábek and Rieder, 2008; Pd-Ag-Te: Vymazalová et al., 2015; Pd-Ag-Se: Vymazalová et al., 2014b; Pd-Hg-Se: Drábek et al., 2014; Pd-Pb-Te: Vymazalová and Drábek, 2011; Pd-Sn-Te: Vymazalová and Drábek, 2010; Pd-Cu-Se: Makovicky and Karup-Møller, in press; Pd-Cu-Fe-S: Karup-Møller et al., 2008; Pd-Ni-Fe-S: Makovicky and Karup-Møller, 2016; Pd-Pt-Sb: Kim, 2009; Pd-Pt-Ni-Te: Helmy et al., 2007; Pt-Hg-Se: Drábek et al., 2012; Rh-Cu-S: Karup-Møller and Makovicky, 2007). Nevertheless, as emphasized by Makovicky (2002) there is still a need to cover additional phase systems important for PGE deposits. Particularly unresearched are systems containing phases that are less common but of mineralogical significance and potentially of an economic importance (e.g., Pd-As-Sn,

Table 10.1 Platinum-Group Minerals (Ideal Formulas)

Name	Formula	Name	Formula	Name	Formula
Anduoite	$RuAs_2$	Ferrorhodsite	$FeRh_2S_4$	Kojonenite[a]	$Pd_{7-x}SnTe_2$
Arsenopalladinite	$Pd_8As_{2.5}Sb_{0.5}$	Froodite	$PdBi_2$	Konderite	$Cu_3PbRh_8S_{16}$
Atheneite	$Pd_2As_{0.75}Hg_{0.25}$	Gaotaiite	Ir_3Te_8	Kotulskite	$PdTe$
Atokite	Pd_3Sn	Genkinite	$(Pt,Pd)_4Sb_3$	Kravtsovite[a]	$PdAg_2S$
Borovskite	Pd_3SbTe_4	Geversite	$PtSb_2$	Laflammeite	$Pd_3Pb_2S_2$
Bortnikovite[a]	Pd_4Cu_3Zn	Hexaferrum	(Fe,Ru,Os,Ir)	Laurite	RuS_2
Bowieite	Rh_2S_3	Hollingworthite	$RhAsS$	Lisiguangite[a]	$CuPtBiS_3$
Braggite	$(Pt,Pd)S$	Hongshiite	$PtCu$	Luberoite	Pt_5Se_4
Cabriite	Pd_2SnCu	Inaglyite	$Cu_3PbIr_8S_{16}$	Lukkulaisvaaraite[a]	$Pd_{14}Ag_2Te_9$
Changchengite	$IrBiS$	Insizwaite	$PtBi_2$	Majakite	$PdNiAs$
Chengdeite	Ir_3Fe	Irarsite	$IrAsS$	Malanite	$CuPt_2S_4$
Cherepanovite	$RhAs$	Iridarsenite	$IrAs_2$	Malyshevite[a]	$PdCuBiS_3$
Chrisstanleyite	$Ag_2Pd_3Se_4$	Iridium	Ir	Marathonite[a]	$Pd_{25}Ge_9$
Coldwellite[a]	Pd_3Ag_2S	Isoferroplatinum	Pt_3Fe	Maslovite	$PtBiTe$
Cooperite	PtS	Isomertieite	$Pd_{11}Sb_2As_2$	Mayingite	$IrBiTe$
Crerarite	$(Pt,Pb)Bi_3S_{4-x}$	Jacutingaite[a]	Pt_2HgSe_3	Menshikovite	$Pd_3Ni_2As_3$
Cuproiridsite	$CuIr_2S_4$	Jagüéite[a]	$Cu_2Pd_3Se_4$	Merenskyite	$PdTe_2$
Cuprorhodsite	$CuRh_2S_4$	Kalungaite[a]	$PdAsSe$	Mertieite I	$Pd_{11}(Sb,As)_4$
Damiaoite	$PtIn_2$	Kashinite	Ir_2S_3	Mertieite II	$Pd_8Sb_{2.5}As_{0.5}$
Daomanite	$PtCuAsS_2$	Keithconnite	$Pd_{20}Te_7$	Miessiite[a]	$Pd_{11}Te_2Se_2$
Erlichmanite	OsS_2	Kharaelakhite	$(Cu,Fe)_4(Pt,Pb)_4NiS_8$	Michenerite	$PdBiTe$
Ferhodsite[a]	$(Fe,Rh,Ni,Ir,Cu,Pt)_9S_8$	Kingstonite[a]	Rh_3S_4	Milotaite[a]	$PdSbSe$
Ferronickelplatinum	$PtFe_{0.5}Ni_{0.5}$	Kitagohaite[a]	Pt_7Cu	Moncheite	$PtTe_2$

(Continued)

Table 10.1 Platinum-Group Minerals (Ideal Formulas) *Continued*

Name	Formula	Name	Formula	Name	Formula
Naldrettite[a]	Pd_2Sb	Polarite	$PdBi$	Tainyrite	$(Pd,Cu)_3Sn$
Nielsenite[a]	$PdCu_3$	Polkanovite	$Rh_{12}As_7$	Tatyanaite	$Pt_9Cu_3Sn_4$
Nigglite	$PtSn$	Potarite	$PdHg$	Telargpalite	$Pd_{2-x}Ag_{1+x}Te$
Norilskite[a]	$(Pd,Ag)_7Pb_4$	Prassoite	$Rh_{17}S_{15}$	Telluropalladinite	Pd_9Te_4
Omeiite	$OsAs_2$	Rhodarsenide	$(Rh,Pd)_2As$	Temagamite	Pd_3HgTe_3
Oosterboschite	$(Cu,Pd)_7Se_5$	Rhodium	Rh	Testibiopalladite	$PdSbTe$
Osarsite	$OsAsS$	Rhodplumsite	$Rh_3Pb_2S_2$	Tetraferroplatinum	$PtFe$
Osmium	Os	Ruarsite	$RuAsS$	Tischendorfite	$Pd_8Hg_3Se_9$
Oulankaite	$Pd_5Cu_4SnTe_2S_2$	Rustenburgite	Pt_3Sn	Tolovkite	$IrSbS$
Padmaite	$PdBiSe$	Ruthenarsenite	$RuAs$	Törmroosite[a]	$Pd_{11}As_2Te_2$
Palarstanide	$Pd_5(Sn,As)_2$	Rutheniridosmine	(Ir,Os,Ru)	Tulameenite	$PtFe_{0.5}Cu_{0.5}$
Palladium	Pd	Ruthenium	Ru	Ungavaite[a]	Pd_4Sb_3
Palladoarsenide	Pd_2As	Shungfengite	$IrTe_2$	Urvantsevite	$Pd(Bi,Pb)_2$
Palladobismutharsenide	$Pd_2As_{0.8}Bi_{0.2}$	Skaergaardite[a]	$PdCu$	Vasilite	$Pd_{16}S_7$
Palladodymite	$(Pd,Rh)_2As$	Sobolevskite	$PdBi$	Verbeekite	$PdSe_2$
Palladgermanide[a]	Pd_2Ge	Sopcheite	$Ag_4Pd_3Te_4$	Vincentite	$(Pd,Pt)_3(As,Sb,Te)$
Palladosilicide[a]	Pd_2Si	Sperrylite	$PtAs_2$	Vymazalováite[a]	$Pd_3Bi_2S_2$
Palladseite	$Pd_{17}Se_{15}$	Stannopalladinite	Pd_5Sn_2Cu	Vysotskite	PdS
Paolovite	Pd_2Sn	Stibiopalladinite	$Pd_{5+x}Sb_{2-x}$	Yixunite	Pt_3In
Pašavaite[a]	$Pd_3Pb_2Te_2$	Stillwaterite	Pd_8As_3	Zaccariniite[a]	$RhNiAs$
Platarsite	$PtAsS$	Stumpflite	$PtSb$	Zvyagintsevite	Pd_3Pb
Platinum	Pt	Sudburyite	$PdSb$		
Plumbopalladinite	Pd_3Pb_2	Sudovikovite	$PtSe_2$		

[a]*New minerals described from 2002 to 2016.*

Table 10.2 Revised Platinum-Group Minerals, Crystallographic Data

Mineral		Crystallographic Data	Unit Cell Parametrs						Reference
Name	Formula	Space group	a (Å)	b (Å)	c (Å)	β (°)	V(Å^3)	Z	
Atheneite	$Pd_2As_{0.75}Hg_{0.25}$	Hexagonal $P6/2m$	6.813		3.4892		140.26	3	Bindi (2010)
Chrisstanleyite	$Ag_2Pd_3Se_4$	Monoclinic $P2_1/c$	5.676	10.342	6.341	114.996	337.3	2	Topa et al. (2006)
Mertieite II	$Pd_8Sb_{2.5}As_{0.5}$	Trigonal $R\bar{3}ch$	7.5172		43.037		2106.1	12	Karimova et al. (in press)
Sopcheite	$Ag_4Pd_3Te_4$	Orthorhombic $Cmca$	12.212	6.138	12.234		917.1	4	Laufek et al. (in press)
Temagamite[a]	Pd_3HgTe_3	Hexagonal $P3m1$	7.8211		17.281		917.8	6	Laufek et al. (2016)
Tischendorfite[a]	$Pd_8Hg_3Se_9$	Orthorhombic $Pmmm$	7.1886	16.8083	6.4762		782.51	2	Laufek et al. (2014)

[a]Crystal structure solved from the synthetic analogue.

Table 10.3 Platinum-Group Minerals, Described Since 2002 and Their Crystallographic Data

| Mineral | | Crystal Structure Data | | Unit Cell Parameters | | | | | | Reference |
Name	Formula		Space Group	a (Å)	b (Å)	c (Å)	β (°)	V (Å³)	Z	
Bortnikovite	Pd_4Cu_3Zn	Tetragonal	$P4/mmm$?	6.00		8.50		306.0	3	Mochalov et al. (2007)
Coldwellite	Pd_3Ag_2S	Cubic	$P4_332$	7.2470				380.61	4	McDonald et al. (2015)
Ferhodsite	$(Fe,Rh,Ni,Ir,Cu,Pt)_9S_8$	Tetragonal	?	10.009		9.840		985.77	2	Begizov (2009)[a]
Jacutingaite	Pt_2HgSe_3	Trigonal	$P\bar{3}m1$	7.3477		5.2955		247.59	2	Vymazalová et al. (2012a)
Jaguéite	$Cu_2Pd_3Se_4$	Monoclinic	$P2_1/c$	5.672	9.910	6.264	115.40	318.1	2	Paar et al. (2004), Topa et al. (2006)
Kalungaite	$PdAsSe$	Cubic	$Pa\bar{3}$	6.089				225.78	4	Botelho et al. (2006)
Kingstonite	Rh_3S_4	Monoclinic	$C2/m$	10.4616	10.7527	6.2648	109.0	666.34	6	Stanley et al. (2005)
Kitagohaite	Pt_7Cu	Cubic	$Fm\bar{3}m$	7.7891				472.57	4	Cabral et al. (2014)
Kojonenite	$Pd_{7-x}SnTe_2$	Tetragonal	$I4/mmm$	4.001		20.929		335.0	2	Stanley and Vymazalová (2015)
Kravtsovite	$PdAg_2S$	Orthorhombic	$Cmcm$	7.9835	5.9265	5.7451		271.82	4	Vymazalová et al. (in press)
Lisiguangite	$CuPtBiS_3$	Orthorhombic	$P2_12_12_1$	7.7152	12.838	4.9248		487.80	4	Yu et al. (2009)
Lukkulaisvaaraite	$Pd_{14}Ag_2Te_9$	Tetragonal	$I4/m$	8.9599		11.822		949.1	2	Vymazalová et al. (2014a)
Malyshevite	$PdCuBiS_3$	Orthorhombic	$Pnam$	7.541	6.4823	11.522		563.204	4	Chernikov et al. (2006)
Marathonite	$Pd_{25}Ge_9$	Trigonal	$P3$	7.391		10.477		495.65	1	McDonald et al. (2016)
Miessiite	$Pd_{11}Te_2Se_2$	Cubic	$Fd\bar{3}m$	12.448				1929.0	8	Kojonen et al. (2007)
Milotaite	$PdSbSe$	Cubic	$P2_13$	6.3181				252.20	4	Paar et al. (2005)
Naldrettite	Pd_2Sb	Orthorhombic	$Cmc2_1$	3.3906	17.5551	6.957		414.097	8	Cabri et al. (2005)

Mineral	Formula	Crystal system	Space group	a	b	c	V	Z	Reference
Nielsenite	$PdCu_3$	Tetragonal	$P4mm$	3.7125		25.62	353.2	4	McDonald et al. (2008)
Norilskite	$(Pd, Ag)_7Pb_4$	Trigonal	$P3_121$	8.9656		17.2801	1202.92	6	Vymazalová et al. (2017)
Palladogermanide	Pd_2Ge	Hexagonal	$P\bar{6}2m$	6.712		3.408	132.96	3	McDonald et al. (2017b)
Palladosilicide	Pd_2Si	Hexagonal	$P\bar{6}2m$	6.496		3.433	125.5	3	Cabri et al. (2015)
Pašavaite	$Pd_3Pb_2Te_2$	Orthorhombic	$Pmmn$	8.599	5.9381	6.3173	322.6	2	Vymazalová et al. (2009)
Skaergaardite	$PdCu$	Cubic	$Pm3m$	3.0014			27.0378	1	Rudashevsky et al. (2004)
Törnroosite	$Pd_{11}As_2Te_2$	Cubic	$Fd\bar{3}m$	12.3530			1885.03	8	Kojonen et al. (2011)
Ungavaite	Pd_4Sb_3	Tetragonal	?	7.7388		24.145	1446.02	8	McDonald et al. (2005)
Vymazalováite	$Pd_3Bi_2S_2$	Cubic	$I2_13$	8.3097			573.79	4	Sluzhenikin et al. (in press)
Zaccariniite	$RhNiAs$	Tetragonal	$P4/nmm$	3.5498		6.1573	77.59	2	Vymazalová et al. (2012b)

[a]No further data available.

Table 10.4 Experimental Data for Ternary Systems and the Corresponding PGM (Ideal Formulae)

System	Mineral Name	Formula	Type of Diagram, at T (°C)	Compositional Range	Reference
Ir-As-S	Irarsite	$IrAsS$	None		
Ir-Bi-S	Changchengite	$IrBiS$	None		
Ir-Bi-Te	Mayingite	$IrBiTe$	None		
Ir-Cu-S	Cuproiridsite	$CuIr_2S_4$	None		
Ir-Sb-S	Tolovkite	$IrSbS$	None		
Os-As-S	Osarsite	$OsAsS$	None		
Pd-Ag-Pb	Norilskite	$(Pd,Ag)_7Pb_4$	Isothermal section, 400	Full	Sarah et al. (1981)
Pd-Ag-S	Coldwellite	Pd_3Ag_2S	Isothermal section, 400, 550	Full	Vymazalová et al., in prep.
	Kravtsovite	$PdAg_2S$	Vertical section, 400–1600	Partial, Ag_2S-Pd	Raub (1954)
Pd-Ag-Se	Chrisstanleyite	$Ag_2Pd_3Se_4$	Isothermal section, 350, 430, 530	Full	Vymazalová et al. (2014b)
Pd-Ag-Te	Lukkulaisvaaraite	$Pd_{14}Ag_2Te_9$	Isothermal section, 350, 450	Full	Vymazalová et al. (2015)
	Sopcheite	$Ag_4Pd_3Te_4$	Vertical section, 0–1000	Partial, $Ag_{22.2}Pd_{66.6}Te_{11.2}$-$Ag_{66.7}Te_{33.3}$	Chernyaev et al. (1968)
	Telargpalite	$Pd_{2-x}Ag_{1+x}Te$			
Pd-As-Bi	Palladobismutharsenide	$Pd_2As_{0.8}Bi_{0.2}$	Isothermal section, 400	Partial, As-Bi- $Bi_{50}Pd_{50}$- $As_{15}Bi_{35}Pd_{50}$-$As_{33}Pd_{67}$-$As_{25}Pd_{75}$- $Bi_{25}Pd_{75}$-Pd	El-Boragy and Schubert (1971a)
			480	Partial, $Pd_{1.97}As_{0.80}Bi_{0.23}$- Pd_2As	Cabri et al. (1976)
Pd-As-Hg	Atheneite	$Pd_2As_{0.75}Hg_{0.25}$	None		
Pd-As-Sb	Arsenopalladinite	$Pd_8As_{2.5}Sb_{0.5}$	Isothermal section, 710	Partial, $PdAs_2$-$Pd_{73}As_{27}$-$Pd_{72}Sb_{28}$- $Pd_{38}Sb_{62}$	Cabri et al. (1975)
	Isomertieite	$Pd_{11}Sb_2As_2$			
	Mertieite I	$Pd_{11}(Sb,As)_4$			
	Mertieite II	$Pd_8Sb_{2.5}As_{0.5}$			
Pd-As-Se	Kalungaite	$PdAsSe$	None		

System	Mineral	Formula	Section	Solid solution / notes	Reference
Pd-As-Sn	Palarstanide	$Pd_5(Sn,As)_2$	Tentative isothermal section, 500	Partial	Pratt (1994)
Pd-As-Te	Törnroosite	$Pd_{11}As_2Te_2$	Isothermal section, 25	Partial, $SnAs$-$Pd_{50}Sn_{50}$-Sn	El-Boragy and Schubert (1971b)
Pd-Bi-Pb	Urvantsevite	$Pd(Bi,Pb)_2$	Isothermal section, 480	Full	Zhuravlev (1976)
Pd-Bi-S	Vymazalováite	$Pd_3Bi_2S_2$	Vertical section, 180–600	Partial, $PdBi_2$, $PdPb_2$	El-Boragy and Schubert (1971b)
Pd-Bi-Se	Padmaite	$PdBiSe$	None		Hoffman and MacLean (1976)
Pd-Bi-Te	Michenerite	$PdBiTe$	None	Partial, Pd-rich corner	Makovicky and Karup-Møller (in press)
Pd-Cu-Se	Jaguéite	$Cu_2Pd_3Se_4$	Isothermal section, 480	Partial, Te-Bi_4Te_3-$Bi_{60}Pd_{40}$-$PdTe$	Evstigneeva and Nekrasov (1980)
	Oosterboschite	$(Cu,Pd)_7Se_5$	Isothermal section, 480	Full	
Pd-Cu-Sn	Cabriite	Pd_2SnCu	Isothermal section, 300, 400, 550, 650	System studied in hydrothermal chloride solutions (Pd-Sn-Cu-HCl) at 300 and 400 °C	Lebrun et al. (2007) based on data from Dobersek and Kosovinc (1989)
	Stannopalladinite	Pd_5Sn_2Cu	None	Partial	Dobersek and Kosovinc (1989)
	Taimyrite	$(Pd,Cu)_3Sn$			
Pd-Cu-Zn	Bortnikovite	Pd_4Cu_3Zn	Tentative isothermal section, 800	Partial	Schubert et al. (1955)
			Experimental points at 350 to 800	Partial, Pd-Cu region containing up to 5 at.% Zn	
Pd-Hg-Se	Tischendorfite	$Pd_8Hg_3Se_9$	Isothermal section, 430	Full	Drábek et al. (2014)
Pd-Hg-Te	Temagamite	Pd_3HgTe_3	Isothermal section, 400	None	
Pd-Ni-As	Majakite	$PdNiAs$	Isothermal section, 450	Partial	Gervilla et al. (1994)

(Continued)

Table 10.4 Experimental Data for Ternary Systems and the Corresponding PGM (Ideal Formulae) *Continued*

System	Mineral Name	Formula	Type of Diagram, at T (°C)	Compositional Range	Reference
	Menshikovite	$Pd_3Ni_2As_3$	Isothermal section, 790	Full	El-Boragy et al. (1984)
Pd-Pb-Te	Pašavaite	$Pd_3Pb_2Te_2$	Isothermal section, 500	Partial, $NiAs_2$-Ni_5As_2-Pd_5As_2-$PdAs_2$	Vymazalová and Drábek (2011)
			Isothermal section, 400	Partial Pd_3Pb-Pb-Te-$PdTe$	El-Boragy and Schubert (1971b)
Pd-Pb-S	Laflammeite	$Pd_3Pb_2S_2$	Isothermal section, 480	Partial, Pd-rich corner	
Pd-Sb-Se	Milotaite	$PdSbSe$	None		
Pd-Sb-Te	Borovskite	Pd_3SbTe_4	None		El-Boragy and Schubert (1971b)
			Isothermal section, 400	Full	
	Testibiopalladite	$PdSbTe$	Isothermal section, 600	Full	Kim and Chao (1991)
			Isothermal section, 800	Partial, Pd-$Pd_{40}Sb_{60}$-$Pd_{40}Te_{60}$	
			Isothermal section, 1000	Partial, $Pd_{60}Sb_{40}$-Pd-$Pd_{60}Te_{40}$	
Pd-Sn-Te	Kojonenite	$Pd_{7-x}SnTe_2$	Isothermal section, 400	Full	Vymazalová and Drábek (2010)
Pd-Te-Se	Miessiite	$Pd_{11}Te_2Se_2$	None		
Rh-Pd-As	Rhodarsenide	$(Rh,Pd)_2As$	Tentative isothermal section, 750	Partial, As-$PdAs_2$-$RhAs_2$	Pratt (1994) based on data from Bennett and Heyding (1966)
	Palladodymite	$(Pd,Rh)_2As$			
Pt-As-S	Platarsite	$PtAsS$	Isothermal section, 1000	Full composition	Skinner et al. (1976)
Pt-Bi-Te	Maslovite	$PtBiTe$	None		
Pt-Cu-Fe	Tulameenite	$PtFe_{0.5}Cu_{0.5}$	Isothermal section, 600, 1000	Full composition	Shahmiri et al. (1985)
Pt-Cu-S	Malanite	$CuPt_2S_4$	None		

System	Mineral	Formula	Experiment	Composition	Reference
Pt-Cu-Sn	Tatyanaite	$Pt_9Cu_3Sn_4$	None		
Pt-Fe-Ni	Ferronickelplatinum	$PtFe_{0.5}Ni_{0.5}$	None		
Pt-Hg-Se	Jacutingaite	Pt_2HgSe_3	Isothermal section, 400	Full composition	Drábek et al. (2012)
Pt-Pd-S	Braggite	$(Pt,Pd)S$	Isothermal section, 800, 1000	Partial composition; Pd-Pt-PtS-PdS	Cabri et al. (1978)
			Isothermal section, 1000	Full composition	Skinner et al. (1976)
Pt-Pd-Sb	Genkinite	$(Pt,Pd)_4Sb_3$	Isothermal section, 600	Partial composition; $Pd_{72}Sb_{28}$-$Pt_{84}Sb_{16}$-Sb	Kim and Chao (1996)
			Isothermal section, 800, 1000	Full composition	
			Isothermal section, 1000	Full	Kim (2009)
Rh-As-Ni	Zaccariniite	$RhNiAs$	None		
Rh-As-S	Hollingworthite	$RhAsS$	None		
Rh-Cu-S	Cuprorhodsite	$CuRh_2S_4$	Isothermal section, 500, 700, 900	Full	Karup-Møller and Makovicky (2007)
			Isothermal section, 540	Partial, $Cu_{2-x}S$-$Rh_{17}S_{15}$-RhS_3-Cu_xRhS_{3+x}-$Cu_{2-x}S$	
Rh-Fe-S	Ferrorhodsite	$FeRh_2S_4$	Isothermal section, 500, 900	Full composition	Makovicky et al. (2002)
			Vertical section 900–1500	Partial composition; $FeS_{1.09}$-FeRh	Bryukvin et al. (1990)
			Vertical section 1050–1350	Partial composition; $FeS_{1.09}$-Rh_2S_3	
			Liquidus projection	Partial composition; Rh-$Rh_{40}S_{60}$-$Fe_{48}S_{52}$-Fe	
Rh-Pb-S	Rhodplumsite	$Rh_3Pb_2S_2$	None		
Ru-As-S	Ruarsite	$RuAsS$	None		

Pd-As-Sb, Pd-As-Hg, Pd-Cu-Sn, Pt-Cu-Sn, Pt-Bi-Te). We summarize the ternary PGM (taking into the account the ideal composition of minerals) and corresponding systems, and refer to data on phase diagrams in the literature (Table 10.4). There are still a large number of systems containing PGE-bearing ternary minerals without a knowledge of ternary phase diagrams and phase relations. As shown in Table 10.4, there is still not enough knowledge on several complete isothermal sections and very few systems have been studied at more than one temperature. Furthermore, some ternary systems also require re-investigation due to new observations within the corresponding binary systems and by the discovery of new minerals.

10.2 EXPERIMENTAL METHODS FOR SYNTHESIS OF PLATINUM-GROUP MINERALS

Experimental studies have various implications for geological processes. One of the aims of experimental synthesis is to determine phase relations under specific conditions with a view to understanding natural processes. The knowledge of phase relations allows prediction of the mineral assemblages stable at natural conditions. Experimental study of thermal stabilities and the solid solutions of PGM help to understand their formation, occurrence, and accumulation in nature, thus are directly applicable to the study of mineral deposits.

The other aim of experimental synthesis of PGE compounds is to obtain single crystals of sufficient size for detailed investigations. Single crystals can be studied in terms of crystal structure determination that is also applicable to the mineralogy of PGM. The knowledge of crystal structures and structural mechanisms of various substitutions within PGM also have implications in mineral processing and geo-metallurgy. Various PGE compounds are also of materials science interest and have industrial applications whereby PGE phases are studied in terms of specific properties (e.g., chemical, electrical, optical, magnetic, semiconductor, etc.).

We describe the preparation methods that can be used for PGE phases and minerals synthesis. In the first part (Section 10.3) we focus on methods resulting in powder products and in the second part (Section 10.4) we discuss suitable methods for producing single crystals. We also provide some examples of crystal growth and synthesis conditions for various PGE phases and minerals that might be applicable for preparation of other PGE compounds.

10.3 POWDER SAMPLES
10.3.1 DRY TECHNIQUE

The silica-glass tube method belongs to the classical experimental methods, so-called dry technique or solid phase synthesis. The method has been described in detail by Kullerud (1971), focusing on sulfide synthesis.

The silica tube method is suitable for synthesis of PGE sulfides, selenides, and tellurides, as well as for other compounds and alloys that are formed at temperatures below 1300°C (when silica starts to recrystallize and may become permeable). Charges are carefully weighed out from the pure elements and placed into the silica-glass tube. The starting materials are either pure elements or presynthesized phases, or combinations. To prevent loss of material to the vapor phase during experiments, the free space in the tubes is reduced by placing closely fitting glass rods over the charge. The charges should be as small as possible, but still having the sufficient material for the product investigation (e.g., polished section, X-ray diffraction). Smaller quantities are easier to homogenize, and another consideration is the high prices of pure platinum-group-element starting reagents. The tubes with the charge are evacuated and subsequently sealed in, e.g., a hydrogen-oxygen flame. The capsules with the charge are then heated in horizontal or vertical furnaces, which allow faster quenches. In order to ensure the homogeneity, the samples are after the first melting/heating opened and the products are finely reground under acetone in an agate mortar. In some cases, the reground samples are pelletized under pressure before being reinserted in silica tubes, evacuated and reheated (e.g., Cabri, 1973; Cabri et al., 1976 (palladobismutharsenide)). In many cases the samples are reground several times during the heating before the equilibrium is attained. Reactions in silica tubes rely on diffusion and can be exceedingly slow, particularly for PGE-rich phases and for phase studies at low temperatures. Therefore, some experiments require long term heating to reach the equilibrium that may take from several months up to a year. After heating the experimental runs are quenched by dropping the capsule in cold water.

In order to obtain sulfides with high sulfur content (e.g., PtS_2), a long silica tube located in a temperature gradient (Pt in a hot and S in a cold zone) can be used to prevent the explosion of the tube that might be caused by the high sulfur vapor pressure.

The run products are suitable for powder X-ray diffraction analyses, and are usually examined in polished sections using reflected light microscopy, electron-scanning microscopy and electron-microprobe techniques. In most cases, the run products are very fine grained and required detailed and precise examination (Fig. 10.1A,B).

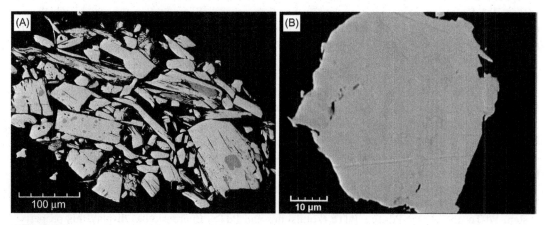

FIGURE 10.1

Back-scattered electron images of experimental products obtained by the dry technique: (A) synthetic jacutingaite (Pt_2HgSe_3) in association with sudovikovite ($PtSe_2$), a quench from 400°C (heated for 71 days); (B) fine grained intergrowths of lukkulaisvaaraite ($Pd_{14}Ag_2Te_9$) with telargpalite ($Pd_{2-x}Ag_{1+x}Te$) and hessite (Ag_2Te), a quench from 350°C (heated for 130 days).

10.4 SINGLE CRYSTALS

Single crystals of PGE phases and minerals are required for crystal structure investigations and for more accurate determination of their properties. In some cases, phase associations obtained in the form of sufficiently large coexisting crystals help the study of phase relations. Crystal growth techniques can in general be divided into congruent and incongruent. Congruent techniques constitute crystal growth from the melt of the desired substance (melt of a similar composition) and are characterized by a $L \rightarrow S$ phase reaction. Incongruent techniques are based on crystal precipitation from liquid or gas solvents. The principal methods are vapor transport, hydrothermal (water solution) and flux techniques. Moreover, techniques can be distinguished based on spontaneous crystallization (when crystals are formed via nucleation) or controlled crystallization (with seed crystals). The silica-glass tube method (dry technique or so-called solid phase synthesis technique) described in Section 10.3 also allows the preparation of micrometer-size crystals.

10.4.1 CONGRUENT TECHNIQUES

There are various congruent techniques for crystal preparation (e.g., Wilke, 1973), and all of them are based on a gradual removal of heat from the liquid-solid boundary; they differ in the way this boundary is shifted. An advantage of incongruent techniques is the possibility of synthesizing incongruently melting crystals and synthesis at lower temperatures.

The Bridgman—Stockbarger technique (or Bridgman technique, Bridgman, 1925) is usually performed in a vertical reaction vessel with a cone-shaped bottom. The vessel with the melt is slowly moved down to the gradient tube furnace, where the temperature in its lower part is lower than in its upper part. Crystallization starts from the bottom of the vessel. The cone shape of the bottom ensures the formation of a single nucleus rather than several. Sometimes a process in which the reaction vessel is cooled in the temperature gradient without moving along the gradient furnace is also called the Bridgman technique.

The Czochralski technique (Czochralski, 1918) is implemented by gradually pulling the cooled seed crystal above the free surface of the melt. The Kyropoulos technique (Kyropoulos, 1926) is similar but differs in that there is no mechanical movement of the seed crystal. The temperature of the melt is slowly decreased so that the crystal gradually grows on the seed forming a half-sphere.

Crystal growth via *the Verneuil technique* is implemented by gradually adding melt droplets to the surface of the growing crystal.

The Pfann technique or zone melting constitutes melting by a movable circular heater of a powder-like starting mixture, which usually is compressed into a cylinderical shape (Pfann, 1966).

Phases suitable for growing crystals from their melt can be identified from the corresponding known phase diagrams by having a common liquidus and solidus maximum. For example, in the system Pd-Te: the phases Pd_8Te_3, PdTe (kotulskite) and $PdTe_2$ (merenskyite) display this property. Most of the congruently melting crystals of chalcogenides and pnictogenides are grown using the Bridgman—Stockbarger technique. For example, crystals of $PdTe_2$, up to 7×16 mm in size, were grown using this method by Lyons et al. (1976). The growth was carried out in a silica-glass reaction vessel, which was moved down at 2.8 mm/hour inside a gradient furnace with the temperature of the hot zone at 800°C. Apparently, crystals of other congruently melting phases can be prepared using a similar technique: $PtAs_2$ (sperrylite), PdBi (polarite), Pd_3Pb (zvyagintsevite), PdSb (sudbur-yite), Pd_2Si (palladosilicide), $PtSb_2$ (geversite), Pt_3Sn (rustenburgite), PtSn (niggliite), Rh_2Se_3 (bowieite), $PtSe_2$ (sudovikovite), $PtTe_2$ (moncheite), and others.

Congruent techniques are mainly used to grow single phase samples at temperatures near the liquidus and therefore in general are not suitable for studies of phase relations.

10.4.2 INCONGRUENT TECHNIQUES

10.4.2.1 Vapor Transport Technique

In the vapor-transport crystal-growth technique (Schafer, 1962), a solid substance interacts with a transport agent via a reversible chemical reaction, forming only gas products which flow to another part of the reaction system with different physical and chemical conditions, where the initial substance is formed again but in a single-crystal state. The main requirements for the transport are a concentration gradient and a reversible transport reaction. Usually the transport agents are halogens or their compounds (e.g., ICl_3, $AlCl_3$, S_2Cl_2). The transport of transition elements (PGE and Fe, Ni, Co) can be facilitated by phosphorus, oxygen, hydrogen and, consequently, water. The vapor transport process is similar to sublimation but it can occur at lower temperatures. To describe the temperature profile of the transport it is convenient to use the $T_1 \rightarrow T_2$ notation where T_1 is the temperature of the charge and T_2 is the temperature of the growth zone.

Crystal growth is usually performed in closed cylindrical reaction vessels made of silica or ordinary glass. A part of the charge (the solid substance) with the transport agent is placed at one end of the vessel. Usually the quantity of the transport agent is orders of magnitude less than that of the charge. In order to create different temperatures at the two ends of the vessel, gradient furnaces or a natural temperature gradient are used. Cylindrical silica ampoules can be used in the vapor transport technique if the hot end of the ampoule is at a temperature not higher than $1000-1200°C$. At higher temperatures silica-glass softens, recrystallizes or reacts with a charge, and silica becomes permeable to gas(es). If the hot end of the ampoule has to be at a temperature greater than $1300°C$, a white-hot wire is soldered into the ampoule.

Campbell et al. (1949) described transport of various metals including iridium, osmium, platinum, rhodium, and ruthenium using CO and halogens as transport agents. The formation of platinum whiskers by decomposition of platinum chloride at $800°C$ was described by Brenner (1956). Schafer (1962) results in the formation of iridium ($1325 \rightarrow 1130°C$) and platinum crystals (temperature in the hot end $= 1500°C$) using oxygen as a transport agent. At similar or lower temperatures, in a flow of oxygen, crystals of PGE oxides such as: OsO_2 (Rogers et al., 1969; Yen et al., 2004), RuO_2, and IrO_2 (Rogers et al., 1969; Horkans and Shafer, 1977) or $Ru_{1-x}Ir_xO_2$ (Georg et al., 1982) can be also grown. Tiny crystals of osmium phosphide OsP_2 were grown at $1000°C$ using iodine (Bugaris et al., 2014) and crystals of palladium arsenides (Pd_3As and Pd_5As) were grown at $500-600°C$ using chlorine by Saini et al. (1964) who noted that the transport does not occur with other halogens.

Crystals of chalcogenides are usually prepared using halogens. Crystals of RuS_2 and $RuSe_2$ were prepared following such an approach by, e.g., Bichsel et al. (1984), Vaterlaus et al. (1985), Fiechter and Kühne (1987). The growth of RuS_2 occurred in the temperature profile $1040 \rightarrow 1020°C$ using a mixture of ICl_3 and S_2Cl_2 as transport agents with an excess of sulfur. As a result, crystals

$4 \times 4 \times 4$ mm in size were obtained in about 30 days. The growth of RuSe$_2$ occurred at $1100 \rightarrow 1070°$C using ICl$_3$; the charge also contained an excess of the chalcogen, and as a result $10 \times 8 \times 5$ mm crystals were grown. Crystals of RuTe$_2$ were grown using ICl$_3$ at $1060 \rightarrow 960°$C over a period of 10 days (Huang et al., 1994) and Colell et al. (1994) synthesized RuS$_2$ crystals containing up to 50 at.% Ir (temperature profile $1100 \rightarrow 1050°$C) using iodine. Similarly, Tsay et al. (1994) grew iron-doped RuS$_2$ crystals using ICl$_3$. Crystals of Rh$_2$S$_3$ were synthesized within a temperature profile from $1000 \rightarrow 1020°$C using bromine (Parthe et al., 1967).

Crystals of PtS$_2$ were grown using a mixture of chlorine and phosphorus (Finley et al., 1974). The charge contained platinum, sulfur, and phosphorus with molar ratios 1:3:1. Chlorine was added in order to create a pressure of 10,000 Pa. The ideal temperature profile was from $800 \rightarrow 740°$C, under which flat hexagonal crystals with sizes up to 5×5 mm were grown after 6 days. At lower temperatures ($700 \rightarrow 670°$C) the transport was insignificant and at higher temperatures ($900 \rightarrow 840°$C) the two-phase association PtS + PtS$_2$ was obtained. Crystals of Pt(S,Se)$_2$, with varying S/Se ratios, in association with crystals of PtTe$_2$, were prepared with phosphorus as the major transport agent, sometimes with addition of chlorine, with temperatures of $850-875°$C at the hot end and $750-690°$C in the cold end (Soled et al., 1975, 1976).

10.4.2.2 Hydrothermal Techniques

Solution and hydrothermal techniques are based on recrystallization of the substance in aqueous solutions at room or higher temperature. At low temperatures, unlike in the vapor transport technique, supersaturation can be reached by gradual evaporation of the water solution; by addition of compounds that lower the solubility of the desired phase (e.g., pH variation); by mixing solutions containing the components of the substance crystallized; electrolysis; transport reactions; and substance exchange on the boundary of two phases. The gradual temperature increase causes an increase in the solubility of most chemical compounds, reaction rates, and diffusion rates. However, the synthesis needs to be done in durable hermetic vessels (autoclaves) at temperatures above 100°C. Usually transport is driven by a temperature gradient and the migration of the phase(s) occur(s) mostly via convection. Often, in order to increase the solubility of the charge, a mineralizer (e.g., KClO$_3$), which is similar to the transport agent in the vapor transport technique, is added to water. Mineralizers do not form separate phases. The transport of chalcogenides can be increased by adding acid or NH$_4$I to the water, as chalcogenides decompose in alkali solutions. Iodides (e.g., NH$_4$I) are the most preferable halogenides because iodine and sulfide ions have the largest difference in ionic radius, which prevents iodine from substituting into the lattice of the growing sulfide.

Most often autoclaves made of stainless steel or titanium alloys are used as laboratory vessels for hydrothermal synthesis, and allow for operating at temperatures up to $\sim 500°$C and pressures up to 2000 atmospheres. The solution is often placed in a Teflon or a noble metal liner to protect the autoclave. Hermetic ampoules made of noble metals or silica glass, and of a size less than the internal volume of the autoclave, can be also used to protect from an aggressive environment. Free external space in this case is filled with water. When soft metal ampoules are used, the water often serves as the pressure medium. Ampoules made of noble metals in some cases are the source of the noble metal. When silica ampoules are used, the external water creates a counterpressure that protects the silica ampoule from destruction. Gas, e.g., CO$_2$ (Rau and Rabenau, 1968) also can be used as the external medium.

Crystals of various metals including platinum were grown hydrothermally by, e.g., Rau and Rabenau (1968). The crystal growth was carried out in silica ampoules in concentrated aqueous solutions of HCl, HBr, or HI in a temperature gradient (with a hot end $420-600°C$). To prevent the ampoules from cracking they were placed in autoclaves together with dry carbon dioxide which created the counter-pressure by evaporation. It was also shown that the addition of oxidizers, e.g., Cl_2, Br_2, H_2O_2 or residual oxygen boosts the metal transport due to the increase in concentration of metal ions dissolved in water. Schwartz et al. (1982) prepared platinum oxide crystals (β-PtO_2) in equilibrium with other compounds (e.g., $CdPt_3O_6$) in sealed platinum ampoules in $KClO_3$ solution by cooling the solution. Crystals of Rh_2S_3 and $Rh_{17}S_{15}$ with sizes up to a hundred microns were prepared by, e.g., Zhang et al. (2009), using rhodium carbonyl $Rh_6(CO)_{16}$ and crystals of sulfur as starting chemicals, grown in Teflon-lined autoclaves at $200-400°C$.

The hydrothermal technique is less commonly used to grow PGE-bearing crystals. However, it has been used to study the conditions of formations as in, e.g., the Pd-Sn-Cu-HCl system by Evstigneeva and Nekrasov (1980).

10.4.2.3 Flux Technique

The flux technique (e.g., Wilke, 1973) is based on the gradual cooling of a multicomponent system. The solubility of the components in the melt decreases with decreasing temperature and this leads to the formation of crystals of certain compounds. Due to the complexity of a multicomponent system the composition of the grown crystals differs from that of the melt. The flux technique is somewhat intermediate between hydrothermal (incongruent) and congruent techniques. The major advantage of the flux technique, in comparison with the hydrothermal technique, is the broad range of temperatures (from $25°C$ to $1500°C$) that can be used for synthesis. The other advantage is the large variety of possible solvents, e.g., various combinations of oxides, salts, hydroxides, and other heteropolar compounds that can be used. Metals with low melting temperature such as Sn, Pb, Bi, Te, In, Hg, eutectic mixtures of lead fluoride or lead oxide, tungsten, molybdenum oxides, or metal halides or alkali metal poly-chalcogenides (e.g., Na_2S_n) are often used as solvents.

In general, the flux method is not very complicated. Specific amounts of the charge and the solvent, in powder form, are placed in a crucible made of inert material. Usually the crucible is tightly closed with a lid to reduce evaporation of the solvent. Then during the shortest possible period of time (several hours) the crucible is heated up to the maximum point. After this, the temperature is quickly decreased by $50-100°C$ to create a supersaturation or to form nuclei of the desired compound, and then the temperature is slowly decreased at $3-5°C$ per hour. Afterwards, at a certain temperature the experimental product is quenched or the melt is mechanically separated from the crystals (decanted). In some cases, the crucible with the charge is not quenched or cooled to the room temperature because the desired phase can precipitate as crystals only within a certain temperature interval. In some cases, more complicated cooling modes are used, such as oscillations in the area where crystallization starts. Crucibles used for crystal growth are made most often from precious metals, corundum, or silica-glass tubes. However, as described for dry synthesis in Section 10.3, the silica-glass softens at temperatures over $1300°C$ and dissolves at temperatures over $600°C$ in PbF_2-PbO melts.

The flux method is quite universal; crystallization may be possible in poorly concentrated solutions as well as in highly concentrated solutions for a single-component composition, almost in the field of crystallization from the melt. When the solution of a desired material shows a low degree of concentration then the method displays similarities with the hydrothermal or vapor-transport techniques. In cases where the composition of the solution is almost identical with the crystallized substances then the flux method is similar to synthesis from the melt of the desired compound, like the Bridgman or Czochralski techniques. Usually the term "flux technique" is used when the amount of compound being crystallized is more than 6%−8% of the total liquid phase. However, the self-flux technique can be also distinguished when the composition of an initial melt is the same as that of the resulting crystals. Supersaturation is most commonly created by a slow cooling or by gradual evaporation of the solvent. In some cases, small crystals are obtained under isothermal conditions.

10.4.3 PGE-BEARING CRYSTALS PREPARED BY FLUX TECHNIQUE

The first crystals of PGE compounds using the flux technique were obtained in the 19th century (e.g., Rossler, 1895; as summarized in Chirvinskii, 1995). For example, dendritic crystals of gold and platinum were prepared by melting Au/Pt with NaCl, $Na_2S_2O_7$ or $Fe_2(SO_4)_3$, $PtAs_2$ (sperrylite), $PtSb_2$ (geversite) and $PtBi_2$ (insizwaite) crystals in lead flux, and crystals of RuS_2 (laurite) and PtS_2 by melting Ru (or Pt) in iron sulfide and borax together at $\sim 900°C$.

Crystals of PGE compounds can be grown in various solvents by gradually decreasing temperature. Reviews by Fisk and Remeika (1989) and Canfield and Fisk (1992) summarize the application of low-temperature melting metals to prepare crystals by the flux method for many chemical compounds including PGE-bearing phases. The growth of UPt_3, $NpPt_3$, PtMnSb, and UIr_3 crystals in bismuth, RRh_4B_4, RIr_2, and UIr_3 in copper, RPt_2 and $YbPt_x$ in lead, and $R_3Rh_4Sn_{13}$ in tin (where R is a rare-earth element) was shown by Fisk and Remeika (1989). Further, crystals of UPt_3, YPd, RBiPt, and $R_3Bi_4Pt_3$ prepared in bismuth; $R_2Pt_4Ga_8$ in gallium, LaPbPt, CePbPt, LaBiPt, CeBiPt, and PrBiPt in lead, and $U_3Sb_4Pt_3$ and $PtSb_2$ in antimony are summarized in Canfield and Fisk (1992). Melts of Al, Ce, Fe, In, Hg, Zn, and Ag can be used to prepare PGE-bearing crystals. Here we provide some examples for crystal growth of PGE-bearing phases and minerals and discuss the experimental conditions.

Crystals of PGE with P, As, Sb, Bi (pnictides) phases can also be prepared by cooling the metal melts. For example, crystals of RhP_3 were grown in tin flux (Odile et al., 1978); the synthesis was done in silica-glass tubes, where tin formed about 85% of the total volume of the charge. Tubes with the charge were heated to 1150°C for some hours, until complete dissolution. Afterwards, the tube with the charge was cooled to 550°C at a rate of 5°C per hour and the experimental products were washed from tin in hydrochloric acid. Similarly, by cooling the tin melt, other PGE-bearing phosphides can be synthesized, as PtP_2, RuP_2, IrP_2, RuP_4, and OsP_4 (Baghdadi et al., 1974; Kaner et al., 1977 Ruehl and Jeitschko, 1982), OsP_2 and $OsAs_2$ or $OsSb_2$, in an excess of antimony (Bugaris et al., 2014). Savilov et al. (2005) prepared the ternary phase $Pd_{7-x}SnTe_2$, an analogue of kojonenite, using a tin flux.

Ternary PGE arsenides can be also synthesized in a metal flux. For example, crystals of $BaRh_2As_2$ were grown in a lead flux (Singh et al., 2008) with an initial mixture of $Ba_{1.1}Rh_2As_{2.1}Pb_{50}$. Due to the presence of barium the synthesis was performed in a glass made of Al_2O_3, sealed in an evacuated silica-glass tube. The mixture was cooled from 1000°C to 500°C at a rate of 5°C per hour. As a result, crystals $1.5 \times 1.5 \times 0.1$ mm^3 in size were obtained. The authors did not mention if lead was present in the resulting crystals.

Further, the flux technique can be applied for PGE oxides. For example, needle-like crystals of $CaIrO_3$ were grown in platinum crucibles with $CaCl_2 + Ca(OH)_2 + Ir$ melt (in molar ratio 10:1:1) by cooling from 827 to 327°C at a cooling rate of 10°C per hour (Sugahara et al., 2008).

The flux method can be also used for crystal growth of PGE chalcogenides. For example, crystals of $CuIr_2S_4$ were grown by cooling a bismuth-based melt (Matsumoto and Nagata, 2000) from an initial temperature of 1100−1000°C to a final temperature of 500°C. The bismuth melt was decanted and the remaining crystals were washed from the residual bismuth in warm nitric acid. The authors also noted that these crystals could not be prepared by the vapor transport technique and using melts based on Cu, Sn, In, S, and Te. Crystals of RuS_2 (analogue of laurite), $RuSe_2$ and $RuTe_2$ in sizes up to 4 mm were grown in bismuth, and tellurium, respectively, using evacuated silica-glass tubes, cooled from 1000°C at a rate of 2°C per hour (Foise et al., 1985). Due to density differences, crystals of Ru chalcogenides grew above the melt. However, crystals grown in bismuth melts may contain some admixture of Bi, up to 3%.

The same approach was also applied by Ezzaouia et al. (1985) to grow crystals of RuS_2 in a tellurium flux by cooling from 1000°C to 850−700°C at rates of 0.7 or 1.4°C per hour. Iridium-rich crystals of RuS_2 (Colell et al., 1994) were prepared from an initial mixture of $Ir_{0.667x}Ru_{1-x}S_2$ with bismuth, in a ratio 1:40, in an evacuated silica-glass tube, heated at 1100°C for a day. Afterwards the experiment was cooled to room temperature at a rate of 2°C per hour, and the crystals were washed in HNO_3. It should be noted that often solid solutions of Ir-bearing crystals grown by cooling are zoned. Iron-doped RuS_2 crystals (sizes up to 3 mm) were prepared by Tsay et al. (1994) using a tellurium flux (100 g of Te to 5 g of charge) cooling from 1000°C to room temperature at a rate of 1−0.7°C per hour. In order to obtain larger crystals, already-presynthetized crystals can be used as seed material. This approach was used by, e.g., Tsay et al. (1995), following Tsay et al. (1994), to grow crystals of RuS_2, $RuSe_2$, and $RuTe_2$ reaching sizes of 15 × 10 × 10 mm.

Os chalcogenides can be prepared using methods similar to the one used for Ru chalcogenides (Ezzaouia et al., 1984). Crystals of OsS_2 were grown in sulfur and tellurium and $OsSe_2$ in tellurium, with ratios of charge to solvent from 1:8 to 1:4 and 1:32 to 1:12, when using tellurium. The silica-glass tube with the charge was heated, in the case of tellurium, at 885°C for 12 hours and cooled to 600°C at a rate of 1°C per hour. For sulfur, the experiment was heated at 650°C and cooled to 300°C at the same rate. The resulting crystals, up to one mm in size, were washed from the tellurium in aqua regia and the sulfur was removed by evaporation (Ezzaouia et al., 1984).

The growth of RuS_2 and OsS_2 crystals using the sulfur flux or $RuTe_2$ and $OsTe_2$ using tellurium flux can be considered as a self-flux technique. Applying a similar approach, crystals of Ir_3Te_8 were grown by cooling tellurium flux from 1000°C to 700°C (Liao et al., 1997) and platinum-doped $IrTe_2$ crystals by cooling from 1160°C to 900°C (Pyon et al., 2013). The chalcogen flux can be applied to crystal growth of those chalcogenides that are in equilibrium with the respective chalcogen. The possibility of preparing only chalcogen-rich substances limits the self-flux technique.

In cases where the solvent has a sufficient partial vapor pressure, then the crystals can be grown by evaporation resulting in crystals of monolayers growing at a constant temperature unlike what occurs during cooling. Zoned crystals are not formed.

The application of evaporation can be nicely demonstrated by the crystal growth of RuS_2, as documented by Fiechter and Kühne (1987), and earlier by Ezzaouia et al. (1983, 1985). A boomerang-shaped silica-glass tube (by oxygen torch) was used as a reaction vessel (Fig. 10.2A). A tellurium or selenium melt containing RuS_2 was placed in the left end of the vessel (growth

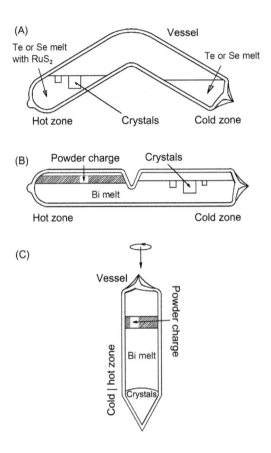

FIGURE 10.2

Reaction vessels showing the growth of RuS_2 crystals: (A) in selenium or tellurium flux by solvent evaporation. The solvent evaporates from the solution in the left part and condensates in the right part; (B) in bismuth flux under gradient conditions, with separated growth and dissolving zones; (C) in the traveling-solvent technique.

Adapted from Fiechter, S., Kühne, H.M., 1987. Crystal Growth of RuX₂ (X = S, Se, Te) by chemical vapour transport and high temperature solution growth. J. Crystal Growth 83(4), 517–522.

zone) at 920°C; the solvent gradually evaporates and condenses at the right end of the ampoule at 900°C. Crystals of RuS_2 grew up to 3 mm in size in 5 days and small crystals of $RuTe_2$ (or $RuSe_2$) were found in the melt. The authors also noted that bismuth is not suitable for this technique due to its low partial vapor pressure. However, they used a bismuth flux under gradient conditions. A sketch of the reaction vessel is shown in Fig. 10.2B. A silica-glass vessel about 14 cm long consists of two parts with a neck located approximately in the middle of the vessel. The neck enables RuS_2 powder to float on the melt in the left part of the vessel and does not get into the right part (the crystallization zone). At the beginning of the experiment the entire vessel was held for a day at 1000°C, then the right part was cooled at a rate 2°C per hour to 500°C. As a result, crystals of RuS_2 were grown with a surface area of up to 15 mm². According to Fiechter and Kühne (1987)

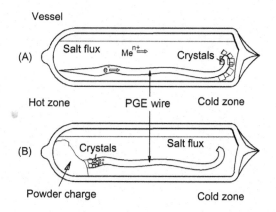

FIGURE 10.3

A reaction vessel for salt-flux synthesis under temperature-gradient conditions for: (A) crystal growth of metals and alloys; (B) crystal growth of Ag-Pd chalcogenides.

this method is the most preferable for Ru chalcogenides. A traveling-solvent technique (Hurle et al., 1967) was applied by Fiechter and Kühne (1987) for RuS_2 crystal growth; a basic outline is shown in Fig. 10.2C. A silica vessel with a conical base (similar to the Bridgeman technique) and a conical tip with a neck on the other end is placed vertically. A hole is made in order that only a single nucleus is formed. In the beginning of the experiment the ampoule contains bismuth melt with RuS_2 powder floating on it. The ampoule is slowly moved down at 0.01 mm/hour into a temperature gradient whereby crystals grow from the base fed by the charge that floats on the bismuth melt.

Gradient conditions can be also applied to obtain some PGE intermetallic phases. For example, crystals (up to 1 mm) of platinum, palladium, and $AuPd_4$ phase were grown in chlorides of alkali metals (Chareev, 2015). A schematic sketch of the synthesis is shown in Fig. 10.3A. The crystal growth was performed in a salt mixture of CsCl/KCl/NaCl, of eutectic composition in silica-glass tubes. A wire made of a noble metal, located along the length of the tube, serves as a source of that metal. The temperature at one end of the tube was about 600°C and in the other end about 50−100°C lower. The wire gradually dissolved at the hot end and formed metal crystals on the wire in the cold part of the tube. The metal ions migrate to the place of crystallization via the salt flux and electrons migrate along a wire. Likewise, using an electrically conducting wire it is possible to grow crystals of Ag-Pd chalcogenides (Chareev et al., 2016). The general scheme of the reaction vessel is shown in Fig. 10.3B. The silica-glass tubes were located in a temperature gradient; the hot end was at 450°C, and the cold end about 50 °C lower. Material transport occurred in a LiCl-RbCl melt of eutectic composition. A charge of approximate $(Ag,Pd)_4X$ (where X was S or Se) composition, was placed at the hot end. A wire or a plate of palladium or silver-palladium alloy was touching the charge. After several weeks, small crystals of $AgPd_3Se$ (Fig. 10.4A), in an equilibrium association with Pd_3Ag_2S (coldwellite) and Pd_4S (Fig. 10.4B) were obtained on the wire close to the charge. This technique can be confidently used to grow crystals only of those substances which are in equilibrium with silver-palladium alloy. This ongoing research, focused on

FIGURE 10.4

Back-scattered electron images showing: (A) AgPd$_3$Se crystals with a recrystallized Pd$_{0.6}$Ag$_{0.4}$ plate in the background; (B) equilibrium association of Pd$_3$Ag$_2$S (coldwellite) and Pd$_4$S crystals.

transport of material under gradient conditions, has shown the possibility to grow PtS$_2$ from FeS$_2$ + Pt charge in the (850 → 780°C) profile in eutectic NaCl/KCl.

10.5 PLATINUM-GROUP MINERALS AND SYNTHETIC ANALOGUES

PGM recognized by the Commission on New Minerals, Nomenclature, and Classification (CNMNC) of the International Mineralogical Association (IMA) up to 2002 were summarized and thoroughly evaluated by Cabri (2002). All the pertinent information on the 109 PGM known at that time can be found in this review. An earlier summary of PGM was given by Cabri (1981a). Since 2002 an additional 27 new PGM have been described after approved by CNMNC IMA; in total there are 136 known PGM in 2016. In this chapter, we provide a brief summary of PGM described since Cabri's overview.

All 136 PGM known in 2016 are listed in Table 10.1. The ideal formulae are given as accepted by Cabri (2002); newly described PGM are marked with an *a*. Formulae for atheneite and mertierite II, based on new studies on the crystal structure and formulaes of telagpalite and keithconnite, are shown based on the study of the Ag-Pd-Te system by Vymazalová et al. (2015). A summary of PGM data with revised or newly determined crystal structures are presented in Table 10.2. The crystal structures of atheneite, chrisstanleyite, mertierite II, and sopcheite were newly determined, and using synthetic analogues for temagamite and tischendorfite. The crystal structure of atheneite was solved from data collected from a crystal from Itabira, Minais Gerais, Brazil; based on these results (Table 10.2), the formula was revised to be Pd$_2$[As$_{0.75}$Hg$_{0.25}$] ($Z = 3$) (Bindi, 2010), instead of the previously reported $(Pd,Hg)_3As$ ($Z = 2$) (Clark et al., 1974). The crystal structure of mertierite II was reinvestigated on crystals from the Kaarrojja river, Inari commune, Polar Finland by Karimova et al. (in press), and yielded a formula of that is in agreement with the suggestion of Cabri (2002), proposing the 5:1 ratio for Sb:As. The crystal structure of sopcheite was solved using single-crystal data from the Lukkulaisvaara intrusion, northern Russian Karelia (Table 10.2; Laufek et al.,in press). Cooperite (PtS), some of whose phase relations were studied by Cabri et al. (1978), was reinvestigated in terms of crystal structure by Rozhdestvina et al. (2016). They found that in addition to the main reflections (corresponding to the known tetragonal cell, $P4_2/mmc$) there are many weak reflections which fit the tetragonal cell ($I4/mmm$) with double parameters. The crystal structures of isomertieite and törnroosite were refined (Karimova et al., 2016). Table 10.3 summarizes new PGM described after 2002.

As pointed out by Cabri (2002) there is still a significant number of PGM that require re-examination, re-definition or additional data for better characterization, particularly in terms of ideal formula and crystal structure. The main reason for the lack of insufficient identification of PGM is their mode of occurrence (as minute inclusions), intergrowths with other PGM, often embedded in base-metal sulfides and in most cases of very rare occurrence. These peculiarities often prevent XRD-based structural study. Therefore, in some cases further studies of synthetic analogues including crystal structure determination of PGM are desirable, ideally with natural minerals being investigated at the same time. Among newly described PGM (Table 10.3) the synthetic analogues were used in the description of the following mineral species: jacutingaite, milotaite, pašavaite, zaccariniite, lukkulaisvaaraite, kojonenite, palladosilicide, and norilskite. The XRD data were collected and crystal-structure studies were performed on their synthetic analogues, using either single crystals or powder samples. Consequently, the identity of a natural phase can be verified by a synthetic sample. The optical properties, physical properties, and chemical composition must be in agreement. In order to prove the structural identity, various methods can be used. The best way is to compare the X-ray diffraction data of a natural sample with its synthetic analogue (e.g., kitagohaite, sudburyite) but the extraction of a suitable grain of the natural sample is not always possible. Therefore, in order to prove the structural identity of natural and synthetic sample Electron Backscatter Diffraction or Raman spectroscopy (e.g., zaccariniite) can be applied. However, Raman spectroscopy has limited success, particularly among PGE alloy phases. Electron Backscatter Diffraction (EBSD), a technique based on the scanning electron microscope, enables sample microstructure to be analyzed, visualized, and quantified; it is a non-destructive method for the study of tiny grains, in situ (in polished sections). The diffraction pattern is characteristic of the crystal structure and orientation in the spot where it was generated. Hence the diffraction pattern can be used to determine the crystal orientation, discriminate between crystallographically different phases, characterize grain boundaries, and provide information about the local crystalline perfection. EBSD was used for the description of jacutingaite, pašavaite, lukkulaisvaaraite, kojonenite, palladosilicide, and norilskite. The EBSD technique can be applied as a supporting tool in order to distinguish various PGM minerals with defined crystal structures, but does not work in every case, depending on the polished characteristics of different minerals.

We tabulate all known PGM in 2016 (Table 10.1) and provide a brief overview on new PGM described from 2002 to 2016, as summarized in Table 10.3. Minerals are listed in alphabetical order; the data were collected from the original sources. All listed minerals were approved by the CNMNC of IMA and resulted in a publication except for ferhodsite (IMA No

2009-056, Begizov). The mineral was accepted by the Commission in 2009 but the description has not yet been published so that insufficient information about the mineral composition and other properties is known. Reflectance data for all listed minerals were measured in the range of λ from 400 to 700 nm (in case of pašavaite and jacutingaite from λ 420 to 700 nm). Herein we present only four wavelengths recommended by COM. The strongest lines from the XRD patterns are given, with a reference to the PDF or ICSD databases, if known. An average of electron-microprobe analyses (EMPA) is given for chemical composition. The experimental aspects for the corresponding compound are discussed for each mineral/synthetic analogue. Minerals marathonite $Pd_{25}Ge_9$ (IMA No 2016-080), palladogermanide Pd_2Ge (IMA No 2016-086), kravtsovite $PdAg_2S$ (IMA No 2016-092) and vymazalováite $Pd_3Bi_2S_2$ (IMA No 2016-105) were accepted when this Chapter was completed therefore a detailed description of these four minerals is not provided.

10.5.1 BORTNIKOVITE PD$_4$CU$_3$ZN

Crystallography: tetragonal, *P4/mmm* ?
   Unit cell: a 6.00, *c* 8.50 Å, *V* 306.0 Å^3, *Z* =3
   X-ray powder diffraction pattern, strongest lines [*d* in Å(I)(*hkl*)]: 3.00(1)(200,112), 2.64(1)(120), 2.36(0.5)(113), 2.13(10)(004, 220), 1.737(1)(132), 1.501(3)(400,224), 1.346(2)(240,332), 1.224(8)(404), 1.161(1)(151), 1.059(4)(440), ICDD PDF2 card 00-060-0492
   Structure: not defined
   Appearance: Forms rims (50−60 μm thick and 50−150 μm long) on isoferroplatinum, intergrowing with titanite and chlorite. Found in a heavy concentrate sample from sediments of the Konder valley.
   Optical properties: White with a slight greyish tint in reflected light; bireflectance, anisotropy, and internal reflections not observed.

Reflectance, [λnm, R %]		
	470	56.9
	546	61.7
	589	63.4
	650	65.4

   Physical Properties: Opaque, metallic luster, steel-white with slight cream tint, poorly malleable
   Hardness: VHN$_{25}$ 367.9 (354−382) kg/mm^2 Mohs
   Cleavage: not observed
   Density: 11.16 g/cm^3 (calc.)
   Chemical Composition (EMPA data, wt.%):
   Pd 58.19, Pt 4.06, Fe 1.41, Cu 27.26, Zn 8.02 $\sum$98.94$(Pd_{3.82}Pt_{0.14})_{\sum3.96}Cu_{3.00}(Zn_{0.86}Fe_{0.18})_{\sum1.04}$ (*n* =10)
   Pd 62.44, Cu 27.97, Zn 9.59 $\sum$100.00 Pd$_4$Cu$_3$Zn—ideal
   Type locality and occurrence: The Konder placer deposit, Ayan-Maya district, Khabarovsky krai, Russia.

Name: For Nikolai S. Bortnikov of the Institute of Geology of Ore Deposits, Petrography, Mineralogy and Geochemistry RAS, in recognition of his contributions to mineralogy and mineral deposits.

Reference: Mochalov et al. (2007), IMA No 2006-027

Note: Further investigations on XRD data and crystal structure is desirable.

Experimental: The phase diagram of the Pd-Cu-Zn system at 400°C and 800°C was assessed tentatively by Lebrun et al. (2007), based on data of Dobersek and Kosovinc (1989). The corresponding synthetic phase is not known. A further, more detailed experimental study is needed.

10.5.2 COLDWELLITE PD_3AG_2S

Crystallography: Cubic, $P4_332$

Unit cell: a 7.2470 Å, *V* 380.61 Å^3, Z = 4

X-ray powder diffraction pattern, strongest lines [*d* in Å(I)(*hkl*)]: 2.427(100)(221), 2.302(38) (310), 2.195(38)(311), 1.4280(44)(510,431), 0.9294(24)(650,643), 0.9208(20)(732,651)

Structure: isostructural with β-Mn

Appearance: Discovered in a high-grade heavy-mineral concentrate from the Marathon deposit. Coldwellite (in size $150 \times 80\ \mu m$), is anhedral, slightly angular with scalloped grain edges with overgrowth of vysotskite (PdS); most coldwellite occur in small grains, sizes $0.5-11\ \mu m$.

Optical properties: In reflected light white with a light pinkish brown tint, isotropic.

Reflectance [λnm, R%]		
	470	41.9
	546	44.9
	589	44.0
	650	45.0

Physical Properties: Opaque, with a metallic luster

Hardness: not measured

Cleavage: not observed

Density: 9.90 g/cm^3 (calc.)

Chemical Composition (EMPA data, wt.%):

Pd 56.10 Fe 0.16 Ag 38.20 S 5.63 $\sum$100.09, $(Pd_{2.99}Fe_{0.02})_{\sum 3.01}Ag_{2.00}S_{0.99}$ (*n* = 23)

Pd 56.30 Ag 38.05 S 5.65 $\sum$100.00, Pd_3Ag_2S—ideal

Type locality and occurrence: The Marathon Cu-PGE-Au deposit, Coldwell Complex, Ontario, Canada (48°48′7″ N, 86°18′55″ W). It was also reported from the PGE-Cr zone of the Birch Lake Deposit, Duluth Complex, Canada (Severson and Hauck, 2003) and in PGE mineralization of the Fedorova-Pana ore node (Subbotin et al., 2012).

Name: After the locality at the Coldwell Complex, Ontario, Canada.

Reference: McDonald et al. (2015), IMA No 2014-045

Experimental: The synthetic analogue occurs as a distinct phase in the ternary system Pd-Ag-S at 400°C and 550°C (Vymazalová et al., in prep). Based on the experimental study (Ag$_2$S-Pd) of Raub (1954) the phase Pd_3Ag_2S is stable up to 940°C.

10.5.3 FERHODSITE (FE,RH,NI,IR,CU,PT)$_9$S$_8$

Crystallography: Tetragonal

Unit cell: *a* 10.009, *c* 9.840 Å, *V* 985.77 Å^3

X-ray powder diffraction pattern, strongest lines [*d* in Å(I)(*hkl*)]: 5.72(50), 3.01(70), 2.81(30), 2.23(100), 1.933(60), 1.772(40), 1.367(3), 1.167(4)

Type locality and occurrence: Nizhny Tagil ultramafic complex, Solovyeva Gora, Alexandrov Log (57°40′ N, 59°39′ W) and the Konder placer, Konder alkaline-ultrabasic massif, Maya river basin, South Yakutia, Russia (57°36′ N, 134°37′ W).

Name: For the main chemical components (Fe, Rh, S).

Reference: Begizov V.D., IMA No 2009-056

Note: Approved by IMA in 2009 but no other data on the mineral have been published to date.

10.5.4 JACUTINGAITE PT$_2$HGSE$_3$

Crystallography: trigonal, P$\overline{3}$m1

Unit cell: *a* 7.3477, *c* 5.2955 Å, *V* 247.59 Å^3, *Z* = 2

X-ray powder diffraction pattern, strongest lines [*d* in Å(I)(*hkl*)]: 5.2917(100)(001), 2.7273(16)(201), 2.4443(10)(012), 2.0349(18)(022), 1.7653(37)(003), 1.3240(11)(004) 1.0449(11)(025)

Structure: The crystal structure was solved and refined from the powder X-ray-diffraction data on synthetic Pt$_2$HgSe$_3$. Isostructural with Pt$_4$Tl$_2$X$_6$ (X = S, Se, or Te); no structural analogue is known as a mineral.

Appearance: Occurs as a single grain (about 50 μm across) in an aggregate (2 mm across) of atheneite, potarite, and hematite, obtained from a heavy-mineral concentrate.

Optical properties: Light gray in reflected light, bireflectance moderate to distinct, a bluish gray to rusty brown pleochroism; anisotropy weak to distinct.

Reflectance [λnm, R$_1$, R$_2$ %]		
470	47.4	51.1
546	48.2	50.5
589	48.0	49.6
650	47.1	47.8

Physical Properties: Opaque, gray with a metallic luster and gray streak, brittle.

Hardness: VHN$_{10}$ 169 (119−245) g/mm^2, Mohs 3½

Cleavage: {001} very good

Density: 10.35 g/cm^3 (calc.), 10.9 (meas. on synth.)

Chemical Composition (EMPA data, wt.%):

Pt 37.30, Pd 5.91, Hg 25.72, Ag 0.16, Cu 0.82 Se 31.48 $\sum$101.39 (Pt$_{1.46}$Pd$_{0.42}$Cu$_{0.10}$Ag$_{0.01}$)$_{\sum1.99}$Hg$_{0.98}$Se$_{3.04}$ (*n* =3)

Pt 46.73, Hg 24.06, Se 28.33 $\sum$99.12 Pt$_{1.97}$Hg$_{1.04}$Se$_{2.99}$ (synth., *n* =7)

Pt 47.14, Hg 24.24, Se 28.62 $\sum$100.00 Pt$_2$HgSe$_3$—ideal

Type locality and occurrence: The Cauê iron-ore deposit, Itabira district, Minas Gerais, Brazil

Name: After the specular hematite-rich vein-type gold mineralization, locally known as "jacutinga," in which the mineral occurs.

Reference: Vymazalová et al. (2012a), IMA No 2100-078

Note: The structural identity of natural and synthetic Pt_2HgSe_3 was confirmed by electron back-scattering diffraction (EBSD).

Experimental: Phase relations in the system Pd-Hg-Se were studied at 400°C (Drábek et al., 2012). Only one ternary phase (jacutingaite) is known in the ternary system. Jacutingaite forms stable assemblages with tiemannite (HgSe) and sudovikovite ($PtSe_2$), sudovikovite and luberoite (Pt_5Se_4), luberoite and platinum, and coexists with PtHg, $PtHg_2$, $PtHg_4$ phases. Below 250°C, the phase HgPt is no longer stable and the assemblage Pt_2HgSe_3 + $PtHg_2$ + (PtHg)ss appears in the system.

10.5.5 JAGUÉITE CU$_2$PD$_3$SE$_4$

Crystallography: monoclinic, $P2_1/c$

Unit cell: a 5.6719, *b* 9.9095, *c* 6.2636 Å, β 115.403°, V 318.0 Å^3, Z = 2

X-ray powder diffraction pattern, strongest lines [*d* in Å(I)(*hkl*)]: 2.759(23),2.676(100)(121), 2.630(64)($\overline{1}$22), 2.508(31)($\overline{2}$02), 2.269(27)(041), 1.950(27)(122), 1.920(36)($\overline{1}$23), 1.866(24)($\overline{2}$41), ICDD PDF2 card -00-057-0615

Structure: Structure solved from a single crystal by Topa et al. (2006), isostructural with chrisstanlyeite.

Appearance: Occurs in lobate aggregates with chrisstanleyite (up to 500 μm in across) embedded in calcite or as inclusions of anhedral grains (up to 50 μm across) in tiemannite and naumannite with common twinning.

Optical properties: Light creamy-yellowish in reflected light, bireflectance weak to moderate, pleochroic from a light buff to a creamy buff, strong anisotropy (rotation tints from brownish to bluish, greenish).

Reflectance, [λnm, R$_1$, R$_2$ %]		air		im	
	470	41.0	50.1	27.0	31.9
	546	44.1	51.8	29.2	33.8
	589	44.6	51.7	29.4	33.7
	650	45.1	52.0	30.2	34.1

Physical Properties: Opaque with a metallic luster and black streak, brittle, uneven fracture.

Hardness: VHN$_{25}$ 612 (464−772) g/mm^2, Mohs 5

Cleavage: not observed

Density: 8.02 g/cm^3 (calc.)

Chemical Composition (EMPA data, wt.%):

Cu 15.7, Ag 1.59 Pd 42.04, Se 40.15 $\sum$99.48 (Cu$_{1.91}$Ag$_{0.11}$)$_{\sum2.02}$Pd$_{3.05}$Se$_{3.93}$ (*n* = 8)

Cu 16.68, Pd 41.88, Se 41.44 $\sum$100.00 Cu$_2$Pd$_3$Se$_4$—ideal

Type locality and occurrence: Selenide mineralization at El Chire, the depression of Jagué La Rioja province, Argentina (28°38.3′ S, 68°44.3′ W). Observed, as an unnamed phase, from the Copper Hills occurrence, East Pilbara region, Western Australia (Nickel, 2002).

Name: After the village of Jagué, the closest settlement to the El Chire mine, La Rioja, Argentina.

Reference: Paar et al. (2004), Topa et al. (2006), IMA No 2002-60

Experimental: The synthetic analogue occurs as a distinct phase in the ternary system Pd-Cu-Se at 300 and 400 °C (Makovicky and Karup-Møller, in press).

10.5.6 KALUNGAITE PDASSE

Crystallography: Cubic, $Pa\bar{3}$

Unit cell: a 6.089 Å, *V* 225.78 Å^3, Z = 4

X-ray powder diffraction pattern, strongest lines [*d* in Å(I)(*hkl*)]: 3.027(75)(002), 1.838(100) (113), 1.172(95)(115,333), 1.077(80)(044,144,334), 0.988(70)(116,235,253), 0.929(90)(335), 0.918 (70)(226), ICDD PDF2 card 00-058-0509

Structure: not defined

Appearance: Occurs as 0.1–0.5 mm platy anhedral aggregates, in association with gold, chalcopyrite, bohdanowiczite, clausthalite, guanajuatite, Pb-Bi-Se-S phase, padmaite, sperrylite, stilbiopalladinite.

Optical properties: In reflected light cream, creamy gray against gold, isotropic, no internal reflections.

Reflectance, [λnm, R %]	470	air 47.5	im 33.3
	546	46.9	32.6
	589	46.8	32.6
	650	48.0	34.0

Physical Properties: Lead gray with metallic luster, black streak, brittle with an uneven fracture

Hardness: VHN_{25} 438 (429-455) g/mm^2, Mohs

Cleavage: not observed

Density: 7.59 g/cm^3 (calc.)

Chemical Composition (EMPA data, wt.%):

Pd 41.32, As 27.49, Bi 0.35, Sb 1.59, Se 27.67, S 1.22 $\sum$99.64 $Pd_{1.01}(As_{0.95}Sb_{0.03}Bi_{0.004})$ $\sum_{0.98}(Se_{0.91}S_{0.10})\sum_{1.01}$ (*n*=8)

Pd 40.88, As 28.78, Se 30.34 $\sum$100.0 PdAsSe—ideal

Type locality and occurrence: The Buraco do Ouro gold mine, Cavalcante town, Goiás State, Brazil (13°47'45" S, 47°27'35" W).

Name: For the Kalunga people, a community of descendants of African slaves living in the surroundings of the mine, Goiás State, Brazil.

Reference: Botelho et al. (2006), IMA No 2004-047

Note: The XRD data of natural sample are in an agreement with the data of the synthetic phase (ICDD PDF2—01-070-8016 card) and would be worth reinvestigation in terms of refinement based on the structural model proposed by Foecker and Jeitschko (2001).

Experimental: The ternary system Pd-As-Se has not been investigated. Nevertheless, the synthetic analogue was studied by Foecker and Jeitschko (2001). They refined the crystal structure of the synthetic PdAsSe from the single crystal XRD data. The phase is cubic, space group $P2_13$, *a* 6.095 Å, *V* 226.40 Å^3, Z 4, and belongs to the ullmannite (NiSbS) type structure.

10.5.7 KINGSTONITE RH₃S₄

Crystallography: monoclinic, *C2/m*

Unit cell: a 10.4616, *b* 10.7527, *c* 6.2648 Å, *β* 109.000°, *V* 666.34 Å^3, *Z* =6

X-ray powder diffraction pattern, [*d* in Å(I)(*hkl*)]: 3.156(100)(310), 3.081(100)($\bar{1}$31), 2.957(90) (002), 2.234(60)(202), 1.941(50)($\bar{2}$23), 1.871(80)($\bar{4}$41), 1.791(90)(060,$\bar{1}$33)

Structure: Structure solved and refined from single crystal. A new structure type.

Appearance: Occurs as subhedral (tabular, elongate) to anhedral inclusions (10−40 µm) in Pt-Fe alloy with isoferroplatinum, tetraferroplatinum, Cu-bearing Pt-Fe alloy, osmium, enriched oxide remnants of osmium, laurite, bowieite, ferrorhodsite, cuprorhodsite.

Optical properties: In reflected light pale slightly brownish gray, weakly pleochroic, weak bireflectance, weak to moderate anisotropy (rotation tints in dull grays and browns).

Reflectance, [λnm, R_1, R_2 %]: 470 air 47.2 48.9 im 33.2 34.7,

546 48.4 50.3 34.3 36.1

589 49.1 50.7 35.0 36.5

650 49.8 51.0 35.6 36.7

Physical Properties: Opaque with a metallic luster and black streak, brittle.

Hardness: VHN$_{25}$ 895 (871−920) g/mm^2, Mohs 6

Cleavage: good, parallel to [001]

Density: 7.52 g/cm^3 (calc.)

Chemical Composition (EMPA data, wt.%):

Rh 46.5, Ir 16.4, Pt 11.2, S 25.6 $\sum$99.7 (Rh$_{2.27}$Ir$_{0.43}$Pt$_{0.29}$)$_{\sum 2.99}$S$_{4.01}$ (*n* =20)

Rh 70.65, S 29.35, $\sum$100.00 Rh₃S₄—ideal

Type locality and occurrence: The Bir Bir river, Wallaga province, Yubdo district, Ethiopia. Also observed as inclusions in heavy-concentrate platinum from diamond placers of the Mayat-Vodorazdel'nyi site in the Anabar river basin, Russia (Airiyants et al., 2014).

Name: For Gordon A. Kingston (b. 1939), in recognition of his contributions to PGE mineralogy and geology of related deposits.

Reference: Stanley et al. (2005), IMA No 1993-46

Experimental: The mineral has a synthetic analogue Rh₃S₄ (Beck and Hilbert, 2000; ICSD No 410813, ICDD PDF2 01-070-5129), stable up to 1130°C. The binary system Rh-S comprises another three binary phases Rh$_{17}$S$_{15}$, Rh$_2$S$_3$ and RhS$_{\sim 3}$ (Predel, 1998). Based on the study of the ternary system Cu-Rh-S by Karup-Møller and Makovicky (2007), the phase Rh₃S₄ dissolves up to 0.6 Cu and 2.8 at.% Cu at 900°C and 700°C, respectively and is not stable at 540°C. The natural mineral may be stabilized by the content of Pt or Ir at lower temperature.

10.5.8 KITAGOHAITE PT₇CU

Crystallography: Cubic, *Fm$\bar{3}$m*

Unit cell: a 7.7891 Å, *V* 472.57 Å^3, *Z* = 4.

X-ray powder diffraction pattern, strongest lines [*d* in Å(I)(*hkl*)]: 2.246(100)(222), 1.948(8) (004), 1.377(77)(044), 1.174(27)(622), 1.123(31)(444), 0.893(13)(662)

Structure: Ca_7Ge type. The XRD data of kitagohaite were refined by the Rietveld method based on the structural model of a synthetic analogue (Schneider and Esch, 1944; Sluiter et al., 2006) (ICSD No 108775, ICDD PDF2 01-074-6150 card).

Appearance: The mineral comes from a heavy-mineral concentrate. It occurs in grains (around 0.5 mm in size), rimmed by hongshiite (PtCu).

Optical properties: In reflected light, white, isotropic.

Reflectance [λnm, R%]		
	470	63.2
	546	66.6
	589	68.2
	650	70.1

Physical Properties: Opaque, greyish white, with a metallic luster and gray streak, malleable.

Hardness: VHN_{100} 217 (206-237) g/mm^2, Mohs 3½

Cleavage: not observed

Density: 19.958 g/cm^3 (calc.)

Chemical Composition (EMPA data, wt.%):

Pt 95.49, Cu 4.78 $\sum$100.26 $Pt_{6.93}Cu_{1.07}$ (n =13)

Pt 95.55, Cu 4.45 $\sum$100.00 Pt_7Cu—ideal

Type locality and occurrence: The Lubero region of North Kivu, Democratic Republic of the Congo.

Name: For the Kitagoha river, the platiniferous river in the Lubero region, Congo.

Reference: Cabral et al. (2014), IMA No 2013-114

Experimental: It has a synthetic analogue $CuPt_7$ (Schneider and Esch, 1944), with an ordered structure whereas above 500°C it becomes disordered, lacking superlattice reflections.

10.5.9 KOJONENITE $PD_{7-X}SNTE_2$ ($0.3 \leq X \leq 0.8$)

Crystallography: tetragonal, *I4/mmm*

Unit cell: a 4.001, *c* 20.929 Å, *V* 335.0 $Å^3$, Z = 2

X-ray powder diffraction pattern, strongest lines [*d* in Å(I)(*hkl*)]: 10.465(29)(002), 2.496(52) (114), 2.1986(100)(116), 2.0930(18)(0010) 2.0025(48)(200), synth. (ICSD PDF2 cards 01-073-5652 and PDF 01-073-9276)

Structure: Structure solved on a synthetic single crystal by Savilov et al. (2005), and confirmed by x-ray powder diffraction data on experimental products (Vymazalová and Drábek, 2010). Isotypic to nickel analogues of $Ni_{7-x}SnQ_2$ (Q = S, Se, Te) (Baranov et al., 2003, 2004), consists of Cu_3Au-like Sn/Pd blocks and NaCl-like Pd/Te slabs.

Appearance: It forms anhedral grains (< 40 µm) in aggregates (up to 100 µm) with kotulskite as inclusions in chalcopyrite and cubanite.

Optical properties: In reflected light slightly pinkish off-white against (kotulskite bright cream), weak bireflectance (visible only on differently oriented adjacent grains), no pleochroism, anisotropy distinct (with rotation tints in shades of dark greenish-brown).

Reflectance, [λnm, R_o, $R_{e'}$ %]	470	nat. 55.0 52.7	synth. 54.8 53.4
	546	58.5 56.5	58.6 56.9
	589	61.0 58.3	61.2 59.4
	650	63.9 60.1	63.8 61.4

Physical Properties: Opaque, metallic luster, brittle, black streak (synth.).

Hardness: not measured

Cleavage: not observed

Density: 10.07 g/cm^3 (calc.)

Chemical Composition (EMPA data, wt.%):

Pd 62.48, Sn 11.74, Te 25.63 $\sum$99.85 $Pd_{5.96}Sn_{1.00}Te_{2.04}$ ($n = 20$), Pd 63.1, Sn 11.7, Te 25.2—simplified formula for natural sample Pd_6SnTe_2.

Pd 63.8 Sn 11.5 Te 24.7 $\sum$100.0 for x= 0.8; Pd 65.6 Sn 10.9 Te 23.5 $\sum$100.0 for x =0.3

$Pd_{7-x}SnTe_2$—ideal based on crystal structure of synthetic material.

Type locality and occurrence: the Stillwater Layered Igneous Intrusion, Stillwater Valley, Montana, USA (45°23′11″ N, 109°53′03″ W). The mineral is known also to occur in Noril'sk ores (Sluzhenikin et al., pers. comm.).

Name: For Kari K. Kojonen (b. 1949) of the Geological Survey of Finland, for his contributions to ore mineralogy.

Reference: Stanley and Vymazalová (2015), IMA No 2013-132

Note: The structural identity between natural and synthetic kojonenite was confirmed by electron backscatter diffraction (EBSD).

Experimental: Phase relations in the system Pd-Sn-Te were determined at 400°C (Vymazalová and Drábek, 2010). At 400°C the system contains three ternary compounds: kojonenite, and phase $Pd_{72}Sn_{16}Te_{12}$ and PdSnTe, not known to occur in nature. Kojonenite forms a stable assemblage with paolovite (Pd$_2$Sn, dissolving up to 4 at.% Te) and the ternary phase $Pd_{72}Sn_{16}Te_{12}$. It also coexists with palladium tellurides kotulskite (PdTe), phase Pd_3Te_2, telluropalladinite (Pd$_9$Te$_4$), and keithconnite (dissolving up to 4 at.% Sn). The upper stability of kojonenite is 596°C where it melts incongruently.

10.5.10 LISIGUANGITE CUPTBIS$_3$

Crystallography: Orthorhombic, $P2_12_12_1$

Unit cell: a 7.7152, *b* 12.838, *c* 4.9248 Å, *V* 487.80 Å^3, *Z* =4

X-ray powder diffraction pattern, strongest lines [*d* in Å(I)(*hkl*)]: 6.40(30)(020), 3.24(80)(031), 3.03(100)(201), 2.27(40)(051), 2.14(50)(250), 1.865(60)(232)

Structure: The crystal structure was solved from a single crystal fragment. Belongs to the lapieite group, Pt-analogue of mückeite (CuNiBiS$_3$) and malyshevite (PdCuBiS$_3$).

Appearance: Occurs as idiomorphic crystals, generally tabular or lamellae {010}, elongated along [100], up to 2 mm long and 0.5 mm wide. Sampled from a heavy mineral concentrate.

Optical properties: In reflected light, bright white with a yellowish tint, anisotropy weak to moderate (blue-greenish to brownish colors and parallel-axial extinction), no internal reflection, parallel extinction.

Reflectance, [λnm, R_1 R_2 %]	470	air 39.2 36.7	im 23.4 22.3
	546	40.3 37.3	23.6 22.6
	589	40.7 37.9	23.6 22.7
	650	40.8 37.9	23.7 22.9

Physical Properties: Opaque, lead-gray with black streak and metallic luster, brittle

Hardness: VHN_{25} 48.3 (46.7−49.8) g/mm^2, Mohs 2½

Cleavage: {010} perfect, {001} distinct, {100} visible

Density: 7.42 g/cm^3 (calc.)

Chemical Composition (EMPA data, wt.%):

Cu 12.98, Pt 30.04, Pd 2.69, Bi 37.65, S 17.55 $\sum$100.91 $Cu_{1.10}(Pt_{0.83}Pd_{0.14})_{\sum0.97}Bi_{0.97}S_{2.96}$ (*n* = 8)

Cu 11.27, Pt 34.60, Bi 37.07, S 17.06 $\sum$100.00 $CuPtBiS_3$—ideal

Type locality and occurrence: Mineral discovered in a PGE-bearing Co-Cu sulfide vein in garnet pyroxenite of the Yanshan Mountains, Chengde Prefecture, Hebei Province, China. The mineral also observed in Cu-Ni-PGE ores in the footwall of the Sudbury Igneous Complex, Canada (Pentek et al., 2013).

Name: After Li Siguang on the 120th anniversary of his birth (1889−1971); prominent Chinese geologist, a founder of geomechanics, advocate of the theory of tectonic systems, one of the pioneers in oil exploration in China.

Reference: Yu et al. (2009), IMA No 2007-003

Experimental: The quaternary system Cu-Pt-Bi-S has not been experimentally studied. The preliminary experimental results of ongoing research confirm the existence of a synthetic analogue stable at 400°C.

10.5.11 LUKKULAISVAARAITE $PD_{14}AG_2TE_9$

Crystallography: tetragonal, *I4/m*

Unit cell: a 8.9599, *c* 11.822 Å, *V* 949.1 $Å^3$, Z = 2.

X-ray powder diffraction pattern, strongest lines [*d* in Å(I)(*hkl*)]: 2.8323(58)(130,310), 2.8088(92) (213), 2.5542(66)(312), 2.4312(41)(321,231), 2.1367(57)(411,141), 2.1015(52)(233,323), 2.0449(100) (314), 2.0031(63)(420,240), 1.9700(30)(006), 1.4049(30)(246,426), 1.3187(36)(543,453) (synth.).

Structure: The crystal structure was solved and refined from the powder X-ray-diffraction data of synthetic $Pd_{14}Ag_2Te_9$. Unique structure type; shows some similarities to that of sopcheite ($Ag_4Pd_3Te_4$) and palladseite ($Pd_{17}Se_{15}$).

Appearance: Occurs as anhedral grains (about 40 μm in diameter) rimmed by tulameenite and randomly accompanied by telargpalite and Bi-rich kotulskite, enclosed in chalcopyrite, in association with millerite, bornite and hematite. It also is observed as tiny grains (5−10 μm) in intergrowths with telargpalite and Bi-rich kotulskite in association with moncheite, tulameenite, hongshiite, and telluropalladinite.

Optical properties: In reflected light, light gray with a brownish tinge, strong bireflectance, light brownish gray to greyish brown pleochroism, distinct to strong anisotropy; exhibits no internal reflections.

Reflectance, [λnm, R_1 R_2 %]		
470	40.9	48.3
546	47.6	56.4
589	52.1	61.0
650	57.5	65.2

Physical Properties: Opaque, gray, with metallic luster, and gray streak, brittle.

Hardness: VHN_{20} 355 (339−371) kg/mm², Mohs 4

Cleavage: not observed

Density: 9.993 g/cm³ (calc.), 9.9 g/cm³ (meas. on synth.)

Chemical Composition (EMPA data, wt.%):

Pd 52.17, Ag 7.03, Te 40.36 $\sum$99.56 $Pd_{14.05}Ag_{1.88}Te_{9.06}$ (*n* = 5)

Pd 52.13, Ag 7.31, Te 40.58 $\sum$100.02 $Pd_{13.99}Ag_{1.93}Te_{9.08}$ (synth. *n* = 9)

Pd 52.20, Ag 7.56, Te 40.24 $\sum$100.00 $Pd_{14}Ag_2Te_9$—ideal

Type locality and occurrence: The Lukkulaisvaara intrusion, northern Russian Karelia, Russia. (66°19′20″ N, 30°49′50″ E). Also observed, as an unnamed phase, in the South Sopcha massif and from the Monchetundra deposit of the Monchegorsk Complex, Kola Peninsula, Russia (Grokhovskaya et al., 2003, 2009).

Name: For the Lukkulaisvaara intrusion in Russian Karelia.

Reference: Vymazalová et al. (2014a), IMA No 2013-115

Note: The structural identity between natural and synthetic $Pd_{14}Ag_2Te_9$ was confirmed by electron backscatter diffraction (EBSD).

Experimental: Phase relations in the system Pd-Ag-Te were determined at 350°C and 450°C (Vymazalová et al., 2015). At 350°C the system contains five ternary compounds: sopcheite ($Pd_3Ag_4Te_4$), lukkulaisvaaraite ($Pd_{14}Ag_2Te_9$), telargpalite ($Pd_{2-x}Ag_{1+x}Te$ 0.09 < *x* < 0.22), and phases $Pd_{7.5-x}Ag_{0.5+x}Te_3$ (0.02 < *x* < 0.83) and $Pd_{2+x}Ag_{2-x}Te$ (0.18 < *x* < 0.24). Lukkulaisvaaraite coexists with kotulskite and sopcheite, hessite and sopcheite, telargpalite and hessite; it forms stable associations with the phase Pd_3Te_2 and kotulskite, and with Pd_3Te_2 and telluropalladinite. It also coexists with a phase $Pd_{7.5-x}Ag_{0.5+x}Te_3$ and telargpalite, with the phase $Pd_{7.5-x}Ag_{0.5+x}Te_3$ and telluropalladinite. At 450°C, sopcheite is no longer stable and the assemblage lukkulaisvaaraite + kotulskite + hessite become stable.

10.5.12 MALYSHEVITE PDCUBIS₃

Crystallography: Orthorhombic, *Pnam*

   Unit cell: a 7.541, b 6.4823, c 11.522 Å, V 563.204 Å³, Z = 4

   X-ray powder diffraction pattern, strongest lines [d in Å(I)(hkl)]: 3.24(4)(020), 2.88(8)(004), 2.52(6)(300), 1.900(10)(304), 1.715(2)(206), 1.672(2)(225), ICDD PDF2 card—00-060-0390.

   Structure: not determined, chemically Pd-analogue of lisiguangite

Appearance: Malyshevite forms rims (1−20 μm) around clausthalite, replacing padmaite.

Optical properties: In reflected light, white, bireflectance moderate, a bright yellow to pale yellow pleochroism; anisotropy weak to distinct (light yellow tints).

Reflectance, [λnm, R₁ R₂ %]	470	34.1	28.7
	546	36.3	33.0
	589	37.0	34.4
	650	37.4	34.6

Physical Properties: Bluish gray, with metallic luster and gray streak.

Hardness: VHN not measured, Mohs 3

Cleavage: not observed

Density: 6.025 g/cm³ (calc.)

Chemical Composition (EMPA data, wt.%):

Pd 20.6, Pt 1.0, Pb 0.8, Bi 42.6, Cu 13.1, Se 2.2, S 19.0 $\sum$99.3 $(Pd_{0.94}Pt_{0.02}Pb_{0.02})_{\sum 0.98}Bi_{0.99}$ $Cu_{1.00}(S_{2.88}Se_{0.14})_{\sum 3.02}$ ($n = 7$)

Pd 22.39, Cu 13.38, Bi 43.99, S 20.24 $\sum$100.00 $PdBiCuS_3$—ideal

Type locality and occurrence: The Srednyaya Padma U-V deposit in Southern Karelia, Russia. The mineral also observed in Cu-Ni-PGE ores in the footwall of the Sudbury Igneous Complex, Canada (Pentek et al., 2013), and in Noril'sk ores (Spiridonov et al., 2015) and is found in intergrowths with its Pt-analogue lisinguaite from its type locality (Yu et al., 2009).

Name: In honor of I.I.Malyshev (1904−1973) and V.I. Malyshev (1927−2002), father (who first discovered the Malyshevskoe—Samotkanskoe Ti deposit) and son, formerly of the All-Russian Scientific-Research Institute of Mineral Resources (VIMS).

Reference: Chernikov et al. (2006), IMA No 2006-012

Note: Further investigations of the crystal structure are needed.

Experimental: The quaternary system Cu-Pd-Bi-S has not been experimentally studied.

10.5.13 MIESSIITE PD₁₁TE₂SE₂

Crystallography: cubic, $Fd\bar{3}m$

Unit cell: a 12.448 Å, *V* 1929.0 Å³, *Z* = 8

X-ray powder diffraction pattern, strongest lines [*d* in Å(I)(*hkl*)]: 2.395(80)(511,333), 2.197 (100)(440), 1.875(25)(622), 1.555(25)(800), 1.305(25)(931), 1.271(30)(844), ICDD PDF2 card—00-059-0323.

Structure: Isostructural with isomertierite ($Pd_{11}As_2Sb_2$) and törnroosite ($Pd_{11}As_2Te_2$).

Appearance: Discovered with placer gold and PGM nuggets, as one grain (483 × 522 μm). Shows a subidiomorphic cubic morphology.

Optical properties: In reflected light, light gray, isotropic.

Reflectance, [λnm, R%]	470	48.88
	546	51.63
	589	53.91
	650	56.82

Physical Properties: Opaque, black with a metallic luster, malleable.

Hardness: VHN₁₀₀ 362 (348-370) g/mm², Mohs 2−2½

Cleavage: not observed
Density: 10.94 g/cm^3 (calc.)
Chemical Composition (EMPA data, wt.%):
Pd 75.17, Te 17.06, Se 9.61 $\sum$101.84 $Pd_{11.02}Te_{2.09}Se_{1.90}$ ($n =16$)
Pd 73.91, Te 16.12, Se 9.97 $\sum$100.00 $Pd_{11}Te_2Se_2$—ideal
Type locality and occurrence: The Miessijoki River in the Lemmenjoki area, Inari commune, Finnish Lapland, Finland (25°21′ N, 68°25′ E).
Name: For the Miessi river, Finnish Lapland (in the Saami language "Miessijohka," where *johka* means river, and *miessi* a reindeer calf).
Reference: Kojonen et al. (2007), IMA No 2006-013
Experimental: The synthetic analogue has not been studied; nor has the Pd-Te-Se system.

10.5.14 MILOTAITE PDSBSE

Crystallography: cubic, $P2_13$
Unit cell: a 6.3181 Å, *V* 252.20 Å^3, $Z = 4$
X-ray powder diffraction pattern, strongest lines [*d* in Å(I)(*hkl*)]: 2.825(100)(201), 1.905(98)(311), 2.579(81)(211), 3.159(53)(200), 2.233(32)(220), 1.752(27)(320), 1.688(25)(312), 1.378(18)(412), ICDD PDF2 card—01-073-3935, synth.
Structure: The crystal structure was determined on a single crystal of synthetic PdSbSe. The structure of PdSbSe is a homeotype of the structure of pyrite (FeS_2) and an isotype of the cubic ordered structure of gersdorffite (NiAsS).
Appearance: Occurs as subhedral grains (less than 25 µm in diameter), embedded in eucairite and tiemannite, randomly intergrown with a graphic intergrowth of bornite and selenian digenite.
Optical properties: In reflected light, white, isotropic.

Reflectance, [λnm, R%]		nat.	synth.
	470	48.6	48.0
	546	47.5	47.1
	589	47.6	47.0
	650	49.0	48.2

Physical Properties: Silvery gray, metallic luster, opaque, brittle, uneven fracture (synth.).
Hardness: VHN$_{100}$ 465 (420−514) g/mm^2 (synth.), Mohs 4½
Cleavage: not observed
Density: 8.09 g/cm^3 (calc.), 7.98-8.23 g/cm^3 (meas. on synth.)
Chemical Composition (EMPA data, wt.%):
Pd 34.17, Cu 0.78, Ag 0.35, Sb 38.03, Se 26.38 $\sum$99.71 $(Pd_{0.98}Cu_{0.04})_{\sum1.02}(Ag_{0.01}Sb_{0.95})_{\sum0.96}Se_{1.02}$ ($n = 5$)
Pd 34.46, Sb 38.86, Se 26.60 $\sum$99.91 $Pd_{0.99}Sb_{0.97}Se_{1.04}$ (synth., $n = 17$)
Pd 34.65, Sb 39.64, Se 25.71 $\sum$100.00 $Pd_1Sb_1Se_1$—ideal
Type locality and occurrence: Předbořice, a low-temperature selenide-bearing uranium mineralization in the Czech Republic.
Name: For Milota Makovicky (b. 1941), University of Copenhagen, in recognition of her investigations of sulfide and sulfoarsenide systems with PGE.

Reference: Paar et al. (2005), IMA No 2003-056

Note: X-ray powder-diffraction pattern derived from the crystal structure refinement. Identity between natural and synthetic PdSbSe confirmed by optical properties, reflectance data and chemical analyses.

Experimental: The ternary system Pd-Sb-Se has not been studied yet. Our preliminary experimental results of on-going research have shown that milotaite forms stable association with palladseite ($Pd_{17}Se_{15}$) and antimonselite (Sb_2Se_3), and coexists with stibiopalladinite (Pd_5Sb_2) at 400°C.

10.5.15 NALDRETTITE PD₂SB

Crystallography: orthorhombic, $Cmc2_1$

Unit cell: a 3.3906, b 17.5551, c 6.957 Å, V 414.097 Å^3, Z = 8

X-ray powder diffraction pattern, strongest lines [d in Å(I)(hkl)]: 2.2454(100)(132), 2.0567(52)(043), 2.0009(40)(152), 1.2842(42)(115), 1.2122(50)(204), 0.8584(56)(1174), ICDD PDF2 card—00-058-0460.

Structure: isostructural with Pd_2As

Appearance: Occurs as anhedral grains (varying in size from 10 to 239 μm), commonly attached or molded to sulfide minerals, also associated with clinochlor, rarely with magnetite.

Optical properties: In reflected light, bright creamy white, weak bireflectance, no pleochroism, distinct anisotropy (rotation tints deep bright blue, lemon-buff, and mauve pale pink); higher reflectance and less yellow against pentladite.

Reflectance, [λnm, R_1 R_2 %]		air		im	
	470	49.0	50.9	35.9	37.6
	546	53.2	55.1	40.3	42.1
	589	55.4	57.5	42.5	44.3
	650	58.5	60.1	45.4	47.2

Physical Properties: Opaque, metallic, irregular fracture.

Hardness: VHN_{50} 393 (358−418) g/mm^2, Mohs 4-5

Cleavage: not observed

Density: 10.694 g/cm^3 (calc.)

Chemical Composition (EMPA data, wt.%):

Pd 63.49, Fe 0.11, Sb 35.75, As 0.31, S 0.02 $\sum$99.68 $(Pd_{1.995}Fe_{0.007})_{\sum2.002}$ $(Sb_{0.982}As_{0.014}S_{0.002})_{\sum0.998}$ (n = 69)

Pd 63.61, Sb 36.39 $\sum$100.00 Pd_2Sb—ideal

Type locality and occurrence: The Mesamax Northwest deposit, Ungava region, Quebéc, Canada. It was also observed in the Tootoo and Mequillon magmatic sulfide deposits, New Quebec Orogen (Liu et al., 2013) and in concentrates of chromitite from the Korydallos area in the Pindos ophiolite complex, NW Greece (Kapsiotis et al., 2010). Observed as an unnamed phase in Cu-Ni sulfide deposits in clinopyroxenite intruding Permian sandy shales and volcanics, NE China and in a serpentinite intrusion in Permian metamorphic rocks, SW China (Cabri, 1981b, UN1974-8), in PGE mineralization in the Alaskan-type intrusive complexes near Fifield, New South Wales, Australia (Johan et al., 1989), in the vein-type Cu-Ni sulfide ores of the Ioko-Dovyren massif, northern Baikal region, Russia (Rudashevsky et al., 2003), and in Noril'sk ores (unpubl. data).

Name: For Anthony J. Naldrett (b. 1933), professor at the University of Toronto, in recognition of his significant contributions to understanding the genesis of PGE deposits.

Reference: Cabri et al. (2005), IMA No. 2004-007

Experimental: The mineral has a synthetic analogue synthesized by Bälz and Schubert (1969), ICDD PDF2 01-074-6150 card, El-Boragy and Schubert (1971b) and Kim and Chao (1996). It melts incongruently at 580°C (Kim and Chao, 1996); below this temperature it coexists with Pd_5Sb_2 or $PdSb$.

10.5.16 NIELSENITE PDCU$_3$

Crystallography: Tetragonal, *P4mm*

Unit cell: a 3.7125, *c* 25.62 Å, *V* 353.2 Å^3, *Z* = 4

X-ray powder diffraction pattern, strongest lines [*d* in Å(I)(*hkl*)]: 2.137(100)(117), 1.8596(70) (200), 1.8337(40)(00$\underline{14}$), 1.3126(60)(220), 1.1188(55)(317), 1.0663(30)(22$\underline{14}$), ICDD PDF2

Structure: considered to be isostructural with synthetic tetragonal $PdCu_3$, space group *P4mm*, but may belong to the *P4/mmm* group

Appearance: Occurs as discrete grains or in sulfide-bearing, droplet-shaped to irregular grains 5–50 μm in size.

Optical properties: In reflected light, bright creamy white, anisotropy not observed (due to grain orientation, parallel to *c*)

Reflectance, [λnm, R′ %]	470	air 57.6	im 47.5
	546	60.85	50.8
	589	62.8	53.0
	650	66.7	57.5

Physical Properties: Steel-gray with a metallic luster, black streak, sectile tenacity

Hardness: not measured

Cleavage: not observed

Density: 9.53 g/cm^3 (calc.)

Chemical Composition (EMPA data, wt.%):

Pd 29.86, Pt 3.08, Au 3.70, Cu 61.96, Fe 0.59, Pb 0.17 $\sum$99.36 $(Pd_{0.862}Au_{0.058}Pt_{0.049}Fe_{0.028}Pb_{0.003})_{\sum1.00}(Cu_{2.996}Fe_{0.004})_{\sum3.00(n=11)}$

Pd 35.82, Cu 64.18 $\sum$100.00 $PdCu_3$—ideal

Type locality and occurrence: The Skaergaard intrusion, Kangerdlugssuaq area, Est Greenland (68°09′55″ N, 31°41′02″ W). Also, found in concentrates of chromitite from the Korydallos area in the Pindos ophiolite complex, NW Greece (Kapsiotis et al., 2010), in magnetite-bearing gabbroic rocks of the Freetown Layered Complex, Sierra Leone (Bowles et al., 2013) and in Au-Cu-Pd-type mineralization in dolerites from Alexandra Land Island of the Franz Josef Archipelago (Sklyarov et al., 2016).

Name: After Troels F.D. Nielsen (b. 1950), a geologist with the Geological Survey of Denmark and Greenland, in recognition of his field work in the Skaergaard intrusion and his mineralogical and metallurgical studies.

Reference: McDonald et al. (2008), IMA No.2004-046

Experimental: According to the study of Karup-Møller et al. (2008) the formation of the phase can be assumed to be a product of low-temperature re-equilibration, shown as a 1D-LPS (one-dimensional long-period superlattice LPS) phase in the Cu-Pd phase diagram formed below 500°C (Subramanian and Laughlin, 1991).

10.5.17 NORILSKITE (PD,AG)$_7$PB$_4$

Crystallography: trigonal, $P3_121$

Unit cell: a 8.9656, *c* 17.2801 Å, *V* 1202.92 Å^3, *Z* = 6

X-ray powder diffraction pattern, strongest lines [d in Å*(I)(hkl)]:* 3.2201(29)(023,203), 2.3130 (91)(026,206), 2.2414(100)(220), 1.6098(28)(046,406), 1.3076(38)(246,462), 1.2942(18)(600), 1.2115(37)(2212,1213), 0.9626(44)(0612,6012).

Structure: Crystallizes in the Ni$_{13}$Ga$_3$Ge$_6$ structure type, related to nickeline. The crystal structure was solved and refined from the powder X-ray-diffraction data of synthetic (Pd,Ag)$_7$Pb$_4$.

Appearance: Forms anhedral grains in aggregates (up to about 400 μm) with polarite, zvyagintsevite, Pd-rich tetra-auricupride, Pd-Pt bearing auricupride, Ag-Au alloys, (Pb,As,Sb) bearing atokite, mayakite, Bi-Pb rich kotulskite and sperrylite in pentlandite, cubanite, and talnakhite.

Optical properties: In reflected light, orange-brownish pink, moderate to strong bireflectance, orange-pink to greyish-pink pleochroism, strong anisotropy (rotation tints from dull yellow to dull blue in partially crossed polars); no internal reflections.

Reflectance, [λnm, R$_o$ R$_{e'}$ %]	470	51.1	48.8
	546	56.8	52.2
	589	59.9	53.5
	650	64.7	55.5

Physical Properties: Gray with metallic luster and gray streak, brittle.

Hardness: VHN$_{20}$ 310 (296−342) g/mm^2, Mohs 4

Cleavage: not observed

Density: 12.99 g/cm^3 (calc.)

Chemical Composition (EMPA data, wt.%):

Pd 44.33, Ag 2.68, Bi 0.33, Pb 52.34 $\sum$99.68 (Pd$_{6.56}$Ag$_{0.39}$)$_{\sum6.97}$(Pb$_{3.97}$Bi$_{0.03}$)$_{\sum4.00}$ (*n* = 16)

Pd 42.95, Ag 3.87, Pb 53.51 $\sum$100.33 (Pd$_{6.25}$Ag$_{0.56}$)$_{\sum6.81}$Pb$_{4.00}$ (synth., *n* = 8)

Pd 43.93, Ag 3.43, Pb 52.64 $\sum$100.00 (Pd$_{6.5}$Ag$_{0.5}$)$_{\sum7}$Pb$_4$—ideal based on crystal structure

Type locality and occurrence: The Mayak mine of the Talnakh deposit, Russia (69°30′20″ N, 88°27′17″ E). Also observed from the Komsomolsky mine of the Talnakh deposit and from the Zapolyarny (Trans-Polar) mine of the Noril'sk I deposit (Sluzhenikin and Mokhov, 2015).

Name: For the Noril'sk district, Russia. Almost half of all known named platinum-group minerals have been reported to occur in the Noril'sk ores.

Reference: Vymazalová et al. (2017), IMA No 2015-008

Note: The structural identity between the natural and synthetic (Pd,Ag)$_7$Pb$_4$ was confirmed by electron back-scattering diffraction (EBSD).

Experimental: The phase was synthesized at 300°C. The system Pd-Ag-Pb was experimentally studied by Sarah et al. (1981) at 400°C. Based on the proposed phase diagram, norilskite forms a stable assemblage with zvyagintsevite (Pd_3Pb) and Pd-Ag alloy, and coexists with Pd-Pb phases Pd_5Pb_3, $Pd_{13}Pb_9$, and $PdPb$.

10.5.18 PALLADOSILICIDE PD$_2$SI

Crystallography: hexagonal, $P\bar{6}2m$

   Unit cell: a 6.496, *c* 3.433 Å, *V* 125.5 Å^3, Z = 3

   X-ray powder diffraction pattern, strongest lines [*d* in Å(I)(*hkl*)]: 2.3658(100)(111), 2.1263(37)(120), 2.1808(34)(021), 3.240(20)(110), 1.8752(19)030), 1.7265(12)(002), 1.3403(11)(122), 1.2089(10)(231) (synth.)

   Structure: Fe_2P type

Appearance: Found in a heavy-mineral concentrate, grains ranging in size from 0.7 to 39.1 μm.

Optical properties: In reflected light bright creamy white, weak bireflectance, weak anisotropy (rotation tints in shades of light blue and olive green)

Reflectance, [λnm, R$_1$ R$_2$ %]			
	470	air 49.6 52.7	im. 36.3 38.6
	546	51.2 53.8	37.6 39.5
	589	51.6 53.7	37.8 39.5
	650	51.7 53.3	37.9 39.3

Physical Properties: Has an anhedral to subhedral habit, metallic luster.

Hardness: not measured

Cleavage: not observed

Density: 9.562-9.753 g/cm^3 (calc.)

Chemical Composition (EMPA data, wt.%):

Si 7.95, Pd 68.56, Ag 1.07, Ni 4.59, Te 0.32 Sb 0.36, As 3.95, Fe 0.64, Pt 1.72, Sn 1.79, Cu 2.18, Rh 2.39$\sum$95.53 $(Pd_{1.657}Ni_{0.201}Cu_{0.088}Rh_{0.06}Fe_{0.029}Ag_{0.026}Pt_{0.023}Sn_{0.039})\sum_{2.123}(Si_{0.728}As_{0.136}Sb_{0.008}Te_{0.006})\sum_{0.878}$ (*n* = 8)—Kapalagulu

Si 10.13, Pd 68.77, Ag 0.33, Ni 5.16, Sb 0.11, As 2.18, Fe 0.35, Pt 4.45, Sn 3.08, Cu 1.62, Rh 3.76 $\sum$99.94 $(Pd_{1.557}Ni_{0.212}Cu_{0.061}Rh_{0.088}Fe_{0.015}Ag_{0.007}Pt_{0.055}Sn_{0.063})\sum_{2.058}(Si_{0.869}As_{0.07}Sb_{0.002})\sum_{0.941}$ (*n* =12)—UG-2

Pd 88.34 Si 11.66 $\sum$100.00 Pd_2Si—ideal

Type locality and occurrence: PGE-chromite horizon of the Kapalagulu Intrusion near eastern shore of Lake Tanganyika, western Tanzania (30°03′51″ E, 5°53′16″ S and 30°05′37″ E, 5°54′26″ S) and the UG-2 chromitite, Bushveld Complex.

Name: For the chemical composition: Pd, Si.

Reference: Cabri et al. (2015), IMA No. 2014-080

Note: The structural identity of natural and synthetic Pd_2Si (Nylund, 1966) was confirmed by electron back-scattering diffraction (EBSD).

Experimental: The synthetic analogue melts congruently at 1330°C. Langer and Wachtel (1981) also observed a metastable state in the system, depending on the rate of cooling.

10.5.19 PAŠAVAITE PD₃PB₂TE₂

Crystallography: orthorhombic, *Pmmn*

Unit cell: a 8.599, *b* 5.9381, *c* 6.3173 Å, *V* 322.6 Å^3, *Z* = 2.

X-ray powder diffraction pattern, strongest lines [*d* in Å(I)(*hkl*)]: 6.3152(34)(001), 3.1572(33) (002), 3.0495(100)(211), 2.5456(63)(202), 2.4424(34)(220), 2.2786(42)(221), 2.1637(71)(022), 2.1496(30)(400), 1.8906(42)(203), 1.5248(31)(422), ICDD PDF2 card—01-077-9121, synth.

Structure: Solved and refined from the powder X-ray diffraction data of syntetic $Pd_3Pb_2Te_2$. Structurally related to shandite ($Ni_3Pb_2S_2$) and parkerite ($Ni_3Bi_2S_2$).

Appearance: Occurs as subhedral grains (less than 20 μm in diameter) embedded in polarite, randomly accompanied by Pd-Pb-Bi-Te phases, sperrylite or intergrown with Au-Ag phases.

Optical properties: In reflected light, pale pink with brownish tinge, strong bireflectance, pleochroic from brownish to light pink, anisotropy distinct to strong, exhibits no internal reflections.

Reflectance, [λnm, R₁ R₂ %]			
	470	nat. 42.4	synth. 49.9
	546	44.6	51.8
	589	45.7	52.2
	650	46.9	52.8

Physical Properties: Opaque, gray with metallic luster and gray streak, brittle (synth.).

Hardness: VHN$_{25}$ 233 (173−281), Mohs 2

Cleavage: {001} weak

Density: 10.18 g/cm^3 (calc.), 9.9 g/cm^3 (meas. on synth.)

Chemical Composition (EMPA data, wt.%):

Pd 31.51, Pb 41.54, Bi 0.19, Te 25.75 ∑98.99 $Pd_{2.96}(Pb_{2.01}Bi_{0.01})Te_{2.02}$ (*n* = 4)

Pd 32.17, Pb 41.78, Te 25.93 ∑99.88 $Pd_{2.99}Pb_{2.00}Te_{2.01}$ (synth., *n* = 7)

Pd 32.28, Pb 41.91, Te 25.81 ∑100.00 $Pd_3Pb_2Te_2$—ideal

Type locality and occurrence: The Talnakh deposit, Noril'sk-Talnakh Ni-Cu camp, Taimyr Autonomous District, Russia.

Name: For Jan Pašava (b. 1957) of the Czech Geological Survey, in recognition of his contributions to the mineralogy and geochemistry of PGE in anoxic environments and other related ore deposits.

Reference: Vymazalová et al. (2009), IMA No 2007-059

Note: The structural identity of natural and synthetic $Pd_3Pb_2Te_2$ was confirmed by electron back-scattering diffraction (EBSD).

Experimental: Phase relations in the system Pd-Pb-Te were determined at 400°C (Vymazalová and Drábek, 2011); the Pd-rich corner at 480°C was explored by El-Boragy and Schubert (1971b). At 400°C the system contains two ternary compounds: pašavaite and $Pd_{71}Pb_8Te_{21}$ (not known to occur in nature). Pašavaite forms a stable assemblage with kotulskite solid-solution (25−30 at.% Pb) and altaite (PbTe). Pašavaite melts at 500°C.

10.5.20 SKAERGAARDITE PDCU

Crystallography: cubic, *Pm3m*

Unit cell: a 3.0014 Å, *V* 27.0378 Å^3, *Z* = 1

X-ray powder diffraction pattern, strongest lines [*d* in Å(I)(*hkl*)]: 2.122(100)(110), 1.5000(20) (200), 1.2254(20)(211), 0.9491(20)(310), 0.8666(10)(222), 0.8021(70)(321), ICDD PDF2 card—00-057-0606.

Structure: CsCl-type. Isostructural with wairauite (CoFe), synthetic CuZn (β-brass) and structuraly related to hongshiite (PtCu).

Appearance: Occurs as droplets, equant grains with rounded outlines, subhedral to euhedral crystals and irregular grains, in size from 2 to 75 μm.

Optical properties: In reflected light, bright creamy white (against to bornite and chalcopyrite), bright white (against digenite and chalcocite), isotropic.

Reflectance, [λnm, R_1 R_2 %]	470	air 58.65	im. 47.4
	546	62.6	51.1
	589	64.1	52.8
	650	65.25	53.95

Physical Properties: Steel gray with a bronze tint, metallic luster, sectile.

Hardness: VHN_{25} 257 (244−267) g/mm^2, Mohs 4−5

Cleavage: not observed

Density: 10.64 g/cm^3 (calc.)

Chemical Composition (EMPA data, wt.%):

Pd 58.94, Pt 1.12, Au 2.23, Cu 29.84, Fe 3.85, Zn 1.46, Sn 1.08, Te 0.28, Pb 0.39$\sum$99.19 $(Pd_{0.967}Au_{0.020}Pt_{0.010})_{\sum0.997}(Cu_{0.820}Fe_{0.120}Zn_{0.039}Sn_{0.016}Te_{0.004}Pb_{0.003})_{\sum1.002}$ (*n* =311)

Cu 37.39 Pd 62.61 $\sum$100.00 CuPd—ideal

Type locality and occurrence: The Skaergaard intrusion, Kangerdlugssuaq area, East Greenland. Also found in a heavy-mineral concentrate from the Marathon deposit, Coldwell Complex, Canada (McDonald et al., 2015). Observed in concentrates of chromitite from the Korydallos area in the Pindos ophiolite complex, NW Greece (Kapsiotis et al., 2010); in Au-Cu-Pd-type mineralization in dolerites from Alexandra Land Island of the Franz Josef Archipelago (Sklyarov et al., 2016); in PGE mineralization in the Kirakkajuppura PGE deposit, Penikat layered complex, Finland (Barkov et al., 2005), and from the J-M reef, Stillwater Complex (Godel and Barnes, 2008).

Name: For the Skaergaard intrusion in East Greenland.

Reference: Rudashevsky et al. (2004), IMA No. 2003-049

Experimental: The transformation from disordered (Cu,Pd) solid solution to ordered CuPd occurs over the range 35−50 at.% Pd at 596°C (Subramanian and Laughlin, 1991).

10.5.21 TÖRNROOSITE $PD_{11}AS_2TE_2$

Crystallography: cubic, $Fd\bar{3}m$

Unit cell: a 12.3530, V 1885.03 Å^3, Z = 8

X-ray powder diffraction pattern, strongest lines [*d* in Å(I)(*hkl*)]: 2.182(100)(440), 2.376(90) (511,333), 1.544(15)(800), 1.862(13)(622), 1.2606(13)(844), 1.608(11)(731,553)

Structure: Isostructural with isomertierite ($Pd_{11}As_2Sb_2$) and miessiite ($Pd_{11}Se_2Te_2$).

Appearance: Observed as an anhedral grain (size 132 x 200 μm), discovered with placer gold and PGM nuggets.

Optical properties: In reflected light yellowish white, isotropic.

Reflectance, [λnm, R%]	470	45.4
	546	51.0
	589	54.1
	650	57.45

Physical Properties: Black, opaque with metallic luster, silvery black streak, malleable.

Hardness: VHN_{25} 519 (509−536) kg/mm^2, Mohs 5.

Cleavage: not reported

Density: 11.205 g/cm^3 (calc.)

Chemical Composition (EMPA data, wt.%):

Pd 72.04, Pt 1.75, Sb 2.13, Sb 0.85, As 8.77, Te 13.15, Bi 0.79 $\sum$99.48 $(Pd_{10.85}Pt_{0.14})_{\sum 10.99}$ $(As_{1.88}Sb_{0.11})_{\sum 1.99}(Te_{1.65}Sn_{0.29}Bi_{0.06})_{\sum 2.00}$ ($n = 10$)

Pd 72.29, As 9.51, Te 16.20 $\sum$100.00 $Pd_{11}As_2Te_2$—ideal

Type locality and occurrence: The Miessijoki River, Lemmenjoki area, Inari Commune, Finnish Lapland, Finland (25°42′33″ N, 68°42′30″ E). Also observed in a heavy-mineral concentrate from the Marathon Cu-PGE-Au deposit, Coldwell Complex, Ontario, Canada (McDonald et al., 2015) and in the sulfide-bearing pegmatoidal pyroxenites of the South Sopcha intrusion of the Monchegorsk Igneous Complex, Kola peninsula, Russia (Grokhovskaya et al., 2012), in PGE mineralization of the Fedorova-Pana ore node (Subbotin et al., 2012) and as an unnamed phase in PGE concentrate from alluvial sediments from the farm Maandagshoek in the eastern Bushveld (Oberthür et al., 2004).

Name: For Ragnar Törnroos (b. 1943), University of Helsinki, for his pioneering research on placer PGM in Finnish Lapland and his contributions to ore mineralogy.

Reference: Kojonen et al. (2011), IMA No 2010-043

Note: The crystal structure of törnroosite was refined on a single crystal sample from the South Sopcha intrusion, Russia by Karimova et al. (2016), the data are in an agreement with those proposed by Kojonen et al. (2011).

Experimental: The synthetic analogue is not known. The system Pd-As-Te was experimentally studied by El-Boragy and Schubert (1971b) at 480°C, but it requires some further re-examination, particularly in the Pd-rich corner, reflecting new observations in the binary systems Pd-Te and Pd-As. The phase assigned as Pd_6AsTe by El-Boragy and Schubert (1971b) could be in fact $Pd_{11}As_2Te_2$.

10.5.22 UNGAVAITE PD$_4$SB$_3$

Crystallography: tetragonal, $P4_12_12$, $P4_122$, $P4_32_12$, $P4_22_12$ or $P4_222$

  Unit cell: a 7.7388, *c* 24.145 Å, *V* 1446.02 $Å^3$, *Z* = 8

  X-ray powder diffraction pattern, strongest lines [*d* in Å(I)(*hkl*)]: 3.008(90)(008), 2.268(100) (134), 2.147(30)(230), 1.9404(60)(400), 1.2043(30)(22$\underline{18}$,452), 1.2002(30)(624), ICDD PDF2 card—00-058-0505.

Structure: not determined

Appearance: Occurs as rare anhedral grains with inclusions of Au-Ag alloy or with attached chalcopyrite and a chlorite-group mineral, from 36 to 116 µm in size.

Optical properties: In reflected light, bright creamy white, weak bireflectance, no pleochroism, weak anisotropy (rotation tints pale blue-gray to deep blue).

Reflectance, [λnm, R_1 R_2 %]	470	air 50.2 50.5	im. 37.6 38.0
	546	55.6 55.9	43.2 43.5
	589	57.9 58.3	45.9 46.3
	650	60.2 60.7	48.1 48.5

Physical Properties: Opaque, dark gray, metallic, malleable.

Hardness: not measured

Cleavage: not observed

Density: 7.264 g/cm^3 (calc.)

Chemical Composition (EMPA data, wt.%):

Pd 54.53, Fe 0.13, Te 0.09, Sb 44.59, Bi 0.42 Hg 0.19 As 0.20 $\sum$100.15 $Pd_{4.062}(Sb_{2.893}Fe_{0.017}Bi_{0.017}Hg_{0.006}Te_{0.005})_{\sum2.938}$ ($n = 16$)

Pd 53.82 Sb 46.18 $\sum$100.00 Pd_4Sb_3—ideal

Type locality and occurrence: The Mesamax Northwest Ni-Cu-Co-PGE deposit in the Cape Smith fold belt of the Ungava region, northern Quebec, Canada (61°34′25″ N, 73°15′36″ W). Also observed in the Tootoo and Mequillon magmatic sulfide deposits, New Quebec Orogen (Liu et al., 2013).

Name: After the locality in the Ungava region, Quebec, Canada.

Reference: McDonald et al. (2005), IMA No. 2004−020

Note: Some further experimental and crystal structure investigations are needed.

Experimental: The synthetic analogue is not known from the experimental studies related to Pd-Sb system (El-Boragy and Schubert, 1971b), nor from studies on ternary systems involving Pd and Sb including those made by Kim and Chao (1991, 1996), Cabri et al. (1975).

10.5.23 ZACCARINIITE RHNiAs

Crystallography: tetragonal, *P4/nmm*

Unit cell: a 3.5498, *c* 6.1573 Å, *V* 77.59 Å^3, *Z* = 2

X-ray powder diffraction pattern, strongest lines [*d* in Å(I)(*hkl*)]: 2.5092(40)(110), 2.3252(100) (111,102), 1.9453(51)(112), 1.7758(80)(103,200), 1.2555(40)(213,220), 1.1044(22)(302,311), 1.0547(23)(312), 0.9730(42)(215). (synth.)

Structure: Cu$_2$Sb type. The crystal structure was refined from the powder XRD data of synthetic RhNiAs using the initial structural model of Roy-Montreuil et al. (1984), ICSD card No187596.

Appearance: Forms anhedral grains (1−20 µm in size), intergrown with garutiite, in association with hexaferrum, Ru-Os-Ir-Fe alloys, and Ru-Os-Ir-Fe oxygenated compounds in chromite.

Optical properties: In reflected light, white with brownish to pinkish tints, has moderate to strong bireflectance, a strong white to pinkish brownish white pleochroism, strong anisotropy (with rotation tints from orange to blue-green); exhibits no internal reflections.

Reflectance, [λnm, R_1 R_2 %]	470	nat. 49.4 52.6	synth. 49.1 53.3
	546	52.4 53.2	53.1 54.1
	589	54.2 53.2	55.2 53.9
	650	56.6 53.3	57.9 53.9

Physical Properties: Opaque, metallic luster, gray streak, brittle.

Hardness: VHN_5 218 (166–286) g/mm^2, Mohs 3½–4

Cleavage: not observed

Density: 10.19 g/cm^3 (calc.), 10.09 g/cm^3 (meas. on synth.)

Chemical Composition (EMPA data, wt.%):

Rh 41.77, Os 0.51, Ir 0.64, Ru 0.46, Pd 0.34, Ni 23.75, Fe 0.53, As 27.84, S 0.10, $\sum$ 96.09 wt. %, $(Rh_{1.01}Os_{0.01}Ir_{0.01}Ru_{0.01}Pd_{0.01})_{\sum 1.05}(Ni_{1.00}Fe_{0.02})_{\sum 1.02}(As_{0.92}S_{0.01})_{\sum 0.93(n=3)}$

Rh 44.57, Ni 24.50, As 31.82 $\sum$100.88 $Rh_{1.02}Ni_{0.98}As_{1.00}$ (synth., $n = 28$)

Rh 43.51, Ni 24.82, As 31.67 $\sum$100.00 RhNiAs—ideal

Type locality and occurrence: The Loma Peguera ophiolitic chromitite, Dominican Republic. It has been observed from several localities worldwide, e.g., the Guli chromitite, Ray-Iz ophiolite complex, Nizhny Tagil, the ophilitic belt in the Koryak-Kamchatska fold region in Russia, the Thetford Mines in Canada, the Vourinos complex in Greece, and the Bushveld complex in South Africa.

Name: For Federica Zaccarini (b. 1962) of the University of Leoben, in recognition of her contributions to PGE mineralogy and related deposits.

Reference: Vymazalová et al. (2012b), IMA No. 2011-086

Note: The structural identity between the natural and synthetic RhNiAs was confirmed by Raman spectroscopy.

Experimental: The phase was synthesized at 800°C; the system Rh-As-Ni has not been experimentally investigated.

ACKNOWLEDGEMENTS

We are very grateful to Louis J. Cabri and Emil Makovicky for their detailed and helpful comments on the manuscript. We are grateful to Sisir K. Mondal and William L. Griffin for the editorial handling and for reading the final manuscript and correcting the English.

REFERENCES

Airiyants, E.V., Zhmodik, S.M., Ivanov, P.O., Belyanin, D.K., Agafonov, L.V., 2014. Mineral inclusions in Fe-Pt solid solution from the alluvial ore occurrences of the Anabar basin (northeastern Siberian Platform). Russian Geol. Geophys. 55 (8), 945–958.

Baghdadi, A., Finley, A., Russo, P., Arnott, R.J., Wold, A., 1974. Crystal growth and characterization of PtP$_2$. J. Less Common Metals 34 (1), 31–38.

Bälz, U., Schubert, K., 1969. Crystal structure of Pd$_2$As and Pd$_2$Sb (r). J. Less-Common Met. 19, 300–304, in German.

Baranov, A.I., Isaeva, A.A., Kloo, L., Popovkin, B.A., 2003. New metal-rich sulfides Ni_6SnS_2 and $Ni_9Sn_2S_2$ with a 2D metal framework: synthesis, crystal structure, and bonding. Inorg. Chem. 42, 6667−6672.

Baranov, A.I., Isaeva, A.A., Kloo, L., Kulbachinskii, V.A., Lunin, R.A., Nikiforov, V.N., et al., 2004. 2D metal slabs in new nickel-tin chalcogenides $Ni_{7-\delta}SnQ_2$ (Q = Se, Te): average crystal and electronic structures, chemical bonding and physical properties. J. Solid State Chem. 177, 3616−3625.

Barkov, A.Y., Fleet, M.E., Martin, R.F., Halkoaho, T.A.A., 2005. New data on "bonanza"-type PGE mineralization in the Kirakkajuppura PGE deposit, Penikat layered complex, Finland. Can. Mineral. 43 (5), 1663−1686.

Beck, J., Hilbert, T., 2000. An "old" rhodiumsulfide with surprising structure − synthesis, crystal structure, and electronic properties of Rh_3S_4. Z. Anorg. Allg. Chem. 626, 72−79.

Bennett, S.L., Heyding, R.D., 1966. Arsenides of the transitions metals. VIII. Some binary and ternary group VIII Diarsenides and their magnetic and electrical properties. Can. J. Chem. 44, 3017−3036.

Berlincourt, L.E., Hummel, H.H., Skinner, B.J., 1981. Phases and phase relations of the platinum-group elements. In: Cabri, L.J. (Ed.), Platinum Group Elements: Mineralogy, Geology, Recovery, CIM Special Vol, 23. pp. 19−45.

Bichsel, R., Levy, F., Berger, H., 1984. Growth and physical properties of RuS_2 single crystals. J. Physics C: Solid State Phys. 17 (1), L19−L21.

Bindi, L., 2010. Atheneite, $Pd_2As_{0.75}Hg_{0.25}$ from Itabira mines, Mina Gerais, Brazil: crystal structure and revision of the chemical formula. Can. Mineral. 48 (5), 1149−1155.

Botelho, N.F., Moura, M.A., Peterson, R.C., Stanley, C.J., Silva, D.V.G., 2006. Kalungaite, PdAsSe, a new platinum-group mineral from the Buraco do Ouro gold mine, Cavalcante, Goiás State, Brazil. Mineral. Mag. 70 (1), 123−130.

Bowles, J.F.W., Prichard, H.M., Suarez, S., Fisher, P., 2013. The first report of platinum-group minerals in magnetite-bearing gabro, Freetown layered complex, Sierra Leone: occurrences and genesis. Can. Mineral. 51 (3), 455−473.

Brenner, S.S., 1956. The growth of whiskers by the reduction of metal salts. Acta Metallurgica 4 (1), 62−74.

Bridgman, P.W., 1925. Certain physical properties of single crystals of tungsten, antimony, bismuth, tellurium, cadmium, zinc, and tin. Proc. Am. Acad. Arts Sci. 60 (6), 305−383.

Bryukvin, V.A., Fishman, B.A., Reznichenko, V.A., Kuoev, V.A., Vasil'eva, N.A., 1990. Inverstigation of the Fe-Rh-S phase diagram in the $Fe-Rh-Rh_2S_3-FeS_{1.09}$ compositional range. Izvestiya Akademii Nauk SSSR. Metally. 2, 23−28, in Russian.

Bugaris, D.E., Malliakas, C.D., Shoemaker, D.P., Do, D.T., Chung, D.Y., Mahanti, S.D., et al., 2014. Crystal Growth and Characterization of the Narrow-Band-Gap Semiconductors $OsPn_2$ (Pn = P, As, Sb). Inorg. Chem. 53 (18), 9959−9968.

Cabral, A.R., Skála, R., Vymazalová, A., Kallistová, A., Lehmann, B., Jedwab, J., et al., 2014. Kitagohaite, Pt_7Cu, a new mineral from the Lubero region, North Kivu, DR Congo. Mineral. Mag. 78 (3), 739−745.

Cabri, L.J., 1973. New data on phase relations in Cu-Fe-S system. Econ. Geol. 68 (4), 443−454.

Cabri, L.J., 1981a. The Platinum-group minerals. In: Cabri, L.J. (Ed.), Platinum Group Elements: Mineralogy, Geology, Recovery, CIM Special Vol, 23. pp. 83−150.

Cabri, L.J., 1981b. Unnamed Platinum-group minerals. In: Cabri, L.J. (Ed.), Platinum Group Elements: Mineralogy, Geology, Recovery, CIM Special Vol, 23. pp. 175−186.

Cabri, L.J., 2002. The Platinum-group minerals. In: Cabri, L.J. (Ed.), The Geology, Geochemistry, Mineralogy and Mineral Beneficiation of Platinum-Group Elements, CIM Special Vol, 54. pp. 177−210.

Cabri, L.J., Laflamme, J.H.G., Stewart, J.M., Rowland, J.F., Chen, T.T., 1975. New data on some palladium arsenides and antimonides. Can. Mineral. 13, 321−335.

Cabri, L.J., Chen, T.T., Stewart, J.M., Gilles, J.H., Laflamme, J.H.G., 1976. Two new palladium-arsenic-bismuth minerals from the Stillwater complex, Montana. Can. Mineral. 14, 410−413.

Cabri, L.J., Laflamme, J.H.G., Stewart, J.M., Turner, K., Skinner, B.J., 1978. On cooperite, braggite, and vysotskite. Am. Mineral. 63, 832–839.

Cabri, L.J., McDonald, A.M., Stanley, C.J., Rudashevsky, N.S., Poirier, G., Durham, B.R., et al., 2005. Naldrettite, Pd_2Sb, a new intermetallic mineral from the Mesamax Northwest deposit, Ungava region, Quebec, Canada. Mineral. Mag. 69 (1), 89–97.

Cabri, L.J., McDonald, A.M., Stanley, C.J., Rudashevsky, N.S., Poirier, G., Wilhelmij, H.R., et al., 2015. Palladosilicide, Pd_2Si, a new mineral from the Kapalagulu Intrusion, Western Tanzania and the Bushveld Complex, South Africa. Mineral. Mag. 79 (2), 295–307.

Campbell, I.E., Powell, C.F., Nowicki, D.H., Gonser, B.W., 1949. The Vapor-Phase Deposition of Refractory Materials I. General Conditions and Apparatus. J. Electrochem. Soc. 96 (5), 318–333.

Canfield, P.C., Fisk, Z., 1992. Growth of single crystals from metallic fluxes. Philosophical Magazine B 65 (6), 1117–1123.

Chareev, D.A., 2015. The low temperature electrochemical growth of iron, nickel and other metallic single crystals from halide eutectic fluxes in a temperature gradient. J. Crystal Growth 429, 63–67.

Chareev, D.A., Volkova, O.S., Geringer, N.V., Koshelev, A.V., Nekrasov, A.N., Osadchii, V.O., et al., 2016. Synthesis of chalcogenide and pnictide crystals in salt melts using a steady-state temperature gradient. Crystallog. Rep. 61 (4), 682–691.

Chernikov, A.A., Chistyakova, N.I., Uvarkina, O.M., Dubinchuk, V.T., Rassulov, V.A., Polekhovsky, Y.S., 2006. Malyshevite - a new mineral from Srednyaya Padma deposit in Southern Karelia. New Data Mineral. 41, 14–17, in Russian.

Chernyaev, I.I., Zheligovskaya, N.N., Borisenkova, M.K., Subbotina, N.A., 1968. The Ag_2Te-Pd System. Russ. J. Inorg. Chem. 13, 848–850, in Russian.

Chirvinskii, P.N., 1995. Artificial Formation of Minerals in the XIX Century. In: Zharikov, V.A. (Ed.), Classics of Science. Nauka, Moscow, p. 510. , in Russian.

Clark, A.M., Criddle, A.J., Fejer, E.E., 1974. Palladium arsenide-antimonids from Itabira, Minais Gerais, Brazil. Mineral. Mag. 39, 528–543.

Colell, H., Alonso-Vante, N., Fiechter, S., Schieck, R., Diesner, K., Henrion, W., et al., 1994. Crystal growth and properties of novel ternary transition metal chalcogenide compounds $Ir_xRu_{1-x}S_2$ (0.005 < x < 0.5). Mater. Res. Bull. 29 (10), 1065–1072.

Czochralski, J., 1918. A new method for the measurement of the crystallization rate of metals. Z. Physik. Chem. 92, 219–221, in German.

Dobersek, M., Kosovinc, I., 1989. The ternary system palladium-copper-zinc. Z. Metallkd. 80, 669–671, in German.

Drábek, M., Rieder, M., 2008. The system Os-Mo-S. Can. Mineral. 46 (5), 1297–1303.

Drábek, M., Vymazalová, A., Cabral, A.R., 2012. The Pt−Hg−Se system at 400°C: phase relations of jacutingaite. Can. Mineral. 50 (2), 441–446.

Drábek, M., Vymazalová, A., Laufek, F., 2014. The system Hg−Pd−Se at 400°C: phase relations involving tischendorfite. Can. Mineral. 52 (2), 763–768.

El-Boragy, M., Ellner, M., Schubert, K., 1984. On Several Phases of the Mixture $NiPd_xAs_x$. Z. Metallkd. 75, 82–85, in German.

El-Boragy, M., Schubert, K., 1971a. Crystal structures of some ternary phases in T-B-B′ systems. Z. Metallkd. 62, 667–675, in German.

El-Boragy, M., Schubert, K., 1971b. On some variants of the NiAs family in mixtures of palladium with B-elements. Z. Metallkd. 62, 314–323, in German.

Evstigneeva, T.L., Nekrasov, I.Y., 1980. Conditions of phase synthesis and phase relations in the systems Pd_3Sn-Cu_3Sn and Pd-Sn-Cu-HCl. In: Zharikov, V.A., Fedkin, V.V. (Eds.), Sketches of Physico-Chemical Petrology, 10. Nauka, Moscow, pp. 20–35. , in Russian.

Ezzaouia, H., Heindl, R., Parsons, R., Tributsch, H., 1983. Visible light photo-oxidation of water with single-crystal RuS_2 electrodes. J. Electroanal. Chem. Interfacial Electrochem. 145 (2), 279−292.

Ezzaouia, H., Heindl, R., Loriers, J., 1984. Synthesis of ruthenium and osmium dichalcogenide single crystals. J. Mater. Sci. Lett. 3 (7), 625−626.

Ezzaouia, H., Foise, J.W., Gorochov, O., 1985. Crystal growth in tellurium fluxes and characterization of RuS_2 single crystals. Mater. Res. Bull. 20 (11), 1353−1358.

Fiechter, S., Kühne, H.M., 1987. Crystal Growth of RuX_2 (X = S, Se, Te) by chemical vapour transport and high temperature solution growth. J. Crystal Growth 83 (4), 517−522.

Finley, A., Schleich, D., Ackerman, J., Soled, S., Wold, A., 1974. Crystal growth and characterization of $Pt_{0.97}S_2$. Mater. Res. Bull. 9 (12), 1655−1659.

Fisk, Z., Remeika, J.P., 1989. Growth of single crystals from molten metal fluxes. Handbook Phys. Chem. Rare Earths 12, 53−70.

Foecker, A.J., Jeitschko, W., 2001. The atomic order of the pnictogen and chalcogen atoms in equiatomic ternary compounds TPnCh (T5 Ni, Pd; Pn 5 P, As, Sb; Ch 5 S, Se, Te). J. Solid State Chem. 162, 69−78.

Foise, J.W., Ezzaouia, H., Gorochov, O., 1985. Crystal growth of p-type RuS_2 using bismuth flux and its photoelectrochemical properties. Mater. Res. Bull. 20 (12), 1421−1425.

Georg, C.A., Triggs, P., Levy, F., 1982. Chemical vapour transport of transition metal oxides (I) crystal growth of RuO_2, IrO_2 and $Ru_{1-x}Ir_xO_2$. Mater. Res. Bull. 17 (1), 105−110.

Gervilla, F., Makovicky, E., Makovicky, M., Rose-Hansen, J., 1994. The system Pd-Ni-As at 790° and 450°C. Econ. Geol. 89, 1630−1639.

Godel, B., Barnes, S.-J., 2008. Image analysis and composition of platinum-group minerals in the J-M reef, Stillwater Complex. Econ. Geol. 103 (3), 637−651.

Grokhovskaya, T.L., Bakaev, G.F., Sholokhnev, V.V., Lapin, M.I., Muravitskaya, G.N., Voitekhovich, V.S., 2003. The PGE Ore Mineralization in the Monchegorsk Igneous Layered Complex (Kola Peninsula, Russia). Geol. Rudnykh Mestorozhdenii 45 (4), 329−352.

Grokhovskaya, T.L., Lapina, M.I., Mokhov, A.V., 2009. Assemblages and genesis of platinum-group minerals in low-sulfide ores of the Monchetundra deposit, Kola peninsula, Russia. Geol. Ore Deposits 51 (9), 467−485.

Grokhovskaya, T.L., Karimova, O.V., Griboedova, I.G., Samoshnikova, L.A., Ivanchenko, V.N., 2012. Geology, mineralogy amd genesis of PGE mineralization in the South Sopcha massif, Monchegorsk complex, Russia. Geol. Ore Deposit 54, 347−369.

Helmy, H.M., Ballhaus, C., Berndt, J., Bockrath, C., Wohlgemuth-Ueberwasser, C., 2007. Formation of Pt, Pd and Ni tellurides: experiments in sulfide-telluride systems. Cont. Mineral. Petrol. 153, 577−591.

Hoffman, E., MacLean, W.H., 1976. Phase relations of michenerite and merenskyite in the Pd-Bi-Te system. Econ. Geol. 71, 1461−1468.

Horkans, J., Shafer, M.W., 1977. An investigation of the electrochemistry of a series of metal dioxides with rutile-type structure: MoO_2, WO_2, ReO_2, RuO_2, OsO_2, and IrO_2. J. Electrochem. Soc. 124 (8), 1202−1207.

Huang, J.K., Huang, Y.S., Yang, T.R., 1994. The preparation and characterization of $RuTe_2$ single crystals. J. Crystal Growth 135 (1), 224−228.

Hurle, D.T.J., Mullin, J.B., Pike, E.R., 1967. Thin alloy zone crystallisation. J. Material. Sci. 2 (1), 46−62.

Johan, Z., Ohnenstetter, M., Slansky, E., Barron, L.M., Suppel, D., 1989. Platinum mineralozation in the Alaskan-type intrusive complexes near Fifield, New-south Wales, Australia Part1. Platinum-group minerals in clinopyroxenites of the Kelvin Grove Prospect, Owendale intrusion. Miner. Petrol. 40 (4), 289−309.

Kaner, R., Castro, C.A., Gruska, R.P., Wold, A., 1977. Preparation and characterization of the platinum metal phosphides RuP_2 and IrP_2. Material. Res. Bull. 12 (12), 1143−1147.

Kapsiotis, A., Grammatikopoulos, T.A., Tsikouras, B., Hatzipanagiotou, K., 2010. Platinum-group mineral characterization in concentrates from high-grade PGE Al-rich chromitites of Korydallos area in the Pindos Ophiolite Complex (NW Greece). Resour. Geol. 60 (2), 178−191.

Karimova O.V., Zolotarev A.A., Evstigneeva T.L., Johanson B., Mertieite-II, $Pd_8Sb_{2.5}As_{0.5}$, crystal structure refinement and formula revision. *Mineral. Mag.* (in press).

Karimova, O.V., Grokhovskaya, T.L., Zolotarev, A.A., Gurzhiy, V.V., 2016. Crystal structure refinement of isomertieite, $Pd_{11}Sb_2As_2$, and törnroosite, $Pd_{11}As_2Te_2$. Can. Mineral. 54 (2), 511−517.

Karup-Møller, S., Makovicky, E., 2007. The system Cu-Rh-S at 900 °C, 700 °C, 540 °C and 500 °C. Can. Mineral. 45 (6), 1535−1542.

Karup-Møller, S., Makovicky, E., Barnes, S.-J., 2008. The metal-rich portions of the phase system Cu-Fe-Pd-S at 1000 °C, 900 °C and 725 °C: implications for mineralization in the Skaergaard intrusion. Mineral. Mag. 72 (4), 941−951.

Kim, W.S., 2009. Synthesis of Pt-Pd antimonide minerals at 1000 °C and the extent of Pt-Pd substitutions. Geosci, J 13 (1), 25−30.

Kim, W.S., Chao, G.Y., 1991. Phase relations in the system Pd-Sb-Te. Can. Mineral. 29, 401−409.

Kim, W., Chao, G.Y., 1996. Phase relations in the system Pd-Pt-Sb. Neues Jahrb. Mineral., Monatsh. 351−364.

Kojonen, K.K., McDonald, A.M., Stanley, C.J., Johanson, B., 2011. Törnroosite, $Pd_{11}As_2Te_2$, a new mineral species related to isomertieite from Miessijoki, Finnish Lapland, Finland. Can. Mineral. 49, 1643−1652.

Kojonen, K.K., Tarkian, M., Roberts, A.C., Tornroos, R., Heidrich, S., 2007. Miessiite, $Pd_{11}Te_2Se_2$, a new mineral species from Miessijoki, Finnish Lapland, Finland. Can. Mineral. 45 (5), 1221−1228.

Kullerud, G., 1971. Experimental techniques in dry sulfide research. In: Ulmer, G.C. (Ed.), Research Techniques for High Pressure and High Temperature,. Spinger-Verlag, New York, pp. 288−315.

Kyropoulos, S., 1926. Ein Verfahren zur Herstellung grosser Kristalle. Z. Anorg. Allgem. Chemie 154 (1), 308−313, in German.

Langer, H., Wachtel, E., 1981. Aufbau und Magnetische Eigenschaften von Palladium Silicium-Legierungen im Konzentrationsintervall 30 bis 100 at. % Si. Z. Metallkd. 72, 769−775.

Laufek, F., Vymazalová, A., Drábek, M., Dušek, M., Navrátil, J., Černošková, E., 2016. The crystal structure of temagamite, Pd_3HgTe_3. Eur. J. Mineral. 28, 825−834.

Laufek, F., Vymazalová, A., Drábek, M., Navrátil, J., Drahokoupil, J., 2014. Synthesis and crystal structure of tischendorfite $Pd_8Hg_3Se_9$. Eur. J. Mineral. 25, 157−162.

Laufek, F., Vymazalová, A., Grokhovskaya,T.L., Plášil, J., Dušek, M., Orsoev, D.A., Kozlov V.V. The Crystal Structure of sopcheite, $Ag_4Pd_3Te_4$, from the Lukkulaisvaara intrusion, northern Russian Karelia, Russia *Eur. J. Mineral.* in press.

Lebrun, N., Cacciamani, G., Cordes, H., Effenberg, G., Ilyenko, S., Schmid-Fetzer, R., 2007. Cu-Pb-Zn (Copper-Lead-Zinc).. In: Effenberg, G., Ilyenko, S. (Eds.), Ternary Alloy Systems, Noble Metal Systems. Landolt-Börnstein - Group IV Physical Chemistry Series, 11B. Springer Berlin Heidelberg, pp. 408−419.

Liao, P.C., Ho, C.H., Huang, Y.S., Tiong, K.K., 1997. Preparation and characterization of pyrite-like single crystal phase in the Ir-Te system. J. Crystal Growth 171 (3), 586−590.

Liu, Y., Mungall, J.E., Ames, D.E., 2013. Hydrothermal redistribution and local enrichment of platinum group elements in the Tootoo and Mequillon magmatic sulfide deposits, South Raglan Trend, Cape Smith Belt, New Quebec Orogen. Econ. Geol. 111 (2), 467−485.

Lyons, A., Schleich, D., Wold, A., 1976. Crystal growth and characterization of $PdTe_2$. Material. Res. Bull. 11 (9), 1155−1159.

Makovicky, E., 2002. Ternary and quaternary phase systems with PGE. In: Cabri, L.J. (Ed.), The Geology, Geochemistry, Mineralogy and Mineral Beneficiation of Platinum-Group Elements, CIM Special Vol, 54. pp. 131−175.

Makovicky, E., Karup-Møller, S. Exploratory studies of the Cu-Pd-Se system from 300 to 650 °C, *Eur. J. Mineral.* in press.

Makovicky, E., Karup-Møller, S., 2016. The Pd−Ni−Fe−S phase system at 550 and 400°C. Can. Mineral. 54 (2), 377−400.

Makovicky, E., Makovicky, M., Rose-Hansen, J., 2002. The system Fe-Rh-S at 900° and 500 °C. Can. Mineral. 40, 519−526.

Massalski, T.B., 1990. Binary Alloy Phase Diagrams. 2nd ed. ASM International, Materials Park, Ohio.

Matsumoto, N., Nagata, S., 2000. Single-crystal growth of sulphospinel $CuIr_2S_4$ from Bi solution. J. Crystal Growth 210 (4), 772−776.

McDonald, A.M., Ames, D.E., Ross, K.C., Kjarsgaard, I.M., Good, D.J., 2016. Marathonite, IMA 2016-080. CNMNC Newsletter No. 34, December 2016, page 1320. Mineral. Mag. 80, 1315−1321.

McDonald, A.M., Cabri, L.J., Stanley, C.J., Rudashevsky, N.S., Poirer, G., Ross, K.C., et al., 2005. Ungavaite, Pd_4Sb_3, a new intermetallic mineral from the Mesamax Northwest Deposit, Ungava Region, Quebec, Canada. Can. Mineral. 43 (5), 1735−1744.

McDonald, A.M., Cabri, L.J., Rudashevsky, N.S., Stanley, C.J., Rudashevsky, V.N., Ross, K.C., 2008. Nielsenite, $PdCu_3$, a new platinum-group intermetallic mineral from the Skaergaard intrusion, Greenland. Can. Mineral. 46, 709−716.

McDonald, A.M., Cabri, L.J., Stanley, C.J., Good, D.J., Redpath, J., Spratt, J., 2015. Coldwellite, Pd_3Ag_2S, a new mineral species from the Marathon deposit, Coldwell Complex, Ontario, Canada. Can. Mineral. 53, 845−857.

McDonald, A.M., Zhe, W., Ames, D.E., Ross, K.C., Kjarsgaard, I.M., Good, D.J., 2017. Palladogermanide, IMA 2016-086. CNMNC Newsletter No. 35, February 2017, page 210. Mineral. Mag. 81, 209−213.

Mochalov, A.G., Tolkachev, M.D., Polekhovsky, Yu.S., Goryacheva, E.M., 2007. Bortnikovite, Pd_4Cu_3Zn, a new mineral species from the unique Konder placer deposit, Khabarovsk krai, Russia. Geol. Ore Deposits 49 (4), 318−327.

Nickel, E.H., 2002. An unusual occurrence of Pd, Pt, Au, Ag and Hg minerals in the Pilbara region of western Australia. Can. Mineral. 40, 419−433.

Nylund, A., 1966. Some notes on the palladium-silicon system. Acta Chem. Scand. 20, 2381−2386.

Oberthür, T., Melcher, F., Gast, L., Wöhrl, C., Lodziak, J., 2004. Detritaǔ platinum-group minerals in rivers draining, the eastern Bushveld Complex, South Africa. Can. Mineral 42, 563−582.

Odile, J.P., Soled, S., Castro, C.A., Wold, A., 1978. Crystal growth and characterization of the transition-metal phosphides copper diphosphide, nickel diphosphide, and rhodium triphosphide. Inorg. Chem. 17 (2), 283−286.

Paar, W.H., Topa, D., Makovicky, E., Sureda, R.J., de Brodtkorb, M.K., Nickel, E.H., et al., 2004. Jagueite, $Cu_2Pd_3Se_4$, a new mineral species from El Chire, La Rioja, Argentina. Can. Mineral. 42, 1745−1755.

Paar, W.H., Topa, D., Makovicky, E., Culetto, F.J., 2005. Milotaite, a new palladium mineral species from Předbořice, Czech Republic. Can. Mineral. 43, 689−694.

Parthe, E., Hohnke, E., Hulliger, F., 1967. A new structure type with octahedron pairs for Rh_2S_3, Rh_2Se_3 and Ir_2S_3. Acta Crystallog. 23 (5), 832−840.

Pentek, A., Molnar, F., Tuba, G., Watkinson, D.H., Jones, P.C., 2013. The significance of partial melting processes in hydrothermal low sulfide Cu-Ni-PGE mineralization within the footwall of the Sudbury Igneous Complex, Ontario, Canada. Econ. Geol. 108 (1), 59−78.

Pfann, W.G., 1966. Zone Melting. second ed. John Wiley & Sons, New York, p. 310.

Pratt, J., 1994. As-Pd-Rh Ternary Phase Diagram Evaluation. In: Effenberg, G. (Ed.), Phase diagrams, crystalographic and thermodynamic data. Ternary Evaluations, MSI Eureka MSI, Materials Science International Services GmbH, Stuttgart.

Predel, B., 1998. Rh-S (Rhodium-Sulfur). Pu-Re-Zn-Zr. Springer Berlin Heidelberg, pp. 1−2.

Pyon, S., Kudo, K., Nohara, M., 2013. Emergence of superconductivity near the structural phase boundary in Pt-doped IrTe$_2$ single crystals. Physica C: Superconductivity 494, 80–84.

Rau, H., Rabenau, A., 1968. Hydrothermal growth of some elements. J. Crystal Growth 3, 417–421.

Raub, E., 1954. Über die Reaktion von Silber-Palladium-Legierungen mit Schwefel bei erhöhter Temperatur. Z. Metallkd. 45, 533–537, in German.

Rogers, D.B., Shannon, R.D., Sleight, A.W., Gillson, J.L., 1969. Crystal chemistry of metal dioxides with rutile-related structures. Inorg. Chem. 8 (4), 841–849.

Rossler, F., 1895. Synthese einiger Erzmineralien und analoger Metallverbindungen durch Auflosen und krystallisiren lassen derselben in geschmolzenen Metallen. Z. Anorg. Chem. 9, 31–77, in German.

Roy-Montreuil, J., Chaudouet, P., Rouault, A., Boursier, D., Senateur, J.O., Fruchart, R., 1984. Analyse de Iordre dans les arseniures MMAs. Ann. Chim. 4, 411–417, in French.

Rozhdestvina, V.I., Udovenko, A.A., Rubanov, S.V., Mudrovskaya, N.V., 2016. Structural investigation of cooperite (PtS) crystals. Crystallog. Rep. 61 (2), 193–202.

Rudashevsky, N.S., Kretser, Yu,L., Orsoev, D.A., Kislov, E.V., 2003. Palladium-platinum mimeralization in copper-mickel vein ores in the Ioko-Dovyren layered massif. Doklady Earth Sci. 391 (6), 858–861, in Russian.

Rudashevsky, N.S., McDonald, A.M., Cabri, L.J., 2004. Skaergaardite, PdCu, a new mineral platinum-group intermetallic mineral from the Skaergaard intrusion, Greenland. Min. Mag. 68 (4), 615–632.

Ruehl, R., Jeitschko, W., 1982. Preparation and crystal structure of dirhenium pentaphosphide, Re$_2$P$_5$, a diamagnetic semiconducting polyphosphide with rhomboidal Re$_4$ clusters. Inorg. Chem. 21 (5), 1886–1891.

Saini, G.S., Calvert, L.D., Heyding, R.D., Taylor, J.B., 1964. Arsenides og the transitions metals: VII. The Palladium-Arsenic system. Can. J. Chem. 42 (3), 620–629.

Sarah, N., Alasafi, K., Schubert, K., 1981. Kristallstruktur von Pd$_{20}$Sn$_{13}$, Pd$_6$AgPb$_4$ und Ni$_{13}$ZnGe$_8$. Z. Metallkd 72 (7), 517–520, in German.

Savilov, S.V., Kuznetsov, A.N., Popovkin, B.A., Khrustalev, V.N., Simon, P., Getzschmann, J., et al., 2005. Synthesis, crystal structure and electronic structure of modulated Pd$_{7-\delta}$SnTe$_2$. Z. Anorg. Allgem. Chem. 631, 293–301.

Schafer, H., 1962. Chemische Transportreaktionen.. Verlag Chemie, MnbH, Weinhelm, Bergstr, p. 182. , in German.

Schneider, A., Esch, U., 1944. Das System Kupfer–Platin. Z. Elektrochem. 50, 290–301, in German.

Schubert, K., Kiefer, B., Wilkens, M., Haufler, R., 1955. On some ordered phases of long period metals. Z. Metallkd. 46, 692–715, in German.

Schwartz, K.B., Gillson, J.L., Shannon, R.D., 1982. Crystal growth of CdPt$_3$O$_6$, MnPt$_3$O$_6$, CoPt$_3$O$_6$ and β-PtO$_2$. J. Crystal Growth 60 (2), 251–254.

Severson, M.J., Hauck, S.A., 2003. Platinum group elements (PGEs) and platinum group minerals (PGMs) in the Duluth Complex. NRRI, University of Minnesota, Duluth, Technical Report, NRRI/TR-2003/37.

Shahmiri, M., Murphy, S., Vaughan, D.J., 1985. Structural and phase equilibria studies in the system Pt-Fe-Cu and the occurrence of tulameenite (Pt$_2$FeCu). Mineral. Mag. 49, 547–554.

Singh, Y., Lee, Y., Nandi, S., Kreyssig, A., Ellern, A., Das, S., et al., 2008. Single-crystal growth and physical properties of the layered arsenide BaRh$_2$As$_2$. Phys. Rev. B 78 (10), 104512.

Skinner, B.J., Luce, F.D., Dill, J.A., Ellis, D.E., Hagan, H.A., Lewis, D.M., et al., 1976. Phase realations in ternary portions of the system Pt-Pd-Fe-As-S. Econ. Geol. 71, 1469–1475.

Sklyarov, E.V., Karyakin, Yu.V., Karmanov, N.S., Tolstykh, N.D., 2016. Platinum-group minerals in dolerites from Alexandra Land Island (Franz Josef Land Archipelago). Russian Geol. Geophys. 57 (5), 834–841.

Sluiter, M.H.F., Colinet, C., Pasturel, A., 2006. Ab initio calculation of the phase stability in Au-Pd and Ag-Pt alloys. Phys. Rev. B 73, 174–204.

Sluzhenikin, S.F., Mokhov, A.V., 2015. Gold and silver in PGE-Cu-Ni and PGE ores of the Noril'sk deposit, Russia. Mineral. Deposita 50, 465–492.

Sluzhenikin S.F., Kozlov V.V., Stanley C.J., Lukashova M.L., Dicks K. Vymazalováite, Pd₃Bi₂S₂, a new mineral from the Noril'sk -Talnakh deposit, Krasnoyarskiy region, Russia. *Mineral. Mag. – in press.*

Soled, S., Wold, A., Gorochov, O., 1975. Crystal growth and characterization of platinum ditelluride. Material. Res. Bull. 10 (8), 831–835.

Soled, S., Wold, A., Gorochov, O., 1976. Crystal growth and characterization of several platinum sulfoselenides. Material. Res. Bull. 11 (8), 927–932.

Spiridonov, E.M., Kulagov, E.A., Serova, A.A., Kulikova, I.M., Korotaeva, N.N., Sereda, E.V., et al., 2015. Genetic Pd, Pt, Au, Ag, and Rh mineralogy in Noril'sk sulfide ores. Geol. Ore Deposits 57 (5), 402–432.

Stanley, C.J., Criddle, A.J., Spratt, J., Roberts, A.C., Szymanski, J.T., Welch, M.D., 2005. Kingstonite, (Rh,Ir, Pt)₃S₄, a new mineral species from Yubdo, Ethiopia. Mineral. Mag. 69 (4), 447–453.

Stanley, C.J., Vymazalova, A., 2015. Kojonenite, a new palladium tin telluride mineral from the Stillwater Layered Igneous Intrusion, Montana, USA. Am. Mineral 100, 447–450.

Subbotin, V.V., Korchagin, A.U., Savchenko, E.E., 2012. Platinum mineralization of the Fedorova-Pana ore node: types of ores, mineral compositions and genetic features. Vestnik of the Kola Science Center of the Russian Academy of Sciences, Apatity 2012 (1), 54–65, in Russian.

Subramanian, P.R., Laughlin, D.E., 1991. Cu-Pd (Copper-Palladium). J. Phase Equilibria 12, 231–243.

Sugahara, M., Yoshiasa, A., Yoneda, A., Hashimoto, T., Sakai, S., Okube, M., et al., 2008. Single-crystal X-ray diffraction study of CaIrO₃. Amer. Mineral. 93 (7), 1148–1152.

Topa, D., Makovicky, E., Balić-Žunić, T., 2006. The crystal structures of jaguéite, Cu₂Pd₃Se₄, and chrisstanleyite, Ag₂Pd₃Se₄. Can. Mineral. 44, 497–505.

Tsay, M.Y., Chen, S.H., Chen, C.S., Huang, Y.S., 1994. Preparation and characterization of iron-doped RuS₂ single crystals. J. Crystal Growth. 144 (1), 91–96.

Tsay, M.Y., Huang, J.K., Chen, C.S., Huang, Y.S., 1995. Crystal growth in tellurium flux and characterization of ruthenium dichalcogenides. Mater. Res. Bull. 30 (1), 85–92.

Vaterlaus, H.P., Bichsel, R., Levy, F., Berger, H., 1985. RuS₂ and RuSe₂ single crystals: a study of phonons, optical absorption and electronic properties. J. Physics C: Solid State Physics 18 (32), 6063.

Vymazalová, A., Drábek, M., 2010. The system Pd-Sn-Te at 400°C and mineralogical implications. II. The ternary phases. Can. Mineral. 48 (5), 1051–1058.

Vymazalová, A., Drábek, M., 2011. The Pd–Pb–Te system: phase relations involving pašavaite and potencial minerals. Can. Mineral. 49 (6), 1679–1686.

Vymazalová, A., Laufek, F., Drábek, M., Haloda, J., Sidorinová, T., Plášil, J., 2009. Pašavaite, Pd₃Pb₂Te₂, a new platinum-group mineral species from Noril'sk-Talnakh Ni-Cu camp, Russia. Can. Mineral 47 (1), 53–62.

Vymazalová, A., Laufek, F., Drábek, M., Cabral, A.R., Haloda, J., Sidorinová, T., et al., 2012a. Jacutingaite, Pt₂HgSe₃, a new platinum-group mineral from the Cauê iron-ore deposit, Itabira District, Minas Gerais, Brazil. Can. Mineral. 50 (2), 431–440.

Vymazalová, A., Laufek, F., Drábek, M., Stanley, C.J., Baker, R.J., Bermejo, R., et al., 2012b. Zaccariniite, RhNiAs, a new platinum-group minera species from Loma Peguera, Dominican Republic. Can. Mineral. 50, 1321–1329.

Vymazalová, A., Chareev, D.A., Kristavchuk, A.V., Laufek, F., Drábek, M., 2014b. The system Ag-Pd-Se: phase relations involving minerals and potential new minerals. Can. Mineral. 52, 77–89.

Vymazalová, A., Grokhovskaya, T.L., Laufek, F., Rassulov, V.I., 2014a. Lukkulaisvaaraite, Pd₁₄Ag₂Te₉, a new mineral from Lukkulaisvaara intrusion, northern Russian Karelia, Russia. Mineral. Mag. 78 (7), 1491–1502.

Vymazalová, A., Laufek, F., Kristavchuk, A.V., Chareev, D.A., Drábek, M., 2015. The system Ag-Pd-Te: phase relations and mineral assemblages. Mineral. Mag. 79 (7), 1813–1832.

Vymazalová, A., Laufek, F., Sluzhenikin, S.F., Stanley, C.J., 2017. Norilskite, (Pd,Ag)$_7$Pb$_4$, a new mineral from Noriľsk -Talnakh deposit, Russia. Mineral. Mag. 81 (3), 531−541.

Vymazalová, A., Laufek, F., Sluzhenikin, S.F., Stanley, C.J., Kozlov, V.V., Chareev, D.A., et al.: Kravtsovite, PdAg$_2$S, a new mineral from Noriľsk -Talnakh deposit, Krasnoyarskiy kray, Russia. *Eur. J. Mineral.* in press.

Wilke, K.T., 1973. Kristallzüchtung. Harri Deutsch Verlag 600, in German.

Yen, P.C., Chen, R.S., Chen, C.C., Huang, Y.S., Tiong, K.K., 2004. Growth and characterization of OsO$_2$ single crystals. J. Crystal Growth 262 (1), 271−276.

Yu, Z., Cheng, F., Ma, H., 2009. Lisiguangite, CuPtBiS$_3$, a New Platinum-Group Mineral from the Yanshan Mountains, Hebei, China. Acta Geologica Sinica 83 (2), 238−244.

Zhang, W., Yanagisawa, K., Kamiya, S., Shou, T., 2009. Phase Controllable Synthesis of Well-Crystallized Rhodium Sulfides by the Hydrothermal Method. Crystal Growth Design 9 (8), 3765−3770.

Zhuravlev, N.N., 1976. Physicochemical investigation of some superconducting alloys in the System Pd-Pb-Bi (Sections PdBi-PdPb and PdBi$_2$-PdPb$_2$). Vestn. Mosk. Univ., Ser. 3: Fiz. Astron 17, 392−396.

Index

Note: Page numbers followed by "*f*" and "*t*" refer to figures and tables, respectively.

Printed in the United States
By Bookmasters